WILD HARVEST

STUDYING SCIENTIFIC ARCHAEOLOGY

Wild harvest:
Plants in the hominin
and pre-agrarian human worlds

edited by
Karen Hardy and Lucy Kubiak-Martens

Oxbow Books
Oxford & Philadelphia

Published in the United Kingdom in 2016 by
OXBOW BOOKS
10 Hythe Bridge Street, Oxford OX1 2EW

and in the United States by
OXBOW BOOKS
1950 Lawrence Road, Havertown, PA 19083

Paperback Edition: ISBN 978-1-78570-123-8
Digital Edition: ISBN 978-1-78570-124-5

A CIP record for this book is available from the British Library

Library of Congress Cataloging-in-Publication Data

Names: Hardy, Karen (Karen Vanessa), editor. | Kubiak-Martens, Lucy, editor.
Title: Wild harvest : plants in the hominin and pre-agrarian human worlds /
    edited by Karen Hardy and Lucy Kubiak-Martens.
Description: Oxford ; Philadelphia : Oxbow Books, [2016] | Series: Studying
    scientific archaeology | Includes bibliographical references and index.
Identifiers: LCCN 2015041341 (print) | LCCN 2015050097 (ebook) | ISBN
    9781785701238 (pbk.) | ISBN 9781785701245 (epub) | ISBN 9781785701252
    (mobi) | ISBN 9781785701269 (pdf)
Subjects: LCSH: Food crops--History. | Wild plants, Edible.
Classification: LCC SB175 .W53 2016 (print) | LCC SB175 (ebook) | DDC
    633--dc23
LC record available at http://lccn.loc.gov/2015041341

Printed in the United Kingdom by Latimer Trend

For a complete list of Oxbow titles, please contact:

UNITED KINGDOM
Oxbow Books
Telephone (01865) 241249, Fax (01865) 794449
Email: oxbow@oxbowbooks.com
www.oxbowbooks.com

UNITED STATES OF AMERICA
Oxbow Books
Telephone (800) 791-9354, Fax (610) 853-9146
Email: queries@casemateacademic.com
www.casemateacademic.com/oxbow

Oxbow Books is part of the Casemate Group

*Front cover: Gathering plant foods on the Mesolithic river-dune site at Yangtze Harbour, the Netherlands.
Artwork by Martin Valkhoff, Rotterdam.
Back cover: (left) The transverse section through the rhizome of the bracken fern (Pteridium aquilinum)
showing parenchyma cells packed with starch granules (polarised light microscope)
(photo: Lucy Kubiak-Martens); (right) Sarcocephalus latifolius.(photo: Mathieu Grèye).*

Contents

List of contributors vii

Dedication: Lydia Zapata Peña xi

Introduction 1
Karen Hardy and Lucy Kubiak-Martens

PART 1

SETTING THE SCENE 17

1. Food carbohydrates from plants...19
   Les Copeland

2. Why protein is not enough: the roles of plants and plant processing
   in delivering the dietary requirements of modern and early *Homo*......................31
   Peter J. Butterworth, Peter R. Ellis and Michele Wollstonecroft

3. An ape's perspective on the origins of medicinal plant use in humans................55
   Michael A. Huffman

4. Plants as raw materials...71
   Karen Hardy

5. Hunter-gatherer plant use in southwest Asia: the path to agriculture.................91
   Amaia Arranz-Otaegui, Juan J. Ibáñez and Lydia Zapata Peña

PART 2

PLANT FOODS, TOOLS AND PEOPLE...111

6. Scanning electron microscopy and starchy food in Mesolithic Europe:
   the importance of roots and tubers in Mesolithic diet..113
   Lucy Kubiak-Martens

7. Tools, use wear and experimentation: extracting plants
   from stone and bone 135
   Annelou van Gijn and Aimée Little

8. Buccal dental microwear as an indicator of diet in modern
    and ancient human populations 155
    Laura Mónica Martínez, Ferran Estebaranz-Sánchez and Alejandro Pérez-Pérez

9. What early human populations ate: the use of phytoliths for identifying
    plant remains in the archaeological record at Olduvai 171
    Rosa María Albert and Irene Esteban

10. Phytolith evidence of the use of plants as food by Late Natufians
    at Raqefet Cave 191
    Robert C. Power, Arlene M. Rosen and Dani Nadel

11. Evidence of plant foods obtained from the dental calculus
    of individuals from a Brazilian shell mound 215
    Célia Helena C. Boyadjian, Sabine Eggers and Rita Scheel-Ybert

12. Stable isotopes and mass spectrometry 241
    Karen Hardy and Stephen Buckley

Part 3

Providing a context: ethnography, ethnobotany, ethnohistory, ethnoarchaeology ... 251

13. Prehistoric fish traps and fishing structures from Zamostje 2,
    Russian European Plain: archaeological and ethnographical contexts 253
    *Ignacio Clemente Conte, Vladimir M. Lozovski, Ermengol Gassiot Ballbè,
    Andrey N. Mazurkevich and Olga V. Lozovskaya*

14. Plants and archaeology in Australia 273
    Sally Brockwell, Janelle Stevenson and Anne Clarke

15. Plentiful scarcity: plant use among Fuegian hunter-gatherers 301
    Marian Berihuete Azorín, Raquel Piqué Huerta and Maria Estela Mansur

16. Ethnobotany in evolutionary perspective: wild plants in diet
    composition and daily use among Hadza hunter-gatherers 319
    Alyssa N. Crittenden

17. Wild edible plant use among the people of Tomboronkoto,
    Kédougou region, Senegal 341
    Mathieu Guèye and Papa Ibra Samb

Index 361

Contributors

Rosa María Albert
ICREA (Institució Catalana de Recerca i
Estudis Avançats), ERAAUB, Departament
de Prehistòria, Historia Antiga I Arqueologia,
Universitat de Barcelona, Montalegre, 6–8,
08001 Barcelona, Spain.
rosamaria.albert@icrea.cat

Amaia Arranz-Otaegui
Department of Cross-Cultural and Regional
Studies, Faculty of Humanities,
University of Copenhagen, Karen Blixens Vej 4,
2300 Copenhagen, Denmark.
kch860@hum.ku.dk

Marian Berihuete Azorín
Institute of Botany,
University of Hohenehim, Garbenstrasse 30,
70599 Stuttgart, Germany.
mberihueteazorin@gmail.com

Célia Helena C. Boyadjian
Laboratório de Antropologia Biológica,
Departamento de Genética e Biologia Evolutiva,
Instituto de Biociências,
Universidade de São Paulo. Rua do Matão, 277,
05508-090, São Paulo, Brazil.
boyadjian.celia@gmail.com

Sally Brockwell
Department of Archaeology and Natural
History, School of Culture, History and
Language, ANU College of Asia and the Pacific,
The Australian National University,
Acton ACT 2601, Australia.
sally.brockwell@anu.edu.au

Stephen Buckley
BioArCh, Department of Archaeology,
University of York, Biology S Block, Wentworth
Way, York YO10 5DD UK.
sb55@york.ac.uk

Peter J. Butterworth
Diabetes & Nutritional Sciences Division,
Biopolymers Group, King's College London,
Franklin-Wilkins Building, 150 Stamford St,
London SE1 9NH, UK.
peter.butterworth@kcl.ac.uk

Anne Clarke
Department of Archaeology,
School of Philosophical and Historical Inquiry,
The University of Sydney, Sydney
NSW 2006, Australia.
annie.clarke@sydney.edu.au

Ignacio Clemente Conte
AGREST, Departament de Arqueologia y
Antropologia, Institució Milà i Fontanals, CSIC, C/
Egipcíaques 15, Barcelona E-08001, Spain.
ignacio@imf.csic.es

Les Copeland
Faculty of Agriculture and Environment,
University of Sydney, NSW 2006, Australia.
les.copeland@sydney.edu.au

Alyssa N. Crittenden
Department of Anthropology; Metabolism,
Anthropometry, and Nutrition Laboratory,
University of Nevada, Las Vegas, USA.
alyssa.crittenden@unlv.edu

SABINE EGGERS
Laboratório de Antropologia Biológica,
Departamento de Genética e Biologia Evolutiva,
Instituto de Biociências, Universidade de São Paulo.
Rua do Matão, 277
05508-090, São Paulo, Brazil.
saeggers@usp.br

PETER R. ELLIS
Diabetes & Nutritional Sciences Division,
Biopolymers Group, King's College London,
Franklin-Wilkins Building, 150 Stamford St,
London SE1 9NH, UK.
peter.r.ellis@kcl.ac.uk

IRENE ESTEBAN
ERAAUB-Department of Prehistory Ancient
History and Archaeology.
University of Barcelona, Montalegre 6–8,
08001 Barcelona, Spain.
irene.esteban.alama@gmail.com

FERRAN ESTEBARANZ-SÁNCHEZ
Secció d'Antropologia, Departament de Biologia
Animal, Facultat de Biologia, Universitat de
Barcelona, Avd. Diagonal 643–645,
08028 Barcelona, Spain.
estebaranz@ub.edu

ERMENGOL GASSIOT BALLBÈ
AGREST, Departament de Prehistòria, Facultat
de Filosofía i Lletres, Universitat Autònoma de
Barcelona, Bellaterra, 08193 Barcelona, Spain.
ermengol.gassiot@uab.cat

MATHIEU GUÈYE
Institut Fondamental d'Afrique Noire,
Département de Botanique et Géologie,
Laboratoire de Botanique, UMI 3189
"Environment, Health and Society",
BP 206 Dakar, Senegal.
mathieu.gueye@ucad.edu.sn

KAREN HARDY
ICREA (Institució Catalana de Recerca i
Estudis Avançats), AGREST, Departament de
Prehistòria, Facultat de Filosofía i Lletres,
Universitat Autònoma de Barcelona, Bellaterra,
08193 Barcelona, Spain.
khardy@icrea.cat

MICHAEL A. HUFFMAN
Primate Research Institute, Kyoto University
41–2 Kanrin, Inuyama Aichi 484-8506, Japan.
huffman@pri.kyoto-u.ac.jp

JUAN J. IBÁÑEZ
AGREST, Departament de Arqueologia y
Antropologia, Institució Milà i Fontanals, CSIC, C/
Egipcíaques 15, Barcelona E-08001, Spain.
ibanezjj@imf.csic.es

LUCY KUBIAK-MARTENS
BIAX Consult, Biological Archaeology &
Environmental Reconstruction, Hogendijk 134,
1506 AL Zaandam, The Netherlands.
kubiak@biax.nl

AIMÉE LITTLE
BioArCh, Department of Archaeology,
University of York, Biology S Block, Wentworth
Way, York YO10 5DD, UK.
aimee.little@york.ac.uk

OLGA V. LOZOVSKAYA
Institute for the History of Material Culture,
Russian Academy of Science, Dvortsovaya
Naberezhnaya 18, St. Petersburg 191186, Russia.

Sergiev Posad State History and Art Museum-
Reserve, av. Krasnoï Armii 144, Sergiev Posad
141310, Russia.
olozamostje@gmail.com

† VLADIMIR M. LOZOVSKI
Institute for the History of Material Culture,
Russian Academy of Science, Dvortsovaya
Naberezhnaya 18, St. Petersburg 191186, Russia.

Sergiev Posad State History and Art Museum-
Reserve, av. Krasnoï Armii 144, Sergiev Posad
141310, Russia.

MARIA ESTELA MANSUR
Centro Austral de Investigaciones Científicas,
Bernardo Houssay 200, 9410 Ushuaia,
Tierra del Fuego, Argentina.
estelamansur@gmail.com

Laura Mónica Martínez
Secció d'Antropologia, Departament de Biologia
Animal, Facultat de Biologia, Universitat de
Barcelona, Avd. Diagonal 643–645,
08028 Barcelona, Spain.
lmartinez@ub.edu

Andrey N. Mazurkevich
State Hermitage Museum: Dvortsovaya
Naberezhnaya 34, St. Petersburg 190000, Russia.
a-mazurkevich@mail.ru

Dani Nadel
Zinman Institute of Archaeology,
University of Haifa,
Haifa 3498838, Israel.
dnadel@research.haifa.ac.il

Alejandro Pérez-Pérez
Secció d'Antropologia, Departament de Biologia
Animal, Facultat de Biologia, Universitat de
Barcelona, Avd. Diagonal 643–645,
08028 Barcelona, Spain.
martinez.perez-perez@ub.edu

Raquel Piqué Huerta
AGREST, Departament de Prehistòria, Facultat
de Filosofía i Lletres, Universitat Autònoma de
Barcelona, 08193 Bellaterra, Barcelona, Spain.
raquel.pique@uab.cat

Robert C. Power
Max Planck Research Group on Plant Foods in
Hominin Dietary Ecology, Max Planck Institute
for Evolutionary Anthropology,
Deutscher Platz 6, 04103,
Leipzig, Germany.
r.c.power@umail.ucc.ie

Arlene M. Rosen
Department of Anthropology, University of
Texas at Austin, SAC 4.102, 2201 Speedway Stop
C3200, Austin TX 78712, USA.
amrosen@austin.utexas.edu

Papa Ibra Samb
Département de Biologie Végétale,
Faculté des Sciences et Technique, UCAD,
BP 5005 Dakar, Senegal.
pisamb@gmail.com

Rita Scheel-Ybert
Laboratório de Arqueobotânica e Paisagem,
Departamento de Antropologia, Museu
Nacional, Universidade Federal do Rio de
Janeiro, Quinta da Boa Vista, São Cristóvão
20940-040 Rio de Janeiro, RJ, Brazil.
scheelybert@mn.ufrj.br

Janelle Stevenson
Department of Archaeology and Natural
History, School of Culture, History and
Language, ANU College of Asia and the Pacific,
The Australian National University,
Acton ACT 2601, Australia.
janelle.stevenson@anu.edu.au

Annelou Van Gijn
Faculty of Archaeology, Material Culture
Studies, Van Steenis gebouw, Einsteinweg 2,
2333 CC Leiden, The Netherlands.
a.l.van.gijn@arch.leidenuniv.nl

Michele Wollstonecroft
UCL Institute of Archaeology,
31–34 Gordon Square,
London WC1H 0PY, UK.
m.wollstonecroft@ucl.ac.uk

† Lydia Zapata Peña
Department of Geography, Prehistory and
Archaeology. University of the Basque Country
– Euskal Herriko Unibertsitatea. Francisco
Tomás y Valiente s/n. 01006,
Vitoria-Gasteiz, Spain.

Dedication

Lydia Zapata Peña
(June 1965–January 2015)

We dedicate this book to the memory of Lydia Zapata. Lydia was an archaeobotanist who achieved much in her life. She received her PhD in 1999 on *Plant Resource Exploitation and the Origins of Agriculture in the Basque Country through the Analysis of Plant Macroremains* at the University of the Basque Country which showed the importance of plants amongst prehistoric communities of northern Iberia. Her research on prehistoric plant use through the analysis of both wood charcoal and seeds established the bases for the development of archaeobotany in the Iberian Peninsula and more specifically in the Basque Country. Throughout her career, Lydia explored not only agrarian communities but also Epipaleolithic and Palaeolithic groups. Her interrupted work in her dear Balzola, Basque Country, highlights her capacity for undertaking extraordinary research. She was also very interested in ethnography and her work comprised studies of hulled wheat growers, plant gatherers and producers of wood charcoal.

Lydia had recently been awarded a highly prestigious ERC (European Research Council) grant for her project 'Paleoplant', an investigation into the use of plants by Palaeolithic and Mesolithic populations in Iberia and North Africa. Though Lydia's untimely death prevented completion of this project some of the research she initiated is being realised. We hope that Lydia's vision and her optimism, enthusiasm and extraordinary capacity for engaging people, will serve as an inspiration and will encourage new researchers to follow in Lydia's footsteps and work towards a better understanding of the use of plants by the pre-agrarian people of the past.

Leonor Peña-Chocarro

Vladimir Lozovski (May 1968–July 2015). As the book was going to press, we heard that Vladimir had passed away. His work at the outstanding site of Zamostje is well known and the chapter he has co-authored in this book will form part of the extended bibliography which demonstrates his lifelong dedication to that site.

Introduction

Karen Hardy and Lucy Kubiak-Martens

Plants are essential to human life. They provide us with food, fuel, raw materials and medicine as they also did in the hominin and pre-agrarian human past; however, the recovery and identification of plant remains particularly from early prehistoric sites can be challenging and time consuming. But dietary reconstruction is incomplete without plants and the need to redress the balance between the role of animals and plants in the study of hunter-gatherer diet has been highlighted before (Mason and Hather 2002). Likewise, plant parts are likely to comprise most raw materials used, though traces of these rarely survive (Adovasio *et al.* 2007). This has contributed to a situation in which there are large gaps in our understanding of the use of plants, both as raw materials and in the diet, particularly for earlier prehistoric periods.

While the fragile nature of plants means that their recovery from non-agrarian sites is more challenging than more durable remains such as animal bone and lithics, there are many ways in which this important subject can be addressed. The development of flotation methods to extract minute carbonised plant remains from archaeological sites was first developed in the 1960s (Struever 1968). Charcoal analysis or anthracology, identifies wood types through microscopic analysis of anatomical structure. It is particularly useful for reconstruction of woody vegetation and for identifying the types of wood used as fuel. The study of carbonised seeds and fruit and nut fragments uses morphological attributes to identify the edible parts of some plants. The accumulated evidence acquired in this way represents much of what we know about human and hominin use of plants before the development of agriculture, and it continues to form the backbone of our developing understanding and awareness of the use of plants in the pre-agrarian world. However, the increasing focus on other methods, including the recovery and analysis of plant microfossils, the use of scanning electron microscopy in plant identification, the development of increasingly sophisticated methods for analysis of use wear and microwear traces on both tools and teeth, and the application of analytical techniques have extended the scope of recovery of information on plant use in pre-agrarian contexts for which there can sometimes be little in the way of macroscopically visible carbonised remains.

An understanding of why and how plants were used for both dietary and functional purposes in pre- and non-agrarian contexts are vast topics in terms of the length of time involved, the numerous approaches that can be taken and the significant gaps in our knowledge that need to be addressed. This book is divided into three parts. *Part 1: Setting the Scene*, comprises five chapters, each providing a broad contextual background for the use and role of plants in pre-agrarian life. This part begins by outlining the significance and properties of dietary carbohydrates from plant sources (Copeland, Chapter 1). This is followed by a detailed explanation of the physiological roles of different types of foods and the essential roles of plant harvesting and processing methods to nutritional value (Butterworth, Ellis and Wollstonecroft, Chapter 2). In Chapter 3, Huffman examines the use of medicinal plants by apes from behavioural, functional and evolutionary perspectives and also highlights the overlap in medicinal plant use by some contemporary humans and chimpanzees. Chapter 4 (Hardy) focuses on the role of plants as raw materials, in particular the development of controlled use of fire and composite technology, two of the most significant technological advances ever made. Chapter 5 (Arranz, Ibanez and Zapata) comprises a detailed summary of all the available evidence for plant use by the last hunter-gatherers in the Levant, from the Last Glacial Maximum (LGM) to the first experiments with plant cultivation at the beginning of the Holocene.

The development of alternative methods for detecting evidence for use of plants has led to significant advances and this is the focus of the second part (*Part 2: Plant Foods, Tools and People*) of the book. The vast time scales of the hunter-gatherer period are particularly reflected in this part, with chapters from across the world covering the Quaternary period from the Early Pleistocene to the mid-Holocene. Part 2 comprises seven chapters; each chapter focuses on a different method used to recover information on plants from pre-agrarian contexts. This part begins with Kubiak-Martens, who uses scanning electron microscopy to identify underground storage organs (USOs) (Chapter 6). Scanning electron microscopy is a valuable but underused method for the identification of root foods and USOs. Hather (1991; 1993) and Mason *et al.* (1994) have long highlighted the need to recognise and identify archaeological parenchyma, the edible USOs including roots, tubers, rhizomes and bulbs. While carbonised remains of these can be abundant, they are frequently unidentified. Chapters 7 and 8 cover tool use wear and dental microwear analyses. These methods, which are based on comparative experimental data, provide dietary and functional information. In Chapter 7, Van Gijn and Little discuss microwear analysis of ground and chipped stone assemblages, and objects made on bone and antler and how this can contribute to a better understanding of the role of plants in pre-agrarian societies. Martínez, Estebaranz and Pérez-Pérez (Chapter 8) discuss how buccal microwear patterns on enamel surfaces display distinct patterns that can be related to dietary habits and ecological conditions during the Plio-Pleistocene at the levels of both species and at times, populations. Chapters 9, 10 and 11 focus on the extraction and identification of plant microfossils from different contexts to explore environmental, dietary and functional perspectives. While not without limitations (Cabanes and Shahack-Gross 2015), these methods can offer direct evidence for plants on sites and from extended time periods for which macro-botanical remains rarely survive. For example, the endurance of phytoliths, which are made from silica or calcium oxalate and therefore do not readily degrade, has increased access to the presence of plants from a wide range of pre-agrarian sites and extended time periods (Albert *et al.* 2009; Piperno 1988; Madella *et al.* 2002; Chapters 9, 10). In Chapter 9, Albert and Esteban combine a number of techniques, including phytolith analysis, plant macroremains, charcoal analysis and FTIR to investigate environmental influence on human evolution at Olduvai Gorge and in South Africa. Chapter 10 (Power, Rosen and Nadel) comprises a study of phytoliths extracted from mortars carved into bedrock at the Late Natufian site of Raqefet. They show that both small-grained grasses and wheat and barley were present while forest resources also continued to be used. Chapter 11 (Boyadjian, Eggers and Scheel-Ybert) explores the use of plants by the inhabitants of Jabuticabeira II, a *sambaqui* (Brazilian shell mound), through extraction of plant microfossils from samples of dental calculus. The last chapter in this part, Chapter 12 (Hardy and Buckley), examines the role of stable isotope analysis in dietary reconstruction and outlines the contribution of carbon isotope analysis to a better understanding of the changes in plant types eaten by some Australopithecine species. This chapter also discusses methods used in the detection and identification of chemical compounds from ancient samples of dental calculus.

In *Part 3: Providing a Context; Ethnography, Ethnobotany, Ethnohistory, Ethnarchaeology,* five chapters investigate the use of wild plants in the diet or as raw materials, from modern or recent ethnographic contexts. In this final part, the very broad scope of plant use among modern or recent hunter-gatherers both in terms of food and material culture, is highlighted. The part begins with a summary of the evidence for the use of plants in fishing technology from the Mesolithic through the later prehistoric periods and into historical and ethnographic contexts in Russia and the Baltic region, and a detailed description of the fish traps and other artefacts related to fishing from the site of Zamostje, Russia (Clemente, Lozovski, Ballbè, Mazurkevich and Lozovskaya; Chapter 13). Chapter 14 (Brockwell, Stevenson and Clarke) provides a synthesis of the evidence for plant use in Australia going back over 40,000 years, in particular from Kakadu National Park, which is enriched by the modern ethnobotanical information which survives among many Aboriginal people. In Chapter 15, Berihuete Azorín, Piqué Huerta and Mansur investigate the use of plants as food by the Selknam, recent hunter-gatherers from Tierra del Fuego. Crittenden (Chapter 16) explores the ethnobotany of dietary plant food, the use of wild plants as extraction and/or processing tools and non-nutritive daily uses of plants among the Hadza hunter-gatherers of Tanzania. Finally, in Chapter 17 (Guèye and Ibra Samb) discuss the use of wild plant foods that are still part of the diet of traditional Malinké agriculturalists from the region of Kédougou, Senegal, the ways these are used and how some of them are disappearing from the diet.

Theories and evidence for the dietary use of plants before agriculture

During the earliest phases of human evolution, dietary reconstruction is based on theoretical models using skeletal and dental morphology as a starting point (e.g. Aiello and Wheeler 1995; Snodgrass *et al.* 2009; Macho 2014). Additionally, Wrangham and collaborators (Laden and Wrangham 2005; Wrangham 2005; Wrangham *et al.* 2009) have argued that late Miocene hominoids were able to expand their ecological niches through exploitation of USOs (underground storage organs) as fallback foods, with far reaching evolutionary implications. This may be supported by the growing evidence that some higher primates, primarily chimpanzees, but also to a lesser extent other species, use tools for a wide range of activities including digging for tubers (Hernandez-Aguilar *et al.* 2007; McGrew 2007; 2010a). The identification of carbon isotopes (C_3 and C_4) has provided some fascinating data on the development of Plio-Pleistocene hominin species, and the different ecological niches they may have occupied (Lee-Thorp *et al.* 2012; see also Chapter 12). There is little in the way of actual evidence of plants from these early periods, and studies such as the use of actualistics to provide environmental proxies (Chapter 9), and the analysis of microwear traces on teeth have been combined with stable isotope analysis in some instances to develop dietary reconstructions (e.g. Grine *et al.* 2012; see also Chapter 8).

There is very little macro-botanical evidence for plants from Lower and Middle Palaeolithic sites; at the 790,000 year old site of Gesher Benot Ya'aqov in Israel, carbonised nut shell fragments from seven species of edible nuts, including wild almond (*Amygdalus communis* ssp. *Microphylla*; *A. korshinskii*), prickly water lily (*Euryale ferox*), Atlantic pistachio (*Pistacia atlantica*), pistachio (*P. vera*), Palestine oak (*Quercus calliprinos*), Mt Tabor oak (*Q. ithaburensis*) and water chestnut (*Trapa natans*) were found together with pitted basalt stones, possibly used for opening the nuts, though their anthropogenic origin is not entirely assured (Goren-Inbar *et al.* 2002; 2004). Archaeobotanical analysis of macro remains of plant material from the Lower Palaeolithic site of Schöningen has revealed a diverse assemblage of wood and edible plant materials. Whether these plants were used for fuel, as food, medicine and/or raw materials, is unknown; however, this large assemblage significantly enhances the evidence for plants in the Lower Palaeolithic (Bigga *et al.* 2015). Many remains of charred legumes were found at Middle Palaeolithic Kebara Cave, Israel (Lev *et al.* 2005) and remains of charred nuts were found at the late Neanderthal site of Gorham's Cave, Gibraltar (Barton 2000). Jones (2009) suggests that higher latitudes created challenging conditions for plant food acquistion due to a combination of fewer edible plants and more complex packaging of the edible parts. He proposes a shift from monocotyledon stems to a broader use of dicotyledon seeds, fruits, roots and tubers, many of which would have required processing to remove toxins and make them edible. With so little in the way of direct evidence, much of the focus of paleo-dietary reconstruction, in particular in the Middle Palaeolithic, is based on stable isotope data (e.g. Bocherons 2009; Richards and Trinkaus 2009). However, stable isotope analysis preferentially identifies protein which means that plants are generally missing from these dietary reconstructions (Chapter 12).

Large scale flotation at the Upper Palaeolithic site of Dolní Věstonice (~26,000 BP) has recovered significant amounts of parenchymatous tissue from hearths which are thought to represent edible roots on the basis of scanning electron microscopy (SEM) (Mason *et al.* 1994; Pryor *et al.* 2013). Towards the latter part of the Upper Palaeolithic, there is an increase in macro-remains of plants recovered from a relatively small number of highly significant sites, including the 19,000 year old site of Ohalo II (Kislev *et al.* 1992; Weiss *et al.* 2004).

Abundant remains of wild nut-grass (*Cyperus rotundus*) tuber, which are thought to have been collected as food, were found at the 18,000 year old site of Wadi Kubbaniya (Hillman 2000; Hillman *et al.* 1989). Interestingly at the later Al Khiday site, 600 miles (*c.* 966 km) to the south, chemical compounds from *C. rotundus* tubers were found in samples of dental calculus from several individuals, confirming consumption of these tubers, albeit at a later date (Buckley *et al.* 2014). Charred remains from occupational deposits at Grotte des Pigeons at Taforalt in Morocco, dated between 15,000 and 13,700 cal BP, provided evidence for harvesting and processing of edible wild plants, including acorns and pine nuts, and other potential plant foods (Humphrey *et al.* 2013). The evidence for food plants from the Early Epipaleolithic to the Early Neolithic in the Near East is examined in Chapter 5; a site that stands out in this regard is Abu Hureyra with its large quantity of carbonised remains, including possible staples such as wild rye (*Secale* spp.), wheat (*Triticum* spp.) and club-rush (*Scirpus maritimus*) (Hillman 2000). Macro-remains of plant food, including seeds, fruits, legumes, nuts and parenchyma, have been found on many late hunter-gatherer sites worldwide (e.g. Aura *et al.* 2005; Bishop *et al.* 2014; Antolin *et al.* in press; Butler 1996; Fairbairn and Weiss and all chapters therein, 2009; Mason and Hather 2002; Mason *et al.* 1994; Ugent *et al.* 1982; Willcox 2002; Kubiak-Martens and Tobolski 2014; Kubiak-Martens *et al.* 2015); the evidence for the Mesolithic in northern Europe is discussed more fully in Chapter 6.

The development of microfossil extraction and identification has expanded the information available on plants from Palaeolithic sites. These include a study that combined tooth wear and stable isotopes with identification of phytoliths from the dental calculus of an *Australopithecus sediba* sample (Henry *et al.* 2012); evidence for starchy food and also chemical compounds identified as essential polyunsaturated fatty linoleic and linolenic acids, most probably from pine nuts, from the 400,000 year old Lower Palaeolithic site of Qesem Cave (Hardy *et al.* 2015), the Middle Palaeolithic site of Amud Cave, Israel where phytoliths provided evidence for edible grass seeds (Madella *et al.* 2002), starch granules found embedded in Neanderthal dental calculus from the Middle Palaeolithic sites of Shanidar and Spy (Henry *et al.* 2011) and evidence for starchy foods and medicinal plants from the dental calculus of Neanderthal individuals from the 49,000 year old site of El Sidrón, Spain (Hardy *et al.* 2012). Revedin *et al.* (2010) have identified evidence for plant processing, based on starch granules extracted from grinding stones from several Gravettian sites. The study of residues and use wear on stone tools has also been used to highlight exploitation of plants from many pre-agrarian sites (e.g. Van Peer *et al.* 2003; Chapter 7).

There is a wide range of information also available on pre-agrarian use of plants outside Europe and the Near East based on the extraction of microfossils. Some notable examples include evidence for consumption of grass seeds during the Middle Stone Age in Mozambique based on starch granules recovered from the surface of flaked and ground stone tools (Mercader 2009); starch and phytolith assemblages suggesting a range of edible species from Niah Cave, Borneo (Barton 2005) and numerous studies suggesting exploitation of a wide range of plants in China and Japan (e.g. Liu *et al.* 2011; Guan *et al.* 2014; Kitagawa and Yasuda 2008). The available evidence for plant use in pre-agrarian periods provides small glimpses of what is likely to have been extensive use of plants based on a deep applied knowledge that was accumulated over very long time periods.

Plant management may have begun long before it becomes apparent in the archaeological record, and there is abundant ethnographic and archaeological evidence from many places across the world including North America (Turner *et al.* 1990; Boyd, 1999; Stephens *et al.* 2007) and Australia (Latz 1995; Hill and Baird 2003; Gott 2005; Bliege Bird *et al.* 2008), that demonstrates resource and environmental management using fire among non-agrarian people. In central Australia some of the most important dietary wild plants, known as fireweeds, required regular burning to maintain their maximum production (Latz 1995). An increase in pollen from fire-resistant taxa dating from around 60,000 years ago at Niah Cave, Borneo, has been suggested to reflect biomass burning either to maintain open hunting areas or habitats rich in wild edible plants (Hunt *et al.* 2007). Elaborate management of plant resources for raw materials and food has also been recorded for the pre-agrarian Jomon period in Japan (Noshiro and Sasaki 2014). The possibility of plant husbandry in the European Mesolithic was first highlighted by Clarke (1976) and developed by Zvelebil (1994) who suggested that both conservational and promotional approaches to plant husbandry were being practised during the Mesolithic in Europe; evidence for human use of fire to manage plant resources in Mesolithic Europe, possibly linked to the production of acorns (Mason 2000) or for general landscape ecology (Innes and Blackford 2003; 2010; Edwards 1990), has been identified. Various scholars have suggested that burning of vegetation may have been part of Mesolithic hunting strategy (e.g. Mellars 1976). Burning would create open spaces in what was otherwise dense forest environments, but more importantly it would allow young shoots to grow which would attract game to the area (e.g. Innes and Blackford 2003). The practice may also have generated more plant foods (e.g. Mellars, 1976; Simmons 1996; Mason 2000). Pollen diagrams associated with Mesolithic sites sometimes display an increase in hazel pollen during a regeneration phase which follows a fire (e.g. Innes *et al.* 2010). The evidence for the deliberate and recurrent burning of reed marshes as proposed for the Early Mesolithic at Star Carr (Hather 1998; Law 1998), and also for the Early and Middle Mesolithic at Rotterdam Yangtze Harbour (Kubiak-Martens *et al.* 2015) are well documented examples of vegetation management in pre-agrarian Europe. Fire may have been used to improve conditions for hunting, to increase the yield of plant foods, or simply to create and maintain access to open water. However, a study to investigate the earlier use of fire for plant management in Europe found no evidence for this by Neanderthals or early anatomically modern humans (AMH) (Daniau *et al.* 2010).

In Chapter 2, Butterworth, Ellis and Wollstonecroft provide a clear explanation of why a diet comprising only animal produce is deleterious. The Inuit are often used as an example of a population that lived on a diet very high in animal produce, including both meat and fat (Hardy 2007). Recently, a gene (*CPT1A*) that is only present in high Arctic populations and appears to be an adaptation to either a high fat diet or the extreme cold has been identified (Clemente *et al.* 2014). This gene enables people to be in a more or less permanent state of ketosis, relying on gluconeogenesis (lipocentric – relying on fatty acids) with the ability to switch to the glucocentric (relying on glucose) mechanism, which is predominant today (Clemente *et al.* 2014). Even so, they are likely to have obtained more carbohydrates than has sometimes been thought; they ate plants including seaweed, tubers and berries, as well as chyme – the partially digested stomach contents of the animals they killed – and they obtained carbohydrates from their meat through their practice of preserving whole animals which enabled proteins to hydrolyse, or ferment, into carbohydrates (Hardy *et al.* 2015).

The hominins considered the most 'carnivorous', largely on the basis of carbon and nitrogen stable isotope analyses, are Neanderthals (Richards and Trinkaus 2009). While they could have overcome the nitrogen excess that a meat-only diet would have given them, a lack of carbohydrates may have left them dependent on gluconeogenesis (see Chapter 2). It is not yet known if Neanderthals had the *CPT1A* gene; however, they did not have access to the large quantities of marine mammal fat (around 50% of the diet; Speth 2010; 2012) which is such an important part of traditional Arctic diet. Extended use of gluconeogenesis without the *CPT1A* gene may have reduced their energy, and therefore their efficiency in hunting, and may well have had implications for their reproductive rate. However, Neanderthals were immensely successful, they lived for around 250,000 years and demonstrated great adaptability through extreme climatic variations; it is highly unlikely that they were able to do this based on an inefficient diet while Hardy (2010) demonstrates the availability of plants in the regions they occupied even during glacial periods. Up to 25% of the diet can be plant-based without featuring in the C/N isotope data (Jones 2009; Chapter 12), and Henry *et al.* (2011) and Hardy *et al.* (2012) have identified the presence of starch granules in a range of samples of Neanderthal dental calculus. It seems likely therefore, that even those living during glacial periods did include some plants in their diet, as the Inuit and other recent cold climate populations have done. If, as is understood, their primary resource was hunted animal produce, it is hard to imagine consistently successful hunting of any kind, from one which involved rapid bursts of speed to endurance running, without sufficient energy and the appropriate food to provide this.

Neanderthals are sometimes lumped together as a species, but they lived in widely different environmental zones, and there has been extensive investigation into what their diet consisted of, some of which has already suggested that it was as varied as would be expected from the widely different latitudes and climatic regimes in which they lived (e.g. Pérez-Pérez *et al.* 2003; El Zaatari 2011). Much remains to be done to fully understand what Neanderthals ate and a combination of many different approaches will hopefully provide further clarification.

The introduction to new and different environments, either through environmental change or movement, requires being able to adapt to new foods (Jones 2009). Possibly the most important feature of plant eating is knowing what to avoid as many plants can be poisonous. The complex suite of knowledge reflected in the ability to self-medicate suggests accumulation over very long timescales (see Chapter 3). By the time evidence for diet is reflected in the archaeological record, the mechanisms of knowledge acquisition and plant usage for dietary and self-medication purposes are likely to have been very well established. The identification of chemical compounds from non-nutritional medicinal plants from the dental calculus of Neanderthals from

El Sidrón (Hardy *et al.* 2012) has provided an example of this complexity of knowledge among Middle Palaeolithic Neanderthals. It is highly unlikely that any hominin species or human population could have evolved without knowing what to eat, what to avoid, how to take care of themselves and how to successfully treat at least some common ailments (Hardy *et al.* 2013; 2016; see also Chapter 3).

Ethnobotanical data from around the world demonstrates the use of wild plants as food and medicine, including at high latitude, permafrost and mountain regions where plant biodiversity can be considerably lower than in higher latitudes (e.g. Nelson 1899; Bogoras 1904–5; Porsild 1953; Rudenko 1961; Bergman *et al.* 2004; Kang *et al.* 2013). Inner Mongolia, for example, which is a region known for its traditions of animal herding and consumption of animal products, has an abundant record of plants integrated into the traditional diet, and also used for tea, as medicine and for raw materials (e.g Huai Khasbagan and Pei 2000; Khasbagan 2008; Khasbagan *et al.* 2005; Khasbagan and Soyolt 2007; Khasbagan and Hui 2011). Recent hunter-gatherers, horticulturalists and traditional farmers retain a very broad knowledge of the plants in their environments and Cordain *et al.* (2000), extracted data from Murdock's (1967) *Ethnographic Atlas* which is based on extensive literature searches of 1267 hunter gatherer societies, to demonstrate dependence on plant foods ranging from 6–15% for tundra areas, to 46–55% for areas of desert grasses and scrubs and tropical grasslands.

The scope of traditional knowledge of plants can be hard to comprehend particularly for modern urban populations; for example in New Guinea in 1976, 1035 plant species, representing 470 genera and 146 families were known to be used; of these, 332 species, 215 genera and 99 families for medicinal purposes (Powell, 1976) while 169 different plant species were used as raw materials among the Wola of highland Papua New Guinea (Sillitoe 1988). Other examples include the !Kung San who could identify and name over 200 plant species and used 105 for food (Lee 1979), while Kuhnlein and Turner (1991) recorded 550 native plants traditionally used as food in Canada. Likewise, Owen (1993) describes the extensive use of plants in the material culture among groups from northern North America. Among the Tzeltal Maya, Mexico, children aged 12 were recently able to name 95% of the 85 plants selected for the study (Zarger and Stepp 2004). Still today almost all human groups living within traditional economic contexts continue to rely on plants for medicine. For example, in 2002 according to the World Health organisation 'over 80% of the population in developing countries depended directly on plants for their medical requirements' (Ssegawa and Kasene 2007, 522). The efficacy of some medicinal plants used by non-industrialised communities is confirmed by a study which linked useful compounds to parallels in treatment in three separate regions of the world (Haris Saslis-Lagoudakis *et al.* 2012). Still today, the collection and use of selected wild plants in particular for use as medicine is widespread in Turkey (Ertüg 2000), and even in southern Europe. Finally, the number of academic journals that incorporate ethnobotanical studies in their remit is a reflection of the abundant knowledge that still survives. Examples include *Economic Botany, Ethnobotany Research and Applications, Journal of Ethnobiology, Journal of Ethnobiology and Ethnomedicine, Journal of Ethnopharmacology, Journal of Medicinal Plants Research*.

The time covered in this book begins at the earliest stages of human evolution. For most of human history, our species and all our ancestral species have relied entirely on their knowledge of the wild, natural, world, including a deep applied knowledge of plants, to survive. The depth of this synergy with plants may perhaps be illustrated by the psychological benefits to urban dwellers of having plants in their environment (Fuller *et al.* 2007), while Sullivan *et al.* (2011) highlight a possible evolutionary reliance on natural medicines: 'Humans love medicinal drugs ... Worldwide, the amount of money spent on medicines annually is growing exponentially and is expected to reach around US$1 trillion in 2012', adding 'pharmophilia evolved as a means to cope with disease and sickness and is mediated through belief-induced neurological and immunological signalling pathways' (Sullivan *et al.* 2011, 572).

What survives of this deep relationship with the natural world is encompassed in the concept of TEK (Traditional Ecological /Environmental Knowledge) 'a cumulative body of knowledge, practice and belief, evolving by adaptive processes and handed down through generations by cultural transmission about the relationship of living beings (including humans) with one another and with their environment' (Berkes 1999, 8), in terms of understanding and use of plants, it is referred to as 'ecological intelligence' by Jones (2009: 173). Inglis (1993) highlights the continuing relevance and importance of TEK, while the beginning of agriculture has been described as 'the break with our past and the incipient loss of traditional ecological knowledge (TEK)' (Society for Ecological Restoration 2015). The implication is that the vast amount of knowledge that was accumulated over huge periods of time began to be eroded once our basic nutritional requirements had been resolved by a small number of plants whose growth, management and production were secured by agricultural practices. In the light of this, TEK represents part of the enduring heritage of our hunter-gatherer past. We need to search for the archaeological evidence and use this information to enrich our knowledge and understanding of the processes involved. Every time we find out something new about plant use before agriculture, it broadens the perspectives on our hunter-gatherer origins while ultimately contributing to the wider picture of the role of plants in our mental and physiological makeup.

Plants are essential for humans and their role in pre-agrarian life has been widely underappreciated (Mason and Hather 2002). The development of a range of new techniques such as the extraction and identification of residues and microfossils, the increasingly sophisticated methods of dental and artefact microwear analysis together with the various microscopic methods of identification, including optical and scanning electron microscopy, analytical techniques, the developing understanding of the physiological role of plants and the role of processing methods are all contributing to a better understanding of the role plants played in the pre-agrarian past. This will ultimately permit fuller and more realistic perspectives on pre-agrarian lifestyle, knowledge, self-medication and diet.

We are indebted to the I+D micinn 2010 (Ministerio de Ciencia e Innovación, Madrid. Spain) (project number HAR2012-35376) and BIAX Consult for contributing towards the publication costs. Front cover artwork was done by Martin Valkhoff.

References

Adovasio, J. M., Soffer, O. and Page, J. 2007. *The Invisible Sex*. New York: Smithsonian Books

Aiello, L. C. and Wheeler, P. 1995. The Expensive-Tissue Hypothesis: the brain and the digestive system in human and primate evolution. *Current Anthropology* 36, 199–221

Albert, R. M., Bamford, M. K. and Cabanes, D. 2009. Palaeoecological significance of palms at Olduvai Gorge, Tanzania, based on phytolith remains. *Quaternary International* 193, 41–8

Antolín, M., Berihuete, A. and Hardy, K. Potential and limitations for the interpretation of botanical macroremains from Mesolithic shell middens. The case of Sand (Applecross, Island of Skye, Scotland). in press *Proceedings of the Meso 2010 conference, Santander September 2010*

Aura, J. E., Carrión, Y., Estrelles, E. and Pérez Jordà, G. 2005. Plant economy of hunter gatherer groups at the end of the last Ice Age: plant macroremains from the cave of SantaMaira (Alacant, Spain) ca. 12000–9000 b.p. *Vegetation History and Archaeobotany* 14(4), 542–50

Barton, H. 2005. The case for rainforest foragers: the starch record at Niah Cave, Sarawak. *Asian Perspectives* 44, 56–72

Barton, R. N. E. 2000. Mousterian hearths and shellfish: Late Neanderthal activities in Gibraltar. In Stringer, C. B., Barton, R. N. E. and Finlayson, J. C. (eds), *Neanderthals on the Edge: 150th Anniversary Conference of the Forbes' Quarry discovery, Gibraltar*, 211–20. Oxford: Oxbow Books

Bergman, I. Östlund, L. and Zackrisson, O. 2004. The use of plants as regular food in ancient subarctic economies: a case study based on Sami use of Scots Pine innerbark. *Arctic Anthropology* 41(1), 1–13

Berkes, F. 1999. *Sacred Ecology: Traditional Ecological Knowledge and Resource Management.* London: Routledge

Bigga, G., Schoch, W. H. and Urban, B. 2015. Paleoenvironment and possibilities of plant exploitation in the Middle Pleistocene ofSchöningen (Germany). Insights from botanical macro-remains and pollen. *Journal of Human Evolution* 89(92–104), http://dx.doi.org/10.1016/j.jhevol.2015.10.005

Bishop, R. R, Church, M. J. and Rowley-Conwy, P. A. 2014. Seeds, fruits and nuts in the Scottish Mesolithic. *Proceedings of the Society of Antiquaries of Scotland* 143, 9–72

Bliege Bird, R., Bird, D. W., Codding, B. F., Parker, C. H. and Jones, J. H. 2008. The "fire stick farming" hypothesis: Australian Aboriginal foraging strategies, biodiversity, and anthropogenic fire mosaics. *Proceedings of the National Academy of Science (USA)* 105, 14796–801

Bocherons, H. 2009. Neanderthal dietary habits: review of the isotopic evidence. In Hublin, J. J. and Richards, M. P. (eds), *The Evolution of Hominin Diets: Integrating Approaches to the Study of Palaeolithic Subsistence*, 241–50. Dordrecht: Springer

Bogoras, W. 1904–5 *The Chukchee*. New York: Memoirs of the Museum of the American Museum of Natural History 2

Boyd, R. T. (ed.) 1999. *Indians, Fire, and the Land*. Corvallis, OR: Oregon State University Press

Buckley, S., Usai, D., Jakob, T., Radini, A. and Hardy, K. 2014. Dental calculus reveals evidence for food, medicine, cooking and plant processing in prehistoric Central Sudan. *PLOS ONE* 9(7), e100808

Butler, A. 1996. Trifolieae and related seeds from archaeological contexts: problems in identification. *Vegetation History and Archaeobotany* 5(1–2), 157–67

Cabanes, D. and Shack Gross, R. 2015. Understanding fossil phytolith preservation: the role of partial dissolution in paleoecology and archaeology. PLoS ONE 10(5): e0125532. doi:10.1371/journal.pone.0125532

Clarke, D. L. 1976. Mesolithic Europe: the economic basis. In Sieveking, G. de G., Longworth, I. H. and Wilson, K. E. (eds), *Problems in Economic and Social Archaeology*, 449–81. London: Duckworth

Clemente, F. J., Cardona, A., Inchley, C. E., Peter, B. M., Jacobs, G., Pagani, L. *et al.* 2014. A selective sweep on a deleterious mutation in the CPT1A gene in Arctic populations. *American Journal of Human Genetics* 95(5), 584–9

Cordain, L., Miller, J. B., Eaton, S. B., Mann, N., Holt, S. H. A. and Speth, J. D. 2000. Plant animal subsistence ratios and macronutrient energy estimations in worldwide hunter gatherer diets. *American Journal of Clinical Nutrition* 71, 682–92

Daniau, A. L., d'Errico, F. and Goñi, M. F. S. 2010.Testing the hypothesis of fire use for ecosystem management by Neanderthal and Upper Palaeolithic modern human populations. *PLOS ONE* 5(2), e9157

Edwards, K. J. 1990. Fire and the Scottish Mesolithic: evidence from microscopic charcoal. In Vermeersch, P. (ed.), *Contribution to the Mesolithic in Europe*, 71–9. Leuven: Leuven University Press

El Zaatari, S., Grine, F. E., Ungar, P. S. and Hublin, J. J. 2011. Ecogeographic variation in Neandertal dietary habits: evidence from occlusal molar microwear texture analysis. *Journal of Human Evolution* 61, 411–24

Ertuğ, F. 2000. An ethnobotanical study in central Anatolia (Turkey). *Economic Botany* 54(2), 155–82.

Fairbairn, A. S. and Weiss, E. 2009. *From Foragers to Farmers: Papers in Honour of Gordon C. Hillman*. Oxford: Oxbow Books

Fuller, R. A., Irvine, K. N., Devine-Wright, P., Warren, P. H. and Gaston, K. J. 2007. Psychological benefits of greenspace increase with biodiversity. *Biology Letters* 3(4), 390–4

Goren-Inbar, N., Alperson, N., Kislev, M. E., Simchoni, O., Melamed, Y., Ben-Nun, A. and Werker, E. 2004. Evidence of hominin control of fire at Gesher Benot Yaaqov, Israel. *Science* 304(5671), 725–7

Goren-Inbar, N., Sharon, G., Melamed, Y. and Kislev, M. 2002. Nuts, nut cracking, and pitted stones at GesherBenotYa'aqov, Israel. *Proceedings of the National Academy of Science (USA)* 99, 2455–60

Gott, B. 2005. Aboriginal fire management in south-eastern Australia: aims and frequency. *Journal of Biogeography* 32(7), 1203–8

Grine, F. E., Sponheimer, M., Ungar, P. S., Lee-Thorp, J. A. and Teaford, M. F. 2012. Dental microwear and stable isotopes inform the palaeoecology of extinct hominins. *American Journal of Physical Anthropology* 148, 285–317

Guan, Y., Pearsall, D. M., Gao, X., Chen, F., Pei, S. and Zhou, Z. 2014. Plant use activities during the Upper Paleolithic in East Eurasia: Evidence from the Shuidonggou Site, Northwest China. *Quaternary International* 347, 74–83

Hardy, B. L. 2010. Climatic variability and plant food distribution in Pleistocene Europe: implications for Neanderthal diet and subsistence. *Quaternary Science Reviews*, 29(5), 662–79

Hardy, K. 2007. Food for thought. Tubers, seeds and starch in hunter gatherer diet. *Mesolithic Miscellany*.18(2). http://www.york.ac.uk/depts/arch/Mesolithic/mmpdf/18.2.pdf

Hardy, K., Buckley, S. and Huffman, M. 2013. Neanderthal self-medication in context. *Antiquity* 87(337), 873–78

Hardy, K., Buckley, S. and Huffman, M. 2016. Doctors, chefs or hominin animals? Non-edible plants and Neanderthals. *Antiquity* 90

Hardy, K., Brand Miller, J., Brown, K. J., Thomas, M. G. and Copeland, L. 2015. The importance of dietary carbohydrate in human evolution. *The Quarterly Review of Biology* 90(3), 251–68 DOI: 10.1086/682587

Hardy, K., Buckley, S., Collins, M. J., Estalrrich, A., Brothwell, D., Copeland, L., García-Tabernero, A., García-Vargas, S., de la Rasilla, M., Lalueza-Fox, C., Huguet, R., Bastir, M., Santamaría, D., Madella, M., Fernández Cortés, A. and Rosas, A. 2012. Neanderthal medics? Evidence for food, cooking and medicinal plants entrapped in dental calculus. *Naturwissenschaften* 99(8), 617–26. *DOI:* 10.1007/s00114-012-0942-0

Hardy, K., Radini, A., Buckley, S., Sarig, R., Copeland, L., Gopher, A. and Barkai, R. 2015. *First direct evidence for Middle Pleistocene diet and environment from dental calculus at Qesem Cave* DOI 10.1016/j.quaint.2015.04.033

Haris Saslis-Lagoudakis, C., Savolainen, V., Williamson, E. M., Forest, F., Wagstaff, S. J., Baral, S. R., Watson, M. F., Pendry, C. A. and Hawkins, J. A. 2012. Phylogenies reveal predictive power of traditional medicine in bioprospecting. *Proceedings of the National Academy of Science (USA)* 109, 15835–40

Hather, J. G. 1991. The identification of charred archaeological remains of vegetative parenchymatous tissue, *Journal of Archaeological Science* 18(6), 661–75

Hather, J. G. 1993. *An Archaeobotanical Guide to Root and Tuber Identification. Volume 1. Europe and South West Asia.* Oxford: Oxbow Monograph 28

Hather, J. G. 1998. Identification of macroscopic charcoal assemblages. In Mellars, P. and Dark, P. (eds), *Star Carr in Context, New Archaeological and Palaeoecological Investigations at the Early Mesolithic Site of Star Carr in North Yorkshire*, 183–96. Cambridge: McDonald Institute Monograph

Henry, A. G., Brooks, A. S. and Piperno, D. R. 2011. Microfossils in calculus demonstrate consumption of plants and cooked foods in Neanderthal diets (Shanidar III, Iraq; Spy I and II, Belgium). *Proceedings of the National Academy of Sciences (USA)* 108(2), 486–91

Henry, A. G., Ungar, P. S., Passey, B. H., Sponheimer, M., Rossouw, L., Bamford, M., Sandberg, P., De Ruiter, D. J. and Berger, L. 2012. The diet of Australopithecus sediba. *Nature* 487(7405), 90–3

Hernandez-Aguilar, R. A., Moore, J. and Pickering, T. R. 2007. Savanna chimpanzees use tools to harvest the underground storage organs of plants. *Proceedings of the National Academy of Sciences(USA)* 104(49), 19210–3

Hill, R. and Baird, A. 2003. Kuku-Yalanji rainforest Aboriginal people and carbohydrate resource management in the wet tropics of Queensland, Australia. *Human Ecology* 31, 27–52

Hillman, G. C. 1989. Late Palaeolithic plant foods from Wadi Kubbaniya in Upper Egypt: dietary diversity, infant weaning, and seasonality in a riverine environment. In Harris, D. R. and Hillman, G. C. (eds), *Foraging and Farming. The Evolution of Plant Exploitation.* London: Unwin Hyman, One World Archaeology

Hillman, G. C., Madeyska, E. and Hather, J. 1989. Wild plant foods and diet at Late Palaeolithic Wadi Kubbaniya: the evidence from charred remains. In Wendorf, F., Schild, R. and Close, A. (eds), *The Prehistory of Wadi Kubbaniya Volume 2. Stratigraphy, Paleoeconomy and Environment*, 162–242. Dallas TX: Southern Methodist University Press

Hillman, G. C. 2000. Overview: the plant based components of subsistence in Abu Hureyra 1 and 2. In Moore, A. M. T., Hillman, G. C. and Legge A. J. (eds), *Village on the Euphrates: From Foraging to Farming at Abu Hureyra*, 416–22. Oxford: Oxford University Press

Huai, Khasbagan H-Y. and Pei S-J. 2000. Wild plants in the diet of Arhorchin Mongol herdsmen in Inner Mongolia. *Economic Botany* 54(4) 528–36

Huffman, M. A. 2003. Animal self-medication and ethno-medicine: exploration and exploitation of the medicinal properties of plants. *Proceedings of the Nutritional Society* 62, 371–81

Humphrey, L. H., De Groote, I., Morales, J., Barton, N., Collcutt, S., Ramsey, C. B. and Bouzougga, A. 2013. Earliest evidence for caries and exploitation of starchy plant foods in Pleistocene hunter-gatherers from Morocco. *Proceedings of the National Academy of Sciences (USA)* 114(3) 954–9

Hunt, C. O., Gilbertson, D. D. and Rushworth, G. 2007. Modern humans in Sarawak, Malaysian Borneo, during Oxygen Isotope Stage 3: palaeoenvironmental evidence from the Great Cave of Niah. *Journal of Archaeological Science* 34, 1953–69

Inglis, J. T. (ed.) 1993. *Traditional Ecological Knowledge: Concepts and Cases.* Ottawa: International Program on Traditional Ecological Knowledge, International Development Research Centre

Innes, J. B. and Blackford, J. J. 2003. The ecology of late Mesolithic woodland disturbances: model testing with fungal spore assemblage data. *Journal of Archaeological Science*, 30(2), 185–94

Innes, J., Blackford, J. and Simmons, I. 2010. Woodland disturbance and possible land-use regimes during the Late Mesolithic in the English uplands: pollen, charcoal and non-pollen palynomorph evidence from Bluewath Beck, North York Moors, UK. *Vegetation History and Archaeobotany* 19(5–6), 439–52

Jones, M. 2009. Moving north: archaeobotanical evidence for plant diet in Middle and Upper Paleolithic Europe. In Hublin, J. J. and Richards, M. P. (eds), *The Evolution of Hominin Diets: Integrating Approaches to the Study of Palaeolithic Subsistence*, 171–80. Dordrecht: Springer

Kang, Y., Łuczaj, Ł., Kang, J. and Zhang, S. 2013. Wild food plants and wild edible fungi in two valleys of the Qinling Mountains (Shaanxi, central China). *Journal of Ethnobiology and Ethnomedicine* 9(1), 26

Khasbagan, S. 2008. Indigenous knowledge for plant species diversity: a case study of wild plants' folk names used by the Mongolians in Ejina desert area, Inner Mongolia, PR China. *Journal of Ethnobiology and Ethnomedicine* 4(1), 2

Khasbagan, Y. and Hui, Z. H. A. O. (2011). Study on Traditional Knowledge of Wild Edible Plants Used by the Mongolians in Xilingol Typical Steppe Area [J].*Plant Diversity and Resources* 2, 018

Khasbagan, S., Man, L., Enhebayar, G. and Hu, W. 2005.Traditional usage of wild plants for food by the Ejina Mongolians and its exploitation and ethnoecological significance. *Journal of Inner Mongolia Normal University (Natural Science Edition)* 34(4), 471–88

Khasbagan, Soyolt. 2007. Ephedra sinicaStapf (Ephedraceae): the fleshy bracts of seed cones used in Mongolian food and its nutritional components. *Economic Botany* 61(2), 192–7

Kislev, M. E., Nadel, D. and Carmi, I. 1992. Epipalaeolithic (19,000) cereal and fruit diet at Ohalo II, Sea of Galilee, Israel. *Review of Palaeobotany and Palynology* 73, 161–6

Kitagawa, J. and Yasuda, Y. 2008. Development and distribution of *Castanea* and *Aesculus* culture during the Jomon Period in Japan. *Quaternary International* 184(1), 41–55

Kubiak-Martens, L. and Tobolski, K. 2014. Late Pleistocene and Early Holocene Vegetation History and Use of Plant Foods in the Middle Vistula River Valley at Całowanie. In Schild, R. (ed.) *Całowanie a Final Paleolithic and Early Mesolithic Site on an Island in the Ancient Vistula Channel*, 333–48. Warsaw: Institute of Archaeology and Ethnology Polish Academy of Sciences

Kubiak-Martens, L., Kooistra, L. I. and Verbruggen, F. 2015. Archaeobotany: landscape reconstruction and plant food subsistence economy on a meso and microscale. In Moree, J. M. and Sier, M. M. (eds), *Interdisciplinary Archaeological Research Programme Maasvlakte 2, Rotterdam, Part 1, Twenty metres deep! The Mesolithic period at the Yangtze Harbour site – Rotterdam Maasvlakte, the Netherlands. Early Holocene landscape development and habitation*, 223–286. Rotterdam: BOORrapporten 566

Kuhnlein, H. V and Turner, N. J. 1991. *Traditional Plant Foods of Canadian Indigenous Peoples. Nutrition, Botany and Use* 8. Philadelphia PA: Gordon and Breach

Laden, G. and Wrangham, R. 2005. The rise of the hominids as an adaptive shift in fallback foods: plant underground storage organs (USOs) and australopith origins. *Journal of Human Evolution* 49(4), 482–98

Latz, P. 1995. *Bush Fires and Bush Tucker. Aboriginal Plant Use on Central Australia*. Alice Springs: IAD Press

Law, C. A. 1998. The use and fire-ecology of reedswamps vegetation. In Mellars, P. and Dark, P. (eds), *Star Carr in Context, New Archaeological and Palaeoecological Investigations at the Early Mesolithic Site of Star Carr in North Yorkshire*, 197–206. Cambridge: McDonald Institute Monograph

Lee, R. B. 1979. *The !Kung San. Men, Women and Work in a Foraging Society*. Cambridge: Cambridge University Press

Lee-Thorp, J., Likius, A., Mackaye, H. T., Vignaud, P., Sponheimer, M. and Brunet, M. 2012. Isotopic evidence for an early shift to C(4) resources by Pliocene hominins in Chad. *Proceedings of the National Academy of Sciences of the United States of America* 109, 20369–72

Lee-Thorp, J. A., Sponheimer, M. and van der Merwe, N. J. 2003. What do stable isotopes tell us about hominid dietary and ecological niches in the Pliocene? *International Journal of Osteoarchaeology*, 13(1–2), 104–13

Lev, E., Kislev, M. and Bar Yosef, O. 2005. Mousterian vegetal food in Kebara Cave, Mt. Carmel. *Journal of Archaeological Science* 32, 475–84

Liu, L., Ge, W., Bestel, S., Jones, D., Shi, J., Song, Y. and Chen, X. 2011. Plant exploitation of the last foragers at Shizitan in the Middle Yellow River Valley China: evidence from grinding stones. *Journal of Archaeological Science* 38(12), 3524–32

Macho, G. A. 2014. Baboon feeding ecology informs the dietary niche of *Paranthropus boisei*. *PLOS ONE* 9(1), e84942

Madella, M., Jones, M. K. Goldberg, P., Goren, Y. and Hovers, E. 2002. The exploitation of plant resources by Neanderthals in Amud Cave (Israel): the evidence from phytolith studies. *Journal of Archaeological Science* 29, 703–19

Mason, S. L. R. 2000. Fire and Mesolithic subsistence-managing oaks for acorns in northwest Europe? *Palaeogeography, Palaeoclimatology, Palaeoecology* 164, 139–50

Mason, S. L. R. and Hather, J. G. (eds) 2002. *Hunter-Gatherer Archaeobotany: Perspectives from the Northern Temperate Zone*. London: Institute of Archaeology, University College London

Mason, S. L. R., Hather, J. G. and Hillman, G. C. 1994. Preliminary investigation of the plant macro-remains from Dolní Věstonice II, and its implications for the role of plant foods in Palaeolithic and Mesolithic Europe. *Antiquity* 68, 48–57

McGrew, W. C. 2007. Savanna chimpanzees dig for food. *Proceedings of the National Academy of Sciences (USA)* 104(49), 19167–8

McGrew, W. C. 2010a. Chimpanzee technology. *Science* 328(5978), 579–80

McGrew, W. C. 2010b. In search of the last common ancestor: new findings on wild chimpanzees. *Philosophical Transactions of the Royal Society B* 365, 3267–76

Mellars, P. 1976. Fire ecology, animal populations and man: a study of some ecological relationships in Prehistory. *Proceedings of the Prehistoric Society* 42, 15–45

Mercader, J. 2009. Mozambican grass seed consumption during the Middle Stone Age. *Science* 326(5960), 1680–3

Murdock, G. P. 1967. Ethnographic Atlas: a summary. *Ethnology* 6, 109–236

Nelson, E. W. 1899. *The Eskimo about Bering Strait*. 18th Annual Report. Washington DC: Bureau of American Ethnology

Noshiro, S. and Sasaki, Y. 2014. Pre-agricultural management of plant resources during the Jomon period in Japan – a sophisticated subsistence system on plant resources. *Journal of Archaeological Science* 42, 93–106

Owen, L. R. 1993. Material worked by hunter and gatherer groups of northern North America: implications for use-wear analysis. In Anderson, P. S., Beyries, S., Otte, M. and Plisson, H. (eds), *Traces et Fonction: Les Gestes Retrouvés. Volume 2. Actes du Colloque International de Liège, 8–9–10 décembre 1990*, 50. Liège: ERAUL

Pérez-Pérez, A., Espurz, V., Bermúdez de Castro, J. M., de Lumley, M. A. and Turbón, D. 2003. Non-occlusal dental microwear variability in a sample of Middle and Late Pleistocene human populations from Europe and the Near East. *Journal of Human Evolution* 44, 497–513

Piperno, D. R. 1988. *Phytolith Analysis: An Archaeological and Geological Perspective*. San Diego CA: Academic Press

Porsild, A. E. 1953. Edible plants of the Arctic. *Arctic* 6(1), 15–34

Powell, J. M. 1976. Ethnobotany. In Paijmans K. (ed.) *New Guinea Vegetation*, 106–83. Canberra: Australia National University Press

Pryor, A. J. E., Steele, M., Jones, M. K., Svoboda, J. and Beresford-Jones, D. G. 2013. Plant foods in the Upper Palaeolithic at DolníVěstonice? Parenchyma redux. *Antiquity*, 87, 971–84

Revedin, A., Aranguren, B., Becattini, R., Longo, L., Marconi, E., Lippi, M. M. and Svoboda, J. 2010. Thirty thousand-year-old evidence of plant food processing. *Proceedings of the National Academy of Sciences (USA)* 107(44), 18815–9

Richards, M. P. and Trinkaus, E. 2009. Isotopic evidence for the diets of European Neanderthals and early modern humans. *Proceedings of the National Academy of Science (USA)* 106, 16034–9.

Rudenko, S. I. 1961. *The Ancient Culture of the Bering Sea and the Eskimo Problem*. Toronto: Archaeological Institute of North America, University of Toronto Press

Sillitoe, P. 1988. *Made in Niugini: Technology in the Highlands of Papua New Guinea*. London: Trustees of the British Museum

Simmons, I. G. 1996. *The Environmental Impact of Later Mesolithic Cultures. The Creation of Moorland Landscape in England and Wales*. Edinburgh: Edinburgh University Press

Society for Ecological Restoration. 2015. *Traditional Ecological Knowledge*. Indigenous Peoples Restoration Network. http://www.ser.org/iprn/traditional-ecological-knowledge

Speth, J. D. 2010. *The Paleoanthropology and Archaeology of Big-game Hunting: Protein, Fat, Or Politics?* Dordrecht: Springer online

Speth, J. D. 2012. Middle Palaeolithic subsistence in the Near East. *Before Farming* 2, 1–45

Snodgrass, J. J., Leonard, W. R. and Robertson, M. L. 2009. The Energetics of Encephalization in Early Hominids. In Hublin, J. J. and Richards, M. P. (eds), *The Evolution of Hominin Diets: Integrating Approaches to the Study of Palaeolithic Subsistence*, 15–30. Dordrecht: Springer

Ssegawa, P. and Kasenene, J. M. 2007. Medicinal plant diversity and uses in the Sango bay area, Southern Uganda. *Journal of Ethnopharmacology* 113(3), 521–40

Stephens, S. L., Martin, R. E. and Clinton, N. E. 2007. Prehistoric fire area and emissions from California's forests, woodlands, shrublands, and grasslands. *Forest Ecology and Management* 251(3), 205–16

Struever, S. 1968. Flotation techniques for the recovery of small-scale archaeological remains. *American Antiquity* 33(3), 353–62

Sullivan, R., Behncke, I. and Purushotham, A. 2010. An evolutionary perspective on the human love of pills, potions and placebo. *EMBO Report* 11, 572–8. London: Macmillan

Turner, N. J., Thompson, L. C., Thompson, M. T. and York, A. Z. 1990. *Ethnobotany. Knowledge and usage of plants by the Thompson Indians of British Columbia*. Victoria: Royal British Columbia Museum Memoir 3

Ugent, D., Pozorski, S. and Pozorski, T. 1982. Archaeological potato tuber remains from the Casma Valley of Peru. *Economic Botany* 36, 182–92

Van Peer, P., Fullagar, R., Stokes, S., Bailey, R., Moeyersons, J., Steenhoudt, F., Geerts, A., Vanderbeken, T., De Dapper, M. and Geus, F. 2003. The Early to Middle Stone Age transition and the emergence of modern human behaviour at site 8-B-11, Sai Island, Sudan. *Journal of Human Evolution* 45, 187–93

Weiss, E., Kislev, M. E., Simchoni, O. and Nadel, D. 2004. Small-grained wild grasses as staple food at the 23,000 year-old site of Ohalo II, Israel. *Economic Botany* 588, 125–34

Willcox, G. 2002. Charred plant remains from a 10th millennium B.P. kitchen at Jerf el Ahmar (Syria). *Vegetation History and Archaeobotany* 11(1–2), 55–60

Wrangham, R. 2005. The delta hypothesis: hominoid ecology and hominin origins. In Lieberman, D. E., Smith, R. J. and Kelley, J. (eds) *Interpreting the Past: Essays on Human, Primate and Mammal Evolution in Honor of David Pilbeam*, 231–42. Boston: Brill

Wrangham, R., Cheney, D., Seyfarth, R. and Sarmiento, E. 2009. Shallow-water habitats as sources of fallback foods for hominins. *American Journal of Physical Anthropology* 140(4), 630–42

Zarger, R. K. and Stepp J. R. 2004. Persistence of Botanical Knowledge among Tzeltal Maya Children. *Current Anthropology* 45(3), 413–8

Zvelebil, M. 1994. Plant use in the Mesolithic and its role in the transition to farming. *Proceedings of the Prehistoric Society* 60, 35–74

PART 1
SETTING THE SCENE

In this volume, we start with two chapters which set out the nutritional context for why humans need plants. Plant foods provide carbohydrates, proteins, oils, minerals, vitamins, and dietary fibre. As highlighted in Chapters 1 and 2, all these plant nutrients are important components of the human diet. The need to balance consumption of meat with plants is emphasised. Carbohydrates, which are essential for human nutrition, may have been eaten raw, but in order to turn many starchy food items, such as roots and tubers and grains of wild grasses, into readily digestible forms, they need to be processed. This raises interesting questions in particular for how early hominins obtained their dietary energy from starch-rich plants, which is discussed in Chapter 2. In this respect, the concept of 'timely dextrous unpacking' which refers to the ways that food sources can be extracted from either soil or their protective plant coverings (for example shells) as well as the ecological knowledge to know when to collect specific items, have been outlined by Jones (2009: 173).

This is followed with a chapter on the extent of applied knowledge of medicinal plants by apes (Chapter 3). This offers a new contextual horizon for understanding the use of some plants by hominins and early human populations. There is an extensive overlap in the use of certain plants for medicinal purposes by some apes and some contemporary human populations. One of the most fascinating aspects of this chapter is the description of self-medication by chimpanzees and the potential implication of this in terms of plant-based self-medication by hominins and early human populations. Chapter 4 outlines some of the pre-agrarian evidence for plant raw materials in early prehistory, and places into context some of the plant-based technological achievements, many of which form the basis for the human world in its broadest sense. The developments of twisting fibre, the art of pitch and bitumen preparation, and the resulting development of composite technology should be considered as major human technological achievements. The final chapter in this part (Chapter 5) lays out the extent of wild plant use from the Epipaleolithic to the Early Neolithic in the Levant and suggests how the use of this broad range of plants eventually led to the development of agriculture here. Among others, two of the best known Epipaleolithic sites of Ohalo II and Abu Hureyra I are discussed. Both sites revealed exceptionally rich assemblages of small-grained grasses and wild cereals and other plant-foods, including possible staples. This chapter provides a contextual background to the early development of agriculture.

1. Food carbohydrates from plants

Les Copeland

This chapter gives an overview of the significance and properties of dietary carbohydrates from plant sources. Carbohydrates are macro-constituents of many foods and essential in the human diet, providing a source of readily metabolisable energy, and dietary fibre and prebiotics necessary for gut microflora and gut health. Carbohydrates occur in diverse forms, which include monosaccharides (simple sugars), oligosaccharides (containing 2–10 monosaccharides), and polysaccharides (complex biopolymers that contain many monosaccharide units). Monosaccharides are chemically highly reactive molecules and hence have a low natural abundance in the free state. They occur mostly as the building blocks of oligo- and polysaccharides and other biological molecules. Sucrose is the most abundant disaccharide in plants, with trehalose, maltose, cellobiose and lactose also naturally occurring disaccharides. Fructo-oligosaccharides and related polymeric fructans, such as inulins and levans, are water-soluble carbohydrates that are abundant in some food plants, although their botanical distribution is limited. Starch is the main reserve polysaccharide in plants, which contributes 35–70% of the energy intake in the modern human diet, as well as being an important source of pre-formed glucose for brain, red blood cells, and reproductive tissues. Non-starch polysaccharides are a diverse group of carbohydrates that are abundant in many plant foods. They occur mostly in plant cell walls, where they have important structural functions. Non-starch polysaccharides, like fructo-oligosaccharides and fructans, are not digested in the upper gut and pass largely undigested into the large intestine as a component of dietary fibre and fermentable substrates for gut microbiota. The amounts and forms of starch and non-starch polysaccharides in modern food plants are likely to differ considerably from their respective ancestral forms.

Today, much of the human diet comes from plant foods in the form of seeds, tubers, roots, fruits, nuts, and leafy vegetation. Plant storage tissues, in particular, contain reserves of carbohydrates, proteins, lipids (oils), minerals and vitamins to support new growth at different stages of plant development, and hence are potentially a good source of nutrition for humans. Many plants also contain biologically active phytochemicals, some of which contribute aroma and taste, and others that are anti-nutritional or hazardous due to their toxicity. Survival skills of pre-agricultural humans would need to have been based on a good understanding of plants in their environment and an ability to select and adapt species for use as foods or medicines. Plant foods are the source of diverse types of carbohydrates, which are essential for human nutrition. Carbohydrates provide the majority of the energy in the human diet and are the main fermentable substrates that nourish the gut microflora.

The aim of this chapter is to provide a concise overview of the significance and properties of dietary carbohydrates from plant sources. The material is presented in the style of an introductory chapter rather than as a fully referenced review article. Suggestions for further reading are included at the end of the chapter.

What are carbohydrates?

Carbohydrates make up about 90% of the dry matter of plants and are the most abundant carbon compounds in Nature. They serve as stores of energy and in structural roles, forming the framework of plant cell walls and external structures. Carbohydrates are macro-constituents of many foods and essential in the human diet, providing a source of readily metabolisable energy, pre-formed glucose for brain, red blood cells, and reproductive tissues, and dietary fibre and prebiotics necessary for gut microflora and gut health.

Carbohydrates are polyhydroxy aldehydes or ketones that have the general formula $C(H_2O)_n$, as the name hydrates of carbon implies. Most carbohydrates are made up of only carbon, hydrogen and oxygen, although some may also contain nitrogen, phosphorus and sulphur atoms. They occur in diverse forms, which include monosaccharides (simple sugars), oligosaccharides (oligomers containing 2–10 monosaccharides), and polysaccharides (complex biopolymers that contain many monosaccharide units).

Sources of dietary carbohydrates

Most plant cells contain small amounts of carbohydrate energy reserves, but some tissues, notably underground storage organs (USOs) and seeds, accumulate large amounts of carbohydrates to support growth and development at certain stages during the life cycle of the plant.

Underground storage organs

USOs may be derived from tap-root tissue (e.g., carrot, turnip), tuberous roots (e.g., potato, cassava), stem tissue (e.g., sweet potato, yams), corms (e.g., taro), or rhizomes (e.g., ginger). The dry matter (DM) content of USOs is usually 15–35% of their mass, of which carbohydrate (mostly as starch) is the major constituent (70–85% of DM). Their bulk and relatively high moisture content mean that USOs have limited portability and storability and are susceptible to spoilage once harvested. However, if left undisturbed in the ground they remain stable until seasonal changes induce mobilisation of stored reserves for new plant growth. As a food source, USOs can be harvested as needed over a period of months, or they can also be dried to increase durability and portability.

Grains and seeds

Grains (seeds) are the foundation of the modern human diet. Statistical data published by bodies such as the Food and Agriculture Organisation (FAO) indicate that total global grain production is about 2.5 billion tonnes/year, of which 40% is used for livestock feeding. The low moisture content of grains and their hard seed coat, which may contain toxic or anti-nutritional constituents, provide protection against the environment and against microbial and insect attack and herbivory. The ability to harvest seeds from plants with a non-shattering rachis, and the storability and transportability of grains, are considered to have been important in the transformation of humans from hunter-gatherers to agriculturalists. The largest source of food grains is cereals (wheat, rice, corn, barley, oats, millets), with the average annual human consumption of cereal

Table 1.1. Approximate carbohydrate, protein and oil composition of food grains (as % of dry matter)

	Grain	Protein	Oil	Carbohydrate
Cereals	Barley	12	3	76
	Corn	10	5	80
	Oats	13	8	66
	Rice	8	2	75
	Rye	12	2	76
	Wheat	12	2	75
Legumes	Broad bean	23	1	56
	Chickpeas	23	5	66
	Garden pea	25	6	52
	Lentils	29	1	67
	Peanut	31	48	12
	Soybean	40	22	33
Oilseeds	Cotton	50	30	12
	Oil palm	9	48	28
	Rape	21	48	19
	Sunflower	30	45	22
Pseudocereals	Quinoa	16	7	74
	Amaranth	15	7	73

Data compiled food composition tables published by the United States Department of Agriculture Agricultural Research Service (ww.ndb.nal.usda.gov)

grains being about 150 kg per capita (www.fao.org). In addition to cereals, other important modern food grains are pulses (peas, beans, chickpeas, lentils), oilseeds (canola, sunflower, soybean, peanut) and niche grains such as quinoa, buckwheat, and amaranth (sometimes referred to as pseudo cereals). Raw grains are unpalatable and indigestible and need to be processed to increase the bioavailability of nutrients for foods and feed. Grains are also an important source of raw materials for industry. The relative amounts of carbohydrate, protein and lipid in grains differ between and within plant species (Table 1.1), and are influenced by the plant genotype and environmental influences during plant growth and post-harvest processing.

Other sources of carbohydrates

Depending on climate and season, some fruits and nuts may provide a good source of dietary carbohydrates (Tiwari *et al.* 2013). For example, chestnuts and acorns contain over 60% of their DM as carbohydrates (mostly as starch), and cashew nuts may contain 25% of DM as starch. Many other nut species used as foods (e.g., almonds, hazelnuts, pine nuts, pistachios) contain large amounts of lipids (45–50% DM), 15–20% protein and 25–30% carbohydrate, of which only a small amount is starch (USDA National Nutrient Database for Standard Reference 2014). Starch is also stored in significant amounts in the bark and xylem tissues of stems and branches of many tree species, although this varies seasonally and with the age of the tree (Pallardy 2008). The pith of some palms (for example, the sago palm) is a rich source of starch, which is extracted commercially.

Monosaccharides

Monosaccharides with 3, 4, 5, 6, or 7 carbons occur naturally and are referred to as trioses, tetraoses, pentoses, hexoses, and heptaoses, respectively. Monosaccharides are chemically highly reactive molecules and hence have a low natural abundance in the free state. They occur mostly as the building blocks of oligo- and polysaccharides and other biological molecules. One of the carbons in the monosaccharide molecule (referred to as the anomeric carbon) can act as a reducing agent in chemical reactions and hence these compounds are referred to as reducing sugars. The anomeric carbon can also form links to other reactive chemical groups. Glucose (commonly abbreviated as Glc), fructose (Fru), galactose (Gal), and mannose (Man) are examples of hexoses that are commonly combined into more complex molecules, whereas the pentoses deoxyribose and ribose are part of the structural backbone of the nucleic acids DNA and RNA, respectively.

An exception to the low natural abundance of free monosccacharides is glucose, which occurs in appreciable amounts in the free form in animals. Humans need to maintain a steady level of glucose in the blood stream as the essential source of biochemical energy to support the normal functioning of a large brain, which alone accounts for 20–25% of basal expenditure of metabolic energy (Fonseca-Azevedo and Herculano-Houzel 2012). Red blood cells have a requirement for glucose, which is also essential for reproductive fitness during pregnancy and lactation. This requirement for glucose is met from a combination of dietary carbohydrate, mobilisation of transient stores of glycogen in the liver, and glucose synthesis *in vivo* from non-carbohydrates in a biochemical process known as gluconeogenesis. At least some of the glucose must be obtained from the diet, as under normal physiological conditions liver glycogen reserves and gluconeogenesis can meet only part of the daily requirement. Dietary glucose is obtained from the digestion of glycemic carbohydrates (also referred to as available carbohydrates) and subsequent absorption from the small intestine into the bloodstream. Because of its chemical reactivity, the concentration of glucose in the bloodstream in healthy individuals is normally maintained within narrow limits by complex physiological mechanisms that balance the need for continual availability of this essential metabolic substrate, with ensuring that its concentration does not exceed a level that can cause harmful effects due its chemical reactivity. Chronic over-loading of glucose into the bloodstream triggers abnormal physiological responses that increase the risk of obesity and diet related diseases such as type 2 diabetes, cardiovascular disease, and certain types of cancers, as discussed in more detail in Chapter 2.

Oligosaccharides

Oligosaccharides contain between two and about ten monosaccharide units joined by glycosidic bonds (the term "glycosidic" is used generically, whereas "glucosidic" refers specifically to a bond involving glucose).

Sucrose, trehalose, maltose, cellobiose, and lactose are naturally occurring disaccharides (i.e., made up of two monosaccharides joined by a glycosidic bond). Sucrose, the most abundant disaccharide in plants, is a product of photosynthetic carbon capture and the main carbon compound translocated in plants during photosynthesis and when reserve carbohydrates are mobilised, for example in germinating seeds or from tubers to support new plant growth. Sucrose occurs in many fruits but accumulates to only a limited extent in most plant species because of its strong water binding ability and effect on water relations in cells; sugar cane and sugar beet are exceptions, as they have specialised tissues for storing reserves as sucrose.

Animals do not synthesise sucrose but can hydrolyse it in the gut by the enzyme sucrase to produce glucose and fructose in equal proportions for absorption into the bloodstream. Fructose can provide energy to muscle, but is metabolised predominantly in the liver in a different way to glucose. Fructose cannot substitute directly for glucose as an energy source for the brain due to its different metabolic biochemistry. Honey, another carbohydrate-rich food source, contains 45% of DM as glucose and 55% as fructose; most of the sucrose in the original nectar collected by bees is hydrolysed into the monosaccharides in the formation of honey (Murray *et al.* 2001).

Trehalose, maltose, and cellobiose are naturally occurring disaccharides made up of two glucoses joined by glucosidic bonds that differ in their stereochemical orientation. Trehalose occurs in insects and fungi (including mushrooms) but is uncommon in plants (Avigad 1990). It has a similar transport role in insects and fungi to sucrose in plants. Trehalose is used as an additive in various food applications. Maltose occurs in plants mainly as a breakdown product of starch, whereas cellobiose is a product of cellulose metabolism. Lactose, a disaccharide made up of glucose and galactose, is the most abundant sugar in milk.

Fructo-oligosaccharides (oligosaccharides formed by the addition of fructosyl units to sucrose) and related polymeric fructans, such as inulins and levans, are water-soluble carbohydrates that are abundant in some food plants, although their botanical distribution is limited. Fructans are transient reserve polysaccharides that accumulate in stems and other vegetative tissues of cereal plants and forage grasses. Fructans can make up 35–50% of the dry weight of stems and leaves of grasses under certain environmental conditions, particularly in cooler climates, and can represent a quantitatively significant component of the diet of grazing animals. The fructan content of cereal grains is usually 1–2%.

Oligosaccharides that contain galactosyl units joined to the glucosyl moiety of sucrose by an $\alpha(1{\rightarrow}6)$ link occur in appreciable amounts in some plant foods. These compounds are known as raffinose oligosaccharides, of which raffinose (Gal-Glc-Fru) is the most abundant. They make up 5–8% of the dry matter of pulse seeds, such as peas, beans and lentils, and about 0.3% of cereal grains.

Humans lack the enzymes to break down raffinose oligosaccharides, fructo-oligosaccharides, and fructans and hence they are not digested in the upper gut. These oligosaccharides pass largely undigested into the large intestine as a component of dietary fibre and fermentable substrates for gut microbiota. Their contribution to metabolisable energy absorbed from the small intestine is small.

Polysaccharides

Polysaccharides are polymeric carbohydrates that contain many monosaccharide units joined by glycosidic bonds. The degree of polymerisation (the average number of monomers per polymer molecule) is often many thousands and even millions. Polysaccharides have biological functions as energy reserves in storage tissues of plants (for example, starch in seeds, storage roots, tubers) and structural roles to provide mechanical strength to cell walls that regulate the size and shape of plant cells. Polysaccharides may be homopolymers made up of a single type of monomer unit, or they may be heteropolymers of different types of monomers. They may have an unbranched or branched structure. The following discussion will consider plant polysaccharides important in foods in two sections – starch and non-starch polysaccharides.

Starch

Starch is the main reserve polysaccharide in plants and an important material for humans. It is a major determinant of yield and quality of modern cereal grains and a macro-constituent of many foods. Starch contributes 35–70% of the energy intake in the modern human diet. The main sources of dietary starch today are foods made from cereal and pulse grains and root crops. Bananas are also a significant source of dietary starch. Approximately 60 million tonnes of starch are extracted annually worldwide from a range of crops for use in manufactured foods, pharmaceuticals, non-edible industrial products, and biofuels (Burrell 2003; Eliasson *et al.* 2013).

Starch occurs in most plant tissues but is found in largest amounts in seeds and storage organs. Starch accounts for 60–70% by weight of cereal grains and a similar proportion of the dry matter of starchy roots and tubers. The low osmotic effect of starch means that large amounts of glucose can be stored without disrupting water relations in cells. Starch occurs naturally as insoluble, semi-crystalline granules, which vary considerably in size (1–100 μm in diameter) and shape (spherical, elongated, lenticular, multi-lobed, or compound) between and within species, and even within the same plant (Jane *et al.* 1994). Starch granules are inherently stable but under the right conditions can be broken down readily by acids or enzymes to release glucose. When viewed under a polarised light microscope, native starch granules show a birefringent Maltese Cross pattern (Fig. 1.1), which is often used as a diagnostic test for starch granules. However, birefringence is a general property of highly organised, semi-crystalline materials and not unique to starch. The moisture content of native starch granules ranges from 10–12% in cereal grains to 14–18% in root and tuber starches (Copeland *et al.* 2009).

Starch granules from wheat, barley, rye, and triticale grains have a characteristic bimodal size distribution (Fig. 1.1). For example, wheat has one population of lenticular-shaped granules, ranging from about 15–40 μm in diameter (known as A granules) and another population of smaller spherical granules ranging in size from approximately 1–10 μm (referred to as B granules). The following section gives a brief account of the morphological

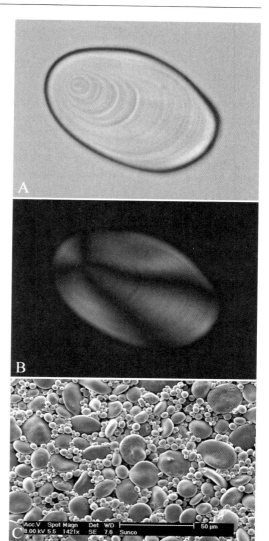

Fig. 1.1. Starch granules. Light micrographs of potato starch granules visualised under brightfield (A) and polarised (B) and SEM image of wheat starch granules (C). Images A and B are from Ek et al. (2012) (with permission), and C is from Salman and Copeland (unpublished)

features and physical, chemical, and functional properties of starch granules, which are described in greater detail in review articles (for example, Copeland *et al.* 2009; Perez and Bertoft 2010).

Starch granules are made up of two polymers of D-glucose: amylose, an essentially linear molecule with mostly α(1→4) links, and the highly branched amylopectin made up of α(1→4) links and up to 5% α(1→6) links. Amylose has a low degree of branching and hence has low water solubility, forms firm gels, and has good film-forming and adhesive properties. Amylose can form complexes with lipids, and binds iodine strongly, which is the basis of methods for detecting and quantifying starch. Complexes between amylose and lipids can significantly modify the properties and functionality of starch in ways that are of interest to the food industry and for human nutrition. Amylopectin has a complex, highly branched molecular architecture in which the placement and length of branches can vary between starches from the same or different species. Amylopectin is a much larger polymer than amylose, but because of its highly branched structure, is more soluble than amylose and forms soft gels and has good food thickening properties. Some starches that contain only amylopectin (known as waxy starches) occur naturally, but there are no known naturally-occurring starches that contain only amylose. The molecular organisation of amylose and amylopectin inside starch granules is highly complex and not fully understood.

Starch granules vary considerably, between and even within plant species, in their physical and chemical properties. This variability is due to the biosynthesis being controlled by multiple genes, which are affected by developmental changes and environmental factors during plant growth (Zeeman *et al.* 2010; Thitisaksakul *et al.* 2011).

Animals do not make starch, but have glycogen, which is also a homopolymer of glucose and an important glucose reserve in muscle and liver cells. Glycogen is structurally analogous to amylopectin, but it has a much higher degree of branching and a more compact, globular shape. More glucose can be packed into the open ended, tree-like structure of amylopectin than into a globular molecule like glycogen.

The functionality of starch in foods and during human digestion is determined by what happens when granules absorb water. During cooking or food processing, water uptake by starch is facilitated by a combination of heating and shear forces. Under these conditions, starch granules undergo an irreversible sequence of events, termed gelatinisation, that result in granules swelling and losing their molecular organisation (Hoover 2010; Wang and Copeland 2013). Gelatinised starch loses its birefringence and is no longer recognisable by the characteristic granular morphology. On cooling, the starch retrogrades gradually into a new semi-crystalline order, which lacks the characteristic morphological features of native granules. Gelatinisation and retrogradation are fundamentally important to the functional properties of starch, including digestibility.

Some starchy foods are eaten raw, for example, nuts, bananas, and certain other fruits. However, most starch consumed by modern humans has undergone some form of processing or cooking and is a mixture of partially gelatinised granules and gelatinised and retrograded starch. The digestibility of starch in the human gut is a subject of great interest for human nutrition because of the implications for diet-related diseases, as discussed in more detail in Chapter 2.

Non-starch polysaccharides
Non-starch polysaccharides are a diverse group of carbohydrates that are abundant in many plant foods. They occur mostly in plant cell walls, where they have important

structural functions. Plant cell walls consist of two phases, a microfibrillar phase composed mainly of cellulose, which is deposited around cells during growth. The cellulose microfibrils are subsequently embedded in a matrix phase consisting of other polysaccharides, proteins, phenolics and lignin, to form secondary, or mature, cell walls. The exterior surfaces of plants are coated with impervious, waxy protective coatings to reduce water loss and pathogen entry. The type and abundance of non-starch polysaccharides vary between and within species and are subject to genotype and environmental influences. Depending on the species, cereal seeds contain between approximately 5% and 25% non-starch polysaccharides, a substantial amount of which are in the outer cell layers of grains. The non-starch polysaccharide content of whole meal flours is much greater than in refined flours. Non-starch polysaccharides are not digested in the upper human gut and pass into the large intestine as substrates for fermentation by the gut microbiota.

Cellulose is the main structural component of plant cell walls and the most abundant carbon compound in Nature. It is a significant component of fruits and vegetables in the human diet. It makes up approximately 98% of cotton fibres, but makes up a smaller fraction of the polysaccharides in the cell walls of most other plants. Cellulose is an unbranched homopolymer of (1–4) linked β–D-glucose units. It is interesting how the two main glucose homopolymers in plants – starch and cellulose – have very different properties and biological roles based on differences in the geometry of the bonds that link the glucosyl units. The α(1–4) glucosidic links in starch produce a polymer that can form helices and pack into semi-crystalline granules, which are stable during storage but can be mobilised and digested readily under the right conditions to serve as a source of energy. In comparison, the β(1–4) glucosidic links of cellulose result in the formation of a fibrous polymer, which is strong, inert and poorly digested. Chitin, the major polysaccharide of insect cuticles and crustacean shells, is a polymer of N-acetyl-glucosamine, which also has close-fitting molecular chains and similar conformation and inertness to cellulose.

Many non-cellulosic polysaccharides occur in the matrix phase of secondary plant cell walls, with the amount and type varying from species to species. These include xylans, arabinoxylans and arabinogalactans, β-glucans, galactomannans and pectins, which can be significant in the diet depending on the types of food eaten. The reader is referred to more detailed sources for further reading on plant cell wall polysaccharides (Doblin *et al.* 2010) and β-glucans and arabinoxylans (Izydorczyk and Dexter 2008; Saeed *et al.* 2011).

β-Glucans are linear homoploymers of glucose in which most of the glucoses are in β(1–4) cellotriose and cellotetraose units joined by β(1–3) bonds. The two types of linkages in β-glucans give the molecule an extended, irregular conformation, which increases the solubility of the polymerand the viscosity of solutions. β–Glucans are abundant in the walls of endosperm cells of cereal grains. They make up 6–8% of endosperm cell walls in oats and barley, and 0.5–2% in wheat and rye. β–Glucans are mobilised during germination by β-glucanases, which has relevance for malting of barley for brewing. Humans lack the glucanases to digest β-glucans, which therefore contribute to dietary fibre. β-Glucans are associated with beneficial health effects, which are thought to be due to a combination of polymer viscosity and the entrapment of molecules, such as cholesterol, in a viscous polymer network in the digesta.

Arabinoxylans have a $\beta(1-4)$ linked D-xylose backbone substituted with α-linked D-arabinose and D-glucuronate residues on some of the xyloses. The arabinoxylan chains may be cross-linked to other arabinoxylan chains and to proteins and lignin, with greater cross-linking decreasing extractability,and solubility. Highly branched arabinoxylans, referred to as pentosans, are abundant in cereal grains. The outer bran layers of wheat grains contain about 35% arabinoxylan, whereas the endosperm and flour usually contain much less arabinoxylans (2–3% of wheat flour, 6–8% of rye flour). Arabinoxylans have very high water holding capacity and form very viscous solutions that can influence the processing and nutritional value of foods. Arabinoxylans may be quantitatively more important in the diet than β-glucans. Several mannose based polysaccharides occur in plant tissues but have a more taxonomically restricted distribution. Galactomannans have a $\beta(1-4)$ linked D-mannan backbone on to which single galactoses are substituted by $\alpha(1-6)$ links. Galactomannans are abundant in seeds of certain legume species, such as guar, carob (locust bean), and fenugreek, and are important food additives as thickeners and gelling agents due to the high viscosity of dilute solutions.

Pectins occur in all terrestrial plants and are closely associated with cellulose and other polysaccharides in cell walls. Pectins also occur in gum exudates, mucilages, and slime secreted from the root caps of plants, where they have protective and lubricating actions. Chemically, pectins are a heterogeneous group of complex acidic polysaccharides that are classified on the basis of properties rather than common structural features. D-Galacturonic acid is the main monosaccharide, as well as L-arabinose, D-galactose, and D-xylose, which make up the various types of polymers that constitute the pectins. Pectins are abundant in many fruits, for example, making up 20–40% of the dry matter of citrus peel and 10–20% of apple pomace. Pectin hydrolysis occurs as part of cell wall softening during fruit ripening. Extracted pectin is used widely in food ingredients as a gelling agent, thickener and stabiliser, especially for making jams and marmalades.

Concluding comments

Domestication and thousands of years of genetic selection and plant improvement has led to modern varieties of food plants that are very different from their ancient wild ancestors. Selection for increased grain yield has likely been accompanied by inadvertent co-selection for increased starch content at the expense of non-starch polysaccharides due to their different densities (1.5 g/cm^3 for starch, compared to 1.35 g/cm^3 for proteins and non-starch polysaccharides, Fischer *et al.* 2004). The amounts and forms of starch and non-starch polysaccharides in modern food plants are likely to differ considerably from their respective ancestral forms.

References

Avigad, G. 1990. Disaccharides. In Dey, P. M. and Harborne, J. B. (eds), *Methods in Plant Biochemistry* 2, 111–88. London: Academic Press

Burrell, M. M. 2003. Starch: the need for improved quality or quantity: an overview. *Journal of Experimental Botany* 54, 451–6

Copeland, L., Blazek, J., Salman, H. and Tang, M. C. M. 2009. Form and functionality of starch. *Food Hydrocolloids* 23, 1527–34

Doblin, M. S., Pettolino, F. and Bacic, A. 2010. Plant cell walls: the skeleton of the plant world. *Functional Plant Biology* 37, 357–81

Ek, K. L., Brand-Miller, J. M. and Copeland, L. 2012. Glycemic effect of potatoes. *Food Chemistry* 133, 1230–40

Eliasson, A-C., Bergenståhl, B., Nilsson, L. and Sjöö, M. 2013. From molecules to products: some aspects of structure–function relationships in cereal starches. *Cereal Chemistry* 90, 326–34

Fischer, H., Polikarpov, I. and Craievich, A. F. 2004. Average protein density is a molecular-weight-dependent function. *Protein Science* 13, 2825–8

Fonseca-Azevedo, K. and Herculano-Houzel, S. 2012. Metabolic constraint imposes trade-off between body size and number of brain neurons in human evolution. *Proceedings of the National Academy of Sciences (USA)* 109, 18571–6

Hoover, R. 2010. The impact of heat-moisture treatment on molecular structures and properties of starches isolated from different botanical sources. *Critical Reviews in Food Science and Nutrition* 50, 835–47

Izydorczyk, M. S. and Dexter, J. E. 2008. Barley β-glucans and arabinoxylans: Molecular structure, physicochemical properties, and uses in food products – a Review. *Food Research International* 41, 850–68

Jane, J-L., Kasemsuwan, T., Leas, S., Zobel, H. and Robyt, J. F. 2004. Anthology of starch granule morphology by scanning electron microscopy. *Starch/Stärke* 46, 121–9

Murray, S. S., Schoeninger, M. J., Bunn, H. T., Pickering, T. R. and Marlett, J. A. 2001. Nutritional composition of some wild plant foods and honey used by Hadza foragers of Tanzania. *Journal of Food Composition and Analysis* 14, 3–13

Myers, A. M., Morell, M. K., James, M. G. and Ball, S. G. 2000. Recent progress toward understanding biosynthesis of the amylopectin crystal. *Plant Physiology* 122, 989–97

Pallardy, S. G. 2010. *Physiology of Woody Plants*, 3rd edn. San Diego: Academic Press

Perez, S. and Bertoft, E. 2010. The molecular structures of starch components and their contribution to the architecture of starch granules: A comprehensive review. *Starch/Stärke* 62, 389–420

Saeed, F., Pasha, I., Anjum, F. M. and Sultan, M. T. 2011. Arabinoxylans and arabinogalactans: a comprehensive treatise. *Critical Reviews in Food Science and Nutrition* 51, 467–76

Thitisaksakul, M., Jiménez, R. C., Arias, M. C. and Beckles, D. M. 2012. Effects of environmental factors on cereal starch biosynthesis and composition. *Journal of Cereal Science* 56, 67–80

Tiwari, B. K., Brunton, N. P. and Brennan, C. (eds). 2013. *Handbook of Plant Food Phytochemicals: sources, stability and extraction*. Chhichester: Wiley

USDA National Nutrient Database for Standard Reference, Release 27. 2014. *Composition of Foods Raw, Processed, Prepared*. www.ndb.nal.usda.gov/ndb/foods

Wang, S. and Copeland, L. 2013. Molecular disassembly of starch granules during gelatinization and its effect on starch digestibility: a review. *Food and Function* 4, 1564–80

Zeeman, S. C., Kossmann, J. and Smith, A. M. 2010. Starch: its metabolism, evolution, and biotechnological modification in plants. *Annual Review of Plant Biology* 61, 209–34

Suggestions for further reading on mono-, oligo- and polysaccharides, and plant biochemistry

BeMiller, J. N. 2007. *Carbohydrate Chemistry for Food Scientists,* 2nd edn. St Paul MN: American Association of Cereal Chemists International

BeMiller, J. and Whistler, R. (eds). 2009. *Starch: Chemistry and Technology,* 3rd edn. London: Academic Press

Buchanan, R. L., Gruissem, W. and Jones, R. L. 2002. *Plant Biochemistry.* Hoboken NJ: Wiley

Delcour, J. A. and Hoseney, C. A. 2010. *Principles of Cereal Science and Technology,* 3rd edn. St Paul MN: American Association of Cereal Chemists International

Gleeson, F. and Chollet, R. 2011. *Plant Biochemistry.* Sudbury: Jones and Bartlett

Heldt, H. W. and Piechulla, B. 2010. *Plant Biochemistry.* San Diego: Academic Press

2. Why protein is not enough: the roles of plants and plant processing in delivering the dietary requirements of modern and early *Homo*

Peter J. Butterworth, Peter R. Ellis and Michele Wollstonecroft

Until recently, serious archaeological discussion about the role of plant foods in the human diet, evolution, and health, and the essential roles of plant harvesting and processing systems in prehistoric economies have been overlooked in favour of meat and hunting. Yet, it is well known that a diet consisting of protein alone has biochemical and physiological consequences that cause ill-health and ultimately death; among the consequences are poisoning of the central nervous system proficiency and impaired renal function, bone de-mineralisation, impaired gut function, and disturbed hormonal production, altogether creating a perfect storm that brings about nausea, discomfort of the digestive system, weight loss, and deterioration in health. This chapter examines these problems, demonstrating that, as well as with the production of potentially toxic ammonia, the most likely health risk in a protein-only diet is what it fails to provide, i.e., adequate production of glucose to satisfy the requirements of glucose-dependent tissues such as the human brain and red blood cells, as well as a range of critical micronutrients, which are readily available from plants. Plants are good sources of carbohydrates, polyunsaturated fatty acids, vitamin C, minerals, phytochemicals (e.g., polyphenols, stanols/sterols), and non-starch polysaccharides (dietary fibre), all of which may contribute to alleviating the adverse consequences listed above. We then examine factors that influence peoples' plant food choices, and conclude that there is interdependence between culinary knowledge, technical choice, and dietary decisions pertaining to edible plants. It is argued that, on a global scale, this interdependence explains much about the different regional historical processes that gave rise to local culinary and husbandry traditions.

The common dietary requirements of present-day humans are thought to have been established in our evolutionary lineage by 150,000–200,000 years ago, prior to the emergence of modern humans (Lindeberg 2009). It is for this reason that the eating habits and biological requirements of our earlier hominin and primate ancestors are of interest in research on human dietary patterns and are investigated by a range of disciplines, including evolutionary anthropology, archaeology, palaeontology, primatology, evolutionary biology, evolutionary medicine, and the food and nutritional sciences. Current evidence indicates that the relative amounts of meat versus plants consumed has varied considerably throughout hominin evolution (Lindeberg 2009). Given an evolutionary history that includes more than 25 million years as anthropoid primates, our ancestral line (Hominoidea) appears to have been primarily herbivorous (Milton 2000). Prior to hominin emergence at *c.* 7 mya, (i.e., the Hominini tribe of the Hominoidea family, comprised of bipedal apes including modern and extinct *Homo* species and our immediate ancestors; see Wood and Richmond 2000), our ancestral hominoids were limited to consuming plants that could be gathered without tools and consumed without processing. These technological limitations evidently restricted them to consuming primarily foliage (rich in several amino acids) and fruits (rich in sugars) for energy and other nutrients. Most plants/plant parts with more concentrated forms of energy, such as carbohydrate-rich underground storage organs (USOs), nuts, and seeds, are inaccessible without the technological ability to extract them (e.g., from the ground or nutshell) or to transform the raw carbohydrates into digestible forms (i.e., through cooking). The diet of our ancestral hominoids thus appears to have been significantly higher in particular amino acids, dietary fibre, and most vitamins and minerals than that of modern humans, and significantly lower in carbohydrates such as starch (Milton 2000; 2002).

Dietary change and shifts in body composition are considered to be the two most important pre-conditions for the split with the apes and subsequent expansion of the hominin brain size (Aiello and Wheeler 1995). Indeed, over the past 3.9 million years the *Homo sapiens* brain increased to three times the size of that of our earliest australopithecine ancestors (Snodgrass *et al.* 2009). Nevertheless, humans and our closest living primate relatives, the great apes, share a number of dietary strategies including omnivory, dietary diversity, dietary flexibility (the use of similar edible resources interchangeably), dietary selection for high quality foods (with high concentrations of energy and nutrients), and the use of food processing tools (Conklin-Brittain *et al.* 1998; Wrangham *et al.* 1999; Hohmann 2009). But the human diet is significantly richer than that of the great apes in terms of the wide range of terrestrial and aquatic plants and animals and birds that we eat, and the forms in which we eat them (Hillman and Wollstonecroft 2014). Major increases in dietary breadth occurred at two critical points in our evolution: the hominin/ape split and the emergence of *Homo* (Sponheimer and Dufour 2009). Notwithstanding the substantial decline in human dietary diversity that followed the introduction of agriculture (Hillman 2004), humans are '… well adapted for lean meat, fish, insects and highly diverse plant foods without being dependent on any particular proportions of plants versus meat' (Lindeberg 2009, 43).

Profound ecological and dietary changes in Plio-Pleistocene hominin behaviours, between 4 and 3.5 mya, have been identified through carbon isotope analyses of fossilised tooth enamel from specimens collected in central, eastern, and southern Africa (Sponheimer and Dufour 2009; Sponheimer *et al.* 2013; Cerling *et al.* 2013b; Wynn *et al.* 2013; see also chapter 12). Despite inter-species differences in the hominin specimens examined, as well as differences in the palaeoenvironments that they inhabited, the carbon isotope studies cited above suggest an overall shift towards C_4 and CAM foods; African C_4 plants include tropical savannah grasses, sedges (Cyperaceae), and other monocotyledons, CAM plants include succulents that are more typical of deserts. Therefore the carbon isotope results suggest a decreasing intake of foods from woodland tree and shrub environments and an increasing intake of savannah grassland resources, e.g., underground storage organs (USOs) of plants, and/or animals that consumed C_4/CAM plants. Significantly, the results of these carbon isotope studies also confirm a continuous expansion of dietary diversity, with hominins consuming foods from a wide range of C_3 as well as C_4/CAM sources. Neither C_4 nor CAM plants are eaten by most other primates (exceptions being extinct *Theropithecus* spp. and some extant baboons, see Cerling *et al.* 2013a)

The role of meat eating in hominin brain expansion is well-established, although when scavenging/hunting for meat actually began and intensified remains in question (Hublin and Richards 2009). The earliest archaeological evidence of tool use is stone tools and cut marks on bone from the Ethiopian sites of Gona and Bouri, dating from *c.* 2.5 mya (Lee-Thorp and Sponheimer 2006). But primate studies otherwise suggest that meat eating, as well as food sharing, hunting, and provisioning, may have been in place earlier, by the last common ancestor of *Homo* and the chimpanzee *(Pan troglodytes)* (Hohmann 2009).

The increasing consumption of high quality (nutrient-dense) foods, such as the concentrated proteins, essential fatty acids, vitamins, and minerals found in meat, is thought to have precipitated the emergence of *Homo* (Laden and Wrangham 2005; Sponheimer and Dufour 2009). Paradoxically, the escalating dietary importance of meat may have promoted a corresponding rise in the diversity of plants consumed. Several authors (Milton 1999; 2000; Sponheimer and Dufour 2009) have argued that, because meat provided adequate nutrients for normal metabolic performance, carbohydrate-rich plants became the main fuel for hominin energy expenditures.

Moreover, Johns (1999) proposed that subsequent to the emergence of *Homo*, anti-oxidant-rich plants such as bitter greens (e.g., plants in the Brassicaceae family), which contain high amounts of polyphenolics and vitamin C, were added to the diet to offset an increasing production of reactive oxidants in the brain due to its continuous expansion and consequential increase in mitochondrial oxidative phosphorylation with its potential for leakage of superoxide radicals.

The idea that fat, rather than protein, was the prime motivation for hominin meat eating was introduced by Speth and Spielman (1983) and supported in later publications by Speth (e.g., 1989; 2004; 2010) and other authors, particularly in a series of papers on human digestion, biochemistry and physiology by Cordain and colleagues (Cordain *et al.* 2000; 2001). Speth and Spielman (1983) argued that starvation can occur if humans are forced to depend on lean meat because there is a ceiling to the amount of calories that humans can safely consume from protein alone. Noting that hominin brain expansion required the ingestion of fatty acids, and that the large-bodied 'prey' of our early hominin African ancestors were lacking in body fat for much of the year, Speth (2004; 2010) further argued that where only lean meat was available, our evolutionary antecedents most likely obtained essential fats and carbohydrates from other resources such as starch-rich edible plant parts, (e.g., USOs) and/or the extraction of marrow and/or fat from animal bones. Indeed, field observations of fresh but abandoned African carnivore kills, where marrow and brains were observed to be the only edible portions remaining, suggest that, through scavenging, our early hominin ancestors would have had access to substantially greater amounts of fats than proteins (Schaller and Lowther 1969, cited in Cordain *et al.* 2001). The fragmenting of bones to obtain marrow dates from the Plio-Pleistocene, but the extraction of bone grease, a more complex process involving pulverising followed by boiling, appears to have begun in the Late Upper Palaeolithic (Munro and Bar-Oz 2005; Speth 2004).

Certainly, a diet consisting of protein alone has potentially lethal biochemical and physiological costs for our species. Among the adverse consequences of a protein-only diet are the poisoning of the central nervous system, and thus a loss in its proficiency, impaired renal function, bone de-mineralisation, impaired gut function, and disturbed hormone production and release, altogether creating a perfect storm that brings about nausea, discomfort of the digestive system, weight loss, and deterioration in health. Seminal biochemical research on this subject was carried out in the 1970s and '80s, most notable being the work of Rudman *et al.* (1973) and Vilstrup (1980). (The complexity of human metabolism and the ethical restrictions now placed on research with human subjects means that direct experimental approaches of the kind described in their papers are significantly more limited today.)

A significant point of the biochemical research that is rarely mentioned in the archaeological literature (an exception being Speth 2010) is that the most likely health risk in a protein-only diet is what this diet fails to provide, i.e., adequate production of glucose to satisfy the requirements of glucose-dependent tissues such as the human brain and red blood cells (see below) as well as a range of critical micronutrients, which are readily available from plants. Plants are a good source of carbohydrates, polyunsaturated fatty acids, vitamin C, minerals, phytochemicals (e.g., polyphenols, stanols/sterols), and non-starch polysaccharides (dietary fibre), all of which may contribute to alleviating the adverse consequences listed above. New research shows that there are many biologically active components in plant foods that are likely to be protective against a range of diseases and helpful in maintaining health and longevity. The beneficial materials include plant sterols/stanols, phenolic compounds (e.g., lignans), glucosinolates (which may explain why bitter species such as the crucifers have potential protective effects in certain cancers), as well as dietary fibre and a broader array of minerals (Liu 2004; Judd and Ellis 2006; Johnson 2013).

Nevertheless, gaining access to nutrients from plants is not always straightforward because many species have physical (e.g., thorns) and/or chemical (e.g., toxins) defences to deter would-be predators. Moreover, the tissue of many edible plants/plant parts (e.g., nut kernels, pulses, and USOs) are comprised of cell walls with chemical and physical properties that promote nutrient encapsulation, hindering nutrient and energy release during chewing and digestion. In-vitro and in-vivo studies of plant microstructure demonstrate that, due to encapsulation, the amount of nutrients contained within a food does not equal the amount that is released, digested, and absorbed by human consumers (Ellis *et al.* 2004; Mandalari *et al.* 2008). Encapsulation slows down or prevents interactions of critical carbohydrates, lipids, and other macro- and micronutrients with digestive substances, resulting in plant foods being excreted without the consumer benefitting from their nutrients. However, encapsulated nutrients may become available following microbial fermentation of plant cell walls in the large intestine, a process that is believed to have positive nutritional benefits.

A survey of plant processing techniques by Johns and Kubo (1988) shows that, on a worldwide scale, humans have great sophistication in their food processing methods and, significantly, that in many cases specific processes are used for particular plants. This is because interspecies differences in the physical and chemical composition of plants mean that there is no 'one size fits all' method of processing. Plants do not respond in identical ways to the same treatment and may require specific techniques or processes to make them edible for humans. In some cases simple processing by washing, boiling, roasting, pulverising, or grinding can remove toxins and promote nutrient release. In other cases a sequence of several techniques (e.g., grinding followed by washing and then cooking) is necessary to transform a plant tissue into an edible form and/or promote nutrient release. But the order in which the techniques are applied can also make a difference to nutrient accessibility.

Altogether, these factors underscore the complexity of the technical, botanical, ecological and culinary skills and knowledge required for plant processing. These issues are not well understood in the archaeological literature despite a range of publications on the subject over the past 20 years (e.g., Yen 1975; 1980; Stahl 1989; Wandsnider 1997; Wrangham *et al.* 1999; Leach 1999; Lyons and D'Andrea 2003; Wollstonecroft *et al.* 2008; 2012; Carmody *et al.* 2011; 2012). For example, carbohydrate-rich USOs are frequently listed among the possible high-quality foods consumed by early hominins (e.g., Laden and Wrangham 2005; Sponheimer and Dufour 2009), but few authors (Cordain *et al.* 2001 being a notable exception) mention that prior to the controlled use of fire, access to energy from starch in otherwise carbohydrate-rich USOs would probably have been limited. To be digested and metabolised by humans, starch must first be transformed physically into a form that will bind with salivary and pancreatic α-amylase and be hydrolysed during the digestive process. The first step in this transformation is gelatinisation prior to consumption, a process that requires the presence of both water and heat (Roder *et al.* 2009). The availability of water during cooking has an important bearing on the physical integrity of the starch granules and its susceptibility to amylolysis. (In some USOs with high levels of internal moisture, heat treatment alone may appear to gelatinise the starch without the addition of exogenous water, but further research is necessary to establish if gelatinisation is sufficiently accomplished under such water-limited conditions.) While little is known about the digestive processes of our early hominin ancestors, it is likely that they obtained relatively low levels of energy by consuming starch-rich plants in the raw form. Some of this energy would be derived potentially from fermentation of starch to short chain fatty acids in the large intestine.

Another problem with perceptions about food processing within the archaeological literature is the pervasive influence of optimal foraging models, which continue to misrepresent the transformative potential of food processing activities. Optimal foraging models typically measure foraging efficiency according to energy (kcal) cost/benefits. Energy investments for food preparation and for the manufacture and maintenance of processing tools are regarded as 'costly' because the returns, i.e., increased access to nutrients, are not factored into these models (see for example Winterhalder and Goland 1997; Barlow and Heck 2002; Bird and O'Connell 2006). Optimal foraging models are problematic because they overlook the ways that food processing can improve human access to energy and other critical nutrients and because they disregard human requirements for other (non-energy) critical nutrients, as well as our need for dietary diversity (Stahl 1989). Models of this type continue to be applied without reference to recent research on the dynamic role of processing in improving nutrient and energy availability, possibly due to assumptions that the elements of processing are already known (Speth 2004).

Thus the aim of the present paper is to clarify several critical issues about human meat and plant consumption that are commonly misunderstood in the archaeological literature. We begin with a review of the biochemistry and digestive and metabolic processes involved in a protein-only diet. We then discuss the role of food processing in human diet and its potential influence on hominin dietary change.

Why protein is not enough

Diets that are high in meat are generally regarded as unhealthy. In societies from developed nations, a high consumption of meat has been shown to be associated with an increased incidence of cardiovascular disease (CVD) (Dimsky 1994) and certain cancers (Talalay and Fahey 2001 and references therein; Sinha *et al.* 2009).

In fact, there are many downsides to just eating meat, but one upside is that the iron in haem is more bioavailable than iron in many sources of plant foods (e.g., grain/ seeds). Also, the literature contains reports of how diets consisting exclusively of lean meat, i.e., of extremely low triacylglycerol (TAG) content and zero carbohydrate, produced symptoms of diarrhoea, nausea, and general feelings of ill-health with loss of appetite in early American explorers (Speth and Spielmann 1983). The condition known as 'rabbit starvation' is reported as being potentially fatal if the period of fat-free and carbohydrate-free nutrition is extended (Hu *et al.* 2000 and references cited therein). The reasons for the failure to thrive on exclusive diets of lean meat have been variously discussed and an obvious target is the propensity for ammonia production accompanying the metabolism of protein-derived amino acids (Rudman *et al.* 1973; Dimsky 1994; Cordain *et al.* 2002). With the known neurotoxicity of ammonia, any condition that exceeds the capacity of systems that convert ammonia to harmless metabolites such as urea is expected to be deleterious to health. Although it is obvious that such a mechanism may be of key importance in attempts to account for a lack of well-being associated with an exclusive diet of lean meat, due consideration has also to be taken of the various biochemical events that categorise the metabolism of a relatively large and persistent consumption of protein. Certain biochemical and physiological effects on adults of diets that are high in protein have been briefly reviewed (Metges and Barth 2000).

Red blood cells and the cells of the kidney medulla are glycolytic tissues and a steady supply of glucose is essential for satisfying their energy requirements. The central nervous system also metabolises glucose to a considerable extent. Skeletal muscle accounts for a major portion of total body mass and can metabolise glucose, free fatty acids (FFA), ketone bodies (acetoacetate and hydroxybutyrate) produced by metabolism of FFA, particularly under conditions of limited carbohydrate availability. Branched chain amino acids are also a source of ketone bodies. With a fat-free diet, availability of FFA is lacking and therefore the energy substrates will be predominantly glucose with small contributions derived from branched chain amino acids and ketone bodies formed only from ketogenic amino acids such as leucine and lysine. The branched chain amino acids, i.e., valine, leucine and isoleucine, are variously metabolised to acetyl-CoA and/or succinyl-CoA. These compounds are intermediates in the tricarboxylic acid (TCA) cycle, which is a central pathway for the oxidative conversion of metabolites with the production of adenosine triphosphate (ATP). ATP and related purine nucleotide triphosphates act as the energy currency of cells and are essential for all energy consuming processes including muscle contraction (see also below). Oxidative metabolism of FFA (β-oxidation) results in production of much acetyl-CoA that enters the TCA cycle. FFA are therefore rich sources of energy in the form of ATP.

An adult brain oxidises about 120 g of glucose per day, which accounts for about 20% of the whole body energy expenditure per day (Frayn 2010, 187–99). The central nervous system metabolises ketone bodies and therefore spares glucose when the latter is in short supply but some glucose may always be required. Thus even allowing for the 'glucose sparing' action of ketone bodies and metabolism of the branched amino acids, the demand for glucose will continue to be great when dietary-derived FFA are not available.

Certain polyunsaturated fatty acids (PUFA) cannot be synthesised by humans and so a dietary source is essential. These essential FFAs include ω-3 and ω-6 PUFA. Food sources such as fish oils that contain linoleic and linolenic acid from which the ω-3 and ω-6 PUFA are derived, are therefore obligatory. A restricted diet of lean meat is unlikely to provide the linoleic and linolenic acids necessary for brain function. ω-3-PUFA also have a role in protection against cardiovascular disease (CVD) by lowering plasma VLDL cholesterol and TAG concentrations. The risk of cardiovascular disease in modern humans rises with age. It is possible that the perceived shorter life-spans of early hominids meant that the development of nutrient-related CVD was less likely to occur.

Glucose supply

In the absence of dietary carbohydrate, essential requirements of glucose have to be met by the biochemical process known as gluconeogenesis (GNG). This pathway results in the synthesis of glucose from so-called glucogenic amino acids as starting material. The genes for characteristic GNG enzymes are expressed in many tissues but only liver and the proximal tubules of kidney (and possibly enterocytes of the ileum) express glucose-6-phosphatase, which is essential for the production of free glucose. Thus, free glucose for export to other tissue and organs, such as the brain, muscle and red blood cells, is produced only by the liver and kidney.

The GNG pathway from pyruvate, for example, is expensive in terms of the direct purine nucleotide triphosphate (i.e., GTP and ATP) requirement, but as it is a reductive process, there is consumption of reducing equivalents (2NADH and $2H^+$) also. Mitochondrial oxidative phosphorylation would potentially produce about five molecules of ATP from the 2NADH plus $2H^+$ and this loss of trinucleotide phosphate has to be added to the direct consumption of ATP in the pathway when considering the overall energy cost of GNG. Glycerol released by the action of lipases on TAG is normally an important substrate for GNG and is slightly less demanding of ATP/GTP, but in a fat-free condition represented by a diet of lean meat, the availability of glycerol is limited.

ATP is the energy currency needed by all cells and tissues. It allows them to maintain their physiological and biological functions which include inter alia, metabolism, synthesis of cellular components such as proteins, membrane structural lipids, DNA, and RNA, and allow movement of metabolites and inorganic ions across cell membranes. For muscle, in addition to these 'housekeeping' requirements, ATP is used to power the contraction process itself. ATP and GTP are inter-convertible and hence the GTP consumption by the GNG pathway is effectively a demand for more ATP. This total demand for ATP is in conflict with the cellular ATP requirements that keep cells alive and provide the power for muscle contraction. Effects of limited ATP availability will impair muscle performance and therefore be expected to have repercussions on the physical activity associated with a hunter-gatherer mode of existence.

If amino acids derived from lean meat are the only source of glucose it is necessary to consider whether the GNG capacity is sufficient to meet the total glucose demand. The GNG pathway is controlled by the substrate supply rate and by hormonal influences on gene expression. Glucagon up-regulates the enzymes and insulin down-regulates them. The activities of certain enzymes are regulated by the levels of key metabolic intermediates such as acetyl-CoA. In the conditions that exist during a diet consisting of lean meat only, it can be argued that GNG enzymes may not be optimally stimulated and therefore unable to produce sufficient glucose to meet the demands of glucose-dependent tissues (see Frayn 2010). Obligate carnivores would normally consume fat as well as protein and so the demand for glucose would be lessened in that the energy fuel for muscles would be largely provided by FFA.

Urea and ammonia production

There has been much speculation in the literature about whether the maximum capacity of biochemical systems that metabolise amino acids may be exceeded in situations associated with abnormally high protein diets (Dimsky 1994; Hue *et al.* 2000; Cordain *et al.* 2002). The amino acids arising from proteolytic digestion of protein are oxidatively deaminated by the combined actions of various aminotransferases and glutamate dehydrogenase. The carbon skeletons are converted variously to TCA cycle intermediates, acetyl-CoA, ketone bodies or glucose (see above) and generated ammonia is converted to urea in the urea (or ornithine) cycle present in the liver (Fig. 2.2). Experiments conducted in rats show that the total activities of all the enzymes in the cycle are directly proportional to the daily protein intake. Increases in protein intake result in up-regulation of the enzymes within four to eight days (Frayn 2010). The levels of a number of aminotransferases are also adjusted upwards in response to an increased load of dietary protein (Evered 1981). This form of long-term regulation is assumed to be a general mechanism applicable to all mammals including humans.

Acute, i.e., more immediate regulation of activity, is suggested to be provided by changes in the concentrations of ammonia, N-acetyl glutamate, ornithine and arginine within liver cells (Newsholm and Leach 2009). N-acetylglutamate is an essential cofactor for mitochondrial carbamoyl phosphate synthetase 1 (CPS1) that catalyses the reaction between NH_4^+ and bicarbonate to form carbamoyl phosphate (CP) which represents the entry point for N into the cycle. The synthetase that produces the essential N-acetylglutamate is up-regulated by dietary protein and is stimulated by arginine (Evered 1981). Since arginine is derived from digested protein, the dietary protein (amino acid) load acts as a feed-forward regulator, (Fig. 2.1).

Not all of the ammonia produced from oxidative deamination reactions is directed towards urea production. Some is required for glutamine synthesis. Glutamine has important metabolic functions including fuel for cells of the immune system and the intestine, and as a donor of NH_4^+ both for synthesis of metabolic intermediates and for pH regulation by the kidneys.

The literature contains instances of where the effectiveness of the ornithine cycle in dealing with large loads of amino acids has been tested (Visek 1979; Vilstrup 1980;

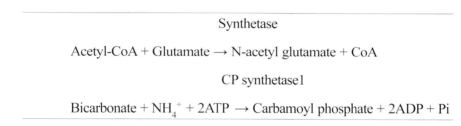

Synthetase

Acetyl-CoA + Glutamate → N-acetyl glutamate + CoA

CP synthetase1

Bicarbonate + NH_4^+ + 2ATP → Carbamoyl phosphate + 2ADP + Pi

Fig. 2.1. Formation of N-acetylglutamate, the activator of carbamoyl phosphate synthetase. The synthetase is stimulated by arginine and its steady state concentration in cells is increased by a large intake of dietary protein

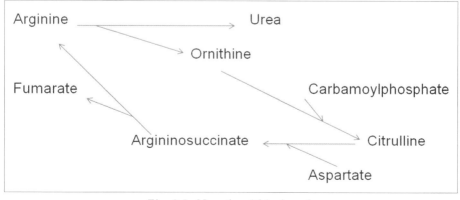

Fig. 2.2. Urea (ornithine) cycle

Metges and Barth 2000). Rudman *et al.* (1973) infused amino acid mixtures into the blood of human subjects and from measurements of urea excretion deduced that the maximum rate of urea synthesis in healthy subjects was about 65 mg urea N/h/kg body weight. This study is widely cited in archaeological literature and has been used by many authors to calculate the maximum dietary protein load that can be consumed without onset of feelings of ill health variously attributed to ammonia toxicity. Scrutiny of the data presented in Rudman *et al.* (1973), however, reveals that in subjects with normal liver function, the resting plasma NH_4^+ concentration was about 23 µM and rose to 30–35 µM during continuous infusion of protein or amino acids for 24 hours. The estimations of maximum rates of urea synthesis, in spite of the relatively small change in plasma ammonia concentration, have been applied to considerations of the protein loads that could be tolerated by humans and inferences drawn of the likely dietary components of hominids. The notion that the capacity of the ornithine cycle can be exceeded implies that the system is saturable, but most biochemistry textbooks (Nelson and Cox 2005; Newsholm and Leach 2009; Frayn 2010) and specialist literature (Vilstrup 1980; Evered 1981) state that provided time is allowed for the adjustment of altered rates of enzyme synthesis, urea production increases in tandem with the amino acid load. Vilstrup (1980) infused subjects with approximately 3 mmol/h/kg body weight of amino acids and found that the rate of urea synthesis was directly proportional to the plasma amino acid concentration which reached 12 mM after 12 hours of infusion. The linear response was indicative of a first order reaction that showed no sign of saturation at the plasma amino acid concentrations reached in the infusion experiment. The author reports however, that after some days of infusion the subjects experienced nausea and several vomited. This is an interesting result because of the reports of nausea, diarrhoea, and appetite loss associated with the so-called rabbit-starvation (see above). In this investigation (Vilstrup 1980), plasma ammonia rose from 19 µM to about 50 µM after 6 hours of infusion, but since urea production can cope with very large amino acid loads, the explanation for the ill health accompanying large intakes of lean meat is unlikely to result from ammonia toxicity alone.

Physiological consequences of excessive consumption of protein

It has been stated that intake of large amounts of protein for prolonged periods can result in chronic renal disease (Brenner *et al.* 1982). The popularity of high protein diets for slimming regimes and for recommended use in patients with type 2 diabetes mellitus (Parker *et al.* 2002; Fine and Feinman 2004; Hulton and Hu 2004) has called for consideration of potential harmful effects of high protein loads on renal function. The kidney responds to a protein load with increased glomerular filtration rate (Metges and Barth 2000) but this seems to be a normal adaptive response and poses no threat in the absence of any underlying renal disease (Hulton and Hu 2004; Martin *et al.* 2005). The metabolism of high protein loads can result in dehydration, metabolic acidosis and increased urinary calcium accompanied by loss of bone mineral (Metges and Barth 2000). The uncertainties that remain about effects on renal physiology have led to a suggestion that nutrient energy from protein consumption should not exceed 50% of the total energy intake, and this could be true for early hominins as well as for modern humans (Speth and Spielman 1989).

Gastrointestinal system

The anatomy of the gut reflects the dietary intake pattern. Carnivores have a relatively short gastrointestinal tract compared with herbivores and omnivores and the transit gut time is correspondingly rapid (Milton 1999). The small intestine of modern humans is longer and the colon is shorter than in extant ape species; an adaptation that is said to be suitable for the digestion of nutritionally dense foods such as meat and dairy products (Milton 1999). Addition of meat and fat to the diet also provides good sources of micronutrients such as vitamin A, zinc, iron, and certain B vitamins. During digestion, iron in meat is also generally more bioavailable than in plant sources. Ascorbate (vitamin C) is not synthesised by humans and apes and must be provided by plant dietary sources. The vitamin is needed for hydroxylation of proline and lysine in collagen (Nelson and Cox 2005) and is a key component of antioxidant defence systems (Padayatty 2003). The collagen malformation accounts for the characteristic symptoms of scurvy. The loss of the biosynthetic pathway for ascorbate occurred presumably before the evolutionary divergence of humans and apes. An extended colon permits the digestion of plant material such as non-starch polysaccharides (dietary fibre) and extant apes feed predominantly on plant-derived material, which will contain significant amounts of non-digestible fibre (Milton 1999).

It is now recognised that the gastrointestinal tract secretes a number of hormones (Stanley *et al.* 2005) including PYY, ghrelin, cholecystokinin, and glucose dependent insulinotropic peptide (GIP) that *inter alia* act on the hypothalamus of the central nervous system to affect appetite and feelings of satiety. In addition glucagon-like peptide 1 (GLP-1) produced in the intestine acts on the pancreas to stimulate insulin release. The release of these intestinal hormones occurs in response to food intake and its macronutrient composition. Adoption of a diet that consists almost entirely of lean meat will have an influence on the hormonal response of the GI tract. Meals that are high in protein content are reported to induce a feeling of satiety (Halton and Hu 2004). Perhaps the loss of appetite and feelings of nausea experienced by those explorers whose food consisted solely of lean meat can be attributed to the effects of the gut hormones released in response to this abnormal diet. Leptin and adiponectin are hormones released by adipose tissue that also act on the hypothalamus and influence appetite and energy expenditure (Stanley *et al.* 2005). An absence of TAG and carbohydrate from the diet will lead to depletion of body fat stored in adipose tissue and affect leptin and adiponectin release. Along with the gut hormones, the wide ranging actions of leptin and adiponectin will add to the general disturbances of the wellbeing of the digestive system accompanying unusual diets.

The intestinal ecosystem contains huge numbers of bacteria that are essential for normal intestinal function and human health (Gibson and Roberfroid 1995; Schell 2002). The total number of bacteria present in the gut is many-fold greater than the total number of cells in an adult human body, with the colon being the region that contains the greatest density of organisms belonging to hundreds of different species (Gibson and Roberfroid 1995). Most of the bacteria are strictly anaerobic and greatly outnumber aerobic organisms (Gibson and Roberfroid 1995). Undigested food material that reaches the colon is fermented by the bacteria to nutrients that may be absorbed into the blood stream and contribute to the total energy capture from food. Of great importance to health however, is the production by bacterial fermentation of short chain FFAs such as acetate, butyrate, and propionate. Butyrate is a major source of fuel for the cells of the colon and is important in maintaining the health of the mucosal cells and is thought to be protective against colorectal cancer (Canani *et al.* 2011).

The main substrates for bacterial fermentation are undigested carbohydrates. These include non-starch polysaccharides originating from plant cell wall materials, starch that resists normal digestion because it remains in native granular form or is encapsulated within plant cell structures and is therefore not easily accessible to α-amylase, plus so-called retrograded starch (Karim *et al.* 2000; Judd and Ellis 2006; Butterworth *et al.* 2011). Interestingly, in feeding experiments performed on humans in the 1920s using certain raw, (i.e., uncooked) starches, the investigators could find little trace of carbohydrate in faeces and therefore concluded that the starch from many of these particular plant sources is completely digestible (Langworthy and Deuel 1922). It is now clear that the combination of normal intestinal digestion and bacterial fermentation accounts for the complete breakdown of starches ingested in many foods.

Protein is also metabolised in the colon. Some of the digestible protein found in the intestines is endogenously produced digestive enzymes (Canani *et al.* 2011). Ingested protein that is resistant to digestion would include collagen and other connective tissue (found in meat gristle) and protein that escaped digestion because of entrapment within a food matrix. If the diet consists primarily of lean meat, virtually all of the bacterial substrates are derived from protein. It is known that the composition of the colonic bacteria is influenced by food intake and there is now considerable promotion of prebiotic food substances that will favour the growth and colonisation of bacteria that are regarded as favourable for healthy colon function (Canani *et al.* 2011). Colonic digestion of protein results in an increase in acidity of the colonic contents. It is known that growth of the 'favourable' bacteria is promoted if conditions are slightly alkaline (Canani *et al.* 2011). The recommended prebiotic substances seem largely to consist of non-digestible oligosaccharides, although all non-digestible carbohydrates, including resistant starch and plant cell wall polysaccharides (dietary fibre) have prebiotic properties to a greater or lesser extent. Protein-derived materials do not seem to be recommended.

The dynamic role of plant food processing in promoting access to energy and other nutrients

The view that food processing was a critical variable in the dietary decisions of prehistoric human populations, introduced into the archaeology literature by Stahl (1989), has only recently begun to receive wider scholarly attention. Stahl discussed how various processing techniques, grinding/pounding/grating, soaking/leaching, fermentation, and heat treatment, can improve the nutritive value of edible plants by physically and chemically transforming the plant tissue and cell contents into substances that are digestible by humans. Following Yen (1975), Stahl proposed that archaeologists recognise processing as an avenue of resource intensification, one that is quite apart from the adoption of agriculture:

> A review of the significant impact that processing can have on the nutritional quality of plant foods indicates that prehistoric populations may have manipulated their diets not only by changing patterns of resource exploitation, but also through modifying the technology of food processing. (Stahl 1989, 186)

Among the most common reasons for processing are: (i) to modify otherwise inedible or toxic plant and animal tissue into forms that are safe and more easily chewed and digested; (ii) to transform a single plant or animal part into variety of food forms, each with a different taste and/or texture; (iii) to make ingredients go further by mixing them into compound foods such as soups, stews, gruels, breads, cakes, beverages; and (iv) to transform fresh harvests into a state that will be preserved during storage. In the case of plants that contain toxins, processing can also provide an alternative to domestication, permitting humans to exploit plant toxins to their advantage during the growing season and circumvent co-evolutionary competition with other organisms (Harborne 1982, cited in Johns and Kubo 1988, 81).

> Although human food procurement is constrained by the same allochemicals (secondary compounds) that make plants unavailable as food for herbivorous animals, processing technology is one means employed by humans for making foods more palatable and less toxic. Selection for genetic changes during domestication risks exposing plants to attack by insects and plant diseases. By eliminating undesirable compounds subsequent to maturation of the plant, we allow chemicals to play their natural role in defence during vegetative and developmental stages, thus insuring a harvest for ourselves. (Johns and Kubo 1988, 81)

More recent anthropological and archaeological research on food processing (Johns and Chapman 1995; Johns 1990; 1999; Wrangham *et al.* 1999; Wollstonecroft *et al.* 2008; Carmody and Wrangham 2009; Carmody *et al.* 2011; 2012; Wollstonecroft 2011) suggests that it has evolutionary significance, having influenced *Homo* dietary selection, physical changes in the gut and digestive chemistry as well as improving species fitness. Innovations in food processing also have ecological implications because they permitted *Homo* to consume a greater share of the energy and nutrients in their environments, e.g., nutrient-rich but otherwise toxic plants (Johns 1990; 1999) and/or otherwise impervious plant tissue (Wollstonecroft 2011). Experimental studies by Carmody *et al.* (2012) show that similar energy gains are obtained with the cooking of meat in addition to improving its safety for consumption.

Previous research on the effects of processing, mastication and digestion on plant tissue (e.g., Brett and Waldron 1996; Waldron *et al.* 1997; 2003; Ellis *et al.* 2004) demonstrates that even the most rudimentary techniques can produce significant energy returns from plants by breaking open cells and exposing nutrients. Grinding, pounding, grating, and heating, for example, can promote nutrient release because they transform the microstructure of plant tissue by disrupting the cell walls, changing nutrient-matrix complexes, and/or transforming tissue substances into more active molecular structures, i.e., increasing nutrient and energy bioaccessibility. Bioaccessibility is defined as the fraction of a nutrient that is released from a food matrix during processing and/or consumption, and its potential availability for absorption in the gastrointestinal tract (Parada and Aguiler 2007; Stahl *et al.* 2002). Food processing can promote the bioaccessibility of macronutrients protein, fats, available carbohydrates, and micronutrients (minerals and vitamins), because it changes the physical or chemical form of plant tissue such that it is more easily and more completely digested (Fennema 1996; Pfannhauser *et al.* 2001). Bioaccessibility is an important factor in bioavailability, which is the rate and proportion of a nutrient that is absorbed by the digestive system to become metabolically available (Verhagen *et al.* 2001; Ellis *et al.* 2004).

A useful identification key for food processing by Johns and Kubo (1988) (following Coursey 1973; see also Johns 1990) provides a range of processing sequences based on seven principal techniques: heating, washing/solubilisation (repeated washing, soaking), fermentation, adsorption (the addition of clay, charcoal, or other substances that bind with toxins so that they are not absorbed during digestion), drying, physical processing (e.g., pounding, pulverizing, grinding, grating), and pH manipulation (e.g., pickling). The timing and place of origins of each of these methods is unknown. Anatomical analyses of australopithecines show that *A. africanus* routinely used pounding/hammering tools (Wood and Richmond 2000). Observational studies of chimpanzee tool uses suggest otherwise, i.e., that pounding was introduced much earlier, possibly by the last common ancestor of Homo and the chimpanzee (Hohmann 2009). Again the earliest actual evidence of tool use is stone tools and cut marks on bone from the Gona and Bouri archaeological sites in Ethiopia, dated to ~2.5 mya (Lee-Thorp and Sponheimer 2006). Firm evidence exists for the controlled use of fire and cooking by 300,000–400,000 years ago (Roebroeks and Villa 2011), but a more controversial date of 790,000 years ago is published for the site of Gesher Benot Ya'aqov in Israel (Goren-Inbar *et al.* 2004). Johns (1990, 76) proposed that adsorption may date from the Middle Stone Age based on charcoal found in Neanderthal coprolites. Fermentation was in use by Predynastic times in Egypt (*c.* 3800 BC) and the Early Dynastic (*c.* 2nd millennium BC) period in Mesopotamia (Samuel 1996; Haaland 2007; Wengrow 2010). Drying by sun or wind may be among the earliest processes but evidence is lacking.

These innovations undoubtedly permitted our ancestors to increase the diversity of food in their diets. The hominin tendency for dietary flexibility, coupled with this ability to modify inedible raw plants into safe and digestible foods further allowed them to move successfully into new environments (Hillman 2004; Jones 2009). Food processing innovations of our Palaeolithic ancestors have thus been linked to evolutionary trends such as changing dietary selection, expanding brain and body sizes, greater longevity and disease prevention as well as increased success in reproduction and the survival of infants (Stahl 1984; 1989; Johns 1990; 1999; Carmody and Wrangham 2009; Wollstonecroft 2011). Advances in the skills and knowledge required to fabricate and utilise food processing tools, the refinement of those tools and methodologies, and associated advances in ecological knowledge for obtaining the foods to be processed, are the types of physical and mental activities attributed with activating *Homo* brain expansion (see Aiello and Wheeler 1995). Altogether these ideas suggest that, through food processing, hominins had an active role in their own evolution. Wollstonecroft (2011) has described this evolutionary process as 'food-processing niche construction', proposing that it acted on human evolution through recursive interactions between culture, environment and biology (human and plant).

The idea that hominins had an active role in their own evolution has been challenged by scholars with more traditional Darwinian perspectives. Based on the position that food processing and other cultural innovations are end products of previous evolutionary change, e.g., increased brain size, advances in cognition and dexterity, opponents argue that it is an adaptation rather than as an initiator of change. In this view, cultural solutions such as food processing can be explained by the enhanced fitness that it induced in those who practiced it (Rindos 1989). Milton (2003, 56; 2000, 665) for example, states that 'such behaviours may well have served to buffer hunter-gatherers' biology from various selective pressures, but that their dietary choices were predetermined by inherited biological requirements for particular nutrients. But regardless of whether food processing is considered to be a critical factor in human evolution or an outcome (adaptation), both sides of the debate agree that it conferred greater fitness on our species.

A recent debate on the effectiveness of cooking versus pounding in energy and nutrient release (Carmody *et al.* 2011; 2012; Wollstonecroft *et al.* 2012) draws attention to significant misunderstandings in the anthropological and archaeological literature about the biology, biochemistry and structure of plants. Here, we wish to emphasise three main points: (i) interspecies differences in functional properties (processing performance) of plant tissue/plant cell wall; (ii.) interspecies differences in cell content, for example the processing performance of starch; and (iii) the point that starch is barely digestible for humans in its native (raw) state and that specific processes, involving heat and water, are necessary to transform it into a digestible form.

Interspecies differences in functional properties (processing performance) of plant tissue/plant cell wall

Interspecies differences in plant processing performance/functional properties means that the effectiveness of individual processing techniques, sequences of techniques and the order in which they are applied can be important in determining nutrient accessibility. Whether or not a processing technique or sequence will deliver nutrient accessibility depends on the inherent, biologically-determined functional properties of the plant in question, i.e., how it responds to the specific technology. In this case, effectiveness refers to both the sensory aspects of the final product as well as the bioavailability of the nutrients that it contains.

Indeed, ethnographic research with societies that continue to use non-mechanised plant processing technology (e.g., Johns 1990; Wandsnider 1997; Leach 1999; Lyons and D'Andrea 2003) shows that the decision to add a particular plant to the diet is highly influenced by peoples' ability to process it with the available technology. These ethnographic observations are confirmed by recent research in the food and nutritional sciences, which show that between-species differences in functional behaviours are significantly influenced by species-specific cell wall properties as well as the species-specific qualities of the cell contents, e.g., starch. (Loh and Breene 1982; Lillford 1991; Brett and Waldron 1996; Fennema 1996; Waldron *et al.* 1997; Pfannhauser *et al.* 2001; Vincent *et al.* 2003; Ellis *et al.* 2004).

The most important variables governing the effectiveness of individual processing techniques in releasing energy and other nutrients from plant tissues are the genetically-inherited physical and chemical properties of the cell wall (Brett and Waldron 1996). Cell wall encapsulation of nutrients is a critical issue in the food and nutritional biosciences because it can impede access to energy and nutrients from raw, and sometimes processed, plant tissue. This tendency cuts across a range of species and plant structural parts, and includes certain nuts, legumes and other seed foods as well as USOs (Brett and Waldron 1996; Noah *et al*, 1998; Ellis *et al.* 2004; Wollstonecroft *et al.* 2008). Indeed, it is now known that the amount of lipid released from raw almonds (*Amygdalus communis* L.) during digestion is limited due to cell wall encapsulation (Ellis *et al.* 2004). Legumes such as chick peas (*Cicer arietinum*) and kidney beans (*Phaseolus vulgaris*) have a low glycaemic index probably mainly because of the effects of the physical barrier of cell walls (Noah *et al.* 1998). These effects include (1) physical inhibition by the cell wall preventing or slowing down the interaction of amylase with the starch substrate, and (2) the restriction of starch swelling/ gelatinisation by the intact cell walls. The degree of the latter would of course depend on the amount of water and the temperature gradient of the cooked plant tissue (see below) and there may be differences between cells at the peripheral surface and in the core tissue. Water restriction is probably more relevant with dried seeds that are being cooked in water compared with materials that have high intrinsic water content, e.g., potatoes and other tubers.

The amount and rate of softening during cooking/grinding/pulverising/grating, etc., depends primarily on whether the innate tendency of the plant cell wall is towards cell rupture or cell separation. Cell rupture occurs when the adhesive forces of the middle lamella that bind the cells together are stronger than the cell wall (Brett and Waldron 1996). With cell rupture, the nutrients are released because the cell is broken open. Thus, the nutrients are more bioaccessible because the intracellular contents are exposed. Cell separation occurs when the cell wall is stronger than the adhesive forces of the middle lamella, so that whole intact cells become detached from each other. The softening of plant tissue during hydrothermal processing and post-harvest ripening is usually the result of cell separation. In the cases of some plants (e.g., legumes), unless the cooked tissue is subjected to further (non-thermal) processing to disrupt the cell wall, the nutrients are likely to remain encapsulated within the cells, are voided in that form, and are therefore unavailable for digestion and assimilation (Parker *et al.* 2003). However, nutrient bioaccessibility is likely to increase during microbial fermentation in the large intestine.

Nuts, for example, are subject to enormous variations in their functional and digestible properties, e.g., in-vitro studies of almond show that it is significantly more digestible in the raw, finely ground almond form than whole blanched or raw states (Mandalari *et al.* 2008.) In cases where a plant requires multiple stages of processing, the order in which the different processing techniques are applied affects bioaccessibility, e.g., chopping followed by steaming is usually more effective in promoting the bioavailability of nutrients in vegetables than steaming followed by chopping (Tydeman *et al.* 2001). In the case of certain root foods such as cassava (*Manihot esculentus*) and tubers in the Cyperaceae family, e.g., water chestnut (*Eleocharis dulcis*), cooking alone, by thermal or hydrothermal processes, can actually promote a toughening of the texture; heat and water are necessary to gelatinise the starch but if some kind of particle reduction (e.g., pulverising, grinding, grating) is not carried out first, the tissue can become even tougher during heating (Parker *et al.* 2003; Wollstonecroft *et al.* 2008). The science behind the processing technology may not be common knowledge but nevertheless these methods are practiced in traditional food processing systems. The survey on global food processing systems by Johns and Kubo (1988) confirms that people who are culturally and geographically unconnected nevertheless choose similar sequences for processing the same plant. A good example is cassava: on a worldwide basis, physical processing such as grating, slicing, or chopping is used before heating, a sequence that is necessary for the safe removal of toxins from the tubers (Johns 1990, 77).

Interspecies differences in the processing performance of starch; why water and heat are necessary to transform it into a digestible form

Starch present as semi-crystalline granules is the main storage form of carbohydrate in plants but its structure and properties differ among plant families and even within related genera and species. These differences explain its behaviour during processing, i.e., the temperature at which it gelatinises is species-specific, and the rate and amount of swelling. For example, phosphate in potato starch potentiates water uptake and thereby promotes sizeable swelling and loss of crystallinity (Slaughter *et al.* 2001). Furthermore, interspecies differences in botanical sources of starch partly explain why it is digested at different rates and to different extents and elicits varying postprandial blood glucose and insulin responses in humans (Seal *et al.* 2003; Warren *et al.* 2011). Because starch occurs in a granular form, to be digested readily and metabolised by humans, it must first be transformed physically into a form that is able to bind with ease to salivary and pancreatic α-amylase and be hydrolysed during the digestive process (Butterworth *et al.* 2011).

Again, the amount of water and the temperature gradient of the cooked plant tissue are critical factors and there may be differences between cells at the peripheral surface and in the middle. Of course water restriction is probably more relevant with dried seeds that are being cooked in water compared with materials that have a high intrinsic water content – e.g., potatoes and other tubers. Nevertheless, in some cases, heat processing alone fails to gelatinise the starch, which may even remain in an almost native state, and is significantly less susceptible to digestion (hydrolysis) by α-amylase in the human gut (Roder *et al.* 2009). In a contemporary example of the importance of hydration, so-called processed foods such as biscuits/wafers and the crust of bread contain birefringent starch granules (indicative of regions of the starch granule that are still crystalline, i.e., they have resisted gelatinisation during food manufacture because of limited water availability (Varianno-Marston *et al.* 1980).

Conclusions

Plants were the primary food source for the greater part of our evolutionary history. After more than 30 million years as herbivorous anthropoid apes, our Plio-Pleistocene ancestors began to scavenge and eventually hunt for meat, thus introducing new types of concentrated proteins and essential fatty acids, and a range of vitamins and minerals into their diets. With the emergence of the hominin line, new patterns of eating were established that focused on (high-quality) foods rich in concentrated nutrients. Significantly, meat eating appears to have promoted, rather than curtailed, increasing diversity in plant consumption, possibly because of the hominin tendency for increasing dietary diversity but, as discussed in this paper, probably to a greater extent because protein and fat alone did not satisfy the basic nutrient requirements and features of a digestive physiology and specific tissue metabolic requirements already shaped by 30 million years of evolution. Many nutrients that are critical for modern humans, and presumably our hominin ancestors, are found only in plants, i.e., available carbohydrates, dietary fibre, polyunsaturated fatty acids, vitamin C, minerals, and phytochemicals such as polyphenols and stanols/sterols. However, due to their predilection for increasing dietary diversity and high quality foods, hominins were confronted with potential dangers of poisoning and/or other natural impediments to obtaining critical nutrients and energy from the plant and animal resources they consumed. In this paper we discussed the toxic effects of a diet of lean meat and the dangers and difficulties posed by interspecies differences in plant chemical and physical properties, such as (undetected) toxins and tough plant tissue. We examined how innovations in food processing, particularly those introduced by the genus *Homo*, provided a means for addressing these problems and concluding that differences in plant processing performance was a critical a factor in *Homo* decisions about dietary selection and technical choice. Ultimately, food processing appears to have stimulated new hominin-plant relationships and possibly different types of hominin-plant relationships with species that were already part of the diet.

Acknowledgements

We thank Gordon Hillman for inspiring this collaboration. We are also grateful to the anonymous reviewer for invaluable comments. PRE and PJB thank the BBSRC (Grant Ref. BB/H004874/1) for their current funding on the role of plant cell walls in regulating macronutrient bioaccessibility; and the Almond Board of California (ABC) for supply of almonds and funding. MW thanks the UCL Institute of Archaeology, particularly Dorian Fuller, for their unflagging encouragement.

References

Aiello, L. C. and Wheeler, P. 1995. The expensive tissue hypothesis: the brain and the digestive system in human primate evolution. *Current Anthropology* 36, 199–221

Bird, D. W. and O'Connell, J. F. 2006. Behavioural ecology and archaeology. *Journal of Archaeological Research* 14, 143–88

Barlow, K. R. and Heck, M. 2002. More on acorn eating during the Natufian: expected patterning in diet and the archaeological record of subsistence. In Mason, S. L. R. and Hather, J. G. (eds), *Hunter-Gatherer Archaeobotany: perspectives from the Northern Temperate Zone*, 128–45. London: UCL Institute of Archaeology

Brett, C. T. and Waldron, K. W. 1996. *Physiology and Biochemistry of Plant Cell Walls.* London: Chapman & Hal

Butterworth, P. J., Warren, F. J. and Ellis, P. R. 2011. Human α-amylase and starch digestion: an interesting marriage. *Starch/Starke* 63, 395–405

Brenner, B. M., Meyer, T. W. and Hostetter, T. H. 1982. Dietary protein intake and the progressive nature of kidney disease: the role of themodynamically mediated glomerular injury in the pathogenesis of progressive glomerular sclerosis in aging, renal ablation and intrinsic renal disease. *New England Journal of Medicine* 307, 652–9

Canani, R. B., Di Constanzo, M., Leone, L., Pefata, M., Meli, R. and Calignano, A. 2011. Potential beneficial effects of butyrate in intestinal and extraintestinal diseases. *World Journal of Gastroenterology* 17, 1519–28

Carmody, R. N. and Wrangham, R. W. 2009. The energetic significance of cooking. *Journal of Human Evolution* 57, 379–91

Carmody, R. N., Weintraub, B. S. and Wrangham, R. W. 2011. Energetic consequences of thermal and nonthermal food processing. *Proceedings of the National Academy of Sciences* (USA) 108, 19199–203

Carmody, R. N., Weintraub, B. S. and Wrangham, R. W. 2012. Reply to Wollstonecroft *et al*.: Cooking increases the bioavailability of starch from diverse plant sources. *Proceedings of the National Academy of Sciences* (USA) 109, E992

Cerling, T. E., Critz, K. L., Jablonski, N. G. and Leakey, M. G. 2013a. Diet of Theropithecus from 4 to 1 Ma in Kenya. *Proceedings of the National Academy of Sciences* (USA), doi:10.1073/pnas.1222571110

Cerling, T. E., Manthi, F. K., Mbua, E. N., Leakey, L. N., Leakey, M. G., Leakey, R. E., Brown, F. H., Grine, F. E., Hart, J. A., Prince Kaleme, Roche, H., Uno, K. T. and Wood, B. A. 2013b. Stable isotope-based diet reconstructions of Turkana Basin Hominins. *Proceedings of the National Academy of Sciences* (USA), doi:10.1073/pnas.1222568110

Conklin-Brittain, N. L, Wrangham, R. W. and Hunt, K. D. 1998. Dietary response of chimpanzees and cercopithicines to seasonal variation in fruit abundance. *International Journal of Primatology* 19, 971–98

Cordain, L., Brand Miller, J., Eaton, S. B., Mann, N. J., Holt, S. H. A. and Speth, J. D. 2000. Plant–animal subsistence ratios and macronutrient energy estimations in worldwide hunter–gatherer diets. *American Journal of Clinical Nutrition* 71(3), 682–92

Cordain, L., Watkins, B. A. and Mann, N. J. 2001. Fatty acid composition and energy density of foods available to African hominids: evolutionary implications for human brain development. In Simopoulos, A. P. and Pavlou, K. N. (eds), *Nutrition and Fitness: metabolic studies in health and disease*, 144–61. Basel: Karger

Cordain, L., Eaton, S. B., Brand Miller, J., Mann, N. and Hill, K. 2002. The paradoxical nature of hunter-gather diets: meat-based yet non-atherogenic. *European Journal of Clinical Nutrition* 56 (Suppl. 1): S42–52

Coursey, D. G. 1973. Cassava as food: toxicity and techonology. In Nestel, B. and MacIntyre, R. (eds), *Chronic Cassava Toxicity*, 27–36. Ottawa: International Development Research Centre

Dimski, D. S. 1994. Ammonia metabolism and the urea cycle: function and clinical implications. *Journal of Veterinary Internal Medicine* 8, 73–8

Ellis, P. R., Kendall, C. W. C, Ren, Y., Parker, C., Pacy, J. F. Waldron K. W. and Jenkins D. J. A. 2004. Role of cell walls in the bioaccessibility of lipids in almond seeds. *American Journal of Clinical Nutrition* 80, 604–13

Evered, D. F. 1981. Advances in amino acid metabolism in mammals. *Biochemical Society Transactions* 9, 159–69

Fennema, O. 1996. *Food Chemistry*. New York: Marcel Dekker

Fine, E. J. and Feinman, R. D. 2004. *Thermodynamics of Weight Loss Diets*. London: Nutrient Metabolism 1: 15.

Frayn, K. N. 2010. *Metabolic Regulation: a human perspective,* 3rd edn. Chichester: Wiley-Blackwell

Gibson, G. R. and Roberfroid, M. B. 1995. Dietary modulation of the human colonic microbiota: introducing the concept of prebiotics. *Journal of Nutrition* 125, 1401–12

Haaland, R. 2007. Porridge and pot, bread and oven: food ways and symbolism in Africa and the Near East from the Neolithic to the Present. *Cambridge Archaeological Journal* 17, 165–82

Halton, T. L. and Hu, F. B. 2004. The effects of high protein diets on thermogenesis, satiety and weight loss: a critical review. *Journal of the American College of Nutrition* 23, 373–85

Hillman, G. C. 2004. *The Rise and Fall of Dietary Diversity*. Paper presented at the 2004 9th meeting of the Society of Economic Botany, University of Kent, Canterbury.

Hillman, G. C. and Wollstonecroft, M. 2014. Dietary diversity: our species-specific dietary adaptation. In Stevens, C. J., Nixon, S., Murray, M. and Fuller, D. Q. (eds) *The Archaeology of African Plant Use*, 37–49. Walnut Creek CA: Left Coast Press

Hohmann, G. 2009. The diets of nonhuman primates: frugivory, food processing and food sharing. Evolution of human diets. In Hublin and Richards (eds) 2009, 1–14

Harborne, J. B. 1982. *Introduction to Ecological Biochemistry*. London: Academic Press

Hublin, J. and Richards, M. P. (eds). 2009. *The Evolution of Hominin Diets: integrating approaches to the study of Palaeolithic subsistence*. Dordrecht: Springer

Hue, F. B., Stampfer, M. J., Manson J. E., Ascherio, A., Spiegelman, D. and Willett, D. C. 2000. Prospective study of major dietary patterns and risk of coronary heart disease in men. *American Journal of Clinical Nutrition* 72, 912–21

Johns, T. 1990. *The Origin of Human Diet and Medicine: chemical ecology*. Tucson AZ: University of Arizona Press

Johns, T. 1999. The chemical ecology of human ingestive behaviours. *Annual Review of Anthropology* 28, 27–50

Johns, T. and Chapman, L. 1995. Phytochemicals ingested in traditional diets and medicines as modulators of energy metabolism. In Arnason, J. T., Mata, R. and Romeo, J. T. (eds), *Phytochemistry of Medicinal Plants*, 161–88. New York: Plenum Press

Johns, T. and Kubo, I. 1988. A survey of traditional methods employed for the detoxification of plant foods. *Journal of Ethnobiology* 8, 81–129

Johnson, I. T. 2013. Phytochemicals and health. In Tiwari, B. K., Brunton, N. P. and Brennan, C. S. (eds), *Handbook of Plant Food Phytochemicals: sources, stability and extraction*, 49–67. Oxford: Wiley

Jones, M. 2009. Moving north: archaeobotanical evidence for plant diet in Middle and Upper Palaeolithic Europe. In Hublin and Richards (eds) 2009, 171–80

Judd, P. A. and Ellis, P. R. 2006. Plant polysaccharides in the prevention and treatment of diabetes mellitus. In Souvamyanath, A. (ed.) *Traditional Medicine for Modern Times*, 257–72. Florida: CRC Press

Karim, A. A., Norziah, M. H. and Scow, C. C. 2000. Methods for the study of starch retrogradation. *Food Chemistry* 71, 9–36

Laden, G. and Wrangham, R. W. 2005. The rise of the hominids as an adaptive shift in fallback foods: plant underground storage organs (USOs) and australopith origins. *Journal of Human Evolution* 49, 482–98

Langworthy, C. J. and Deuel, H. J. 1922. Digestibility of raw rice, arrowroot, canna, cassava, tree-fern and potato starches. *Journal of Biological Chemistry*, 52, 251–61

Leach, H. M. 1999. Food processing technology: its role in inhibiting or promoting change in staple foods. In Gosden G. and Hather, J. G. (eds), *The Prehistory of Food: appetites for change*, 129–38. London: Routledge

Lee-Thorp, J. and Sponheimer, M. 2006. Contributions of biochemistry to understanding hominin dietary ecology. *Yearbook of Physical Anthropology* 49, 131–48

Lindeberg, S. 2009. Modern human physiology with respect to evolutionary adaptations that relate to diet in the past. In Hublin and Richards (eds) 2009, 43–57

Lindeberg, S. 2010. *Food and Western Disease: health and nutrition from an evolutionary perspective.* Chichester: Wiley-Blackwell

Liu, R. H. 2004. Potential Synergy of phytochemicals in cancer prevention: mechanism of action. *Journal of Nutrition* 134, 3479S–85S

Loh, J. and Breene, W. M. 1982. Between-species differences in fracturability loss: comparison of the thermal behaviour of pectin and cell wall substances in potato and Chinese waterchestnut. *Journal of Texture Studies* 13, 381–96

Lyons, D. and D'Andrea, A. C. 2003. Griddles, ovens and agricultural origins: an ethnoarchaeological study of bread baking in Highland Ethiopia. *American Anthropologist* 105, 515–30

Mandalari, G., Faulks, R. M., Rich, G. T., Lo Turco, V., Picout, D. R., Lo Curto, R. B., Bisgnano, G., Dugo, P., Dugo, G., Waldron, K. W., Ellis, P. R. and Wickham, M. S. J. 2008. Release of protein, lipid, and vitamin E from almond seeds during digestion. *Journal of Agricultural and Food Chemistry* 56, 3409–16

Martin, W. F., Armstrong, L. E. and Rodriguez, N. R. 2005. Dietary protein intake and renal function. *Nutrient Metabolism* 2, 25–33

Metges, C. C. and Barth, C. A. 2000. Metabolic consequences of a high dietary-protein intake in adulthood: Assessment of the available evidence. *Journal of Nutrition* 130, 886–9

Milton, K. 1999. A hypothesis to explain the role of meat-eating in human evolution. *Evolutionary Anthropology* 8, 11–21

Milton, K. 2000. Back to basics: why foods of wild primates have relevance for modern human health. *Nutrition* 16, 480–3

Milton, K. 2002. Hunter-gatherer diets: wild foods signal relief from diseases of affluence. In Ungar, P. S. and Teaford, M. F. (eds), *Human Diet: its origin and evolution*, 111–22. Westport VA: Bergin and Garvey

Milton, K. 2003. The critical role played by animal source foods in humans *(Homo)* evolution. *Journal of Nutrition* 133, 3886s–3892s

Munro, N. D. and Bar-Oz, G. 2005. Gazelle bone fat processing the Levantine Epipalaeolithic. *Journal of Archaeological Science* 32, 223–39

Nelson, D. L. and Cox, M. M. 2005a. *Lehninger Principles of Biochemistry*, 4th edn. New York: W. H. Freeman

Newsholme, E. A. and Leech, T. R. 2009. *Functional Biochemistry in Health and Disease*, 215–6. Chichester: Wiley-Blackwell

Noah, L., Guillon, F., Bouchet, B., Buleon, A., Molis, C., Gratas, M. and Champ, M. 1998. Digestion of Carbohydrate from White Beans (*Phaseolus vulgaris* L.) in Healthy Humans. *Journal of Nutrition* 128, 977–85

Padayatty, S. J., Katz, A., Wang, Y., Eck, P., Kwon, O., Lee, J.-H., Chen, S., Corpe, C., Dutta, A., Dutta, S. K. and Levine, M. 2003. Vitamin C as an antioxidant: evaluation of its role in disease prevention. *Journal of the American College of Nutrition* 22, 18–35

Parada, J. and Aguilera, J. M. 2007. Food microstructure affects the bioavailability of several nutrients. *Journal of Food Science* 72, 21–32

Parker, B., Noakes, M., Luscombe, N. and Clifton, P. 2002. Effect of a high protein diet on glycemic control and lipid levels in type 2 diabetes. *Diabetes Care* 25, 425–30

Parker, C. C., Parker, M. L., Smith, A. C. and Waldron, K. W. 2003. Thermal stability of texture in Chinese water chestnut may be dependent on 8.8′-diferulic acid. *Journal of Agricultural and Food Chemistry* 51, 2034–2039

Pfannhauser, W., Fenwick, G. R. and Khokhar, S. 2001. *Biologically-active Phytochemicals in Food: analysis, metabolism, bioavailability and function.* Cambridge: Royal Society of Chemists

Rindos, D. 1989. Darwinism and its role in the explanation of domestication. In Harris, D. R. and Hillman, G. C. (eds), *Foraging and Farming: the evolution of plant exploitation*, 27–41. London: Unwin Hyman

Roder, N., Gerard, C.,Verel, A., Bogracheva, T. Y., Hedley, C. L., Ellis, P. R. and Butterworth, P. J. 2009. Factors affecting the action of α-amylase on wheat starch: Effects of water availability. An enzymic and structural study. *Food Chemistry* 113, 471–8

Roebroeks, W. and Villa, P. 2011. On the earliest evidence for habitual use of fire in Europe. *Proceedings of the National Academy of Sciences* (USA) 108, 5209–14

Rudman, D., DiFulco, T. J., Galambos, J. T., Smith, R. B., Salam, A. A. and Warren, W. D. 1973. Maximal rates of excretions and synthesis of urea in normal and cirrhotic subjects. *Journal of Clinical Investigation* 52, 2241–9

Samuel, D. 1996. Investigation of ancient Egyptian baking and brewing methods by correlative microscopy. *Science* 273, 488–90

Schaller, G. B. and Lowther, B. R. 1969. The relevance of carnivore behaviour to the study of early hominids. *Southwest Journal of Anthropology* 25, 307–41

Schell, M. A. *et al.* 2002. The genome sequence of Bifidobacterium longum reflects its adaptation to the human gastrointestinal tract. *Proceedings of the National Academy of Sciences* (USA) 99, 14422–7

Seal, C. J., Daly, M. E., Thomas, L. C., Bal, W., Birkett, A. M., Jeffcoat, R. and Mathers, J. C. 2003. Postprandial carbohydrate metabolism in healthy subjects and those with type 2 diabetes fed starches with slow and rapid hydrolysis rates determined in vitro. *British Journal of Nutrition* 90, 853–64

Sinha, R., Cross A. J., Graubard, B. I., Leitzmann, M. F. and Schatzkin, A. 2009. Meat intake and mortality: a prospective study of over half a million people. *Archives of Internal Medicine* 169, 562–71

Slaughter, S. L., Ellis, P. R. and Butterworth, P. J. 2001. An investigation of the action of porcine pancreatic α-amylase on native and gelatinised starches. *Biochimica et Biophysica Acta* 1525, 26–9

Snodgrass, J. J., Leonard, W. R. and Robertson, M. L. 2009. The energetics of encaphalization in early hominids. In Hublin and Richards (eds) 2009, 15–30

Speth, J. D. 1989. Early hominid hunting and scavenging: the role of meat as an energy source. *Journal of Human Evolution* 18, 329–43

Speth, J. 2004. *The emergence of bone boiling: why it matters.* Unpublished paper presented at the 69th Annual Meeting of the Society for American Archaeology, April 3, 2004, Montreal.

Speth, J. D. 2010. The paleoanthropology and archaeology of big game hunting. Protein, fat or politics? In Eekens, J. (ed.), *Interdisciplinary Contributions to Archaeology*, 1–223. Dordrecht: Springer

Speth, J. D. and Spielman, K. A. 1983. Energy source, protein metabolism, and hunter-gatherer subsistence strategies. *Journal of Anthropological Archaeology* 2, 1–31

Sponheimer, M. and Dufour, D. L. 2009. Increased dietary breadth in early hominim evolution: Revisiting arguments and evidence with a focus on biogeochemical contributions. In Hublin and Richards (eds) 2009, 229–40

Sponheimer, M., Alemseged, Z., Cerling, T. E., Grine, F. E., Kimbel, W. H., Leakey, M. G. Lee-Thorp, J. A., Manthi, F. K., Reed, K. E., Wood, B. A. and Wynn, J. G. 2013. Isotopic evidence of early hominin diets. *Publication of the National Academy of Science (USA)*. doi:10.1073/pnas.1222579110.

Stahl, A. B. 1984. Hominid dietary selection before fire. *Current Anthropology* 25, 151–68

Stahl, A. B. 1989. Plant-food processing: implications for dietary quality. In Harris, D. R. and Hillman, G. C. (eds), *Foraging and Farming: the evolution of plant exploitation*, pp. 171–96. London: Unwin Hyman

Stahl, W., van den Berg, H., Arthur, J., Bast, A., Dainty, J., Faulks, R. M., Gartner, C., Haenen, G., Hollman, P., Holst, B., Kelly, F. J., Polidori, M. C., Rice-Evans, C., Southon, S., van Vliet, T., Vina-Ribes, J., Williamson, G. and Astley, S. B. 2002. Bioavailability and metabolism. *Molecular Aspects of Medicine* 23, 39–100

Stanley, S., Wynne, K., McGowan, B. and Bloom, S. 2005. Hormonal regulation of food intake. *Physiology Reviews* 85, 1131–58

Tydeman, E., Wickham, M., Faulks, R., Parker, M., Waldron, K., Fillery-Travis, A. and Gidley, M. 2001. Carotene release from carrot during digestion is modulated by plant structure. In Pfannhauser *et al.* (eds) 2001, 429–32

Verhagen, H., Coolen, S., Duchateau, G., Mathot, J. and Mulder, T. 2001. Bioanalysis and biomarkers: the tool and the goal. In Pfannhauser *et al.* (eds) 2001, 125–30

Verriano-Marston, Ke, V., Huang, G. and Ponte, J. 1980. Comparison of methods to determine starch gelatinization in bakery foods. *Cereal Chemistry* 57, 242–8

Vilstrup, H. 1980. Synthesis of urea after stimulation with amino acids: relation to liver function. *Gut* 21, 990–5

Visek, W. J. 1979. Ammonia metabolism, urea cycle capacity and their biochemical assessment. *Nutrition Reviews* 37, 273–82

Waldron, K. W., Smith, A. C., Parr A. J., Ng, A. and Parker, M. L. 1997. New approaches to understanding and controlling cell separation in relation to fruit and vegetable texture. *Trends in Food Science and Technology* 8, 213–21

Waldron, K. W., Parker, M. and Smith, A. C. 2003. Plant cell walls and food quality. *Comprehensive Reviews in Food Science and Food Safety* 2, 101–19

Wandsnider, L. 1997. The roasted and the boiled: Food composition and heat treatment with special emphasis on pit-hearth cooking. *Journal of Anthropological Archaeology* 16, 1–48

Warren, F. J., Royall, P. G., Gaisford, S., Butterworth, P. J. and Ellis, P. R. 2011. Binding interactions of α-amylase with starch granules: The influence of supramolecular structure and surface area. *Carbohydrate Polymers* 86, 1038–47

Wengrow, D. 2010. *What Makes Civilization? The Ancient Near East and the Future of the West.* Oxford: Oxford University Press

Winterhalder, B. and Goland, C. 1997. An evolutionary ecology perspective on diet choice, risk, and plant domestication. In Gremillion, K. J. (ed.), *People, Plants and Landscapes: studies in palaeoethnobotany*, 123–60. Tucaloosa AL: University of Alabama Press

Wollstonecroft, M. 2011. Investigating the role of food processing in human evolution: a niche construction approach. *Journal of Archaeological and Anthropological Sciences* 3, 141–50

Wollstonecroft, M., Ellis, P. R., Hillman, G. C. and Fuller, D. Q. 2008. Advancements in plant food processing in the Near Eastern Epipalaeolithic and implications for improved edibility and nutrient bioaccessibility: an experimental assessment of sea club-rush (*Bolboschoenus maritimus* (L.) Palla). *Vegetation History and Archaeobotany* 17 (Suppl. 1), S19–S27

Wollstonecroft, M. M., Ellis, P. R., Hillman, G. C., Fuller, D. Q. and Butterworth, P. J. 2012. A calorie is not necessarily a calorie: technical choice, nutrient bioaccessibility, and interspecies differences of edible plants. *Proceedings of the National Academy of Sciences USA* 109: E991

Wood, B. and Richmond, B. G. 2000. Human evolution: taxonomy and paleobiology. *Journal of Anatomy* 196, 19–60

Wrangham, R., Jones, J. H., Laden, G., Pilbeam, D. and Conklin-Brittain, N. L. 1999. The raw and the stolen: cooking and the ecology of human origins. *Current Anthropology* 40, 567–94

Wynn, J. G., Sponheimer, M., Kimbel, W. H., Alemseged, Z., Reed, K., Bedaso, Z. K. and Wilson, J. N. 2013. Diet of *Australopithecus afarensis* from the Pliocene Hadar Formation, Ethiopia. *Proceedings of the National Academy of Sciences* (USA) 110,10495–500

Yen, D. E. 1975. Indigenous food processing in Oceania. In Arnott, M. L. (ed.), *Gastronomy, the Anthropology of Food Habits*, 147–68. The Hague: Mouton

Yen, D. E. 1980. Food crops. In Ward, R. G. and Proctor, A. (eds), *South Pacific Agriculture: hoices and constraints*, 172–234. Manila: Asian Development Bank

3. An ape's perspective on the origins of medicinal plant use in humans

Michael A. Huffman

A growing body of literature generated from the study of the medicinal use of plants in animals, chimpanzees in particular, has opened up a new avenue for exploring the origins of our own species' medicinal behaviour. In the African great apes (chimpanzees, bonobos, gorillas), the seasonal consumption of 'medicinal food' plants with both nutritional and bioactive properties, the limited ingestion of pharmacologically active plant parts, or non-nutritional consumption of plants, have been shown to protect or treat individuals suffering from parasite infections and/or related symptoms. There is great overlap in the use of particular medicinal plants by these apes and contemporary humans dependent largely upon ethnomedicinal practices for their health care. These similarities are influenced not only by a shared evolutionary history, but also through the widespread cultural practice of humans to observe the behaviour of sick animals to find new sources and uses of medicinal plants. This chapter reviews the evidence from these different fields of investigation and discusses them in the context of biological and cultural evolution of medicinal plant use in humans.

Why apes?

Among non-human primates, chimpanzees are our closest living relatives. Chimpanzees are often used as models of human evolution, providing insights into the physiological, biological, psychological, and behavioural origins of the human condition. But what can they possibly teach us about medicinal plant use in extant traditional human societies or for that matter in our prehistoric ancestors? A growing body of literature generated from the study of medicinal plant use in animals has opened up a new avenue for considering the origins of our own species' medicinal behaviour.

While the evidence for medicinal plant use throughout the animal kingdom is growing (e.g., Engel 2002; Huffman 1997; 2001; Singer *et al.* 2009), a substantial amount of this evidence comes from long-term observations of chimpanzees, gorillas and, more recently, other monkey species. As a result of this research, many similarities in the selection of medicinal plants by humans and animals have been brought to light, which would suggest that we are driven by similar physiological mechanisms that respond to similar diseases and or environmental pressures. Interestingly, there is also evidence for the widespread practice of humans traditionally seeking medicinal knowledge from the observation of the behaviour of sick animals.

Both our shared evolutionary past, manifest in physiology and behaviour, and dependence upon other species as a source of food and medicinal knowledge, provide a bio-cultural rationale for looking to our ape ancestors for insights in the prehistoric plant use behaviour of humans. The aim of this chapter is three-fold: 1) to review the diversity of plant use in the context of health maintenance and self-medication in chimpanzees and other primates, 2) to describe the overlap in the non-human primate and human pharmacopeia, and 3) to provide examples for how some of this overlap may in fact be tied to experience based cultural belief that animals are a source of medicinal knowledge. In doing so, I propose that prehistoric humans and our hominin ancestors possessed self-medicated behaviour.

The dimensions of health maintenance and self-medication in primates

Behavioural strategies for health maintenance in animals typically form what Hart (1990) calls 'the front line defence' against disease caused by nematodes, microparasites (protozoa, bacteria, viruses, fungi), ectoparasites, and biting insects. I have argued that behavioural strategies against parasite infection and related symptoms come into play when physiological adaptations are insufficient and behavioural avoidance, limited contact with, or a direct response to illness is warranted (i.e., Huffman and Caton 2001). From either perspective, the maintenance of behavioural strategies that ensure basic survival and thus indirectly or directly enhance reproductive fitness must be considered universally adaptive to all animals, including humans.

The need to self-medicate is assumed to be a response to homeostatic challenges faced by an individual (Huffman and Caton 2001; Forbey *et al.* 2009; Foitova *et al.* 2009). The various forms of self-medicative or health maintenance behaviours are likely operated by a combination of both innate mechanisms, individual learning, and cultural transmission from generation to generation (Huffman and Hirata 2003; 2004; Huffman *et al.* 2010). At the ultimate level, behavioural propensities to perform such actions or basic dietary preferences may be selected for their direct benefits to health maintenance and increased reproductive fitness.

At our current level of understanding, health maintenance and self-medicative behaviours can be classified into roughly four different levels (Huffman 2011):

1) 'sick behaviours': lethargy, depression, anorexia, reduction in grooming, behavioural fever (self-induced rise in body temperature, e.g., sun basking), basking behaviour (*sic.* Hart 1988);
2) optimal avoidance or reduction of the possibility for disease transmission: avoidance of faeces contaminated food, water, substrates, etc. (e.g., Freeland 1980);
3) the dietary selection of items with a preventative or health maintenance affect: items eaten routinely in small amounts or on a limited basis (Huffman 1997; Krief *et al.* 2006; MacIntosh and Huffman 2010; Huffman and MacIntosh 2012);

4) ingestion of a substance for the curative treatment of a disease or the symptoms thereof: use of toxic or otherwise biologically active items at low frequency or in small amounts, having little or no nutritional value (e.g., Huffman and Seifu 1989; Huffman *et al* 1993).

Avoidance of disease transmission

Water-borne diseases pose a threat to primates directly dependent on streams and ponds as their main water source. Hamadryas baboons (*Papio hamadryas*) living near the city of Taif, Saudi Arabia, are known to dig drinking holes in the sand directly adjacent to the stagnant alga-tainted watering sites of livestock. They patiently wait for the filtered water to seep through the sand (Huffman 2011). This is a common behaviour of this species at many locations (e.g., Nelson 1960). Gelada baboons (*Theropithecus gelada*) living in the Semien Highlands of Ethiopia, often drink from the cliff springs where they spend their nights, instead of down in the river beds. In addition to the obvious benefits of filtering out unwanted slime, dirt, and debris, this is likely an efficient way for baboons to avoid faecal contamination and water transmitted parasites such as the blood fluke (*Schistosoma mansoni*) found in stagnant ponds and slow running streams in these areas (Huffman 2011).

Hausfater and Mead (1982) described the routine changing of sleeping grove sites by yellow baboon (*Papio cynocephalus*) in Kenya's Amboseli National Park as a strategy for avoiding parasite infection via restricted contact with infective ova and larvae of nematodes (*Oesophagostomum*, *Stongyloides*, and *Trichostrongylus*). Grey-cheeked mangabey (*Cercocebus albigena*) defecate throughout the day, and appear not to be able to avoid, or even try to avoid, contaminated vegetation that they later used for feeding, travelling, or sleeping (Freeland 1980). Freeland proposed an alternative way that these monkeys may be avoiding infection from intestinal protozoa. He showed that *C. albigena* ranged more widely during dry periods than rainy periods, when faecal material was likely to stay longer on the leaves at Ngogo in the Kibale forest, Uganda. In this study, neither food density nor an inhibition to travel in the rain could explain this difference (Freeland 1980). Olupot *et al.* (1997) found that 10 km away in Kanyawara, this species travelled significantly longer distances during rainy periods than dry season, and that fruit density did influence travel patterns.

Differences in food distribution and availability between sites were given as likely reasons for the difference, suggesting that either food takes priority over risk to parasite infection or risk of infection may vary between even neighbouring habitats (Olupot *et al.* 1997). Other ecological factors are also known to affect parasite pressure, challenging the homeostasis of the individual. Risk to parasite infection to monkeys (red tail guenons, red colobus and black and white colobus) living in the same region of Uganda differs significantly between neighbouring logged and unlogged habitats (Gillespie *et al.* 2005). In chimpanzees (*Pan troglodytes*) in different habitats of East Africa, risk to parasite infection varies between years at the same site and between sites based on rainfall patterns (Huffman *et al.* 1997; 2009).

'Medicinal foods' of apes and their medicinal value to humans

The difference between food and medicine is often difficult to detect, even in humans. The concept of medicinal foods first introduced by Etkin and Ross (1982) to anthropology is supported by Johns (1990), who believes these non-nutritional components, once part of our diet, have now been replaced with herbal medicine and modern pharmaceuticals. Within traditional human societies worldwide, there is still much overlap between food and medicinal items.

Supplementing a nutritional diet with non-nutritional bioactive elements is another important part of disease control and prevention in animals too (Huffman 1997; Huffman *et al.* 1998). Janzen (1978) first suggested the possibility that their incidental ingestion by non-human primates might help to combat parasite infection. Lozano (1991) first proposed that parasites have a significant impact on the foraging decisions of animals, comparable to that of the considerations of basic nutritional requirements. Indeed, the diets of several non-human primates investigated thus far reveal that a part of their food intake comprises medicinal foods with anti-parasitic potential.

Laboratory analysis and ethnomedicinal/pharmacological literature searches were conducted on the possible pharmacologically based antiparasitic activity of recognised plant foods in the diet of the Mahale M group chimpanzees in Tanzania (e.g., Ohigashi *et al.* 1991a; 1991b; Koshimizu *et al.* 1993; Ohigashi 1994; Huffman *et al.* 1998). From a total of 172 native plant foods recorded to be eaten by chimpanzees in M group (Nishida and Uehara 1983), 43 species were found to be used to treat parasitic or gastrointestinal related illnesses traditionally by humans in Africa (Huffman *et al.* 1998). In 16 of these species, the same plant part(s) was utilised by both humans and chimpanzees, and 33% (20/63 plant parts) of the specific food items ingested from them by chimpanzees corresponded to the parts utilised in ethnomedicine for the treatment of intestinal nematodes, dysentery, malaria, colic, diarrhoea, and/or as an antiseptic. Chimpanzees ate these items only occasionally and typically in small amounts, but significantly more frequently during rainy season months when parasite reinfection of some species is highest (Huffman *et al.* 1998). In particular, this coincides with the identified period of reinfection by a species of nodular worm (*Oesophogostomum stepanostomum*; Huffman *et al.* 1997), associated with two therapeutic forms of self-medicative behaviour in chimpanzees at Mahale (see below).

Similar trends have been reported among two other chimpanzee study sites in Uganda. In the Kibale chimpanzees, isolation and testing of the pharmacological properties of their diet by Krief and colleagues (2005) found that in the 163 food items tested, 30% possessed some form of mild bioactivity; i.e., antibiotic, antifungal. A sub-set of 117 of these food species cross-referenced with the ethnomedicinal literature revealed that 41% were also used in traditional African medicine. During a one-year study of chimpanzees in the Budongo forest, Uganda, 41 plant food species were recorded (Pebsworth *et al.* 2006). Of these species, 34% are reported in the ethnomedicinal literature as being used in the treatment of parasite infections and related illness. In a comparison of the plant food diets of these two groups (Kibale and Budongo), separated by 200 km, 24 plants used in traditional medicine were recognised, from which eight species possessed known pharmacological properties that could have aided in the treatment for some of the symptoms or illnesses identified in the particular chimpanzees at the time of ingestion (Pebsworth *et al.* 2006). Interestingly, even though the flora overlapped at these two sites, there was evidence for differences in plant selection and use, suggesting unique medicinal cultures in the two groups, with one group using a different part of the same plant with different medicinal properties or both groups using different plants with similar medicinal properties.

With regards to the medicinal properties of the gorilla diet, a few studies have begun to investigate this in some detail. The first of these studies, conducted by Cousins and Huffman (2002), catalogued the ethnomedicinal-pharmacological properties available in the literature for the plant food species reported in the western lowland gorilla, eastern lowland gorilla, and the mountain gorilla. A broader search of potential and known bioactive properties was made, revealing the ingestion of items with such diverse activities as central nervous system stimulants, cardiotonics, hallucinogens, treatments for respiratory ailments, as well as antiparasitic, antifungal, antibacterial and antiviral properties. The tips of the young leaves of *Thomandersia laurifolia* (T. Anders. ex Benth.) Baill. (Acanthaceae) are on rare occasion chewed by western lowland gorillas in the Ndoki forest of northern Congo, and the local hunter-gatherer inhabitants us these leaves as a treatment for parasites and fever (S. Kuroda, pers. comm.). Weak anti-schistosomal activity was noted in crude leaf extracts of the species (Ohigashi 1995).

Curative treatment with medicinal plants

At present, the most convincing evidence of therapeutic self-medication in non-human primates comes from chimpanzees. The hypothesis developed from these investigations is that certain behaviours aid in the control of intestinal parasites and provide relief from related gastrointestinal upset. The two most clearly documented and described behaviours are bitter pith chewing and leaf swallowing. Among the African great apes, one or both of these two proposed therapeutic behaviours have been documented from at least 16 sites and 25 communities at locations spanning their entire geographical distribution (Huffman 2001; McLennan and Huffman 2012), and the number is growing as the work continues (Fig. 3.1).

The first step taken towards demonstrating therapeutic self-medication is identifying the disease potentially being treated. In a longitudinal investigation of the intestinal parasite fauna of chimpanzees conducted on the M group at Mahale (Huffman *et al.* 1997), individuals were monitored over time to detect weekly, monthly, and yearly changes in infection levels. Among all group members monitored in 1991–1992 and 1993–1994, a significant seasonal difference was recognised only for individuals infected by nodular worms (Huffman *et al.* 1997). Among all nematode parasite species detected, nodular worm infections were associated significantly more frequently with bitter pith chewing and leaf-swallowing than other parasites.

This species of nodular worm (*Oesophagostomum stephanastomum*) is the most hazardous species of nodular worm found in the great apes (Brack 1987). Nodular worms produce abdominal pain along with bowel irritation and diarrhoea (Brack 1987), symptoms observed in individuals engaged in leaf-swallowing behaviour in chimpanzees infected with this nematode (Huffman and Seifu 1989; Huffman *et al.*

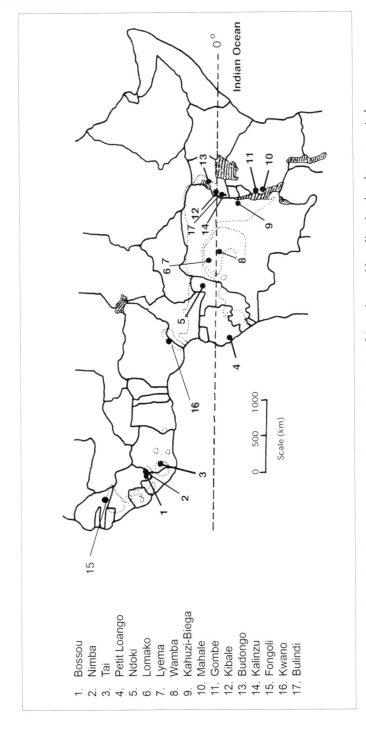

1. Bossou
2. Nimba
3. Tai
4. Petit Loango
5. Ndoki
6. Lomako
7. Lyema
8. Wamba
9. Kahuzi-Biega
10. Mahale
11. Gombe
12. Kibale
13. Budongo
14. Kalinzu
15. Fongoli
16. Kwano
17. Bulindi

Fig. 3.1. Distribution of great ape study sites in Africa where self-medication has been reported

1993; 1996b). Repeated infection occurs in the wild and causes significant complications including secondary bacterial infection, diarrhoea, severe abdominal pain, weight loss, and weakness resulting in high mortality (Brack 1987). Infections caused by other parasites found in M group are not usually serious and may go unnoticed in mild to moderate cases (Brack 1987). The serious effects of nodular worms on the host, however, suggest that it is a sufficiently serious homeostatic challenge to induce self-medicative behaviours.

The second step towards demonstrating the therapeutic action of plant use is to distinguish its use from that of everyday food items and provide evidence for recovery from symptoms identified with the illness associated with its use. This was first demonstrated in animals by the example of bitter pith chewing in chimpanzees.

Bitter pith chewing

Bitter pith chewing has been proposed to aid in the control of intestinal nematode infection, via pharmacological action, and relief from gastrointestinal upset (Huffman 1997). The hypothesis that bitter pith chewing has medicinal value for chimpanzees was first proposed from detailed behavioural observations, and parasitological and phytochemical analyses of patently ill individuals' ingesting *Vernonia amygdalina* (Compositae) at Mahale (Huffman and Seifu 1989; Huffman *et al.* 1993).

When ingesting the pith from young shoots, the outer bark and leaves are carefully removed to chew on the exposed pith, from which they extract the bitter juice and spit out the fibrous remains. The amount of pith ingested in a single bout is relatively small, ranging from portions of 5–120 × 1 cm. The entire process, depending on the amount ingested, takes anywhere from less than 1 to 8 minutes (Huffman 1997). Often, mature conspecifics in proximity to individuals chewing *Vernonia* bitter pith show no interest in ingesting it (Huffman and Seifu 1989; Huffman *et al.* 1997).

Within the home range of M group, *V. amygdalina* is neither abundant nor evenly distributed and usually occurs singly along or near streams. Long-term chimpanzee feeding records from Mahale show that *V. amygdalina* is used in all months except June and October (late dry season), demonstrating its year-round availability. Despite this, bitter pith chewing is highly seasonal and rare (Huffman 1997). In all cases observed by myself and colleagues (Huffman *et al.* 1997) between 1987 and 1993, at the time of use, ill health was evidenced by the presence of diarrhoea, malaise, and nematode infection. Recovery from these symptoms within 20–24 hours has also been reported in detail twice (Huffman and Seifu 1989; Huffman *et al.* 1993). This is quite remarkable given that this recovery time is comparable to that of local human inhabitants, the Tongwe, who use cold concoctions of this plant as a treatment for malaria, intestinal parasites, diarrhoea, and stomach upset. In one case, the intensity of the infection of an ill chimpanzee could be measured and was found to have dropped from 130 eggs per gram (EPG) to 15 within 20 hours. This was quite unusual when compared to seven other individuals with nodular worm infections monitored over the same period. Their nodular worm EPG levels conversely increased over time (Huffman *et al.* 1993; 1997).

Ethnomedicinal reports of this plant's activity and pharmacological analysis of compounds extracted from the specific plants used by the observed sick animals further demonstrated the therapeutic value of this form of self-medication. For numerous African ethnic groups, a concoction made from *V. amygdalina* is prescribed treatment for malarial fever, schistosomiasis, amoebic dysentery, several other intestinal parasites, and stomach aches (Dalziel 1937; Watt and Breyer-Brandwijk 1962; Burkill 1985). Phytochemical analysis of *V. amygdalina* samples collected at Mahale in 1989 and 1991 by our group from plants used by sick chimpanzees revealed the presence of two major classes of bioactive compounds. From this, a total of four known sesquiterpene lactones (vernodalin, vernolide, hydroxyvernolide, vernodalol), seven new stigmastane-type steroid glucosides (vernonioside A1–A4, B1–B3) and two freely occurring aglycones of these glucosides (vernoniol A1, B1) were isolated (Ohigashi *et al.* 1991a; Jisaka *et al.* 1992a; 1992b; 1993 a; 1993b).

The sesquiterpene lactones present in *V. amygdalina*, are well known in many species of *Vernonia* for their anthelmintic, antiamoebic, antitumor, and antibiotic properties (e.g., Toubiana and Gaudemer 1967; Kupchan *et al.* 1969; Asaka *et al.* 1977; Gasquet *et al.* 1985; Jisaka *et al.* 1992a; 1992b; Koshimizu *et al.* 1993).

Our in vitro tests on the antischistosomal activity of the pith's most abundant steroid glucoside (vernonioside B1) and sesquiterpene lactone (vernodalin) showed significant inhibition of egg laying capacity consistent with the observed decline in nodular worm EPG level 20 hours after a sick adult female in M group ingested *V. amygdalina* pith (Jisaka *et al.* 1992b; Huffman *et al.* 1993). In vitro tests on a K1 multi-drug resistant strain of *Plasmodium falciparum* (produces falciparum malaria) showed significant plasmodicidal activity (Ohigashi *et al.* 1994). In total, the evidence from parasitological, pharmacological, and ethnomedicinal observations is substantial and lends support to the hypothesis that bitter pith chewing is a therapeutic form of self-medication stimulated by, and controlling nodule worm infection. Ethnomedicinal use of the plant is also widespread and revealing of the plant's medicinal potential (Table 3.1).

Leaf swallowing
Leaf swallowing is proposed to control parasite infections of whole adult nodular worms (Huffman *et al.* 1996; Huffman and Caton 2001) or tapeworm (*Bertiella studeri*) proglottid fragments (Wrangham 1995; Huffman *et al.* 2009). The mode of parasite control is the physical expulsion of parasites via the self-induced increase in gut motility caused by swallowing indigestible rough, hispid leaves whole (Huffman 1997). In total, the leaves of over 40 different plant species are now confirmed to be swallowed by apes across Africa (Huffman 2011; Fig. 3.1).

The behaviour is quite distinct and discernible from the normal eating of leaves. The distal half of these leaves are selected one at a time, folded by tongue and palate as they are slowly pulled into the mouth and then individually swallowed whole. The leaves' roughness makes them difficult to swallow, so folding them with the tongue and palate before swallowing is a necessary part of ingestion. An individual may swallow anywhere from 1–100 leaves in one sitting and may do so more than once in a day and over several consecutive days (Huffman 1997). Leaves are typically swallowed within the first few hours after leaving the sleeping nest, and or before the first meal of the day, i.e., on an empty stomach (Huffman 1997). The species selected for leaf swallowing are all characterised by rough hispid surfaces, composed of silicate hairs that act to inhibit digestion, inducing the GI tract to flush out the leaves, with parasites, within 6 hours of ingestion (Huffman and Caton 2001).

Table 3.1. Some ethnomedicinal uses of Vernonia amygdalina *in Africa*

Application	Plant part used	Region used: comments
General intestinal upsets:		
enteritis	root, seeds	Nigeria
constipation	leaves, sap	Nigeria, Tanzania, Ethiopia: as a laxative
diarrhoea	stem, root-bark, leaves	W. Africa, Zaire
stomach upset	stem, root-bark, leaves	Angola, Ethiopia
Parasitosis:		
schistosomiasis	root, bark, fruit	Zimbabwe, Mozambique, Nigeria: sometimes mixed with *Vigna sinensis*
malaria substitute	root, stem-bark, leaves	E. Africa, Angola, Guinea, Nigeria, Ethiopia: a quinine substitute
trematode infection	root, leaves	E. Africa: treatment for children used as a suppository
amoebic dysentary	root-bark	S. Africa
ringworm infections	leaves	Nigeria: ringworm & other unidentified epidermal
unspecified	leaves	Nigeria: prophylactic treatment for nursing infants, passed through mother's milk
	root, seeds	Nigeria: worms
	leaves	W. Africa: crushed in water & given to horses as a vermifuge, livestock fodder supplement for treating worms
	leaves	Ghana: purgative
tonic food	leaves	Cameroon, Nigeria: boiled or soaked in cold water prepared as soup or as vegetable fried with meat ; 'n'dole', 'fatefate', 'mayemaye', leaves sold in markets & cultivated in home gardens
Other ailments:		
amenorrhoea	root	Zimbabwe
coughing	leaf	Ghana, Nigeria, Tanzania
diabetes	all bitter parts	Nigeria
fever	leaves	Tanzania, Kenya, Uganda, Congo-Kishasa: leaves squeezed & juice taken
gonorrhoea	roots	Ivory Coast: taken with *Rauwolfa vomitoria*
'heart weakness'	root	W. Africa: vernonine is a cardiotonic glycoside comparable to digitalin
lack of appetite	leaf	W. Africa: leaves soaked in cold water to remove bitter & then boiled in soup
pneumonia	leaf	Ivory Coast: taken with *Argemone maxicana* or used in a bath
rheumatism	stem, root-bark	Nigeria
scurvy	leaves	Sierra Leone, Nigeria, W. Cameroon: leaves sold in markets & cultivated in home gardens
General hygiene:		
dentrifice	twig, stick	Nigeria: chew stick for cleansing & dental caries
disinfectant	not given	Ethiopia
soap	stems	Uganda

Sources: Abebe 1987; Akah and Okafor 1990; Burkill 1985; Dalziel 1937; Irvine 1961; Kokwaro 1976; Muanza *et al.* 1993; Nyazema 1987; Palgrave 1983; Watt and Breyer-Branwijk 1962; Huffman, pers. unpubl. data from interviews in Uganda and Tanzania

Animal origins of human traditional medicine

Evolutionary roots of medicine

In the plant world, a common line of defence against herbivores is to produce a variety of toxic secondary compounds such as sesquiterpenes, alkaloids, and saponins, which inhibit or reduce predation by animals (e.g., Swain 1979; Howe and Westley 1988). At some point in their co-evolutionary history, likely starting with the arthropods, animals began to take advantage of the plant kingdom's protective chemical arsenal to protect themselves from predators and parasites, enhancing their own reproductive fitness (see Blum 1981). For example, adult danaine butterflies of both sexes utilise pyrrolizidine alkaloids for defence against predators and males have also been shown to depend on it as a precursor for the biosynthesis of a pheromone component needed for courtship (Boppré 1978; 1984). The monarch butterfly is reported to feed on *Asclepias* species containing cardiac glucosides that make birds sick, conditioning them not to feed on the species (Brower 1969).

Recent systematic studies in the laboratory based on naturalistic observations have demonstrated that woolly bear caterpillar instar larvae (*Grammia incorrupta*; Lepidoptera: Arctiidae) infected by lethal endoparasites (tacinid flies) parasitoid wasp (sp.) eggs, that ingest a high level of plant toxins (pyrrolizidine alkaloids) are more likely to survive parasitisation by tachinid flies (Singer *et al*. 2009). Such three-tropic level interactions do not require higher-level cognitive abilities, and can be considered a precursor to self-medication in the higher vertebrates.

Cultural roots of medicine and the observation of sick animals

Throughout the history of humankind, people have looked to animals for sources of herbal medicines and narcotic stimulation (e.g., Brander 1931; Riesenberg 1948). Anecdotal reports of the possible use of plants as medicine by wild animals such as the elephant, civet, jackal, and rhinoceros are abundant (e.g., Huffman 2007).

Dr Jaquinto, the trusted physician to Queen Ann, wife of James I in 17th century England, is said to have made systematic observations of domestic sheep foraging in the marshes of Essex, which led to his discovery of a successful cure for consumption (Wilson 1962). Recent controlled studies on sheep have demonstrated their ability to learn to associate the curative properties of items paired with the experimentally induced gastrointestinal discomfort, and then select the appropriate 'cure' for that specific discomfort (Villalba *et al*. 2006). Studies like these have been able to demonstrate what naturalistic studies on other species are unable to do due to ethical constraints, going a long way in revealing the breadth of self-medicative abilities in animals.

Tabernanth iboga (*Apocynacea*) contains several indole alkaloids, and is used as a powerful stimulant and aphrodisiac in many secret religious societies in Gabon (Harrison 1968). Harrison speculated that because of the widespread reports from local people of bush pigs, porcupine, and gorilla going into wild frenzies after digging up and ingesting the roots, they probably learned about these peculiar properties of the plant from watching the animals' behaviour. The most active principle, found in the root, is called ibogaine and is shown to affect the central nervous system and the cardiovascular system. Two other active known compounds in the plant are tabernanthine and iboluteine. The stimulating effects are similar to caffeine (Dubois 1955). The sloth bear and local people of central India are noted to become intoxicated from eating the fermented Madhuca flowers (Brander 1931) and reindeer and the indigenous Lapps consume fly agaric mushrooms known for their intoxicating effects (Phillips 1981).

Tongwe stories on the animal origins of medicine

Mohamedi Seifu Kalunde passed away in October of 2012. He was a member of the WaTongwe ethnic group of western Tanzania and a senior game officer in the Mahale Mountains National Park. He was also a practicing traditional healer, from a family with a long lineage of healers before him. Between 1985 and 2005, I had the great privilege of his friendship, and to have him not only as my field guide following chimpanzees, of which he was an expert, but also as a very generous mentor regarding the medicinal plant tradition of his people. In many societies, traditional healers rely on wisdom passed down from generation to generation. The Tongwe are no exception, but they have another method, that went undocumented until Mohamedi shared that information with me. It was several years into our 20-year collaboration, after we had elucidated the details of bitter pith chewing and leaf swallowing, that he revealed the role of observing self-medicating animals in the acquisition of plant medicines used by his family and the Tongwe in general (Huffman 2007). Mohamedi's elders credit the observation of sick elephants, porcupines, bush pigs, and other animals for the treatment of dysentery, stomach upset, sexually transmitted infections, and many other ailments (Huffman 2007).

In 2003, I was able to witness first-hand the acquisition of a new medicine by Mohamedi as a consequence of our observations of the use of a plant by sick chimpanzees suffering from parasite infection. The plant, *Trema orientalis*, is a common pioneer tree species found in recently cleared forests or fields. It is one of the plants whose rough leaves are swallowed whole by chimpanzees for the expulsion of the same parasite that cause the illness relieved after the use of *V. amygdalina* by chimpanzees at Mahale. For chimpanzees, swallowing of the rough leaves of *T. orientalis* on an empty stomach without chewing them induces temporary diarrhoea that helps to physically purge nematodes from the chimpanzees' intestinal tract. Mohamedi on the other hand, prepared a water extract from the crushed leaves of the tree and tried it to see whether the concoction would stop diarrhoea. He found the extract to be quite effective and both Mohamedi and his mother Joha Kasante used it on their patients. Joha was at first quite skeptical that it would be of any medicinal value. She knew the plant well, but considered it only as a source of firewood and building material.

Minding the gap, medicinal plant use in hominins?

The fundamentals of associating the medicinal properties of a plant by its taste, smell, and texture have their roots deep in our primate history. Therefore, our earliest hominin or *Homo* ancestors can be expected to have exhibited similarities in plant selection criteria with both existing apes and modern humans. While the early fossil record provides no direct evidence for the finer subtleties of feeding behaviour, similarities in plants selected by the African great apes and humans for treating parasitic infection and other illnesses is tantalizing evidence for the possible range of self-medicative behaviours possibly exhibited by early hominins. Further investigation of this could occur through extraction and analysis of microfossil samples from plants with known anti-parasitic activity recovered from early habitation sites (e.g., Hardy *et al.* 2012).

A major turn of events in the evolution of medicine is likely to have come about in early humans with the advent of language to share and pass on detailed experiences about plant properties and their effect against disease. Another major event in human history is considered to have been the attainment of food preparation and detoxification technologies, which allowed humans to exploit a wider range of plant life as food (Wrangham *et al.* 1999). Johns (1990) argues that it was this turning point that may actually have increased our dependence on plant secondary compounds because of their disappearance from the daily diet. In this way, perhaps a greater specialisation of plant use specifically as medicine came about. Furthermore, with the skilled use of fire to boil, steam, vaporise, condense, or otherwise extract useful secondary plant compounds from plants, a greater variety of uses for these compounds were developed.

Acknowledgements

I dedicate this chapter to the memory of the late Mohamedi Seifu Kalunde, a modest man of immense wisdom and humanity.

References

Asaka, Y., Kubota, T. and Kulkarni, A. B. 1977. Studies on a bitter principle from *Vernonia anthelmintica*. *Phytochemistry* 16, 1838–9

Boppre, M. 1978. Chemical communication, plant relationships, and mimicry in the evolution of danaid butterflies. *Entomologia Experimentalis et Applicata* 24, 264–77

Boppre, M. 1984. Redefining "Pharmacophagy". *Journal of Chemical Ecology* 10, 1151–4

Blum, M. S. 1981. *Chemical Defenses of Arthropods*. New York: Academic Press

Brack, M. 1987. *Agents Transmissible from Simians to Man*. Berlin: Springer-Verlag

Brander, A. A. D. 1931. *Wild Animals in Central India*. London: Edward Arnold

Brower, L. P. 1969. Ecological chemistry. *Scientific American* 22, 22–9

Burkill, H. M. 1985. *The Useful Plants of West Tropical Africa* 2nd edn. Vol 1. Kew: Royal Botanical Gardens

Cousins, D. and Huffman, M. A. 2002. Medicinal properties in the diet of gorillas – an ethnopharmacological evaluation. *African Study Monographs* 23, 65–89

Dalziel, J. M. 1937. The useful plants of west tropical Africa. In Hutchinson, J. and Dalziel, J. M. (eds), *Appendix to Flora of West Tropical Africa*. London: Whitefriars Press

Dubois, L. 1955. Tabernanthe Iboga Baillon. *Bulletin Agricole du Congo Belge* 46, 805–29

Engel, C. 2002. *Wild Health*. Boston: Houghton Mifflin

Etkin, N. L. and Ross, P. J. 1982. Food as medicine and medicine as food: an adaptive framework for the interpretation of plant utilization among the Hausa of Northern Nigeria. *Social Science & Medicine* 16, 1559–73

Foitova, I., Huffman, M. A., Wisnu, N. and Olšanský, M. 2009. Parasites and their effect on orangutan health. In Wish, S. A., Utami, S. S., Setia, T. M. and van Schaik, C. P. (eds), *Orangutans – Ecology, Evolution, Behaviour and Conservation*, 157–69. Oxford: Oxford University Press

Forbey, J., Harvey, A., Huffman, M. A., Provenza, F., Sullivan, R. and Tasdemir, D. 2009. Exploitation of secondary metabolites by animals: A behavioural response to homeostatic challenges. *Integrative and Comparative Biology* 49, 314–28

Freeland, W. J. 1980. Mangabey (*Cercocebus albigena*) movement patterns in relation to food availability and fecal contamination. *Ecology* 61, 1297–1303

Gasquet, M., Bamba, D., Babadjamian, A., Balansard, G., Timon-David, P. and Metzger, J. 1985. Action amoebicide et anthelminthique du vernolide et de l'hydroxyvernolide isoles des feuilles de *Vernonia colorata* (Willd.) Drake. *European Journal of Medical Chemical Theory* 2, 111–15

Gillespie, T. R., Chapman, C. A. and Greiner, E. C. 2005. Effects of logging on gastrointestinal parasite infections and infection risk in African primates. *Journal of Applied Ecology* 42, 699–707

Hardy, K. S., Buckley, M. J., Collins, A., Estalrrich, D., Brothwell, L., Copeland, A., García-Tabernero, S., García-Vargas, M., de la Rasilla, C., Lalueza-Fox, R., Huguet, M., Bastir, D., Santamaría, M., Madella, A. and Fernández Cortés, A. Rosas. 2012. Neanderthal medics? Evidence for food, cooking and medicinal plants entrapped in dental calculus. *Naturwissenschaften* 99, 617–26

Harrison, G. P. 1968. *Tabernanthe iboga*: an African narcotic plant of social importance. *Economic Botany* 23, 174–84

Hart, B. L. 1988. Biological basis of the behavior of sick animals. *Neuroscience & Biobehavioral Reviews* 12, 123–37

Hart, B. L. 1990. Behavioral adaptations to pathogens and parasites: five strategies. *Neuroscience & Biobehavioral Reviews* 14, 273–9

Hausfater, G. and Meade, B. J. 1982. Alternation of sleeping groves by yellow baboons (*Papio cynocephalus*) as a strategy for parasite avoidance. *Primates* 23, 287–97

Howe, H. F. and Westley, L. C. 1988. *Ecological Relationships of Plants and Animals*. Oxford: Oxford University Press

Huffman, M. A. 1997. Current evidence for self-medication in primates: a multidisciplinary perspective. *Yearbook of Physical Anthropology* 40, 171–200

Huffman, M. A. 2001. Self-medicative behavior in the African Great Apes: An evolutionary perspective into the origins of human traditional medicine. *BioScience* 51, 651–61

Huffman, M. A. 2007. Animals as sources of medicine in traditional societies. In Bekof, M. (ed.), *Encyclopedia of Human-Animal Relationships: a global exploration of our connections with animals*. Vol. 2, 434–41. Westport CT: Greenwood Press

Huffman, M. A. and Caton, J. M. 2001. Self-induced increase of gut motility and the control of parasitic infections in wild chimpanzees. *International Journal of Primatology* 22, 329–46

Huffman, M. A. and Hirata, S. 2003. Biological and Ecological foundations of primate behavioural traditions. In Fragaszy, D. M. and Perry, S. (eds), *The Biology of Traditions*, 267–96. Cambridge: Cambridge University Press

Huffman, M. A. and Hirata, S. 2004. An experimental study of leaf swallowing in captive chimpanzees – insights into the origin of a self-medicative behaviour and the role of social learning. *Primates* 45, 113–18

Huffman, M. A. and MacIntosh A. J. J. 2012. Plant-food diet of the Arashiyama Japanese macaques and its potential medicinal value. In Leca, J.-B., Huffman, M. A., and Vasey, P. L. (eds), *The Monkeys of Stormy Mountain: 60 years of primatological research on the Japanese macaques of Arashiyama*, 356–431. Cambridge: Cambridge University Press

Huffman, M. A. and Seifu, M. 1989. Observations on the illness and consumption of a possibly medicinal plant *Vernonia amygdalina* by a wild chimpanzee in the Mahale Mountains, Tanzania. *Primates* 30, 51–63

Huffman, M. A., Spiezio, C., Sgaravatti, A. and Leca, J-B. 2010. Option biased learning involved in the acquisition and transmission of leaf swallowing behaviour in chimpanzees (*Pan troglodytes*)? *Animal Cognition* 13, 871–80

Huffman, M. A., Gotoh, S., Izutsu, D., Koshimizu, K. and Kalunde, M. S. 1993. Further observations on the use of the medicinal plant, *Vernonia amygdalina* (Del) by a wild chimpanzee, its possible effect on parasite load, and its phytochemistry. *African Study Monographs* 14, 227–40

Huffman, M. A., Gotoh, S. Turner, L. A., Hamai, M. and Yoshida, K. 1997. Seasonal trends in intestinal nematode infection and medicinal plant use among chimpanzees in the Mahale Mountains National Park, Tanzania. *Primates* 38(2), 111–25

Huffman, M. A., Ohigashi H., Kawanaka, M., Page, J. E., Kirby G. C., Gasquet, M., Murakami, A. and Koshimizu, K. 1998. African great ape self-medication: A new paradigm for treating parasite disease with natural medicines? In Ebizuka Y. (ed.), *Towards Natural Medicine Research in the 21st Century*, 113–23. Amsterdam: Elsevier Science

Huffman, M. A., Page, J. E., Sukhdeo, M. V. K., Gotoh, S., Kalunde, M. S., Chandrasiri, T. and Towers, G. H. N. 1996a. Leaf-swallowing by chimpanzees, a behavioral adaptation for the control of strongyle nematode infections. *International Journal of Primatolology* 17(4), 475–503

Huffman, M. A., Koshimizu, K. and Ohigashi, H. 1996b. Ethnobotany and zoopharmacognosy of *Vernonia amygdalina*, a medicinal plant used by humans and chimpanzees. In Caligari, P. D. S. and Hind, D. J. N. (eds), *Compositae: Biology & Utilization* Vol. 2, 351–60. Kew: Royal Botanical Gardens

Huffman, M. A., Pebsworth, P., Bakuneeta, C., Gotoh, S. and Bardi, M. 2009. Macro-habitat comparison of host-parasite ecology in two populations of chimpanzees in the Budongo forest, Uganda and the Mahale Mountains, Tanzania. In Huffman, M. A. and Chapman, C. (eds), *Primate Parasite Ecology: the dynamics of host-parasite relationships*, 311–30. Cambridge: Cambridge University Press

Janzen, D. H. 1978. Complications in interpreting the chemical defenses of trees against tropical arboreal plant-eating vertebrates. In Montgomery, G. G. (ed.), *The Ecology of Arboreal Folivores*, 73–84. Washington DC: Smithsonian Institute Press

Jisaka, M., Kawanaka, M., Sugiyama, H., Takegawa, K., Huffman, M. A., Ohigashi, H. and Koshimizu, K. 1992a. Antischistosomal activities of sesquiterpene lactones and steroid glucosides from *Vernonia amygdalina*, possibly used by wild chimpanzees against parasite-related diseases. *Bioscience, Biotechnology & Biochemistry* 56, 845–6

Jisaka, M., Ohigashi, H., Takagaki, T., Nozaki, H., Tada, T., Hirota, M., Irie, R., Huffman, M. A., Nishida, T., Kaji, M. and Koshimizu, K. 1992b. Bitter steroid glucosides, vernoniosides A1, A2, and A3 and related B1 from a possible medicinal plant *Vernonia amygdalina*, used by wild chimpanzees. *Tetrahedron* 48, 625–32

Jisaka, M., Ohigashi, H., Takegawa, K., Hirota, M., Irie, R., Huffman, M. A. and Koshimizu, K. 1993a. Steroid glucosides from *Vernonia amygdalina*, a possible chimpanzee medicinal plant. *Phytochemistry* 34, 409–13

Jisaka, M., Ohigashi, H., Takegawa, K., Huffman, M. A. and Koshimizu, K. 1993b. Antitumoral and antimicrobial activities of bitter sesquiterpene lactones of *Vernonia amygdalina*, a possible medicinal plant used by wild chimpanzee. *Bioscience, Biotechnology, Biochemistry* 57(5), 833–4

Johns, T. 1990. *With Bitter Herbs Shall They Eat It: chemical ecology and the origin of human diet and medicine.* Tucson AZ: University of Arizona Press

Koshimizu, K., Ohigashi, H., Huffman, M. A., Nishida, T. and Takasaki, H. 1993. Physiological activities and the active constituents of potentially medicinal plants used by wild chimpanzees of the Mahale Mountains, Tanzania. *International Journal of Primatology* 14, 345–56

Krief, S., Hladik, C. M. and Haxaire, C. 2005. Ethnomedicinal and bioactive properties of plants ingested by wild chimpanzees in Uganda. *Journal of Etnopharmacology* 101, 1–15

Krief, S., Huffman, M. A., Sévenet, T., Guillot, J., Hladik, C.-M., Grellier, P., Loiseau, M. and Wrangham, R. W. 2006. Bioactive properties of plant species ingested by chimpanzees (*Pan troglodytes schweinfurthii*) in the Kibale National Park, Uganda. *American Journal of Primatology* 68, 51–71

Kupchan, S. M., Hemingway, R. J., Karim, A. and Werner, D. 1969. Tumor Inhibitors XLVII. vernodalin and vernomygdin, two new cytotoxic sesquiterpene lactones from *Vernonia amygdalina* Del. *Journal of Organic Chemistry* 34, 3908–11

Lozano, G. A. 1991. Optimal foraging theory: A possible role for parasites. *Oikos* 60, 391–5

MacIntosh, A. J. J. and Huffman, M. A. 2010. Towards understanding the role of diet in host-parasite interactions in the case of Japanese macaques, In Nakagawa, F., Nakamichi, M. and Sugiura, H. (eds), *The Japanese Macaques*, 323–44. Tokyo: Springer

McLennan, M. R. and Huffman, M. A. 2012. High frequency of leaf swallowing and its relationship to parasite expulsion in "village" chimpanzees at Bulindi, Uganda. *American Journal of Primatology* 74, 642–50

Nelson, G. S. 1960. Schistosome infections as zoonoses in Africa. *Transcripts of the Royal Society of Tropical Medicine & Hygiene* 54, 301–14

Nishida, T. and Uehara, S. 1983. Natural diet of chimpanzees (*Pan troglodytes schweinfurthii*): long term record from the Mahale Mountains, Tanzania. *African Studies Monographs* 3, 109–30

Ohigashi, H. 1995. *Plants Used Medicinally by Primates in the Wild and their Physiologically Active Constituents.* Report to the Ministry of Science, Education and Culture for 1994 Grant-in-Aid for Scientific Research (No. 06303012)

Ohigashi, H., Huffman, M. A., Izutsu, D., Koshimizu, K., Kawanaka, M., Sugiyama, H., Kirby, G. C., Warhurst, D. C., Allen, D., Wright, C. W., Phillipson, J. D., Timmon-David, P., Delmas, F., Elias, R. and Balansard, G. 1994. Toward the chemical ecology of medicinal plant use in chimpanzees: the case of *Vernonia amygdalina*, a plant used by wild chimpanzees possibly for parasite-related diseases. *Journal of Chemical Ecology* 20, 541–53

Ohigashi, H., Jisaka, M., Takagaki, T., Nozaki, H., Tada, T., Huffman, M. A., Nishida, T., Kaji, M. and Koshimizu, K. 1991a. Bitter principle and a related steroid glucoside from *Vernonia amygdalina*, a possible medicinal plant for wild chimpanzees. *Agricultural & Biological Chemistry* 55, 1201–3

Ohigashi, H., Takagaki, T., Koshimizu, K. Nishida, T., Huffman, M. A., Takasaki, H., Jato, J. and Muanza, D. N. 1991b. Biological activities of plant extracts from tropical Africa. *African Studies Monographs* 12, 201–10

Olupot, W., Chapman, C. A, Waser, P. M. and Isabirye-Basuta, G. 1997. Mangabey (*Cercocebus albigena*) ranging patterns in relation to fruit availability and the risk of parasite infection in Kibale National Park, Uganda. *American Journal of Primatology* 43, 65–78

Pebsworth, P., Krief, S. and Huffman, M. A. 2006. The role of diet in self-medication among chimpanzees in the Sonso and Kanyawara communitites, Uganda. In Newton-Fisher, N. E., Notman, H., Reynolds, V. and Paterson, J. (eds), *Primates of Western Uganda*, 105–33. New York: Springer

Phillips, R. 1981. *Mushrooms and Other Fungi of Great Britain and Europe*. London: Pan

Riesenberg, S. H. 1948. Magic and medicine in Ponape. *Southwest Journal of Anthropology* 4, 406–29

Singer, M. J., Mace, K. C. and Bernays, E. A. 2009. Self-medication as adaptive plasticity: increased ingestion of plant toxins by parasitized caterpillars. *PLOS ONE* 4, e4796. doi:10.1371/journal. pone.0004796

Swain, T. 1979. Tanins and lignins. In Rosenthal, G. A. and Janzen, D. H. (eds), *Herbivores: their interactions with secondary plant metabolites*, 657–82. New York: Academic Press

Toubiana, R. and Gaudemer, A. 1967. Structure du vernolide, nouvel ester sesquiterpique isole de *Vernonia colorata*. *Tetrahedron Letters* 14, 1333–6

Villalba, J. J., Provenza, F. and Shaw, R. 2006. Sheep self-medicate when challenged with illness-inducing foods. *Animal Behaviour* 71, 1131–9

Watt, J. M. and Breyer-Brandwijk, M. G. 1962. *The Medicinal and Poisonous Plants of Southern and East Africa*. Edinburgh: Livingstone

Wilson, E. 1962. Forward. In Dick, O. L. (ed.), *Aubrey's Brief Lives*, xix. Ann Arbor MI: Ann Arbor Paperbacks

Wrangham, R. W., Jones, J. H., Laden, G., Pilbeam, D. and Conklin-Brittain, N. 1999. The raw and the stolen: cooking and the ecology of human origins. *Current Anthropology* 40, 567–94

Wrangham, R. W. 1995. Leaf-swallowing by chimpanzees, and its relation to tapeworm infection. *American Journal of Primatology* 37, 297–303

4. Plants as raw materials

Karen Hardy

The use of plants as raw materials is likely to have existed throughout the whole of the human past. Usable plant components consist of leaves, bark, wood, fibres and vines, sticks, saps and resins, nut and large seed shells and a wide range of plant secondary compounds. These were all used in an apparently unlimited way as raw materials to manufacture all manner of items. Examples of the way multiple lines of evidence can be used to reconstruct past use of plant materials, even when primary evidence is lacking, include the use of plant materials in oral hygiene activities, the manufacture of pitch, twisted fibre technology and woodworking.

Introduction

In the pre-agrarian world, people are likely to have had a deep understanding of their environments and exploited all the available resources within them. Other than stone and some minerals, this largely comprised material from terrestrial and marine plants and animals. Animal products, such as bone and antler and shell can survive deep into the archaeological record; however, plant materials, which may well have a heritage far greater even than stone, survive in only rare cases and under certain conditions. Usable plant components consist of leaves which can be pounded to provide glue or pigment or, as demonstrated by some chimpanzees, as a sponge (McGrew 2010a), bark, wood, fibres and vines, sticks, saps and resins, nut and large seed shells and plant secondary compounds; these can be used in an apparently limitless way as raw materials to manufacture all manner of items. The way raw materials were used, which ranged from whole, for example ostrich shells (Texier *et al.* 2010), to the twisting and adding that created string and cord, the extraction of pitch and resin from trees, the delicate and complex technique of creating bitumen from these and the development of composite technology in which different materials are combined to create new objects, demonstrates the extraordinary creativity and skill displayed by hominins and prehistoric humans.

The central role of plants in the pre-agrarian world is highlighted by Adovasio *et al.* (2007) who point out that in archaeological contexts with exceptional survival of plant materials, fibre artefacts outnumber stone tools by a factor of 20 to 1, while in anaerobic conditions 95% of all artefacts are either made of wood or fibre. Cordage has been described as 'the unseen weapon that allowed the human race to conquer the earth' (Barber, 1994) and there is little doubt that the technological development of joining fibres together through twisting and adding, was one of the most significant breakthroughs in the development of humankind (Hardy 2008).

But where did it all start? Unlike stone, where some of the first traces of use and modification can still be seen, the earliest uses of plant materials as tools or as raw materials, is lost in time. Jane Goodall was the first to record the use of items as tools by chimpanzees (Goodall 1964). Subsequently, there have been many studies that demonstrate a wide range of tools and tool using behaviour among chimpanzee and other higher primate groups; many of these are summarised in McGrew (1992; 2004) and Haslam (2014). These include the use of unmodified sticks, twigs and leaves as tools (McGrew 2010a) for tasks such as perforating termite mounds (Sanz and Morgan 2007) and leaf sponges to obtain drinking water (McGrew 2010b). Chimpanzees have also demonstrated complexity in tool use, incorporating careful selection of certain types of twigs depending on characteristics such as the aggressiveness of the ants targeted (Humle and Matsuzawa 2002), use of tool sets and tool composites involving two or more tools used in an ordered sequence (McGrew 2010b) or simultaneously (Koops *et al.* 2015), while bonobos have been recorded modifying tree branches to create tools (Roffman *et al.* 2015). An idea of the potential time depth of plant use in the human lineage is provided by McGrew (2010a), who suggests that it should be assumed that anything a chimpanzee can do could also have been done by the Last Common Ancestor (LCA) 6–7 million years ago. Haslam (2014) points out that some modern observed tool using activities may have developed more recently; however, we take McGrew's point and assume that though some observed behaviours may be recent, the abilities these activities demonstrate will have been within the capabilities of the LCA. It is therefore very likely that the use of unmodified materials as tools has always existed in the hominin/human past.

In this chapter some of the archaeological evidence for early use of plants as raw materials is examined. The survival of actual plant remains linked to raw materials is very rare in the pre-agrarian world and becomes increasingly so further back in time. In some cases use of plants is demonstrated through secondary evidence. Examples include early evidence for marine transport and shells with wear traces that demonstrate they have been strung or threaded. Ethnographic evidence can also provide useful insights, particularly if activities are repeated over a broad range of unconnected areas. One such example, comprising multiple lines of indirect evidence, is the probable use of plant materials in oral hygiene activities.

Oral hygiene

The use of plant materials in oral hygiene activities is likely to be as old as the probable use of probing sticks. Dental hygiene activities, including the use of sticks as tooth picks, have been widely observed among non-human primates including chimpanzees (McGrew and Tutin 1973), bonobos (Ingmanson 1996) and orang utans (Russon *et al.*

2009) while long-tailed macaques and Japanese macaques have been recorded using hair to floss between their teeth (Watanabe *et al.* 2007; Leca *et al.* 2010). Indirect evidence for toothpicking, in the form of interproximal grooves which are thought to be the result of oral hygiene activities (Brothwell 1963), and possibly tooth picking (Uberlaker *et al.* 1969; Ungar *et al.* 2001), is widespread and can be found across the geographical and temporal spectra in prehistoric populations (Hinton 1981; Lukacs and Pastor 1988; Ryan and Johansen 1989; Bonfiglioli *et al.* 2004; Eshed *et al.* 2006; Molnar 2008; Lozano *et al.* 2009; Volpato *et al.* 2012). Interproximal grooves have been found on samples from *Homo habilis* and all subsequent hominin species (Puech and Gianfarani 1988) and are abundant among Neanderthal populations (Frayer *et al.* 1987; Formicola 1988; Villa and Giacobini,1995; Urbanowski *et al.* 2010; Estalrrich *et al.* 2011; Dąbrowski *et al.* 2013; Lozano *et al.* 2013). Materials suggested as tooth picks include wood, bone, sinew and grass (Eckhardt and Piermarini 1988; Brown and Molnar 1990; Hlusko 2003). Pieces of tar, most frequently birch bark but also lumps of pine resin, have been found with human tooth impressions in them on some Scandinavian Mesolithic sites. It has been suggested that these were chewed, possibly as a disinfectant and remedy for toothache (Regnell *et al.* 1995; Aveling and Heron 1999). Ethnographic records document that chewing resin was a common practice in northern Scandinavia in historical times (Eidlitz 1969) and the use of numerous plants in oral hygiene activities has been widely recorded ethnographically (Almas 2002), in many cases the plants have anti-microbial properties (Elujoba *et al.* 2005; Idu *et al.* 2009; Jose *et al.* 2011). For example, pitch of the Sitka spruce (*Picea sitchensis*), which is native to the northern part of the American Pacific coastline, was chewed as a remedy for toothache among Native American populations of the Pacific North West Coast (Pojar and MacKinnon 1994). In the Almor District, North India, 17 plant species from 15 different families, all with medicinal qualities, have been recorded in oral hygiene activities (Sharma and Joshi 2010) while chewing sticks are very common in many places in Africa. Though no physical remains survive to prove the use of plant materials in oral hygiene activities, a combination of their recorded use among higher primates, a pattern of behaviour that produced characteristic tooth use wear patterns, thought to be linked to the use of toothpicks identified on multiple species over 2 million years, widespread ethnographic information demonstrating use of plants and the complexity of knowledge of plant properties used in oral hygiene activities in societies with traditional material culture today, all combine to suggest plants were very likely to have been widely used in this way among pre-agrarian people.

Fire and wood

The timing of the development of controlled use of fire is hotly debated. Wrangham and colleagues have argued that cooking began around 1.8 million years ago (Wrangham *et al.* 1999; Wrangham and Conklin-Brittain 2003; Wrangham 2007; 2009; Wrangham and Carmody 2010) while Roebrocks and Vila (2011) argue that there is no secure evidence for habitual use of fire in Europe until 300–400,000 years ago. Fire can be used for heat, light, protection, cooking, food smoking and land management (Hardy *et al.* 2016; see also Introduction) and is considered crucial in human evolution not only in terms of the

Fig. 4.1. Roasting shellfish, Diakhanor, Saloum Delta, Senegal

improvements in food efficiency, but also social development (Wiessner 2014). Access to fuel has always been an essential requirement since the controlled use of fire began and among hunter-gatherers its local depletion may have been one reason to move location (Winterhalder 2001; Bishop and Plew 2015).

The earliest tentative evidence for human use of fire at around 1.6 million years ago comes from a number of sites including Swartkrans (Brain and Sillent 1988), Chesowanja, Kenya (Gowlett *et al.* 1981) and Koobi Fora (Bellomo 1994; Rowlett 2000). At Wonderwerk Cave, South Africa, micromorphological evidence revealed burnt bone and ashed plant remains representing possibly grasses, leaves and brush material in an Acheulean context dated to approximately one million years ago (Berna *et al.* 2012). The site of Gesher Benot Ya'aqov, Israel, dates to around 790,000 BP and here, charcoal, plant remains and burned micro-artefacts were found in concentrations that have been proposed as evidence for hearths (Alperson-Afil 2008). However, one of the earliest confirmed use of fire is at Qesem Cave, Israel and dates to 400,000 years ago (Karkanas *et al.* 2007) while a multi-used heath, at the same site dates from 300,000 BP (Shahack-Gross *et al.* 2014). Bolomor Cave has some of the earliest hearths in southern Europe dating to 225–240,000 years ago (Fernández Peris *et al.* 2012) while the evidence from the Middle Palaeolithic onwards is widespread (Finlayson *et al.* 2001; Madella *et al.* 2002; Badal *et al.* 2011; Burjachs *et al.* 2011; Roebroeks and Vila 2011; Albert *et al.* 2012; Aldeias *et al.* 2012). One of the problems in understanding the earliest evidence for fire is that it may not be linked to a domestic location and may be hard to trace archaeologically (Scherjon *et al.* 2015). In the Saloum Delta, Senegal, for example, fire circles where shellfish are roasted using dry brushwood are located just above the shoreline, well away from the village. It is not clear whether they would be detectable in an archaeological future (Hardy *et al.* 2016) (Fig. 4.1). Likewise, Crittenden (chapter 16, 327) highlights the use of ephemeral fires for rapid cooking of tubers among the Hadza, for which long term evidence is unlikely to survive.

While fuel for fire can come from a range of materials including animal dung and bones, wood was a major source of fuel when it was available. Different woods have different properties when burning, including providing for example a brighter light or greater warmth (Austin 2009) or smoke for hide smoking (Henry and Théry-Parisot 2014). The identification of wood is conducted using a number of different methods, including identification of carbonised wood fragments (e.g. Austin 2009; Théry-Parisot *et al.* 2010) and phytoliths (e.g. Albert *et al.* 2000) and this provides a major source of cultural and paleoenvironmental information.

Very few examples of woodworking survive from the early Palaeolithic periods; however, the few fragments there are, demonstrate this ability deep in the Lower Palaeolithic. Dominguez-Rodrigo *et al.* (2001) claim evidence for woodworking based on the detection of adhering phytoliths on Acheulian tools dated to 1.7–1.5 million years ago while small fragments of pine (*Pinus* sp.) with traces of modification and burning were found at the Acheulean site of Torralba (Carbonell and Castro-Curel 1992). Wooden items include a piece of wood with possible modification marks from Gesher Benot Ya'aqov in Israel dated to 790,000 years ago (Goren Inbar *et al.* 2002) and a spear tip made on yew (*Taxus baccata*) dated to 450,000 years ago from Clacton-on-Sea, England (Oakley *et al.* 1977). Three spears made on common silver fir (*Abies alba*) wood, were found at Schöningen, Germany (Thieme 1997); these date to between 337–300,000 years ago (Richter and Krbetschek 2015). A spear point from Lehringen, dating to 120–125,000 years ago has been attributed to Neanderthals (Langley *et al.* 2008). Several wooden artefacts, some of which were shaped in the form of a scoop, were found at Abri Romanic, Catalunya (Spain) which dates to between 45,000 and 49,000 years ago (Carbonell and Castro-Curel 1992). Wooden digging sticks have been widely recorded ethnographically for harvesting tubers (Nelson 1899; Rudenko 1961; Vincent 1985; Hardy and Sillitoe 2003; Gott 1982) and they are likely to be very ancient tools though there is little in the way of direct evidence for them in earlier Palaeolithic periods.

More broadly, while phytoliths demonstrate use of wood as fuel at the Middle Palaeolithic site of Amud Cave (70,000–55,000 BP), evidence for grasses here was interpreted as food, fuel and bedding (Madella *et al.* 2002). A range of plants including sedges, monocotyledons, and aromatic leaves with insecticidal properties, have been identified as bedding at Sidubu Cave, South Africa, dating to 77,000 years ago (Wadley *et al.* 2011) and evidence for a hut and bedding somewhat later, at around 19,000 years ago, at Ohalo II, Israel (Nadel and Werker 1999; Nadel *et al.* 2004). A range of artefacts and materials, including digging sticks, a poison applicator, traces of resins thought to be part of an adhesive mixture, and ricinoleic acid suggestive of the poisonous castor seeds, were found at Border Cave, South Africa, and dated to 40–25,000 years ago, together with a range of other shell and bone artefacts, suggesting an elaborate material culture based on a wide range of raw materials from the start of the Upper Palaeolithic (D'Errico *et al.* 2012). Some Mesolithic and Neolithic sites have outstanding survival of a wide range of wooden artefacts; examples include the decorated skis and bows from Vis I, Russia (Burov 1990).

Composite technology

The development of composite technology, which involves connecting two items together to make a third item, represents an extraordinary conceptual and technological achievement. In the Palaeolithic and Mesolithic, there were only three ways to join two items together, through the use of resin based adhesives, such as tar or pitch or bitumen, through binding with fibres, or through woodworking techniques that permit slotting of one piece of wood into another. Chapter 13 discusses one example of woodworking skills, the evidence for wooden fish traps and fences, from Russia and the Baltic region from the Mesolithic and into later prehistory.

Tar and bitumen

The manufacture of tar and bitumen represents one of the earliest technological processes for which there is evidence. Resins (or resinous components) and pitch extracted from some plants, including conifer and birch bark wood, have adhesive qualities. In some cases, these can be extracted to create tar or bitumen; however, its manufacture is a relatively complex smouldering process, which requires a specific temperature range and exclusion of oxygen. From a chemical point of view this oxygen-free smouldering process (or burning with no flame) represents destructive dry distillation (known as pyrolysis). The liquid or semi-liquid tar is the initial product of this technique; if further heated this leads to a thickening process which would result in obtaining more viscous residuum, which is the wood or bark pitch (Beck *et al.* 1997; Koller *et al.* 2001).

The earliest evidence for the use of resin-based pitch is from a Mid-Pleistocene site in Italy (Mazza *et al.* 2006) where chemical evidence for resins of Betulaceae origin was found on a series of flint flakes. Numerous samples of ochre mixed with plant gum were found on stone tools, suggesting use as hafting material, from Sidubu, South Africa, dated to approximately 70,000 years ago (Wadley *et al.* 2009). Chemical evidence has identified resins (Hardy *et al.* 2012) and a conifer fragment (Radini *et al.* 2016) from a sample of dental calculus from the 49,000 year old Neanderthal site of El Sidrón in association with unusual dental wear, suggest chewing, possibly to soften this material prior to use (Estalrrich and Rosas 2013), though chewing resins as a disinfectant and remedy for toothache has also been suggested (Aveling and Heron 1999; Regnell *et al.* 1995). Two pieces of birch bark pitch, one of which contained a fingerprint impression and the imprint of a stone tool and the structure of wood cells was also identified at Konigsaue, Germany, dating to a minimum age of 48,500–43,000 years ago (Koller *et al.* 2001; Grünberg 2002). Bitumen, an analogous and alternative product to tree-based pitch, which was found to have been highly heated, has been found adhering to stone tools at Umm el Tlel, dated to around 40,000 years ago (Boeda *et al.* 1996). Functional analysis studies detected use wear traces which are indicative of hafting in the early Middle Palaeolithic (Rots 2013). Chemical evidence for bitumen has also been identified at the Middle-Upper Palaeolithic site of Gura Cheii-Râşnov Cave, Romania (Cârciumaru *et al.* 2012). At the Early Mesolithic site at Tłokowo in north-eastern Poland, a bone point with a set of 16 flint insets was found. All of the flints were fixed with birch bark tar into grooves carved into the bone point (Sułgostowska 1997). Hundreds of hearth-pits,

Fig. 4.2. Applying a waterproofing agent to the base of a wooden canoe, Saloum Delta, Senegal. Traditionally, the inner bark of Detarium senegalense is dried and milled then mixed with baobab leaves (Adansonia digitata), and ground in water. This mixture is then spread using fingers into the cracks and slots left between the planks of the canoes. Today, a mixture using polystyrene mixed with diesel is also used. This is heated and when it liquefies, baobab tree sawdust is added

possibly used for extraction of pine wood tar, were found at various Mesolithic sites in the Netherlands, including Hattemerbroek, Dronten-N23, Tunnel-Drontermeer and Scheemda-Scheemderzwaag (Kubiak-Martens et al. 2011; 2012). A piece of birch bark tar with teeth imprints from the Middle Neolithic site at Schipluiden on the Dutch North Sea coast, shows that fats or plant oil and some beeswax were added to it (Van Gijn and Boon 2006), suggesting it may have been a binding agent.

Related to this, but still poorly understood, is the extraction of plant oil. Dogwood (Cornus sanguinea) was available to pre-agrarian people in Europe and the seeds and fruit stones contain up to 50% of non-volatile oils, which can be used as fuel (for example in lamps), or to impregnate wood or leather. Dogwood fruit stones are frequently encountered in large numbers, and often in a fragmented and partly charred state at south Scandinavian Mesolithic sites, suggesting that oil had been extracted from them (Regnell et al. 1995; Regnell 2012). Many crushed fruit-stones were recently found at the

Mesolithic river dune site at Yangtze Harbour in the Netherlands; some fruit-stones were charred as well as broken, which suggests that the dogwood fruit-stones and possibly also the seeds, may have been used for oil extraction (Kubiak-Martens *et al.* 2015).

Though tar and pitch were clearly widely used, the manufacturing methods of tar extraction are unknown; experimental work suggests it can be achieved using a birch bark container set below a specifically constructed fire. In early prehistory, tar and pitch are likely to have been used as adhesives for joining or fixing lithic implements and stone tools to their wooden, bone or antler hafts. Being insoluble in water, tar may also have been used for waterproofing objects such as canoes, leather clothes and shoes, and for improving the durability of nets for hunting (Fig. 4.2).

Fibres and twisting

Fibres, and their construction into strings and cordage, are probably one the most essential of all artefactual materials. Still today, fibres are used in everything from making clothes to cable-stayed bridges; string and fibres are so ubiquitous and familiar that what they actually meant, in terms of material culture and the likely constant demand on the lives of the manufacturers, has been all but forgotten (Hardy and Sillitoe 2003; Hardy 2008). Plant fibres are likely to be amongst the earliest material items used, though when or how the use of simple fibres developed into twisting and combining

Fig. 4.3. String making, Saloum Delta, Senegal

fibres into string, cord and rope is lost in time. Plant fibres survive very rarely, and fragments of twisted fibre only appear in the archaeological record relatively recently; however, fibre technology is considered to be a cultural universal, that is, something that is a feature of every known human group (Brown, 1991) (Fig. 4.3).

One way to determine the early use of fibre is through the presence of perforated objects as these need to be hung or tied in some way. Beads require fine strings that are strong enough to hold them; this is likely to mean that fibres have been twisted and extended using multiple threads. There are a number of Lower and Middle Palaeolithic sites that have items that may have been strung or tied (Hardy 2008), including Acheulean sites with naturally perforated *Porosphaera globularis*. These small fossil sponges have been found on some Acheulean sites in England and northern France (Rigaud *et al.* 2009). Though the human use of these beads remains uncertain, Bednarik (2005) suggests evidence for artificial enlarging and of use wear around the holes. Rigaud *et al.* (2009) consider the evidence is somewhat ambiguous; however, they do not discount it, and suggest that more data is needed. A study of use wear traces detected on eagle talons from the 120,000 year old Neanderthal site of Krapina suggest these represent evidence for stringing into a necklace (Radovčić *et al.* 2015). Vanhaeren *et al.* (2006) record perforated shells which they believe are beads dating to between 100,000 and 125,000 BP. Perforated beads with clear evidence for having been strung, date to around 75,000 years ago from Blombos Cave (Henshilwood *et al.* 2004), and Border Cave (D'Errico *et al.* 2012), South Africa. By 40,000 years ago, shell beads occur in large numbers on some sites, for example Uçagızlı Cave in Turkey (Kuhn *et al.* 2001) and Ksar'Akil in Lebanon, while ostrich shell beads are found in many MSA sites in east Africa from 40,000 years ago (Ambrose 1998). The Upper Palaeolithic site of Kostenki (approximately 36,000 BP) has a range of perforated items including shell beads and animal teeth (Sinitsyn 2003). From this time onwards perforated beads are found on many sites worldwide.

Another type of secondary evidence comprises eyed needles; the earliest of these known so far date to the mid-40th milennium BP, from Russian sites including Kostënki and Denisova Cave (Gilligan 2010) while needles and other artefacts thought to relate to weaving are widely recorded from the early 30th millennium BP (Soffer *et al.* 2001; Soffer 2004). The human burials at Sungir, Russia (22,000 BP), contained many thousands of beads, thought to have been sewn into close-fitting clothing (Adovasio *et al.* 2007). Several Upper Palaeolithic sites including Dolní Věstonice, Pavlov, Gonnersdorf, Kostenki I and Zaraisk, have impressions of complex items of woven material, including 'dressed' figurines, clay fragments and other materials (Adovasio *et al.* 1996; Soffer 2004; Soffer *et al.* 2000). Though the plant types are not identified as to species, the woven material on the 'dressed' figurines was identified visually by Soffer (2004) as bast fibre. Adovasio *et al.* (2007) suggest nets were used in hunting as well as for fishing and collecting, construction of fibres and clothing, bags, nets, snares and virtually all other fishing gear, bows and arrows, harpoons, containers, footwear, carrying gear, lashings for houses, boats, jewellery and personal adornment. There is a range of secondary evidence from Mesolithic sites for the use of fibre, in the form of harpoons, needles, or bodkins and perforated shell beads (Albrethsen and Brinch Peterson 1976; Mellars 1987; Mordant and Mordant 1992; Soffer 2004; Hardy 2009). Likewise, indirect evidence such as, for example, use of a catch-all method of fishing, demonstrates the use of nets or traps (Parks and Barratt 2009). String bags lined with grass or moss can be used for gathering small items though other raw materials such as birch bark and hide could also be used to make containers. The detection of use-wear traces on tools is also used to identify evidence of weaving and net making (Soffer 2004).

In terms of actual fragments of ancient twisted fibre, possible plant fibres have also been identified on a stone tool at the Middle Palaeolithic site of La Grotte du Portel. Small flax fibre fragments, including some that were coloured and twisted, were detected in samples of clay from Dzudzuana Cave, Georgia (36,000–31,000 BP (Kvavadze *et al.* 2009). Three fragments of fibre were found at Ohalo II in Israel (19,000 BP; Nadel *et al.* 1994) and a piece of cord made from twisted fibre was found at Lascaux (17,000 BP) (Leroi Gourhan 1982). In the Mesolithic, the evidence for twisted plant fibres is wider, and fragments of fibres are known from many sites in the Baltic region and Russia (Hardy 2007, see also chapter 13). At Friesack in Germany, many fragments of nets, ropes and thousands of fragments of twisted bast fibre string were dated to the Early Boreal period, *c.* 7000 BC; several bone points, which have been considered as hafting material, also had narrow strips of bast fibre twisted around them (Gramsch 1992). In Denmark, fragments of textiles and rope were found at Tybrind Vig; these constituted an important feature of this Ertebølle site (Andersen 1985; Jørgensen 2013). Various plant materials were identified including willow (*Salix*) bast and possibly poplar (*Populus*) bast, as well as grass stems (Gramineae). Willow bast was found both as fragments of string and as woven textile fragments. It appears that the grass stems were mixed with willow bast in the textile pieces. Some pieces of Tybrind Vig textile are so tight they appears to be the remains of clothing (Jørgensen 2013).

The oldest fishing net yet known was found at Antrea (then Finland, now Kamennogorsk, Russia), and dates to 8300 BC (9310±120 BP). It was constructed from willow bast, contained knots, and was found with associated artefacts including bark weights (Äyräpää 1950; Miettinen *et al.* 2003 and chapter 13). Other net fragments include one from the Estonian Mesolithic site of Siiversti. Again this was made from twisted bast fibre, and the sample had been woven using the knotless netting or nålebinding technique (Clark 1952). At the Mesolithic site of Forstermoor in Germany, bast string was found on a small bow, while at the Mesolithic sites of Vinkelmore and Fippenborg, string was found wrapped round arrows, perhaps to hold feathers; other sites also have remains of string but it is not clear whether these are made of twisted bast fibre of some other material (Mertens 2000). Over 50 knots tied on plant fibre string were found at the Mesolithic site of Zamostje in Russia in association with a wide range of evidence for fishing including fibre made from bulrush (*Scirpus lacustris*) (see Clemente *et al.* chapter 13). Numerous twisted plant fibre fragments, in some cases knotted, were found at the Mesolithic site of Vis I (Burov 1998). Several Danish Mesolithic and early Neolithic sites have produced evidence of twisted plant fibre string. These include Skoldnaes, Derjø (Skaarup 1982), Møllegabet II (Skaarup and Grøn 2004) and Sigersdal Mose (Bender Jorgensen 1986). String and textile fragments have been discovered at many other Mesolithic and Neolithic sites across northern Europe and Russia, Estonia, Latvia and Lithuania (e.g. Vogt 1937; Burov 1967; Indreko 1967; chapter 13). Danish sites include Tulstrup Mose where coarse fibres were identified as lime (*Tilia*) bast and worked using twined weave (Becker 1947). At Ulkestrup Mose a bone point was found with the shaft attached by twisted bast fibre, most likely lime. At the early Neolithic site of Bolkilde, three textile fragments woven using the knotless netting (nålebinding) technique and a plaited piece were found beside the skeleton which suggests that they might have been clothing (Bennike *et al.* 1986). At Kongsted Lyng remains of string were found that had held a pot and were identified as either willow, poplar or rowan bast (Jørgensen 2013). More broadly, many Neolithic sites have also produced large numbers of inner bast fragments, these include Sārnate in Latvia where the remains of a net was found as well as stone sinkers and floats with bast fibre wrapped round them, a fragment of knotted netting from Abora I, Latvia (Bērziņš 2006), a net and a household mat made of twisted lime bast from Šventoji, Lithuania (Rimantiené 2005). At the underwater Neolithic site of Šerteya II, the extraordinary survival of its organic remains offers an insight into the extent of plant raw materials used at the time. These include two piece of rope that had two different materials twisted together, lime-bast and bilberry roots. Lime bast cordage has a history of use in Scandinavia that began in the Mesolithic and continued into the early 20th century (Myking *et al.* 2005). Several net fragments were also found; these were made from bilberry root, viburnum and willow root respectively (Mazurkevich *et al.* 2010). Chapter 13 describes the extensive and intricate wood and fibre work linked to the manufacture of fishing equipment. It appears that a range of basts from different trees were used to make fibres at this time though it is not clear whether different species were used for different purposes. Ethnographic evidence (Sillitoe 1988; Piqué *et al.* 2016) suggests different materials were selected for specific purposes on the basis of strength, colour and coarseness.

Ethnographically, raw materials used for making string include bast fibres from birch, juniper and pine; willow and heather twigs, nettle in particular makes durable, strong and fine cloth though the earliest evidence for the use of nettle appears later in the archaeological record. Animal based raw materials were also used widely (Thomson 1936; Emmons 1991) and the use of sinews and hide is known ethnographically from the Kalahari (Lee 1979) and the Arctic (e.g. Garth Taylor 1974); however, fibres from spruce and willow root were also used to bind materials together in the high Arctic (Nelson 1899). Use of inner bark from pine has been recorded ethnographically among the Saami who used it as wrapping and to manufacture string, until the early 20th century (Zackrisson *et al.* 2000).

There are four conceptual and technological phases required in order to reach the stage of constructing items solely out of twisted fibre. First, the concept of using natural plant fibres as a material, and the technological use of it, to hold or bind, had to be achieved. Secondly, an understanding of the raw material properties is required; thirdly, the technique of twisting fibres together and adding new fibres in, to create string or cord of unlimited length and a much greater degree of strength, had to be developed. Finally, the development of technologies such as looping and weaving that enabled artefacts to be manufactured solely from manipulation of string developed. The development of fibre technologies represent extraordinary conceptual, technological and functional achievements and to have reached the level of sophistication in networking observed in the Upper Palaeolithic suggests a very deep ancestry.

Conclusion

This chapter has focused on a small number of key plant-based technologies and uses of plant materials. Other technologies not included here are fishing equipment (covered in Clemente *et al.*, chapter 13), and the role of socially embedded items such as the string bags of highland Papua New Guinea (MacKenzie 1991). Though there is little direct evidence for plants as raw materials in the pre-agrarian record, there is sufficient, together with the broader secondary evidence, to demonstrate their fundamental technological importance in pre-agrarian life, while ethnographic evidence gives an idea of the extent of use of plants in construction of items of material culture in recent non-industrialised societies (e.g. Sillitoe 1988).

The technological developments discussed here, including the controlled use of fire, twisting fibre to create string, preparation of tar and bitumen and the resulting development of composite technology, cannot be overestimated and should take their place as major human achievements. It goes without saying that without the use of plants, the human world as we know it today would not exist.

Acknowlegements

Thanks to Lucy Kubiak-Martens for providing some extra information from northern Europe on basts and tar extraction. All figures were taken as part of "Project for the study, enhancement, preservation and spreading of the cultural heritage in the Saloum Delta (Senegal): 5000 years of history". Funded by the Spanish Agency for Cooperation and Development (AECID) (grant no. C/032099/10, 2011).

References

Adovasio, J. M., Soffer, O. and Page, J. 2007. *The Invisible Sex*. New York: Smithsonian Books

Adovasio, J. M., Soffer, O. and Kléma, B. 1996. Upper Palaeolithic fibre technology: Interlaced woven finds from Pavlov I, Czech Republic, *c.* 26,000 years ago. *Antiquity* 70(269), 526–34

Albert, R. M., Berna, F. and Goldberg, P. 2012. Insights on Neanderthal fire use at Kebara Cave (Israel) through high resolution study of prehistoric combustion features: Evidence from phytoliths and thin sections. *Quaternary International* 247, 278–293

Albrethsen, S. E. and Brinch Petersen, E. 1976. Excavation of a Mesolithic cemetery at Vedbeck, Denmark. *Acta Archaeologica* 47, 1–28

Albert, R. M., Weiner, S., Bar-Yosef, O. and Meignen, L. 2000. Phytoliths in the Middle Palaeolithic deposits of Kebara Cave, Mt Carmel, Israel: study of the plant materials used for fuel and other purposes. *Journal of Archaeological Science*, 27(10), 931–47

Aldeias, V., Goldberg, P., Sandgathe, D., Berna, F., Dibble, H. L., Mcpherron, S. P., Turk, A. and Rezek, Z. 2012. Evidence for Neandertal use of fire at Roc de Marsal (France). *Journal of Archaeological Science* 39(7), 2414–23

Almas, K. 2002. The effect of Salvadora persica extract (miswak) and chlorhexidine gluconate on human dentin: a SEM study. *Journal of Contemporary Dental Practice* 3(3), 27–35

Alperson-Afil, N. 2008. Continual fire-making by Hominins at Gesher Benot Ya'aqov, Israel. *Quaternary Science Reviews* 27, 1733–39

Ambrose, S. H. 1998. Chronology of the Later Stone Age and Food Production in East Africa. *Journal of Archaeological Science* 25, 377–92

Andersen, S. H. 1985. Tybring Vig. A preliminary report on a submerged Ertebølle settlement. *Journal of Danish Archaeology* 4, 52–69

Austin, P. 2009. The wood charcoal macro-remains from Mesolithic midden deposits at Sand, Applecross. In Hardy, K. and Wickham-Jones, C. R. (eds) 2009. *Mesolithic and Later Sites Around the Inner Sound, Scotland: the Scotland's First Settlers project 1998–2004. Scottish Archaeological Internet Report* 31. www.sair.org.uk

Aveling, E. M. and Heron, C. 1999. Chewing tar in the early Holocene: an archaeological and ethnographic evaluation. *Antiquity* 73, 579–84

Äyräpää A. 1950. Die ältesten steinzeitlichen Funde aus Finnland. *Acta Archaeologica* 21, 1–43

Badal, E., Villaverde, V. and Zilhão, J. 2011. The fire of Iberian Neanderthals. Wood charcoal from three new Mousterian sites in the Iberian Peninsula. *SAGVNTVM* Extra 11, 77–78

Barber, E. W. 1994. *Women's Work: The First 20,000 years. Women, Cloth and Society in Early Times.* New York: W. W. Norton

Beck, W. C., Stout, E. C. and Jänne, P. A. 1997. The pyrotechnology of pine tar and pitch inferred from quantitative analyses by gas chromatography-mass spectrometry and carbon-13 nuclear magnetic resonance spectrometry. In Breziński, W. and Piotrowski, W. (eds), *Proceedings of the First International Symposium on Wood Tar and Pitch.* Warsaw: State Archaeological Museum, 181–92

Becker, C. J. 1947. *Mosefundne lerkar fra yngre Stenalder.* Copenhagen: Aarbøger for nordisk Oldkyndighed og Histoire

Bednarik, R. G. 2005. Middle Pleistocene beads and symbolism. *Anthropos*, 537–52

Bellomo, R. V. 1994. Methods of determining early hominid behavioral activities associated with the controlled use of fire at Fxjj-20 Main, Koobi-Fora, Kenya. *Journal of Human Evolution* 27, 173–95

Bennike, P., Ebbesen, K., Jørgensen, L. B. and Rowley-Conwy, P. 1986. The bog find from Sigersdal: human sacrifice in the early Neolithic. *Journal of Danish Archaeology* 5(1), 85–115

Bender Jørgensen, L. 1986. The string from Sigersdal Mose. *Journal of Danish Archaeology* 5, 105–6

Berna, F., Goldberg, P., Horwitz, L. K., Brink, J., Holt, S., Bamford, M. and Chazan, M. 2012. Microstratigraphic evidence of *in situ* fire in the Acheulean strata of Wonderwerk Cave, Northern Cape province, South Africa. *Proceedings of the National Academy of Science (USA)*109, E1215–20

Bērziņš, V. 2006. Net fishing gear from Sārnate Neolithic site, Latvia. In Vesa-Pekka, H. (ed.), *People, Material Culture and Environment in the North.* Proceedings of the 22nd Nordic Archaeological Conference. University of Oulu, 18–23 August 2004, 150–158. Oulu: *Studia humaniora ouluensia 1*

Bishop, M. and Plew, M. G. 2016. Fuel exploitation as a factor in Shoshone winter mobility. *North American Archaeologist* 37(1) 3–19

Boëda, E., Connan, J., Dessort, D., Muhesen, S., Mercier, N., Valladas, H. and Tisnérat, N. 1996. Bitumen as a hafting material on Middle Palaeolithic artefacts. *Nature* 380(6572), 336–338.

Bonfiglioli, B., Mariotti, V., Facchini, F., Belcastro, M. G. and Condemi, S. 2004. Masticatory and non-masticatory dental modifications in the epipalaeolithic necropolis of Taforalt (Morocco). *International Journal of Osteoarchaeology* 14(6), 448–56

Brain, C. K. and Sillent, A. 1988. Evidence from the Swartkrans cave for the earliest use of fire. *Nature* 336, 464–6

Brothwell, D. 1963. The macroscopic dental pathology of some earlier human populations. In Brothwell, D. R. (ed.), *Dental Anthropology*, 27–88. Oxford: Pergamon Press

Brown, D. E. 1991. *Human Universals.* New York: McGraw Hill

Brown, T. and Molnar, S. 1990. Interproximal grooving and task activity in Australia. *American Journal of Physical Anthropology* 81, 545–53

Burjachs, F., Allué, E., Blain, H. A., Expósito, I., López-García, J. M. and Rivals, F. 2011. Neanderthals palaeoecology from Abric Romaní (Capellades, NE Iberian Peninsula). *Quaternary International* 247 (C), 26–37

Burov, G. M. 1967. *Drevnij Sindor.* Moscow: Nauka

Burov, G. M. 1990. Some Mesolithic wooden artifacts from the site of Vis I in the European north east of the USSR. In Bonsall, C. (ed.), *The Mesolithic in Europe. Papers Presented at the Third International Symposium, Edinburgh, 1985*, 391–401. Edinburgh: John Donald

Burov, G. M. 1998. The use of vegetable materials in the Mesolithic of Northeast Europe. In Zvelebil, M., Dennell, R. and Domanska, L. (eds), *Harvesting the Sea, Farming the Forest. The Emergence of Neolithic Societies in the Baltic Region*, 53–63. Sheffield: Sheffield Archaeological Monograph 10

Carbonell, E. and Castro-Curel, Z. 1992. Palaeolithic wooden artefacts from the Abric Romani (Capellades, Barcelona, Spain). *Journal of Archaeological Science* 19(6), 707–19

Cârciumaru, M., Ion, R. M., Niţu, E. C. and Ştefănescu, R. 2012. New evidence of adhesive as hafting material on Middle and Upper Palaeolithic artefacts from Gura Cheii-Râşnov Cave (Romania). *Journal of Archaeological Science* 39(7), 1942–50

Clark, J. D. G. 1952. *Prehistoric Europe: the Economic Basis.* London: Methuen

Courty, M. A., Carbonell, E., Vallverdú Poch, J. and Banerjee, R. 2012. Microstratigraphic and multi-analytical evidence for advanced Neanderthal pyrotechnology at Abric Romaní (Capellades, Spain). *Quaternary International* 247, 294–312

Dąbrowski, P., Nowaczewska, W., Stringer, C. B., Compton, T., Kruszyński, R., Nadachowski, A., Stefaniak, K. and Urbanowski, M. 2013. A Neanderthal lower molar from Stajnia Cave, Poland. *HOMO-Journal of Comparative Human Biology* 64(2), 89–103

D'Errico, F., Backwell, L., Villa, P., Degano, I., Lucejko, J. J., Bamford, M. K., Higham, T., Colombini, M. P. and Beaumont, P. B. 2012. Early evidence of San material culture represented by organic artifacts from Border Cave, South Africa. *Proceedings of the National Academy of Sciences (USA)* 109(33), 13214–9

Dominguez-Rodrigo, M., Serrallonga, J., Juan-Tresserras, J., Alcala, L. and Luque, L. 2001. Woodworking activities by early humans: a plant residue analysis on Acheulian stone tools from Peninj (Tanzania). *Journal of Human Evolution* 40(4), 289–99

Eckhardt, R. B. and Piermarini, A. L. 1988. Interproximal grooving of teeth: additional evidence and interpretation. *Current Anthropology* 49(4), 663–71

Eidlitz, K. 1969. Food and emergency food in the circumpolar area. *Studia Ethnographica Uppsaliensis* 32, 1–175

Elujoba, A. A., Odeleye, O. M. and Ogunyemi, C. M. 2005. Review-Traditional medicine development for medical and dental primary health care delivery system in Africa. *African Journal of Traditional and Complementary Medicines* 2(1), 46–51

Emmons, G. T. 1991. *The Tlingit Indians.* Washington DC: University of Washington Press

Eshed, V., Gopher, A. and Hershkovitz, I. 2006. Tooth wear and dental pathology at the advent of agriculture: new evidence from the Levant. *American Journal of Physical Anthropology* 130(2), 145–59

Estalrrich, A. and Rosas, A. 2013. Handedness in Neandertals from the El Sidrón (Asturias, Spain): Evidence from Instrumental Striations with Ontogenetic Inferences. *PLOS ONE* 8(5), e62797. doi:10.1371/journal.pone.0062797

Estalrrich, A., Rosas, A., García Vargas, S., García Tabernero, A., Santamaría, D. and de la Rasilla, M. 2011. Brief communication: subvertical grooves on interproximal wear facets from the El Sidron (Asturias, Spain) Neandertal dental sample. *American Journal of Physical Anthropology* 144(1), 154–61

Fernández Peris, J., González, V. B., Blasco, R., Cuartero, F., Fluck, H., Sañudo, P. and Verdasco, C. 2012. The earliest evidence of hearths in Southern Europe: the case of Bolomor Cave (Valencia, Spain). *Quaternary International* 247, 267–77

Finlayson, C., Barton, R. and Stringer, C. 2001. The Gibraltar Neanderthals and their extinction. Les Premiers Hommes Modernes de la Péninsule Ibérique. *Trabalhos de Arqueologia* 17, 117–22

Formicola, V. 1988. Interproximal grooving of teeth: additional evidence and interpretation. *Current Anthropology* 663–71

Frayer, D. W. and Russell, M. D. 1987. Artificial grooves on the Krapina Neanderthal teeth. *American Journal of Physical Anthropology* 74(3), 393–405

Garth Taylor, J. 1974. *Netsilik Eskimo Material Culture. The Roald Amundsen Collection from King William Island*. Oslo: Norwegian Research Council for Science and the Humanities

Gilligan, I. 2010. The prehistoric development of clothing: archaeological implications of a thermal model. *Journal of Archaeological Method and Theory* 17(1), 15–80

Goodall, J. 1964. Tool-using and aimed throwing in a community of free-living chimpanzees. *Nature* 201, 1264–6

Goren-Inbar, N., Werker, E. and Feibel, C. S. 2002.*The Acheulian site of Gesher Benot Ya'aqov, Israel. The Wood Assemblage*. Oxford: Oxbow Books

Gott, B. 1982. The ecology of root use by the Aborigines of Southern Australia. *Archaeology in Oceania* 17, 59–67

Gowlett, J. A. J., Harris, J. W. K., Walton, D. and Wood, B. A. 1981. Early archaeological sites, hominid remains and traces of fire from Chesowanja, Kenya. *Nature* 294, 125–9

Gramsch, B. 1992. Friesack Mesolithic Wetlands. In Coles, B. (ed.), *The Wetland Revolution in Prehistory*. Exeter: Prehistoric Society/WARP. Occasional Paper 6

Grünberg, J. 2002. Middle Palaeolithic birch-bark pitch. *Antiquity* 76(291), 15–16

Hardy, K. 2007. Where would we be without string? Evidence for the use, manufacture and role of string in the Upper Palaeolithic and Mesolithic of northern Europe. In Beugnier, V. and Crombé, P. (eds), *Préhistoire et ethnographie du travail des plantes Actes de la Table ronde de l'Université de Gandt (Belgique)*, 9–22. Oxford: British Archaeological Report 1718

Hardy, K. 2008. Prehistoric String Theory. How twisted fibres helped to shape the world. *Antiquity* 82, 271–80

Hardy, K. 2009. Worked and modified shell from Sand. In Hardy, K. and Wickham-Jones, C. R. (eds), *Mesolithic and Later Sites Around the Inner Sound, Scotland: the Scotland's First Settlers project 1998–2004. Scottish Archaeological Internet Report 31*. www.sair.org.uk

Hardy, K., Buckley, S., Collins, M. J., Estalrrich, A., Brothwell, D., Copeland, L., García-Tabernero, A., García-Vargas, S., de la Rasilla, M., Lalueza-Fox, C., Huguet, R., Bastir, M., Santamaría, D., Madella, M., Fernández Cortés, A. and Rosas, A. 2012. Neanderthal medics? Evidence for food, cooking, and medicinal plants entrapped in dental calculus. *Naturwissenschaften* 99(8), 617–26

Hardy, K., Camara, A., Piqué, R., Dioh, E., Guèye, M., Diaw Diadhiou, H., Faye, M. and Carré, M. 2016. Shellfishing and shell middens in the Saloum Delta, Senegal. *Journal of Anthropological Archaeology* 41, 19–32. DOI 10.1016/j.jaa.2015.11.001

Hardy, K., Radini, A., Buckley, S., Sarig, R., Copeland, L., Gopher, A. and Barkai, R. 2015. First direct evidence for Middle Pleistocene diet and environment from dental calculus at Qesem Cave. *Quaternary International DOI 10.1016/j.quaint.2015.04.033*

Hardy, K. and Sillitoe, P. 2003. Material Perspectives: Stone Tool Use and Material Culture in Papua New Guinea. *Internet Archaeology* 14. http://intarch.ac.uk/journal/issue14

Haslam, M. 2014. On the tool use behavior of the bonobo-chimpanzee last common ancestor, and the origins of hominine stone tool use. *American Journal of Primatology* 76(10), 910–8

Henry, Au. and Théry-Parisot, I. 2014. From Evenk campfires to prehistoric hearths: charcoal analysis as a tool for identifying the use of rotten wood as fuel. *Journal of Archaeological Science* 52, 321–36

Henshilwood, C., d'Errico, F., Vanhaeren, M., Van Niekerk, K. and Jacobs, Z. 2004. Middle Stone Age shell beads from South Africa. *Science* 304(5669), 404

Hinton, R. J. 1981. Form and patterning of anterior tooth wear among aboriginal human groups. *American Journal of Physical Anthropology* 54(4), 555–64

Hlusko, L. J. 2003. The Oldest Hominid Habit? Experimental Evidence for Toothpicking with Grass Stalks 1. *Current Anthropology* 44(5), 738–41

Humle, T. and Matsuzawa, T. 2002. Ant dipping among the chimpanzees at Bossou, Guinea, and comparisons with other sites. *American Journal of Primatology* 58, 133–48.

Idu, M., Umweni, A. A., Odaro, T. and Ojelede, L. 2009. Ethnobotanical Plants Used for Oral Healthcare Among the Esan Tribe of Edo State, Nigeria. *Ethnobotanical Leaflets* 4, 15. http://opensiuc.lib.siu.edu/ebl/vol2009/iss4/15

Indreko, R. 1967. *Die Mittlere Steinzeit im Estland*. Stockholm: Almqvist snf Wiksells

Ingmanson, E. J. 1996. Tool-using behavior in wild *Pan paniscus*: social and ecological considerations. In Russon, A. E., Bard, K. A. and Parker, S. T. (eds), *Reaching into Thought: The Minds of the Great Apes*, 190–210. New York: Cambridge University Press

Jørgensen, L. B. 2013. The textile remains from Tybrind Vig. In Andersen, S. H. (ed.), *Tybrind Vig Submerged Mesolithic Settlement in Denmark*, 393–400. Højbjerg: Jutland Archaeological Society, Moesgård Museum

Jose, M., Sharma, B. B., Shantaram, M. and Ahmed, S. A. 2011. Ethnomedicinal herbs used in oral health and hygiene in coastal Dakshina Kannada. *Journal of Oral Health Community Dentistry* 5, 107–11

Karkanas, P., Shahack-Gross, R., Ayalon, A., Bar-Matthews, M., Barkai, R., Frumkin, A., Gopher, A. and Stiner, M. C. 2007. Evidence for habitual use of fire at the end of the Lower Paleolithic: site-formation processes at Qesem Cave, Israel. *Journal of Human Evolution* 53, 197–212

Koller, J., Baumer, U. and Mania, D. 2001. High-tech in the middle Palaeolithic: Neanderthal-manufactured pitch identified. *European Journal of Archaeology* 4, 385–97

Koops, K., Schöning, C., McGrew, W. C. and Matsuzawa, T. 2015. Chimpanzees prey on army ants at Seringbara, Nimba Mountains, Guinea: Predation patterns and tool use characteristics. *American Journal of Primatology* 77, 319–29

Kubiak-Martens, L., Kooistra, L. I. and Langer, J. J. 2011. Mesolithische teerproductie in Hattemerbroek. In Lohof, E., Hamburg, T. and Flamman, J. (eds), *Steentijd opgespoord. Archeologisch onderzoek in het tracé van de Hanzelijn-Oude Land*, 497–512. Leiden/Amersfoort: Archol rapport 138/ADC rapport 2576

Kubiak-Martens, L., Langer, J. J. and Kooistra, L. I. 2012. Plantenresten en teer in haardkuilen. In Hamburg, T., Müller, A. and Quadflieg, B. (eds) *Mesolithisch Swifterbant. Mesolithisch gebruik van een duin ten zuiden van Swifterbant (8300-5000 v. Chr.). Een archeologische opgraving in het tracé van de N23/N307, provincie Flevoland*, 341–60. Leiden/Amersfoort: Archol rapport 174/ADC rapport 3250

Kubiak-Martens, L., Kooistra, L. I. and Verbruggen, F. 2015. Archaeobotany: landscape reconstruction and plant food subsistence economy on a meso and microscale. In Moree, J. M. and Sier, M. M. (eds), *Interdisciplinary Archaeological Research Programme Maasvlakte 2, Rotterdam. Part 1*, 223–86. Rotterdam: BOORrapporten 566

Kuhn, S. L., Stiner, M. C., Reese, D. S. and Güleç, E. 2001. Ornaments of the earliest Upper Paleolithic: New insights from the Levant. *Proceedings of the National Academy of Sciences (USA)* 98, 7641–6

Kvavadze, E., Bar-Yosef, O., Belfer-Cohen, A., Boaretto, E., Jakeli, N., Matskevich, Z. and Meshveliani, T. 2009. 30,000-year-old wild flax fibers. *Science* 325(5946), 1359–1359

Langley, M. C., Clarkson, C. and Ulm, S. 2008. Behavioural complexity in Eurasian Neanderthal populations: a chronological examination of the archaeological evidence. *Cambridge Archaeological Journal* 18(3), 289–307

Lee, R. B. 1979. *The !Kung San: Men, Women and Work in a Foraging Society*. Cambridge: Cambridge University Press

Leca, J-B., Gunst, N. and Huffman, M. A. 2010. The first case of dental flossing by a Japanese macaque (Macaca fuscata): implications for the determinants of behavioral innovation and the constraints on social transmission. *Primates* 51 (1), 13–22

Leroi Gourhan, A. 1982. The archaeology of Lascaux Cave. *Scientific American* 246(6), 80–8.

Lozano, M., Bermúdez de Castro, J. M., Carbonell, E. and Arsuaga, J. L. 2008. Non-masticatory uses of anterior teeth of Sima de los Huesos individuals (Sierra de Atapuerca, Spain). *Journal of Human Evolution* 55(4), 713–28

Lozano, M., Mosquera, M., de Castro, J. M. B., Arsuaga, J. L.and Carbonell, E. 2009. Right handedness of *Homo heidelbergensis* from Sima de los Huesos (Atapuerca, Spain) 500,000 years ago. *Evolution and Human Behavior* 30(5), 369–76

Lozano, M., Subirà, M. E., Aparicio, J., Lorenzo, C. and Gómez-Merino, G. 2013. Toothpicking and periodontal disease in a Neanderthal specimen from Cova Foradà Site (Valencia, Spain). *PLOS ONE* 8(10), e76852

Lukacs, J. R. and Pastor, R. F. 1988. Activity-induced patterns of dental abrasion in prehistoric Pakistan: Evidence from Mehrgarh and Harappa. *American Journal of Physical Anthropology* 76(3), 377–98

MacKenzie, M. 1991. *Androgynous Objects: String Bags and Gender in Central New Guinea.* Chur: Harwood Academic

Madella, M., Jones, M. K., Goldberg, P., Goren, Y. and Hovers, E. 2002. The exploitation of plant resources by Neanderthals in Amud Cave (Israel): the evidence from phytolith studies. *Journal of Archaeological Science* 29(7), 703–19

Mazurkevich, A., Dolbunova, E., Maigrot, Y. and Hook, D. 2010. The results of underwater excavations at Serteya II, and research into pile-dwellings in northwest Russia. *Archaeologica Baltica* 14, 47–64

Mazza, P. P. A., Martini, F., Sala, B., Magi, M., Colombini, M. P., Giachi, G., Landucci, F., Lemorini, C., Modugno, F. and Ribechini, E. 2006. A new Palaeolithic discovery: tar-hafted stone tools in a European Mid-Pleistocene bone-bearing bed. *Journal of Archaeological Science* 33(9), 1310–8

McGrew, W. C. 1992. *Chimpanzee Material Culture: Implications for Human Evolution.* Cambridge: Cambridge University Press

McGrew, W. C. 2004. The *Cultured Chimpanzee: Reflections on Cultural Primatology.* Cambridge: Cambridge University Press

McGrew, W. C. 2010a. In search of the last common ancestor: new findings on wild chimpanzees. *Philosophical Transactions of the Royal Society B,* 365, 3267–76

McGrew, W. C. 2010b. Chimpanzee technology. *Science* 328, 579–80

McGrew, W. C. and Tutin, C. E. G. 1973. Chimpanzee Tool Use in Dental Grooming. *Nature* 241, 477–8

Mellars, P. 1987. *Excavations on Oronsay. Prehistoric Human Ecology on a Small Island.* Edinburgh: Edinburgh University Press

Mertens, E-M. 2000. *Linde, Ulme, Hasel, Zur Verwendung von Pflanzen für Jagd- und Fischfanggeräte im Mesolithikum Dänemarks und Schleswig-Holsteins.* Berlin: Prehistorische Zeitschrift 75

Miettinen, A., Sarmaja-Korjonen, K., Sonninen, E., Jungner, H., Lempiäinen, T., Ylikoski, K. and Carpelan, C. 2003 (1998). The Palaeoenvironment of the Antrea Net Find. Karelian Isthmus. *Stone Age Studies in 2003,* 71–87

Molnar, P. 2008. Dental wear and oral pathology: possible evidence and consequences of habitual use of teeth in a Swedish Neolithic sample. *American Journal of Physical Anthropology* 136(4), 423–31

Mordant, D. and Mordant, C. 1992. Noyen-sur-Seine: A Mesolithic Waterside Settlement. In Coles B. (ed.) *The Wetland Revolution in Prehistory.* Exeter: Prehistoric Society, WARP. Occasional Paper 6

Myking, T., Hertzberg, A. and Skrøppa, T. 2005. History, manufacture and properties of lime bast cordage in northern Europe. *Forestry* 78(1), 65–71

Nadel, D., Danin, A., Werker, E., Schick, T., Kislev, M. E. and Stewart, K. 1994. 19,000-year-old twisted fibers from Ohalo II. *Current Anthropology* 35(4), 451–8

Nadel, D. and Werker, E. 1999. The oldest ever brush hut plant remains from Ohalo II, Jordan Valley, Israel (19,000 BP). *Antiquity* 73(282), 755–64

Nadel, D., Weiss, E., Simchoni, O., Tsatskin, A., Danin, A. and Kislev, M. 2004. Stone Age hut in Israel yields world's oldest evidence of bedding. *Proceedings of the National Academy of Science (USA)* 101(17), 6821–6

Nelson, E. W. 1899. *The Eskimo about Bering Strait.* Washington DC: Smithsonian Institution Press

Oakley, K. P., Andrews, P., Keeley, L. H. and Clark, J. D. 1977. A reappraisal of the Clacton spear point. *Proceedings of the Prehistoric Society* 43, 13–30

Parks, R. and Barrett, J. 2009. The Zooarchaeology of Sand. In Hardy, K. and Wickham-Jones, C. R. (eds) *Mesolithic and Later Sites Around the Inner Sound, Scotland: the Scotland's First Settlers project 1998–2004. Scottish Archaeological Internet Report 31.* www.sair.org.uk

Piqué, R., Gueye, M., Hardy, K., Camara, A. and Dioh, E. 2015. Not just shellfish: Wild terrestrial resource use among the people of the Saloum Delta, Senegal. In Biagetti, S. and Lugli, F. *The Intangible Elements of Culture in Ethnoarchaeological Research.* Dordrecht: Springer

Pojar, J. and Mackinnon, A. 1994. *Plants of the Pacific Northwest Coast: Washington, Oregon, British Columbia, and Alaska.* Redmond WA: Lone Pine

Puech, P. F. and Gianfarani, F. 1988. Interproximal grooving of teeth: additional evidence and interpretation. *Current Anthropology* 29(4), 663–71

Radini, A., Buckley, S., Rosas, A., Estalrrich, A., de la Rasilla, M. and Hardy, K. 2016. Neanderthals and Trees: Non-edible conifer fibres found in Neanderthal dental calculus suggests extra-masticatory activity. *Antiquity*

Radovčić, D., Sršen, A. O., Radovčić, J. and Frayer, D. W. 2015. Evidence for Neandertal Jewelry: Modified White-Tailed Eagle Claws at Krapina. *PLOS ONE* 10(3), e0119802 doi:10.1371/journal.pone.0119802

Regnell, M. 2012. Plant subsistence and environment at the Mesolithic site Tågerup, southern Sweden: new insights on the "Nut Age". *Vegetation History and Archaeobotany* 21(1), 1–16

Regnell, M., Gaillard, M. J., Bartholin, T. S. and Karsten, P. 1995. Reconstruction of environment and history of plant use during the late Mesolithic (Ertebølle culture) at the inland settlement of Bökeberg III, southern Sweden. *Vegetation History and Archaeobotany* 4, 67–91

Richter, D. and Krbetschek, M. 2015. The age of the Lower Palaeolithic occupation at Schöningen. *Journal of Human Evolution* 89, 46–56. http://dx.doi.org/10.1016/j.jhevol.2015.06.003

Rigaud, S., d'Errico, F., Vanhaeren, M. and Neumann, C. 2009. Critical reassessment of putative Acheulean *Porosphaera globularis* beads. *Journal of Archaeological Science* 36(1), 25–34

Rimantiené, R. 2005. *Akmens Amžiaus žvejai prie Pajūrio Lagūnos.* Lietuvos Nacionalinis Muziejus. Vilnius.

Roebroeks, W. and Villa, P. 2011. On the earliest evidence for habitual use of fire in Europe. *Proceedings of the National Academy of Science (USA)* 108(13), 5209–14

Rots, V. 2013. Insights into early Middle Palaeolithic tool use and hafting in Western Europe. The functional analysis of level IIa of the early Middle Palaeolithic site of Biache-Saint-Vaast (France). *Journal of Archaeological Science* 40(1), 497–506

Rowlett, R. M. 2000. Fire Control by *Homo Erectus* in East Africa and Asia. *Acta Anthropologica Sinica* 19, 198–208

Rudenko, S. I. 1961. *The Ancient Culture of the Bering Sea and the Eskimo Problem.* Toronto: Archaeological Institute of North America. University of Toronto Press

Russon, A. E., van Schaik, C. P., Kuncoro, P., Ferisa, A., Handayani, D. P. and van Noordwijk, M. A. 2009. Innovation and intelligence in orangutans. In Wich, S.A., Utami Atmoko, S.S. and Mitra Setia, T. (eds), *Orangutans: Geographic Variation in Behavioral Ecology and Conservation*, 279–98. New York: Oxford University Press

Ryan, A. S. and Johanson, D. C. 1989. Anterior dental microwear in Australopithecus afarensis: comparisons with human and nonhuman primates. *Journal of Human Evolution* 18(3), 235–68

Sanz, C. M. and Morgan, D. B. 2007. Chimpanzee tool technology in the Goualougo Triangle, Republic of Congo. *Journal of Human Evolution* 52(4), 420–33

Scherjon, F., Bakels, C., MacDonald, K. and Roebroeks, W. 2015. Burning the land: an ethnographic study of off-site fire use by current and historically documented foragers and implications for the interpretation of past fire practices in the landscape. Free content contains supplements. *Current Anthropology* 56(3) 299–326

Shahack-Gross, R., Berna, F., Karkanas, P., Lemorini, C., Gopher, A. and Barkai, R., 2014. Evidence for the repeated use of a central hearth at Middle Pleistocene (300 ky ago) Qesem Cave, Israel. *Journal of Archaeological Science* 44, 12e21

Sharma, V. and Joshi, B. D. 2010. Traditional medicines used for dental health care amongst the local people of Almora district of Central Himalaya in India. *Asian Journal of Traditional Medicines* 5, 177–121

Sillitoe, P. 1988. *Made in Niugini. Technology in the Highlands of Papua New Guinea.* London: British Museum Publications

Sinitsyn, A. A. 2003. A Palaeolithic 'Pompeii' at Kostenki, Russia. *Antiquity* 77, 9–14

Skaarup, J. 1982. Sites 7, Skjoldnaes & 8 Derjø. Recent excavations and discoveries. *Journal of Danish Archaeology* 1, 166–7

Skaarup, J. and Grøn, O. 2004. *Møllegabet II. A submerged Mesolithic Settlement in Southern Denmark.* Oxford:British Archaeological Report S1328

Soffer, O. 2004. Recovering perishable technologies through use wear on tools: preliminary evidence for Upper Palaeolithic weaving and net making. *Current Anthropology* 45(3), 407–13

Soffer, O. and Adovasio, J. M. 2010. The Roles of Perishable Technologies in Upper Palaeolithic Lives. *The Magdalenian Household: Unraveling Domesticity*, 235–44. Institute for European and Mediterranean Archaeology Distinguished Monograph Series SUNY PRESS

Soffer, O, Adovasio, J. M. and Hyland, D. C. 2000. The 'Venus' figurines. textiles, basketry, gender and status in the Upper Palaeolithic. *Current Anthropology* 41(4), 511–37

Soffer, O, Adovasio, J. M. and Hyland, D. C. 2001. Perishable technologies and invisible people: nets, baskets and 'Venus' wear ca. 26,000BP. In Purdy, B. (ed.), *Enduring Records: The Environmental and Cultural Heritage*. Oxford: Oxbow Books

Sułgostowska, Z., 1997. Examples of the application of wood tar during the Mesolithic on Polish Territory. In Breziński, W. and Piotrowski, W. (eds), *Proceedings of the First International Symposium on Wood Tar and Pitch*, 19–23. Warsaw: State Archaeological Museum

Texier, P.-J., Porraz, G., Parkington, J., Rigaud, J.P., Poggenpoel, C., Miller, C., Tribolo, C. Cartwright, C., Coudenneau, A., Klein, R., Steele, T. and Verna, C. 2010. A Howiesons Poort tradition of engraving ostrich eggshell containers dated to 60,000 years ago at Diepkloof Rock Shelter, South Africa. *Proceedings of the National Academy of Sciences* 107(14), 6180–5

Théry-Parisot, I., Chabal, L. and Chrzavzez, J. 2010. Anthracology and taphonomy, from wood gathering to charcoal analysis. A review of the taphonomic processes modifying charcoal assemblages, in archaeological contexts. *Palaeogeography, Palaeoclimatology, Palaeoecology* 291(1), 142–53

Thieme, H. 1997. Lower Palaeolithic hunting spears from Germany. *Nature* 385 (6619), 807–10

Thomson, D. F. 1936. Notes on Some Bone and Stone Implements from North Queensland. *Journal of the Royal Anthropological Institute of Great Britain and Northern Ireland* 66, 71–4

Ubelaker, D. H., Phenice, T. W. and Bass, W. M. 1969. Artificial interproximal grooving of the teeth in American Indians. *American Journal of Physical Anthropology* 30, 45–50

Ungar, P. S., Grine, F. E., Teaford, M. F. and Pérez-Pérez, A. 2001. A review of interproximal wear grooves on fossil hominin teeth with new evidence from Olduvai Gorge. *Archives of Oral Biology* 46(4), 285–92

Urbanowski, M., Socha, P., Dąbrowski, P., Nowaczewska, W., Sadakierska-Chudy, A., Dobosz, T., Stefaniak, K. and Nadachowski, A. 2010. The first Neanderthal tooth found North of the Carpathian Mountains. *Naturwissenschaften* 97(4), 411–5

Vanhaeren, M., d'Errico, F., Stringer, C., James, S. L., Todd, J. A. and Mienis, H. K. 2006. Middle Paleolithic shell beads in Israel and Algeria. *Science* 312 (5781), 1785–8

Van Gijn, A. L. and Boon, J. 2006. Birch bark tar. In Louwe Kooijmans, L. P. and Jongste, P. F. B. (eds) *Schipluiden, a Neolithic Settlement on the Dutch North Sea coast c. 3500 cal BC*, 339–352. *Analecta Praehistorica Leidensia* 37/38

Villa, G., Giacobini, G. 1995. Subvertical grooves of interproximal facets in Neandertal posterior teeth. *American Journal of Physical Anthropology* 96(1), 51–62

Vincent, A. S. 1985. Plant foods in savanna environments: a preliminary report of tubers eaten by the Hadza of Northern Tanzania. *World Archaeology* 17(2), 131–48

Vogt, E. 1937. *Geflechte und Gewebe der Steinzeit*. Basel: Birkhauser

Volpato, V., Macchiarelli, R., Guatelli-Steinberg, D., Fiore, I., Bondioli, L. and Frayer, D. 2012. Hand to Mouth in a Neandertal: Right-Handedness in Regourdou 1. *PLOS ONE* 7(8), e43949. doi:10.1371/journal.pone.0043949

Wadley, L., Sievers, C., Bamford, M., Goldberg, P., Berna, F. and Miller, C. 2011. Middle Stone Age Bedding Construction and Settlement Patterns at Sibudu, South Africa. *Science* 334 (6061), 1388–91

Wadley, L., Hodgskiss, T. and Grant, M. 2009. Implications for complex cognition from the hafting of tools with compound adhesives in the Middle Stone Age, South Africa. *Proceedings of the National Academy of Sciences (USA)* 106, 9590–4

Watanabe, K., Urasopon, N. and Malaivijitnond, S. 2007. Long-tailed macaques use human hair as dental floss. *American Journal of Primatology* 69(8), 940–4

Wiessner, P. W. 2014. Embers of society: firelight talk among the Ju/'hoansi Bushmen. *Proceedings of the National Academy of Sciences USA* 111(39), 14027e14035

Winterhalder, B. 2001. The behavioural ecology of hunter-gatherers. In Panter-Brick, C., Layton, A. H. and Rowley-Conwy, P. (eds), *Hunter-Gatherers: An Interdisciplinary Perspective* 13, 12–38. Cambridge: Cambridge University Press

Wrangham, R. W. 2007. The cooking enigma. In Ungar, P. S. (ed.), *Evolution of the Human Diet: the Known, the Unknown, and the Unknowable* 308–23. Oxford: Oxford University Press

Wrangham, R. W. 2009. *Catching Fire: How Cooking Made Us Human*. New York: Basic Books

Wrangham, R. W. and Carmody, R. 2010. Human adaptation to the control of fire. *Evolutionary Anthropology* 19, 187–199

Wrangham, R. W. and Conklin-Brittain, N. L. 2003. Cooking as a biological trait. *Comparative Biochemistry and Physiology Part A: Molecular and Integrative Physiology* 136, 35–46

Wrangham, R. W., Jones, J. H., Laden, G., Pilbeam, D. and Conklin-Brittain, N. L. 1999. The raw and the stolen: cooking and the ecology of human origins. *Current Anthropology* 40, 567–94

Zackrisson, O., Östlund, L., Korhonen, O. and Bergman, I. 2000. The ancient use of *Pinus sylvestris* L. (Scots pine) inner bark by Sami people in northern Sweden, related to cultural and ecological factors. *Vegetation History and Archaeobotany* 9 (2), 99–109

5. Hunter-gatherer plant use in southwest Asia: the path to agriculture

Amaia Arranz-Otaegui, Juan J. Ibáñez and Lydia Zapata-Peña

This paper focuses on plant use by the last hunter-gatherers in the Levant, from the Last Glacial Maximum (LGM) to the first experiments with plant cultivation at the beginning of the Holocene. This review of Epipaleolithic and Early Neolithic plant use summarises available archaeobotanical and technological data. Information for the Early Epipalaeolithic, especially from the site of Ohalo II, shows that, from the LGM, humans had access to exceptionally rich plant-food staples that included small-grained grasses and wild barley (Hordeum spontaneum) and wild wheat (Triticum dicoccoides). Grasses seem to have been the staple plant foods but other plants were also present: wild pulses, acorns, almonds, pistachios, wild olives, fruits, and berries. Grinding and pounding stone tools were in use at this time for processing plant resources. During the Late Epipaleolithic (Natufian) period plant use intensified, as can be seen in the site of Abu Hureyra. The seed assemblage from Abu Hureyra I may have included more than 120 food types comprising possible staples such as the grain of wild rye (Secale spp.) and wheat (Triticum spp.), feather grasses (Stipa and Stipagrostis spp.), club-rush (Scirpus maritimus), Euphrates knotgrass (Polygonum corrigioloides), small-seeded grasses, and wild shrubby chenopods (Atriplex spp. and others). The presence in Natufian sites of tools with glossy edges that were used for harvesting cereals, and the widespread nature of mortars suggest that cereals were a more common food. During the Pre-Pottery Neolithic A (PPNA), the first experiments with cultivation of morphologically wild cereals, and also probably of legumes, took place. This involved cereals such as wild emmer (T. dicoccoides), wild einkorn (T. boeoticum), wild barley (Hordeum spontaneum) and wild oat (Avena sterilis), and pulses such as rambling vetch (Vicia peregrina) and probably others. Human manipulation of plant resources opened the path to domestication with the first evidence found during the Early Pre-Pottery Neolithic B (EPPNB). However, the exploitation of wild plants continued to be important for these societies, as is suggested by the admixture of plant exploitation strategies during most of the PPN period and the late establishment of crop 'packages' during the Late PPNB.

The Levant, encompassing areas of the modern countries of Jordan, Israel, the Palestinian Authority, Syria, Lebanon, Turkey, Iraq, and Iran, remains one of the earliest and best-known foci of early agriculture. This innovation was the end result of accumulated experience and widespread use of wild plants, which took place among the last communities of hunter-gatherers who exploited a rich, diverse and shifting environment, during the long period between the Last Glacial Maximum (LGM) and the beginning of the Holocene.

This huge region is bounded by the Mediterranean Sea to the west, the Taurus and Zagros Mountains in the north, the Euphrates River Valley in the north-east, and the Syro-Arabian, Negev and Sinai Deserts in the east, south-east, and south, measuring 1100 km from north to south and 250–350 km from east to west (Simmons 2007). The main geographical regions, which are distributed on a north–south axis, are, from west to east, the coastal Mediterranean belt, the Levantine corridor, following the Jordan, Litani, and Orontes Rivers and the Jordan–Syrian Plateau. These areas roughly correspond to three climatic and vegetation areas; Mediterranean forest, steppe, and desert. The Mediterranean zone, in the west, is the richest, with over 100 edible fruits, seeds, leaves, and tubers (Zohary 1973), while this richness decreases towards the east as the precipitation rate drops drastically. Moisture is the most important ecological variable and its shift during the last 23,000 years has conditioned the geographical distribution of plant resources.

In this chapter we aim to summarise the available evidence on wild plant use from the Early Epipaleolithic to the Pre-Pottery Neolithic by analysing archaeobotanical data together with tools and devices related to plant exploitation and processing. We do not intend to conduct an in depth discussion on the origins of cultivation and plant domestication, which has been subject of a number of recent publications (among others Fuller 2007; Kilian *et al.* 2007; 2010; Allaby *et al.* 2008; Fuller *et al.* 2010; 2011; Purugganan and Fuller 2011; Bar-Yosef 2011; Zeder 2011; Heun *et al.* 2012; Asouti and Fuller 2013; Willcox 2013), but rather to present and evaluate the main archaeobotanical data currently available.

Early and Middle Epipaleolithic 21,000–12,700 BC

The palaeoclimatic records of the LGM (21,000–17,000 cal BC) in the Eastern Mediterranean and Levant suggest that the region was generally cooler and more arid than at present (Robinson *et al.* 2006). The LGM was followed by the cool and dry Heinrich I event (around 14,500 BC) as suggested by the decrease on oceanic salinity and surface sea temperatures, and increase in δ18 O levels (Maher *et al.* 2011 and references therein). Around 13,000 BC, a sudden climatic amelioration occurred, known as the Bølling-Allerød interstadial. This was a period of increasing climatic seasonality in temperature and precipitation (Felis *et al.* 2004).

The Early (21,000–15,500 cal BC) and Middle Epipaleolithic (15,500–12,700 cal BC) have been distinguished by their lithic industries; the former is characterised by a dominance of non-geometric bladelet microliths and the latter by flint tools of geometric forms, mainly rectangles and trapezes (Goring-Morris *et al.* 2009). Sites are composed of a small number (two or three) of habitation structures, whose remains appear as semi-subterranean depressions with multi-stratified living floors, as in Ohalo II, Ein Gev I or Kharaneh IV (Goring-Morris and Belfer-Cohen 2008; Maher *et al.* 2012).

Ground stone tools, which first appeared in the Upper Paleolithic (Bar-Yosef 1998), are present in larger quantities in the Early/Middle Epipaleolithic in both domestic and burial contexts. In addition, new types of these tools, such as deep vessels and mortars (Maher *et al.* 2012), also appear at this time suggesting a growing relevance of plant processing activities.

One of the problems faced by archaeobotanists working in this period is the low density of plant remains preserved in archaeological sites. This contrasts with the wealth of information we find for later periods such as the Pre-Pottery Neolithic. The reasons for this may lie in the differences in plant processing activities, such as the consumption of raw plants or the roasting of tubers and roots that can be preserved as unidentifiable vascular plant material (Colledge 2001). In addition to preservation biases, parching and cooking of grain crops were routine activities, which in later PPN sites resulted in larger amounts of burnt debris in comparison with pre-agrarian sites (Colledge and Conolly 2010, 128–9). Therefore, in spite of this being one of the areas in the world where archaeobotanists have been working longest, information regarding plant exploitation during this period is still patchy and more materials and well-sampled sites are needed.

The exception to the scarcity of archaeological plant material is Ohalo II, a site located by the shore of Lake Kinneret where excavations have revealed the remains of six brush huts, concentrations of hearths, a grave, a pit, and a dumping zone (Fig. 5.1). The fisher-hunter-gatherer site was occupied *c.* 23,000 years ago, during the LGM, on the shore of a small lake with tamarisk and willow trees growing by the shore and an open woodland further away. Charcoal fragments show that tamarisk (*Tamarix*), willow (*Salix*), and oak (*Quercus ithaburensis*) branches together with smaller pieces of other taxa were used for the construction of the wall and roofs of at least one of the huts (Nadel *et al.* 2004). All hut remains from the camp were burned and large quantities of charred debris accumulated. The site was quickly submerged by water, and fine silts and clay sealed the plant remains quickly after abandonment and prevented movement, allowing for exceptional preservation (for a comprehensive view see different authors in Nadel 2002) which has even permitted studies of intrasite spatial organisation with the plant remains (Weiss *et al.* 2008). The site demonstrates that a wide range of plants were used by these early Epipaleolithic groups. The plant assemblage includes >90,000 plant remains belonging to 142 taxa – at least 30 of them with economic value – from Floor II of Hut 1 alone. The largest component of the assemblage is the grasses, *c.* 19,000 remains (Weiss *et al.* 2004). Small-grained Gramineae include the brome *Bromus pseudobrachystachys/tigridis* – the most abundant with *c.* 11,000 remains –, alkali grass (*Puccinellia* cf. *convoluta*), bladder/creeping foxtail (*Alopecurus utriculatus/arundinaceus*), smooth barley (*Hordeum glaucum*), and seaside/Mediterranean barley (*Hordeum marinum/hystrix*). Wild cereals have also been identified: wild barley (*Hordeum spontaneum* with *c.* 2500 remains) and wild emmer wheat (*Triticum dicoccoides*). Grasses seem to have been the staple plant food at Ohalo since they represent >90% of the edible seeds retrieved, and suggests that this type of plant may have been appreciated by different hunter-gatherers and farming groups (Weiss *et al.* 2004, 9552; 2005). Besides grasses, other plants were also present in Ohalo II: wild pulses, acorns (*Quercus*), wild almonds (*Amygdalus*), wild pistachios (*Pistacia*), wild olives (*Olea*), fruits, and berries such as Christ's thorn

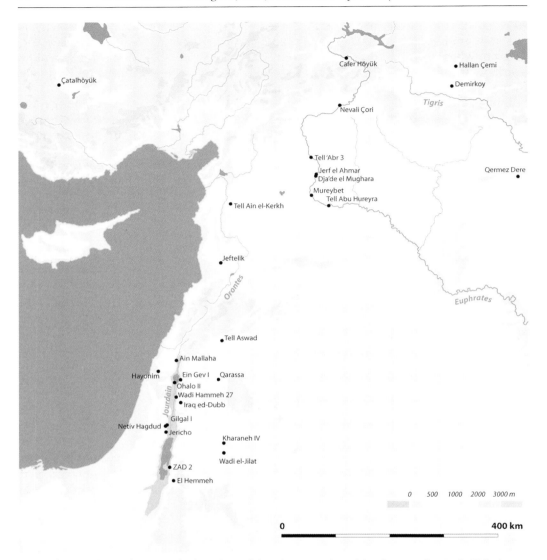

Fig. 5.1. Map showing the location of the sites mentioned in the text (by Luis Teira)

(*Paliurus spina-christi*), raspberry (*Rubus*), wild fig (*Ficus*), wild grape (*Vitis*), plants from the borage family (Boraginaceae), and sunflower family (Compositae), which means that storable high fat content foods were consumed routinely. The storage of *Rubus* berries has also been suggested, which might imply long term planning, and the presence of medicinal plants such as holy thistle (*Silybum marianum*), mallow (*Malva parviflora*), and Indian melilot (*Melilotus indicus*) or *Adonis* spp., has also been suggested, although their use is impossible to prove through archaeobotanical analysis alone (Weiss *et al.*

2008, 2404). The recovery of a grinding stone in Floor II of Hut 1 permitted the analysis of starch granules. This study points to grass seed processing, notably of barley and possibly wheat. Food and medicinal plants cluster around this tool and the oven-like hearth nearby has been interpreted as a possible structure for baking the dough made from wild grain flour (Piperno *et al.* 2004).

Interpretation of Ohalo II strongly suggests that the camp was occupied for most or all of the year taking into account the gathered plant foods, gazelle teeth, and seasonal birds. Bioarchaeological remains indicate a broad-spectrum strategy with an emphasis on gazelle hunting but also evidence for other mammals such as fallow deer, wild pig, red deer, aurochs, and wild goat (Rabinovich 2002), together with the remains of many species of water and field-dwelling birds (Simmons 2002). Fishing was also an extremely important activity; *c.* 10,000 fish remains of Cyprinidae and Cichlidae were recovered in Hut 1 alone (Zohar 2002). On the basis of ethnobotanical analogues, the high quantities of fully mature grass grains that were found in Ohalo II may indicate that small-grained grasses were used as staple foods and that broad spectrum strategies and features that are assumed to be typical of late complex hunter-gatherer groups, were already present several millennia earlier in south-west Asia (Weiss *et al.* 2004).

Ohalo II is extremely important as it demonstrates how much our interpretations regarding bioarchaeological material and in particular plant material are biased by preservation. Other sites from the Early and Middle Epipalaeolithic in south-west Asia do not preserve plant material so well but they are equally important in that they contribute not only to an overall perspective of plant use but also to the taphonomic issues involved. In the north, the only insight into the period – although this includes dates that extend into the Late Epipaleolithic – comes from Öküzini Cave (18,200–11,800 BC) in Anatolia (Martinoli 2004; Martinoli and Jacomet 2004). Here, fruits and nuts such as wild almond (*Amygdalus*) and fragments of fruit flesh from wild pear (*Pyrus*) dominate the samples, together together with wild pistachio (*Pistacia*), wild grape (*Vitis sylvestris*), rose (*Rosa*), and indeterminate amorphous remains that could be fragments of tubers, roots, or bulbs. On the basis of seasonality indices of plants and animals, it is suggested that the site was occupied at least during the late summer, with a wide access to carbohydrate and fat rich plants. In Wadi Jilat 6 (19,500 BC), 19 taxa from *c.* 300 items have been identified although detailed identification of these has not been possible (Colledge 2013, 364). The highest number of seeds corresponds to Chenopodiaceae family that includes perennial shrub species common in steppe-desert environments. Other taxa are *Stipa*, *Atriplex*, and *Verbascum*, and these could have been present nearby while the presence of Cyperaceae seeds indicates marshy environments. The plants could have been used for a wide range of purposes; grass seeds and species of both Chenopodiaceae and Cyperaceae could have been harvested for human consumption. The young leaves of *Atriplex* are edible and the fleshy part and roots of *Verbascum* have medicinal properties. The perennial, woody chenopods are known to have been used for fuel and sedges may have been collected for their fibrous stems (Colledge 2001, 144).

Late Epipaleolithic (12,700–9800 BC)

Around 13,000 BC, the climate improved as the Bølling–Allerød climatic regime became established. In the improved climatic conditions of the Late Epipaleolithic, the Early Natufian has been linked to a more sedentary way of life, as suggested by the presence of dwellings with stone foundations, the occurrence of large sites and the first evidence of possible storage practices (Bar-Yosef 1998; Belfer-Cohen and Bar-Yosef 2000; Valla 2008). Nevertheless, settlement types vary from very small campsites around 15 m² to base camps of more than 1000 m². These base camps, such as Ain Mallaha or Wadi Hammeh 27 (Edwards 2013), are characterised by large semi-subterranean dwellings (7–15 m in diameter), with stone built foundations and probable superstructures made from plant material. Base camps are restricted to the Mediterranean forest belt, from Israel in the south to the central Syrian coast in the north, including the site of Jeftelik (Rodríguez Rodríguez *et al.* 2013; Ibáñez *et al.* 2013).

The wet and warmer climate resulted in optimal growing conditions and hunter-gatherers had access to an exceptionally diverse plant-food resource base that included a range of productive high calorie staples including fruits and nuts of *Quercus, Pistacia,* and Rosaceae and seeds of annual herbaceous taxa (Hillman 2000, 397). Despite the relative number of Early Natufian sites, the knowledge about the use of plant remains during this time is limited by the dearth of archaeobotanical studies. One exception would be Wadi Hammeh 27, located in Jordan, in which several taxa could have been used as sources of food: the grains of grasses (at least wild barley, *Hordeum spontaneum*), medium sized pulses (at least *Lens*), small legumes of the Trifolieae tribe (e.g., *Trifolium, Medicago, Melilotus, Trigonella* spp.), *Pistacia* fruits, and possibly the leaves and seeds of *Malva* (Colledge 2013). Also in the southern Levant, Hayonim Cave is the other Early Natufian site with plant remains; here, wild barley (*Hordeum spontaneoum*), almond (*Amygdalus communis*), pea (cf. *Pisum*), and lupin (*Lupinus pilosus*) were identified (Hopf and Bar-Yosef 1987).

Natufian lithic industries are dominated by microlith segments, which were mounted into projectiles used for hunting (Ibáñez *et al.* 2008). Among lithic tools, glossed flakes and blades are present for the first time and were mostly used for harvesting wild cereals (Anderson 1992). Reaping with sickles, which is a harvesting technique that maximises the yield obtained per unit of surface, was carried out when the ears were not completely ripe, to avoid grain loss (Hillman and Davies 1990). Bedrock mortars, portable mortars, bowls of various types, cupholes, mullers, and pestles, are documented in large quantities in base-camp sites (Bar-Yosef 1998). Eighty-four bedrock mortars have been documented in Qarassa 3 (south Syria), where ten contemporaneous huts are aligned to form an arch on the top of a basaltic hill (Terradas *et al.* 2013; Ibáñez *et al.* 2013). Use-wear analysis carried out on one of them showed that a stone pounding tool was used in the mortar, with a complex pounding and twisting movement, while residue analysis (phytoliths and starch) of another two strongly suggests that they were intended for cereal processing (Terradas *et al.* 2013). Use-wear analysis on grinding tools from Ain Mallaha, Hayonim Cave, and Hayonim Terrace has shown that the Natufians used them for a variety of tasks, including hide working, mineral grinding, and the processing of legumes, and cereals (Dubreil 2004). Interestingly, the Early Natufian shows a higher percentage of mortars and pestles relative to handstones/grinding slabs, while during the Late Natufian more handstones/grinding tools were made; these were most often used for legume and cereal processing (Dubreil 2004). This shift from pounding to grinding tools is a trend that intensifies in the Pre-Pottery Neolithic. The increasing use of these tools could have maximised the calorific returns from a diverse range of plant species (Wright 1993; 1994; Colledge and Conolly 2010, 136).

Somewhat later, *c.* 10,700–9500 BC, the cool and dry conditions of the Younger Dryas (YD) set in (Alley 2000). This somewhat more challenging climate could have led to the need for higher mobility during the Late Natufian (Rosen and Rivera-Collazo 2012), which is manifested in the presence of smaller and often more ephemeral dwellings (Goring-Morris and Belfer-Cohen 2008). Some scholars have considered this renewed stress as a trigger which led to the start of plant cultivation (Hillman *et al.* 2001), while others have suggested that it led to a need to expand dietary breadth with the presence of increased numbers of non-cereal wild plant remains in archaeobotanical samples at this time (Colledge and Conolly 2010). However, some authors claim that the impact of the YD may have been less harsh in some regions (Bottema 1995; Lev-Yadun and Weinstein-Evron 2005; Stein *et al.* 2010; Laggunt *et al.* 2011), which is consistent with the wood charcoal taxa present at several archaeological sites at this time (Noy 1988; Liphschitz and Noy 1991; Baruch and Goring-Morris 1997; Deckers *et al.* 2009).

The three Late Epipaleolithic phases defined in Tell Abu Hureyra remain the best published and most comprehensive archaeobotanical study for this period in the region with *c.* 31,000 remains and 95 different taxa retrieved through extensive flotation (Hillman *et al.* 1989; Hillman 2000). The site was located in a strategic area with water availability at the conjunction of two major resource zones, namely, riverine forest and woodland-steppe; a third, park-woodland, was also close. Hillman describes subzones with wild ryes, wheats and feather-grasses (*Stipa* and *Stipagrostis* spp.), which would support many other plant food resources (Hillman 2000, 349). Archaeobotanical data and their changing proportions have been summarised in 16 ecological groups on the basis of habitat, taxonomy and potential economic value (Hillman 2000, 341–8; Colledge and Conolly 2010, 130). The seed assemblage from Abu Hureyra Phase I may have included more than 120 food types including possible staples such as the grains of wild wheats and ryes (*Triticum* and *Secale* spp.), feather grasses (*Stipa* and *Stipagrostis* spp.), club-rush (*Scirpus maritimus*), Euphrates knotgrass (*Polygonum corrigioloides*), small-seeded grasses, and wild shrub-chenopods (*Atriplex* spp. and others). Ethnoecological modelling (as proposed by Hillman in 1989) allows our knowledge on past availability to be combined with ethnographic models of plant-food preferences amongst present human groups who have access to similar resources and suggests other plant foods that were most probably used as additional staples such as acorns (*Quercus* spp.), almonds (*Amygdalus*), and a range of wild underground storage organs. Altogether, the number of plant-food species consumed in this site could exceed 250 (Hillman 2000, 397).

Also during the Late Natufian from the Euphrates Valley, only 20 km away from Tell Abu Hureyra, Mureybet Phase I has provided a significant assemblage of plant remains (31 taxa). The most frequent edible plants are *Polygonum, Bolboschoenus maritimus*, and *Asparagus*, wild fruits from *Capparis* and *Pistacia atlantica* have also been found. During this period, wild cereals (barley and einkorn) and pulses (lentils and pea) are rare (van Zeist and Bakker-Heeres 1984; Willcox 2008). This was also found at the burnt building from Dederiyeh Cave in northwest Syria (Tanno *et al.* 2013). Preliminary analysis of the remains has yielded over 12,000 charred seeds primarily comprising *Pistacia* and *Amygdalus* nutshells but also feather grass, significant numbers of *Ziziphora* seeds along with some wheat, barley, and pulses such as lentils (*Lens*) and bitter vetch (*Vicia ervilia*). The full publication of the analyses will no doubt contribute to a better understanding of plant exploitation during such a key time period.

Pre-Pottery Neolithic (9800–7000 BC)

Since K. Kenyon's excavation at Jericho (Kenyon 1981), Early Neolithic in the Levant has been divided into the Pre-Pottery Neolithic A (PPNA; 9800–8700 BC) and the Pre-Pottery Neolithic B (PPNB; 8500–7000 BC). The Pre-Pottery Neolithic developed within the onset of the Holocene and the Pre-Boreal around 9600 BC, although the earliest PPNA (the Khiamian period) still coincided with the Younger Dryas (Evin and Stordeur 2008). Proxies such as pollen, isotopes, fluvial, and soil records point to a warm and wet phase, the wettest in the last 25,000 years (among others Rossignol-Strick 1995; 1999; Goodfriend 1999; Gvirtzman and Wieder 2001; McLaren *et al.* 2004; Robinson *et al.* 2006; Roberts *et al.* 2008). During the Pre-Pottery Neolithic, the development of bigger villages with more elaborate pluri-cellular buildings and the first non-domestic monumental and collective structures occurred, showing the shift to increased sedentism and a more complex social organisation (Cauvin 1994; Kuijt 2000; Goring-Morris and Belfer-Cohen 2008; Stordeur 2013).

The first evidence for cultivation of morphologically wild plants, commonly referred to as pre-domestication cultivation (Weiss *et al.* 2006; Fuller 2007; Willcox *et al.* 2008), have been found during the PPNA. Factors such as the presence of weed floras (Colledge 1998; 2001; Willcox 2012a) or the gradual increase in seed size (Colledge 2001; Willcox 2004; Fuller 2007; Willcox *et al.* 2008) have been considered to suggest pre-domestication cultivation of cereals and pulses in a number of sites throughout the Levant (Table 5.1). However, regional differences exist. While barley is primarily exploited in the southern Levant at sites such as ZAD 2 (Edwards *et al.* 2004; Meadows 2004), Gilgal I (Weiss *et al.* 2006), el-Hemmeh (White and Makarewicz 2011), and Iraq ed-Dubb (Colledge 2001), in the northern Levant it is less common and the pre-domestication cultivation of barley has only been claimed at Jerf el Ahmar (Willcox *et al.* 2008; Stordeur and Willcox 2009; Willcox 2012b) and somewhat later, around 8500 BC, at the Iranian sites of Chogha Golan and Chia Sabz (Riehl *et al.* 2012). Instead in the Euphrates area, two-grained einkorn/rye remains predominate and appear to be cultivated at Jerf el Ahmar (Willcox 2004; Stordeur and Willcox 2009) as well as at Mureybet (van Zeist and Bakker-Heeres 1984; Colledge 1998; Willcox 2008) and Tell `Abr (Stordeur and Willcox 2009, 695). Some 'dead-end' cultivars such as oat (*Avena sterilis*) and possibly fig (*Ficus carica*) at Gilgal I (Weiss *et al.* 2006; Kislev *et al.* 2006) and rambling vetch (*Vicia peregrina*) at Netiv Hagdug (Melamed *et al.* 2008) have also been identified. Pulse cultivation seemed more difficult to demonstrate and has been suggested in fewer sites on the basis of seed size, abundance or the presence of sites beyond current distribution of wild relatives (see Table 5.1).

Table 5.1. *Summary of the main Pre-Pottery Neolithic sites with evidence of pre-domestication cultivation and reasons given by authors to support it*

Area	Site	Date ka cal BC	Taxa	Main arguments given	References
	Dhra´	9.7–9.4	barley?	storage evidence	Kuijt and Finlayson 2009
	Gilgal I	9.5–9.1	fig	presence of parthenocarpic fig	Kislev 2006
			barley and oat	abundance of seeds, storage evidence, arable flora	Weiss et al. 2006
	Iraq ed-Dubb	9.7–8.8	barley	arable flora, domestic type seed size	Colledge 2001; see also Fuller 2007
	Netiv Hagdug	9.3–8.8	barley, lentil, rambling vetch	arable flora, abundance	Weiss et al. 2006; Melamed et al. 2008
Southern Levant	ZAD 2	9.2–8.8	barley, pulses?	domestic type seed size	Edwards et al. 2004
	el-Hemmeh	9.1–8.7	barley	domestic type seed size, arable flora	White and Makarewicz 2011
	Jericho I	9.1–8.3	barley and emmer, chickpea	domestic type seed size, location of the site beyond natural pulse habitat	Hopf 1983; see review by Asouti and Fuller 2012; 2013
	Tell ´Abr 3	9.5–9.2	rye/2g einkorn	location of the site beyond natural cereal habitat	Willcox 2008
Northern Levant and southeast	Jerf el Ahmar	9.4–8.7	barley, rye/2g einkorn, lentil	gradual seed size increase, arable flora, reduction in small gathered seeds of non-founder plants, location of the site beyond natural cereal habitat, gradual adoption of founder crops	Willcox 2004; 2012b; Willcox et al. 2008; Stordeour and Willcox 2009; Willcox and Stordeour 2012
Turkey	Mureybet (I–III)	9.7–8.5	rye/2g einkorn, barley	location of the site beyond wild cereal habitat, arable flora	van Zeist and Bakker-Heeres 1984, Colledge 1998
	Dja´de	8.7–8.3	barley, rye/2g einkorn, emmer, chick-pea, faba bean, lentil	domestic type seed size, arable flora, location of the site beyond natural wild cereal and pulse habitat	Willcox 2004; 2012b; Willcox et al. 2008
	Çayönü (RP, GP, ChH)	8.6–8.2	einkorn, emmer, pea?	domestic type seed size, arable flora	Van Zeist and de Roller 1994; see review by Asouti and Fuller 2013
Iran	Chogha Golan	8.7–7.7	barley	adoption of the founder crops, gradual seed size increase	Riehl et al. 2012
	Chia Sabz	8.4–7.7	barley	adoption of the founder crops, gradual seed size increase	Riehl et al. 2012

A trend towards increasing use of cereals and pulses and a lesser use of wild plants can be observed in this period. Weiss *et al.* (2004) showed that, starting in the Upper Palaeolithic, a gradual decrease in the use of small seeded grasses occurred with considerably lower values from the PPNA onwards. This tendency is also confirmed at sites such as Mureybet (Phase III) where wild cereals comprise 74% of the assemblage (van Zeist and Bakker-Heeres 1984; Savard *et al.* 2006) and at Jerf el Ahmar with a marked decrease in *Polygonum/Rumex* during late levels (Willcox *et al.* 2008).

Related to the growing importance of cereals and legumes, the first large-scale grain storage installations are found at this time. At the site of Dhra` in Jordan (Kuijt and Finlayson 2009) a circular shaped installation with raised floors was built to store barley remains. This is not an isolated case, the presence of other contemporaneous cereal storage structures at sites such as Gilgal I, Netiv Hagdug or Jericho (Kuijt and Finlayson 2009) and in other geographical regions such as the northern Levant (Stordeur and Willcox 2009; Willcox and Stordeur 2012), illustrate the emergence of new strategies developed to minimise risk of food shortages by storing food resources. Furthermore, the presence of communal storage structures suggests that food gathering or production would have entailed communal participation (Willcox and Stordeur 2012), although food preparation and consumption could have been based in individual households (Asouti and Fuller 2013).

The intensification in cereal exploitation is not only attested by the presence of pre-domestication cultivation and storage structures. Lithic industries also show increased use of sickle blades, and ground stone tools are an important component of the PPNA tool kit (Wright 1993; Ibáñez *et al.* 2008). Besides, new materials employed in the construction of PPNA buildings emerge at this time, such as pisé (daub), which could constitute approximately one-third cereal wastes (Hopf 1983; Willcox and Fornite 1999; White and Makarewicz 2011). On the other hand, the presence of clean cereal stores in close relationship with symbolic elements such as aurochs skulls or bucrania (Stordeur 2000; Stordeur and Willcox 2009; Willcox and Stordeur 2012) suggests that at least some plant related activities could have been integrated into the new imaginary and ritual practices emerging at this time and undertaken in specific communal structures (Asouti and Fuller 2013).

Despite this evidence, and although most of the research done on the use of plant remains has been centred on the exploitation of cereals and pulses, during the PPNA, the use and eventual cultivation of cereals and pulses was combined with the gathering of other wild plants such as small-seeded legumes, sedges, knotgrass, small-seeded grasses, and nuts and fruits (pistachio, almond, and fig among others) that were also known from previous periods (Colledge 2001; Edwards *et al.* 2004; Willcox *et al.* 2009). The consumption of wild plants is evidenced at Jerf el Ahmar, with the presence of two cakes made of ground *Brassica/Sinapis* seeds on a saddle quern (Willcox 2002). Moreover, in contrast to the long-held view, which maintains the importance of cereals in pre-agrarian sedentary societies, the evidence suggests that cereals and pulses were not major components of the subsistence base in every PPNA site throughout the Levant (cf. Savard *et al.* 2006). For example, sites in the upper Tigris zone such as M'lefaat or Qermez Dere have produced large amounts of small grasses and at Hallan Çemi and Demirkoy wild plants such as *Bolboschoenus maritimus*, *Rumex/Polygonum*, Brassicacea,

Chenopodium, or *Taeniatherum* predominate whilst few barley and einkorn remains have been found (Savard *et al.* 2006). At Körtik Tepe, located in south-east Turkey, seeds of wild relatives account for less than 6% of the plant assemblage (Riehl *et al.* 2012). This diversity in the archaeobotanical record, including sites with a predominance of cereals and pulses and others with evidence for more diverse wild plant exploitation, highlights the complexity of subsistence strategies during the PPNA.

During the Early Pre-Pottery Neolithic B (EPPNB 8700–8200 BC) the diffusion of pluri-cellular rectangular architecture, which originated in the northern Levant during the PPNA, spreads to the southern Levant (among others, Coqueugniot 1998; 1999; 2000; Özdogan 1995; 1999; Ibañez *et al.* 2010). This includes more elaborate household installations in which new materials such as mud bricks were used (Özdogan 1995; 1999) as well as the construction of specific structures destined for death (Coqueugniot 1998; Özdogan 1999).

These changes coincided with the appearance of the first morphologically domestic cereals (Tanno and Willcox 2006b; 2012) and the first evidence for animal domestication (Peters *et al.* 2005). Plant domestication is mainly identified by the elimination or reduction of natural seed dispersal (brittle rachis) (Zohary and Hopf 2000; Nesbitt 2002; Fuller 2007). Harvesting with sickles may have played a major role in cereal domestication, as it unwittingly provoked selection of mutant individuals among populations of wild cereals which, through successive sowing of the selected seeds and the avoidance of interbreeding with wild cereals, would have led to domestication (Hillman and Davies 1990). Nevertheless, evidence for cereal domestication is limited to a few sites such as Tell Aswad and possibly Nevali Çori, whilst at contemporary Dja'de wild-type wheat and barley rachis predominate. The material from Çayönü would benefit from re-examination and quantification using new criteria (Tanno and Willcox 2006b; 2012). Therefore, it is likely that plant domestication was a protracted process since domesticated rachis emerge gradually over time (Fuller 2007; Purugganan and Fuller 2011). However, in order to assess whether each domesticated species originated from a single 'core area' *sensu* Lev-Yadun *et al.* (2000), more archaeobotanical data from new EPPNB sites along with genetic studies are needed.

Apart from cereals, there is extensive evidence for the exploitation and consumption of pulses and their importance in the diet. For example at Dja'de, new cultivars such as faba bean (*Vicia faba*) and chickpea (*Cicer arietinum*) are found for the first time far from their natural habitat (Willcox *et al.* 2008). At Nevali Çori, the percentages of pulses (Pasternak 1998) and the palaeodietary stable isotopes confirm the importance of this type of plant foods in the diet of the inhabitants (Lösch *et al.* 2006). At Tell Ain el-Kerkh pulses such as faba bean (*Vicia faba*) and chickpea (*Cicer arietinum*) are dominant (Tanno and Willcox 2006a) whilst lentil domestication has been claimed at Tell Aswad on the basis of the size increase, which occurs during the period of occupation of the site (Fuller *et al.* 2012).

Nevertheless, it is clear that wild plants other than cereals and pulses would have continued to constitute an important part of the subsistence strategies. For example, at Nevali Çori the proportion of seeds from Brassicaceae and willow-leaved sunrose (*Helianthemum salicifolium*) was larger than the proportion of einkorn grains (Pasternak 1998), whilst at Dja'de, the number of seeds from wild grasses such as *Hordeum murinum/bulbosum* and *Taeniatherum* was superior to those from cereals such as emmer and one-grained einkorn (Willcox *et al.* 2008). At Tell Qarassa North in southern Syria, concentrations of *Tolpis virgata*, which is nowadays a well-known edible green in regions such as Italy and Greece, were found in refuse and pit contexts (Arranz-Otaegui unpublished data). Therefore, the continuing exploitation of wild plants and fruits along with cereals and pulses that is observed throughout the PPNA, is still evident during the EPPNB.

The appearance of new cereal species such as the free threshing wheat (*T. aestivum/durum*) during the Middle Pre-Pottery Neolithic B (MPPNB, 8200–7500 BC), together with the increased reliance on domesticated crops and animals undoubtedly changed the economic bases of the Neolithic communities, and eventually led to societies which were based predominantly on farming. At the same time, a suite of agricultural tools appeared, including curved sickles (Ibáñez *et al.* 1998) and tools for intensive hand tillage, such as the cattle bone hoes at Beidha and Basta (Stordeur 1999) and the ground stone hoes found at Tell Halula (Ibáñez *et al.* 2007, 159), Çayönü (Davis 1982, 108) and sites in the Zagros (Hole *et al.* 1969, 189–92). However, in some areas these improvements took a long time to become established. While in the northern Levant and Anatolia, most assemblages normally contain more than six potential cultivars (i.e., Cafer Höyük: de Moulins 1997) in the southern Levant, the archaeobotanical record does not display a homogeneous 'package' of potential crops (Kislev 1985; Kislev and Bar-Yosef 1988; Nesbitt 2002; Asouti and Fuller 2012) and at some sites intensive exploitation of wild cereals continued at this time (Colledge 2001). These differences between regions continued until the end of the LPPNB period (*c.* 7000 BC), at which point well established crop packages became dominant throughout the region (Asouti and Fuller 2012). Nevertheless, it is important to remember that although crops gained importance over time, wild plant foods continued to be exploited; this can be seen at the well-established farming community of Çatalhöyük where stores of Cruciferae seeds were found; these may have been used for oil production (Fairbairn *et al.* 2007).

Conclusions

Overall, archaeobotanical evidence suggests an extended process of gathering a wide range of wild plants around at least from 21,000 years BC, as seen at Ohalo II. This is followed by the first evidence of cultivation at around 9400 years BC, during the PPNA period; the first signs of morphologically domesticated cereals around 8500 BC in the Early PPNB, and the widespread presence of domesticates around 7500 BC in the MPPNB, although wild plant foods continued to be a component of the diet.

During the Epipalaeolithic, plant assemblages from archaeological sites in the Levant are not rich by comparison to Pre-Pottery Neolithic and agrarian sites. In addition to a low number of archaeological sites with published archaeobotanical reports, analyses are also conditioned by poor preservation and low taxonomic diversity. Very important exceptions are Ohalo II and Tell Abu Hureyra I, which evidence an intensity of exploitation of wild plants during the Epipalaeolithic. In Ohalo II grasses, including small-grained grasses – the most abundant – and wild cereals (*Hordeum spontaneum* and *Triticum dicoccoides*) seem be the staple plant food and the seed assemblage from Abu Hureyra Phase I, Late Epipalaeolithic, may have included more than 120 food types including possible staples such as the grains of wild rye (*Secale*) and wheat (*Triticum* spp.), feather grasses (*Stipa* and *Stipagrostis* spp.), club-rush (*Scirpus maritimus*), Euphrates knotgrass (*Polygonum corrigioloides*), small-seeded grasses, and wild shrubby chenopods (*Atriplex* spp. and others). However, the early dates suggested for the broad spectrum strategies of plant exploitation need to be viewed with caution due to the large gaps that still exist in the archaeobotanical record for the very long period 21,000–12,000 BC.

The start of food production activities, which may have begun with the cultivation of morphologically wild plants during the PPNA and domesticates from EPPNB onwards, entailed a considerable change in the subsistence strategies of the time. Besides, the presence of storage structures and specialised lithic tools suggest an increasing focus on a small number of domesticated species in place of a wide range of wild plants found in earlier periods. In this way, PPNA and Early PPNB sites can be regarded as transitional between earlier Epipalaeolithic complex hunter-gatherers and later Middle to Late PPNB agricultural villages.

During the Middle Pre-Pottery Neolithic B, agricultural practices became firmly established, with the exploitation of new cereal species such as free threshing wheat (*T. aestivum/durum*), the increased reliance on domesticated crops and the appearance of soil tilling tools. However, it is not until the Late PPNB that well established crop packages became dominant among plant resources across the Levant, at the beginning of the widespread expansion of agriculture to Eurasia (Harris 1996). In fact, to see this process of the advent of agriculture in the southwest Asia in a uni-directional perspective would be simplistic. Halfway points between exclusively plant gatherers and 'true' farmers would have existed during the entire Pre-Pottery Neolithic. This can be seen in the complexity of the archaeobotanical evidence which show different strategies, some focused predominantly on cereals and pulses and others on a wider range of plant foods such as small-seeded grasses, knotgrass or sea club-rush. Nevertheless, wild plant foods would probably have continued to be a significant part of subsistence strategies, as can be seen at sites such as Çatalhöyük, well beyond the establishment of agricultural practice.

References

Allaby, R. G., Fuller, D. Q. and Brown, T. A. 2008. The genetic expectations of a protracted model for the origins of domesticated crops. *Proceedings of the National Academy of Science (USA)* 105, 13982–6

Alley, R. B. 2000. The Younger Dryas cold interval as viewed from central Greenland. *Quaternary Science Reviews* 19(1–5), 213–26

Anderson, P. C. 1992. Experimental cultivation, harvest and threshing of wild cereals and their relevance for interpreting the use of Epipaleolithic and Neolithic artifacts. In Anderson, P. C. (ed.), *Préhistoire de l'Agriculture: Nouvelles Approches Expérimentales et Ethnographiques*, 179–209. Paris: CRNS Monographie du CRA 6.

Asouti, E. and Fuller, D. Q. 2012. From foraging to farming in the southern Levant: the development of Epipalaeolithic and Pre-Pottery Neolithic plant management strategies. *Vegetation History & Archaeobotany* 21, 149–62

Asouti, E. and Fuller, D. Q. 2013. A contextual approach to the emergence of agriculture in southwest Asia: reconstructing Early Neolithic plant-food production. *Current Anthropology* 54(3), 299–345

Baruch, U. and Goring-Morris, N. 1997. The arboreal vegetation of the Central Negev Highlands, Israel, at the end of the Pleistocene: evidence from archaeological charred wood remains. *Vegetation History & Archaeobotany* 6, 249–59

Bar-Yosef, O. 1998. The Natufian Culture in the Levant: threshold to the origins of agriculture. *Evolutionary Anthropology* 6, 159–77

Bar-Yosef, O. 2011. The origins of agriculture: new data, new ideas. *Current Anthropology* 52(S4), S175–93

Belfer-Cohen, A. and Bar-Yosef, O. 2000. Early sedentism in the Near East – a bumpy ride to village life. In Kuijt (ed.) 2000, 19–37

Bottema, S. 1995. The Younger Dryas in the eastern Mediterranean. *Quaternary Science Reviews* 14(9), 883–91

Cauvin, J. 1994. *Naissance des Divinités, Naissance de l'Agriculture: La Révolution des Symboles au Néolithique* (2 edition augmentée et corrigée parue en 1997). Paris: CNRS Éditions

Coqueugniot, É. 1998. Dja'de El Mughara (moyen Euphrate): un village Néolithique dans son environnement naturel à la veille de la domestication. In Fortin, M. and Aurenche, O. (eds), *Espace Naturel, Espace Habité en Syrie du Nord (10ᵉ–2ᵉ millénaires av. J.-C.)*, 109–14. Lyons: Travaux de la Maison de l'Orient 28

Coqueugniot, É. 1999. Tell Dja'de El Mughara. In Del Olmo Lette, G. and Montero Fenollós, J. L. (eds), *Archaeology of the Upper Syrian Euphrates: the Tishrin Dam area*, 41–55. Barcelona: Editorial Ausa

Coqueugniot, É. 2000. Dja'de (Syrie): un village Néolithique à la veille de la domestication (second moitié du 9ᵉ millénaire av. J.-C.). In Guilaine, J. (ed.), *Premiers Paysans du Monde: Naissances des Agricultures*, 63–79. Paris: Errance

Colledge, S. 1998. Identifying pre-domestication cultivation using multivariate analysis. In Damania, A. B., Valkoun, J., Willcox, G. and Qualset, C. O. (eds), *The Origins of Agriculture and Crop Domestication*, 121–31. Aleppo: ICARDA

Colledge, S. 2001. *Plant Exploitation on Epipalaeolithic and Early Neolithic Sites in the Levant*. Oxford: British Archaeological Report S986

Colledge, S. 2013. Plant remains and archaeobotanical analysis. In Edwards, P. C. (ed.), *Wadi Hammeh 27, an Early Natufian Settlement at Pella in Jordan*, 353–66. Leiden: Brill

Colledge, S. and Conolly, J. 2010. Reassessing the evidence for the cultivation of wild crops during the Younger Dryas at Tell Abu Hureyra, Syria. *Environmental Archaeology* 15(2), 124–38

Davis, M. K. 1982. The Çayönü ground stone. In Braidwood, L. and Braidwood, R. J. (eds), *Prehistoric Village Archaeology in South-Eastern Turkey*, 73–174. Oxford: British Archaeological Report S138

Deckers, K., Riehl, S., Jenkins, E., Rosen, A., Dodonov, A., Simakova, A. N. and Conard, N. J. 2009. Vegetation development and human occupation in the Damascus region of southwestern Syria from the Late Pleistocene to Holocene. *Vegetation History & Archaeobotany* 18, 329–40

de Moulins, D. 1997. *Agricultural Changes at Euphrates and Steppe Sites in the Mid-8th to the 6th Millennium B.C.* Oxford: British Archaeological Report S683

Dubreuil, L. 2004. Long-term trends in Natufian subsistence: a use-wear analysis of ground stone tools. *Journal of Archaeological Science* 31(11), 1613–29

Edwards, P. (ed.) 2013. *Wadi Hammeh 27: an Early Natufian settlement at Pella in Jordan*. Leiden: Culture and History of the Ancient Near East 59

Edwards, P. C., Meadows, J., Sayej, G. and Westaway, M. 2004. From the PPNA to the PPNB: new views from the southern Levant after excavations at Zahrat adh-Dhra 2 in Jordan. *Paléorient* 30(2), 21–60

Evin, J. and Stordeur, D. 2008. Chronostratigraphie de Mureybet. Apport des datations radiocarbone. In Ibáñez, J. J. (ed.), *Le Site Néolithique de Tell Mureybet (Syrie du Nord)* I, 24–30. Oxford: British Archaeological Report S1843

Fairbairn, A., Martinoli, D., Butler, A. and Hillman, G. 2007. Wild plant seed storage at Neolithic Çatalhöyük East, Turkey. *Vegetation History & Archaeobotany* 16, 467–79

Felis, T., Lohmann, G., Kuhnert, H., Lorenz, A. J., Scholz, D., Pätzold, J., Al-Rousan, S. A. and Al-Moghrabi, S. M. 2004. Increased seasonality in Middle East temperatures during the last interglacial period. *Nature* 429, 164–8

Fuller, D. Q. 2007. Contrasting patterns in crop domestication and domestication rates: recent archaeobotanical insights from the Old World. *Annals of Botany* 100(5), 903–24

Fuller, D. Q., Allaby, R. G. and Stevens, C. 2010. Domestication as innovation: the entanglement of techniques, technology and chance in the domestication of cereal crops. *World Archaeology* 42(1), 13–28

Fuller, D. Q., Asouti, E. and Purugganan, M. D. 2012. Cultivation as slow evolutionary entanglement: comparative data on rate and sequence of domestication. *Vegetation History & Archaeobotany* 21, 131–45

Fuller, D. Q., Willcox, G. and Allaby, R. G. 2011. Cultivation and domestication had multiple origins: arguments against the core area hypothesis for the origins of agriculture in the Near East. *World Archaeology* 43(4), 628–52

Goodfriend, G. A. 1999. Terrestrial stable isotope records of Late Quaternary paleoclimates in the Eastern Mediterranean region. *Quaternary Science Reviews* 18, 501–13

Goring-Morris, A. N. and Belfer-Cohen, A. 2008. A roof over one's head: developments in Near Eastern residential architecture across the Epipalaeolithic–Neolithic transition. In Bocquet-Appel, J. P. and Bar-Yosef, O. (eds), *The Neolithic Demographic Transition and its Consequences*, 239–286. New York: Springer

Goring-Morris, N., Hovers, E. and Belfer-Cohen, A. 2009. The dynamics of Pleistocene and Early Holocene settlement patterns and human adaptations in the Levant: an overview. In Shea, J. J. and Lieberman, D. E. (eds), *Transitions in Prehistory. Essays in Honor of Ofer Bar-Yosef*, 185–252. Oxford: Oxbow Books/American School of Prehistoric Research

Gvirtzman, G. C. and Wieder, M. 2001. Climate of the last 53,000 years in the eastern Mediterranean, based on soil sequence stratigraphy in the coastal plain of Israel. *Quaternary Science Reviews* 20(18), 1827–49

Harris, D. R. (ed.) 1996. *The Origins and Spread of Agriculture and Pastoralism in Eurasia*. London: University College London Press

Heun, M., Abbo, S., Lev-Yadun, S. and Gopher, A. 2012. A critical review of the protracted domestication model for Near-Eastern founder crops: linear regression, long-distance gene flow, archaeological and archaeobotanical evidence. *Journal of Experimental Botany* 63, 4333–41

Hillman, G. C. 1989. Late Palaeolithic plant foods from Wadi Kubbaniya in Upper Egypt: dietary diversity, infant weaning, and seasonality in a riverine environment. In Harris, D. R. and Hillman, G. C. (eds), *Foraging and Farming. The Evolution of Plant Exploitation*, 207–39. London: Unwin Hyman

Hillman, G. C. 2000. Abu Hureyra 1: the Epipaleolithic. In Moore, A. M. T., Hillman, G. C. and Legge, A. J. (eds), *Village on the Euphrates*, 327–99. Oxford: Oxford University Press

Hillman, G. C. and Davies, M. S. 1990. Measured domestication rates in wild wheats and barley under primitive cultivation, and their archaeological implications. *Journal of World Prehistory* 4, 157–222

Hillman, G. C., Colledge, S. M. and Harris, D. R. 1989. Plant food economy during the Epipaleolithic Period at Tell Abu Hureyra, Syria: dietary diversity, seasonality and modes of exploitation. In Harris, D. R. and Hillman, G. C. (eds), *Foraging and Farming. The Evolution of Plant Exploitation*, 240–68. London: Unwin Hyman

Hillman, G. C., Hedges, R., Moore, A., Colledge, S. and Pettitt, P. 2001. New evidence of lateglacial cereal cultivation at Abu Hureyra on the Euphrates. *Holocene* 11(4), 383–93

Hole, F., Flannery, K. and Neely, J., 1969. *Prehistory and Human Ecology of the Deh Luran Plain: an early village sequence from Khuzistan, Iran*. Ann Arbor MI: University of Michigan Museum of Anthropology Memoir 1

Hopf, M. 1983. Appendix B: Jericho plant remains. In Kenyon, K. M. and Holland, T. A. (eds), *Excavations at Jericho 5. The Pottery Phases of the Tell and Other Finds*, 576–621. London: British School of Archaeology in Jerusalem

Hopf, M. and Bar-Yosef, O. 1987. Plant remains from Hayonim Cave, western Galilee. *Paléorient* 13(1), 117–20

Ibáñez, J. J., González Urquijo, J. E., Palomo, A. and Ferrer, A. 1998. Pre-Pottery Neolithic A and Pre-Pottery Neolithic B lithic agricultural tools on the Middle Euphrates: the sites of Mureybet and Tell Halula. In Damania, A. B., Valkoun, J., Willcox, G. and Qualset, C. O. (eds), *The Origins of Agriculture and Crop Domestication*, 132–44. Aleppo: International Center for Agricultural Research in Dry Areas

Ibáñez, J. J., González-Urquijo, J. E. and Rodríguez, A. 2007. The evolution of technology during the PPN in the middle Euphrates: a view from use-wear analysis of lithic tools. In Astruc, L., Binder, D. and Briois, F. (eds), *Systèmes Techniques et Communautés du Néolithique Précéramique*, 153–6. Antibes: Éditions APDCA

Ibáñez, J. J., González-Urquijo, J. E. and Rodríguez, A. 2008. Analyse fonctionnelle de l'outillage lithique de Mureybet. In Ibáñez, J. J. (ed.), *Le Site Néolithique de Tell Mureybet (Syrie du Nord). En Hommage a Jacques Cauvin*, 363–406. Oxford: British Archaeological Report S1843

Ibáñez, J. J., Balbo. A., Braemer, F., Gourichon, L., Iriarte, E., Santana, J. and Zapata, L. 2010. The Early PPNB levels of Tell Qarassa North (Sweida, southern Syria). *Antiquity Project Gallery* 84, 325

Ibáñez, J. J., Terradas, X., Armendáriz, A., González-Urquijo, J. E., Teira, L., Braemer, F., Gourichon, L., Haïdar-Boustani, M. and Rodríguez-Rodríguez, A. 2013. Nouvelles données sur les architectures des sites Natoufiens de Jeftelik et Qarassa 3 (Syrie centro-occidentalle et du sud). In Montero Fenellós, J. L. (ed.), *Du village Néolithique a la ville Syro-Mésopotamienne*, 9–33. Ferrol: Bibliotheca Euphratica 1

Kenyon, K. 1981. *Excavations at Jericho. III: The Architecture and Stratigraphy of the Tell*. London: British School of Archaeology in Jerusalem

Kilian, B., Özkan, H., Walther, A., Kohl, J., Dagan, T., Salamini, F. and Martin, W. 2007. Molecular diversity at 18 loci in 321 wild and 92 domesticate lines reveal no reduction of nucleotide diversity during *Triticum monococcum* (einkorn) domestication: implications for the origin of agriculture. *Molecular Biology & Evolution* 24, 2657–68

Kilian, B., Martin, W. and Salamini, F. 2010. Genetic diversity, evolution and domestication of wheat and barley in the Fertile Crescent. In Glaubrecht, M. (ed.), *Evolution in Action*, 137–66. Heidelberg: Springer

Kislev, M. E. 1985. Early Neolithic Horsebean from Yiftah'el, Israel. *Science* 228, 319–20

Kislev, M. E. and Bar-Yosef, O. 1988. The legumes: the earliest domesticated plants in the Near East? *Current Anthropology* 29, 175–9

Kislev, M. E., Hartmann, A. and Bar-Yosef, O. 2006. Early domesticated fig in the Jordan Valley. *Science* 312, 1372–4

Kuijt, I. (ed.) 2000. *Life in Neolithic Farming Communities. Social Organization, Identity and Differentiation*. New York: Kluwer Academic/Plenum

Kuijt, I. and Finlayson, B. 2009. Evidence for food storage and predomestication granaries 11,000 years ago in the Jordan valley. *Proceedings of the National Academy of Sciences (USA)* 106(27), 10966–70

Laggunt, D., Almogi-Labin A., Bar-Matthews, M. and Weistein-Evron, M. 2011. Vegetation and climate changes in the south eastern Mediterranean during the Last Glacial–Interglacial cycle (86 ka): new marine pollen record. *Quaternary Science Reviews* 30, 3960–72

Lev-Yadun, S. and Weinstein-Evron, M. 2005. Modeling the influence of wood use by the Natufians of el-Wad on the forest of Mount Carmel. *Journal of the Israel Prehistoric Society* 35, 285–98

Lev-Yadun, S., Gopher, A. and Abbo, S. 2000. The cradle of Agriculture. *Science* 288, 1602–3

Liphschitz, N. and Noy, T. 1991. Vegetational landscape and the macroclimate of the Gilgal Region during the Natufian and Pre-Pottery Neolithic A. *Journal of the Israel Prehistoric Society* 24, 59–63

Lösch, S., Grupe, G. and Peters, J. 2006. Stable isotopes and dietary adaptations in human and animals at Pre-Pottery Neolithic Nevalı Çori, SE-Anatolia. *American Journal of Physical Anthropology* 131, 181–93

Maher, L., Banning, E. B. and Chazan, M. 2011. Oasis or mirage? Assessing the role of abrupt climate change in the prehistory of the southern Levant. *Cambridge Archaeological Journal* 21(1), 1–29

Maher, L., Richter, T., Macdonald, D., Jones, M., Martin, L. and Stock, J. T. 2012. Twenty thousand-year-old huts at a hunter-gatherer settlement in eastern Jordan. *PLOS-ONE* 7(2): e31447. doi:10.1371/journal.pone.0031447

Martinoli, D. 2004. Food plant use, temporal changes and site seasonality at Epipalaeolithic Öküzini and Karain B Caves, southwest Anatolia, Turkey. *Paléorient* 30(2), 61–80

Martinoli, D. and Jacomet, S. 2004. Identifying endocarp remains and exploring their use at Epipalaeolithic Öküzini in southwest Anatolia, Turkey. *Vegetation History & Archaeobotany* 13, 45–54

McLaren, S. J., Gilbertson, D. D., Grattan, J. P., Hunt, C. O., Duller, G. A. T. and Barker, G. A. 2004. Quaternary palaeogeomorphologic evolution of the Wadi Faynan area, southern Jordan. *Palaeogeography, Palaeoclimatology, Palaeoecology* 205, 131–54

Meadows, J. 2004. *The Earliest Farmers? Archaeobotanical Research at Pre-Pottery Neolithic Sites in Jordan*, 321–30. amman: *Studies in the History and Archaeology of Jordan* 8

Melamed, Y., Plitzmann, U. and Kislev, M. E. 2008. *Vicia peregrina*: an edible early Neolithic legume. *Vegetation History & Archaeobotany* 17 (suppl. 1), S29–34

Nadel, D. 2002. Ohalo II. *A 23,000-year-old fisher-hunter-gatherers' camp on the shore of the Sea of Galilee.* Haifa: Reuben and Edith Hecth Museum, University of Haifa

Nadel, D., Weiss, E., Simchoni, O., Tsatskin, A., Danin, A. and Kislev, M. 2004. Stone Age hut in Israel yields world's oldest evidence of bedding. *Proceedings of the National Academy of Sciences (USA)* 101(17), 6821–6

Nesbitt, M. 2002. When and where did domesticated cereals first occur in Southwest Asia. In Cappers, R. T. J. and Bottema, S. (eds), *The Dawn of Farming in the Near East*, 113–32. Berlin: Studies in Early Near Eastern Production, Subsistence, and Environment 6

Noy, T. (with contribution by M. Kislev) 1988. Gilgal I, An Early Village in the Jordan Valley. *Israel Museum Journal* 2, 113–14

Özdoğan, A. 1995. Life at Çayönü during the Pre-Pottery Neolithic period (according to the artefactual assemblage). In Faculty of Letters, University of Istanbul Section of Prehistory (ed.), *Readings in Prehistory – Studies presented to Halet* Çambel, 70–100. Istanbul: Graphis

Özdoğan, A. 1999. Çayönü. In Özdoğan, M. and Başgelen, N. (eds), *Neolithic in Turkey*, 35–63. İstanbul: Arkeoloji ve Sanat Yayınları

Pasternak, R. 1998. Investigation of botanical remains from Nevalı Çori, PPNB, Turkey: a short interim report. In Damania, A. B., Valkoun, J., Willcox, G. and Qualset, C. O. (eds), *The Origins of Agriculture and Crop Domestication*, 170–7. Aleppo: International Center for Agricultural Research in Dry Areas

Peters, J., von den Driesch, A. and Helmer, D. 2005. The Upper Euphrates-Tigris basin, cradle of agropastoralism? In Vigne, J. D., Peters, J. and Helmer, D. (eds), *The First Steps of Animal Domestication*, 96–124. Oxford: Oxbow Books

Piperno, D. R., Weiss, E., Holst I. and Nadel, D. 2004. Processing of wild cereal grains in the Upper Palaeolithic revealed by starch grain analysis. *Nature* 430, 670–3

Purugganan, M. D. and Fuller, D. Q. 2011. Archaeological data reveal slow rates of evolution during plant domestication. *Evolution* 65, 171–83

Rabinovich, R. 2002. The mammal bones: environment, food and tools. In Nadel, N. (ed.) *Ohalo II. A 23,000-Year-Old Fisher-Hunter-Gatherers' Camp on the Shore of the Sea of Galilee*. Haifa: Reuben and Edith Hecth Museum, University of Haifa, pp. 24–27.

Riehl, S., Benz, M., Conard, N., Deckers, K., Fazeli, H., Zeidi, M. 2012. The modalities of plant use in three PPNA sites of the northern and eastern Fertile Crescent - A preliminary report. *Vegetation History and Archaeobotany* 21, 95–106.

Roberts, N., Jones, M. D., Benkaddour, A., Eastwood, W. J., Filippi, M. L., Frogley, M. R., Lamb, H. F., Leng, M. J., Reed, J. M., Stein, M., Stevens, L., Valero-Garces, B., Zanchetta, G. 2008. Stable isotope records of Late Quaternary climate and hydrology from Mediterranean lakes: the ISOMED synthesis. *Quaternary Science Reviews* 27, 2426–2441.

Robinson, S. A., Black, S., Sellwood, B. W. and Valdes, P. J., 2006. A review of palaeoclimates and palaeoenvironments in the Levant and eastern Mediterranean from 25,000 to 5000 years BP: setting the environmental background for the evolution of human civilization. *Quaternary Science Reviews* 25, 1517–1541.

Rodríguez Rodríguez, A., Haïdar-Boustani, M., González Urquijo, J. E., Ibáñez, J. J., Al-Maqdissi, M., Terradas, X. and Zapata, L. (2013). The Early Natufian site of Jeftelik (Homs Gap, Syria). In Bar-Yosef, O.,Valla, F. (eds) *Natufian Foragers in the Levant*. International Monographs in Prehistory, Ann Arbor, Michigan, pp. 61–72.

Rosen, A. and Rivera-Collazo, I., 2012. Climate change, adaptive cycles, and the persistence of foraging economies during the late Pleistocene/Holocene transition in the Levant. *Proceedings of the National Academy of Sciences (USA)* 109(10), 3640–3645.

Rossignol-Strick, M., 1995. Sealand correlation of pollen records in the eastern Mediterranean for the glacial-interglacial transition: biostratigraphy versus radiometric time-scale. *Quaternary Science Reviews* 14, 893–915.

Rossignol-Strick, M., 1999. The Holocene climatic optimum and pollen records of Sapropel 1 in the eastern Mediterranean, 9000–6000 BP. *Quaternary Science Reviews* 18, 515–530.

Savard, M., Nesbitt, M. and Jones, M. K., 2006. The role of wild grasses in subsistence and sedentism: new evidence from the northern Fertile Crescent. *World Archaeology* 38(2), 179–196.

Simmons, A. H. 2007. *The Neolithic Revolution in the Near East: Transforming the Human Landscape*. The University of Arizona Press, Tucson, Arizona.

Simmons, T., 2002. The Birds from Ohalo II. In Nadel (ed.) 2002, 32–6

Stein, M., Torfstein, A., Gavrieli, I. and Yechieli, Y. 2010. Abrupt aridities and salt deposition in the post-glacial Dead Sea and their North Atlantic connection. *Quaternary Science Reviews* 29, 567–75

Stordeur, D. 1999. Néolithisation et outillage osseux: la révolution a-t-elle eu lieu? In Julien, M., Averbouh, A., Ramseyer, D., Bellier, C., Buisson, D., Cattelain, P., Patou-Mathis, M. and Provenzano, N. (eds), *Préhistoire d'Os: Recueil d'Etudes sur l'Industrie Osseuse Préhistorique*, 261–72. Aix-en-Provence: Publications de l'Université de Provence

Stordeur, D. 2000. Jerf el Ahmar et l'émergence du Néolithique au Proche Orient. In Guilaine, J. (ed.), *Premiers Paysans du Monde: Naissance des Agricultures*, 33–60. Paris: Errance

Stordeur, D. 2013. Les villages et l'organisation des groupes au Néolithique précéramique A. L'exemple de Jerf el-Ahmar, Syrie du Nord. In Montero Fenellós, J-L. (ed.), *Du village Néolithique à la Ville Syro-Mésopotamienne*, 35–54. Ferrol: Bibliotheca Euphratica 1

Stordeur, D. and Willcox, G. 2009. Indices de culture et d'utilisation des céréales à Jerf el Ahmar. In Collectif (ed.), *De Mediterranée et d'ailleurs ... Mélanges offerts à Jean Guilaine*, 693–710. Toulouse: Archives d'Ecologie Préhistorique

Tanno, K. I. and Willcox, G. 2006a. The origins of cultivation of *Cicer arietinum* L. and *Vicia faba* L.: early finds from north west Syria (Tell el-Kerkh, late 10th millennium BP). *Vegetation History & Archaeobotany* 15, 197–204

Tanno, K. I. and Willcox. G. 2006b. How fast was wild wheat domesticated? *Science* 311, 1886

Tanno, K. I. and Willcox, G. 2012. Distinguishing wild and domestic wheat and barley spikelets from early Holocene sites in the Near East. *Vegetation History & Archaeobotany* 21, 107–15

Tanno, K. I., Willcox, G., Muhesen, S., Nishiaki, Y., Kanjo, Y. and Akazawa, T. 2013. Preliminary results from analyses of charred plant remains from a burnt Natufian building at Dederiyeh Cave in northwest Syria. In Bar-Yosef, O. and Valla, F. (eds), *Natufian Foragers in the Levant*, 83–7. Ann Arbor MI: International Monographs in Prehistory

Terradas, X., Ibáñez, J. J., Braemer, F., Gourichon L. and Teira, L. C. 2013. The Natufian occupations of Qarassa 3, Sweida, Southern Syria. In Bar-Yosef, O. and Valla, F. (eds), *Natufian Foragers in the Levant*, 45–60. Ann Arbor MI: International Monographs in Prehistory

Terradas, X., Ibáñez, J. J, Braemer, F., Hardy, K., Iriarte, E., Madella, M., Ortega, D., Radini, A. and Teira, L. C. 2013. Natufian bedrock mortars at Qarassa 3: preliminary results from an interdisciplinary methodology. In Borrell, F., Ibáñez, J. J. and Molist, M. (eds), *Stone Tools in Transition: from hunter-gatherers to farming societies in the Near East*, 441–56. Barcelona: Universitat Autònoma de Barcelona

Valla, F. 2008. *L'Homme et l'Habitat. L'Invention de la Maison durant la Préhistoire.* Paris: CNRS Éditions

van Zeist, W. A. and Bakker-Heeres, J. H. 1984. Archaeobotanical Studies in the Levant, 3. Late-Paleolithic Mureybit. *Palaeohistoria* 26, 171–99

van Zeist, W. and de Roller, G. J. 1994. The plant husbandry of aceramic Çayönü, SE Turkey. *Palaeohistoria* 33/4, 65–96

Weiss, E., Wetterstrom, W., Nadel, D. and Bar-Yosef, O. 2004. The broad spectrum revisited: Evidence from plant remains. *Proceedings of the National Academy of Sciences (USA)* 101(26), 9551–5

Weiss, E., Kislev, M. E., Simchoni, O. and Nadel, D. 2005. Small-grained wild grasses as staple food at the 23,000 year old site of Ohalo II, Israel. *Economic Botany* 58, 125–34

Weiss, E., Kislev, M. E. and Hartmann A. 2006. Autonomous cultivation before domestication. *Science* 312, 1608–10

Weiss, E., Kislev, M. E., Simchoni, O., Nadel, D. and Tschauner, H. 2008. Plant food preparation area on an Upper Paleolithic brush hut floor at Ohalo II, Israel. *Journal of Archaeological Science* 35, 2400–14

White, C. E. and Makarewicz, C. A. 2011. Harvesting practices and early Neolithic barley cultivation at el-Hemmeh, Jordan. *Vegetation History & Archaeobotany* 21, 85–94

Willcox, G. 2002. Charred plant remains from a 10th millennium B.P. kitchen at Jerf el Ahmar (Syria). *Vegetation History & Archaeobotany* 11, 55–60

Willcox, G. 2004. Measuring grain size and identifying Near Eastern cereal domestication: evidence from the Euphrates valley. *Journal of Archaeological Science* 31, 145–50

Willcox, G. 2005. The distribution, natural habitats and availability of wild cereals in relation to their domestication in the Near East: multiple events, multiple centres. *Vegetation History & Archaeobotany* 14, 534–41

Willcox, G. 2008. Les nouvelles données archéobotaniques de Mureybet et la Néolithisation du Moyen Euphrate. In Ibáñez, J. (ed.), *Le Site Neolithique de Tell Mureybet (Syrie du Nord), en Hommage à Jacques Cauvin*, 103–14. Oxford: British Archaeological Report S1843

Willcox, G. 2012a. Searching for the origins of arable weeds in the Near East. *Vegetation History & Archaeobotany* 21(2), 163–7

Willcox, G. 2012b. Pre-domestic cultivation during the late Pleistocene and early Holocene in the northern Levant. In Gepts, P., Famula, T. R., Bettinger, R. L., Brush, S. B., Damania, A. B., McGuire, P. E. and Qualset, C. O. (eds), *Biodiversity in Agriculture: domestication, evolution, and sustainability*, 92–109. Cambridge: Cambridge University Press

Willcox, G. 2013. The roots of cultivation in southwestern Asia. *Science* 341(6141), 39–40

Willcox, G. and Fornite, S. 1999. Impressions of wild cereal chaff in pisé from the tenth millennium at Jerf el Ahmar and Mureybet: northern Syria. *Vegetation History & Archaeobotany* 8, 2–24

Willcox, G. and Stordeur, D. 2012. Large-scale cereal processing before domestication during the tenth millennium cal BC in northern Syria. *Antiquity* 86, 99–114

Willcox, G., Fornite, S. and Herveux, L. 2008. Early Holocene cultivation before domestication in northern Syria. *Vegetation History & Archaeobotany* 17, 313–25

Willcox, G., Buxo, R. and Herveux, L. 2009. Late Pleistocene and Early Holocene climate and the beginnings of cultivation in northern Syria. *Holocene* 19, 151–8

Wright, K. I. 1993. Early Holocene ground stone assemblages in the Levant. *Levant* 25, 93–110

Wright, K. I. 1994. Ground-stone tools and hunter-gatherer subsistence in southwest Asia: implications for the transition to farming. *American Antiquity* 59, 238–63

Zeder, A. M. 2011. The origins of agriculture in the Near East. *Current Anthropology* 52(S4), S221–35

Zohar, I. 2002. Fish and fishing at Ohalo II. In Nadel (ed.) 2002, 65

Zohary, M. 1973. *Geobotanical Foundations of the Middle East*. Stuttgart: G. Fischer

Zohary, D. and Hopf, M. 2000. *Domestication of Plants in the Old World*. New York: Oxford University Press

PART 2
PLANT FOODS, TOOLS AND PEOPLE

This part examines approaches to the study of pre-agrarian plant food remains. With little in the way of recognisable carbonised material available, methods of study cover direct evidence for plants, including both chemical and physical evidence, as well as indirect methods including dental and tool microwear analyses. This part covers a wide time span from early hominins to late hunter-gatherers. The first chapter covers the identification of roots and tubers from European Mesolithic contexts using scanning electron microscopy (Chapter 6). This chapter contributes to the discussion on Mesolithic subsistence and diet by presenting examples of archaeological parenchyma remains recovered across temperate Europe. It highlights how a broad diversity of plant species were used for their starchy roots and tubers, and the broad range of ecological zones that were explored by Mesolithic people in their search for vegetative and non-vegetative starchy foods.

This is followed by two chapters in which microwear traces are studied. A chapter on tool wear (Chapter 7) in which tools that imply food preparation, including grinding tools and lithic artefacts, as well as the morphologies of the tools themselves, are used to reconstruct the use of plants, both as food and raw materials. The examples of wear traces of plant-working on tools involved in subsistence and craft activities are drawn from Mesolithic sites in north-west Europe along with experimental examples.

This is followed by a chapter on buccal dental wear analysis (Chapter 8) which explores the ways in which microwear traces are used to reconstruct diet and paleo-ecological conditions. The most significant and diagnostic buccal enamel microwear patterns are presented in this chapter and the way in which these are used to reconstruct early hominin diet based on both buccal and occlusal microwear analyses, is outlined. A low abrasive diet is suggested for early hominins, in contrast to early *Homo* diet which would have involved chewing on hard or tough food, and might have included consumption of fruits and seeds. The results of buccal microwear analyses are compared with signals from stable isotope analysis.

The next three chapters focus on microfossil retrieval and identification. Each chapter offers an example of a different way microfossil evidence can be used to explore early human diet. In Chapter 9, phytoliths are used to infer what was available to the hominin population at Olduvai Gorge. Using phytoliths extracted from archaeological soil samples, the vegetation at the time the hominin populations were present in Olduvai

Gorge was reconstructed. This suggests a more humid landscape than previously thought, comprising grasses, sedges and palms. All would have provided plant foods for early hominins present in the area. Phytoliths are also used to trace the use of plants as food by the Natufian hunter-gatherers in the Southern Levant (Chapter 10). Here, the phytoliths were analysed from bedrock mortars, human burials and cave floors and show that small-grained grasses and wild cereals were processed and consumed. Chapter 11 covers the extraction of plant microfossils, including phytoliths and starch granules, from samples of dental calculus from a Brazilian shell mound. A broad range of root foods, including arum, yams and sweet potato have been tentatively identified; though further evidence is required, the authors suggest this may indicate evidence for early horticulture in the area.

The final chapter (Chapter 12) focuses on stable isotopes and the use of mass spectrometry in dietary reconstruction. Stable isotopes have been used extensively to examine ancient diet, both in terms of the type of plants (C_3 and C_4) and also in reconstruction of dietary trends using carbon and nitrogen ratios, though these tend to focus on the more dominant protein evidence. However, carbon isotopes have been used to identify the changes in dietary regime in the late Pliocene; the identification of a dominant signal for C_4 has led to the suggestion that some early hominin groups were able to survive on savannah resources, possibly tubers over 3 million years ago.

6. Scanning electron microscopy and starchy food in Mesolithic Europe: the importance of roots and tubers in Mesolithic diet

Lucy Kubiak-Martens

Studies of Mesolithic subsistence diet in temperate Europe have focused principally on animal and fish resources. This is mainly due to the abundance of bone remains and artefacts associated with hunting and fishing found at Mesolithic sites. Although there were some early attempts to emphasise the importance of plant foods during the Mesolithic, the lack of archaeobotanical evidence, or at least the limited range of encountered species, prevented a direct assessment of the proportions between the animal and plant food components in Mesolithic diet. Even when the recovery of plant remains was part of archaeological research it often resulted in a rather limited spectrum of plant foods, often mainly hazelnuts, at some sites complemented by acorns, water chestnut, and fleshy fruits and berries. Recently, the deployment of scanning electron microscope techniques to identify charred remains of vegetative plant tissue derived from underground storage organs, also known as storage parenchyma, has shown that starchy root foods, including true roots, tubers, rhizomes and bulbs of various plant species, are among the food resources that contributed substantially to the Mesolithic diet.

The examples of starchy foods discovered in the last two decades and presented here have considerable implications for the way in which we view the plant component of Mesolithic diet. There are clear indications that starchy foods were frequently gathered, implying that starch was a significant dietary energy source in Mesolithic Europe. A broad range of plant species was used and many ecological zones were explored by Mesolithic groups in their search for vegetative and non-vegetative starchy foods. The finds of charred archaeological parenchyma from Mesolithic sites will therefore continue to hold our interest. Although it is difficult, and perhaps still too early, to estimate the complex proportion between animal protein and plant food (starchy foods in particular), it is clear that a more balanced view of Mesolithic diet is emerging from archaeological sites.

The Mesolithic period in north-western Europe extends from *c.* 8800 to 4300 cal BC. In southern Scandinavia, where it is referred to as the Late Mesolithic Ertebølle Culture it continued until *c.* 3900 cal BC (Price 1991). Compared to the preceding Late Pleistocene, the Early Holocene was a period with a warmer climate which witnessed some major changes in the vegetation cover of temperate Europe, from pine and pine–birch dominated forests during the Pre-Boreal and early Boreal to the development of mixed deciduous forests during the early Atlantic, with trees such as oak, hazel, elm, lime, and ash. Terrestrial animals such as red deer, wild pig, roe deer, and aurochs and various small fur-bearing mammals replaced the reindeer and soon became important game animals for Mesolithic people. In addition, a wide range of fish from both marine and freshwater habitats was incorporated into the Mesolithic diet. The study of Mesolithic subsistence diet has long focused on these animal and fish resources, even though there were early attempts to emphasise the importance of plant foods in hunter-gatherer diet (Clarke 1976; Price 1989; Zvelebil 1994). David Clarke, in his work on Mesolithic Europe (1976), even suggested that the edible biomass of Mesolithic Europe has been concentrated especially on resources such as roots and tubers. The lack of archaeobotanical evidence, however, prevented a direct assessment of the importance of the plant foods, not only in Clarke's model of Mesolithic Europe but in hunter-gatherer subsistence diet as a whole. One of the main reasons why the plant component has largely been neglected (or overlooked) on hunter-gatherer sites is the belief that plant remains are rare in such early sites or even do not survive at all. This assumption was mainly based on the great abundance of animal remains, while the plant remains seemed to be archaeologically much less visible. Even when the recovery of plant remains was incorporated into archaeological research, the focus was often limited to one category of botanical remains: hazelnuts. This strategy resulted in the formation of a rather limited spectrum of plant foods, comprising hazelnuts, accompanied at some sites, by acorns, water chestnut and various fruits and berries (Price 1989; Zvelebil 1994). This spectrum remained incomplete as long as one of the most frequently consumed carbohydrate in the human diet – starch – was not included in the elements of the hunter-gatherer subsistence diet.

Fortunately the situation has improved since, particularly due to the increasing interest among archaeologists involved in more complete investigations of hunter-gatherer sites. Archaeobotanists are increasingly invited to excavations of hunter-gatherer sites so that we can join the projects in their early stages and be involved in archaeobotanical research, including sampling.

However, the real breakthrough in the archaeobotany of hunter-gatherers was the introduction of a scanning electron microscope (SEM) technique for the identification of charred remains of vegetative plant tissue derived from underground storage organs, also known as parenchyma. This new method for identification of archaeological parenchyma was introduced and developed by Jon Hather in the early 1990s at the University College London (Hather 1991; 1993; 2000). The identification is based on the examination of the internal anatomy of charred parenchymatous tissue; it is only successful if the vascular tissue is preserved together with the parenchyma cells.

In this chapter, we contribute to the discussion on Mesolithic subsistence and diet by presenting examples of archaeological vegetative and non-vegetative parenchyma remains recovered across temperate Europe over almost two decades of working with this method.

Starch and its chances for preservation in the archaeological record

In terms of nutritional value, roots and tubers and other underground storage organs are considered to be an important energy source in the human diet, mainly because of their high concentrations of starch as well as other carbohydrates and sugars. Starch is largely concentrated within the parenchyma tissue of these vegetative storage organs (Fig. 6.1a). Starch however can also be found in non-vegetative plant foods such as seeds, grains and nuts (see Chapter 1).

Starch is a polysaccharide carbohydrate composed of a large number of glucose units (predominantly amylose and amylopectin) which are joined together to form starch granules (Brown and Poon 2005; see also Chapter 1). As organic micro-remains, however, starch granules have a much smaller chance of being preserved in the archaeological record than, for example, pollen and seeds. Starch granules are readily decomposed by both biological and chemical processes, if not eaten first by bacteria and other soil micro-organisms. Still, some of the starch granules do survive in archaeological assemblages. In certain stable conditions, starch granules remain embedded within residual material and can survive for an extended period of time. For example, starch granules were recently successfully recovered from dental calculus of various Neanderthal individuals (Henry *et al.* 2011; Hardy *et al.* 2012). In some conditions, starches can survive better if they are carbonised. Isolated charred granules, however, are difficult to be identified to species or even taxon level. A few examples of charred starch granules were found in archaeological parenchyma from an Early Mesolithic site at Całowanie in

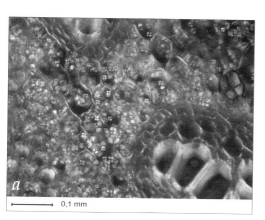

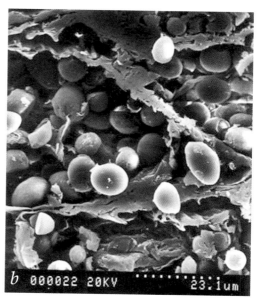

Fig. 6.1. a) The transverse section through the rhizome of the bracken fern (Pteridium aquilinum) *showing parenchyma cells packed with starch granules (polarised light microscope). Bracken fern is an example of a highly starchy root food indigenous to temperate Europe with a long history of use as plant food; b) Charred elliptical to spherical starch granules (unidentified) embedded within charred archaeological parenchyma from Całowanie, Early Mesolithic in Poland (SEM microscope) (photos: L. Kubiak-Martens)*

Poland and were morphologically identified as such under the SEM microscope (Fig. 6.1b). Unfortunately, no further identification was possible. Starch granules were also recently extracted from carbonised organic residues embedded in Late Mesolithic Ertebølle and Early Neolithic ceramic vessels (Saul *et al.* 2012).

Availability of starchy food in Mesolithic temperate Europe

There is little doubt that the most abundant and readily available source of starch in the increasingly forested environment of Mesolithic temperate Europe was found in the form of roots and tubers. Individual groups of hunter-gatherers probably used 20–30 species of edible roots in the course of their annual rounds (Mears and Hillman 2007). This category of plant foods also represented a significant part of traditional diet of recent northern hunter-gatherers. The ethnographic literature suggests that nearly 40 species of various tubers, corms, bulbs, rhizomes, and true roots were used as food (Kuhnlein and Turner 1991).

Carbohydrates in root foods can be present in a variety of forms, some of which may not always be readily digestible by humans and may therefore need preparation before they can be consumed. Many members of the Ranunculaceae family, for example, lesser celandine (*Ranunculus ficaria*, syn. *Ficaria verna*), contain toxins which must, at a minimum, be reduced, usually by boiling or by another heat treatment, before the tubers can be used as food (see Mason and Hather 2000 for discussion; see also Chapter 2). Some other indigenous root foods, for example, wild onion (*Allium* spp.) bulbs, contain inulin as a carbohydrate compound and need to be cooked to improve their palatability and digestibility. Other food plants, for example, horsetail (*Equisetum* spp.) tubers, can either be eaten raw, so that their sweet coconut-like taste can be appreciated (author's personal experience), or they can be cooked.

Although many roots and tubers can be consumed throughout the greater part of the year, their highest concentration of starch occurs at the end of the leaf-growing season, before new shoots appear, between autumn and early spring (Kuhnlein and Turner 1991). Still, for Mesolithic people, root foods probably represented one of the few foods that were available year round, even though their starch content varied through the year. A few examples of edible rhizomes of marsh plants that Mesolithic people would probably have had access to throughout the year are rhizomes of perennials such as bulrush (*Typha latifolia*), club-rushes (*Schoenoplectus* spp.), and common reed (*Phragmites australis*).

The examples of non-vegetative starchy foods in Mesolithic Europe would have included acorns of oak, inner bark tissue of birch and pine, seeds of water lilies (*Nymphaea* spp. and *Nuphar lutea*), seeds of fat hen (*Chenopodium album*), seeds of various orache species (*Atriplex* spp.), and grains of floating sweet grass (*Glyceria fluitans*). Even though the fruits and berries would have been of little calorific value to Mesolithic people, they would have provided an important nutrient in the form of fruit sugar (a simple monosaccharides or fructose) which is the most basic unit of carbohydrates and, of course, they would have been a good source of vitamin C.

Archaeological parenchyma – preservation and identification

The remains of vegetative parenchyma are fragile and therefore have a much lower chance of being preserved in the archaeological record than, for example, seeds or nutshells. This is due to the fact that vegetative tissues are often rich in water and oil and are therefore highly susceptible to damage when exposed to fire (Hather 1993). They also fracture easily; this can occur both during excavation, and then later in the laboratory where botanical soil samples are processed for the recovery of plant remains. The specimens of parenchyma tissues which survive both processes often consist of small fragments, which increase the difficulty in identification.

The remains of archaeological parenchyma are often round and contain either regular or irregular patterns of cavities. Parenchyma remains are often distinguished from wood charcoal at the sorting stage using a binocular incident light microscope. In parenchyma, the cells appear spherical, or more or less isodiametric, as opposed to being elongated like vessels in wood charcoal (Hather 1993). Nevertheless, badly deteriorated wood charcoal can be quite easily confused with parenchyma, especially to the untrained eye. The technique developed by Jon Hather for the identification of archaeological parenchyma is based mainly on examination of the internal anatomy of charred tissues; it therefore requires the use of a SEM microscope (Hather 1991; 1993; 2000). Identification of charred parenchyma to species or taxon level is possible only if the vascular tissue is preserved together with parenchyma cells. No identification can be performed if only isolated fragments of archaeological parenchyma tissue are preserved. Intact preservation of complete roots or tubers is very rare. If it occurs, it is more likely to do so in the case of small vegetative organs. Tubers of lesser celandine (*Ranunculus ficaria*), for example, often survive intact (Bakels 1988; Mason and Hather 2000; Bakels and van Beurden 2001).

Archaeological evidence for starchy root foods in the Mesolithic diet in temperate Europe

The application of new SEM techniques in the identification of archaeological parenchyma has found its expression in a number of systematic studies of archaeobotanical evidence principally from European Mesolithic sites. These studies have all shown that root foods, including true roots, tubers, rhizomes, and bulbs of various species, contributed significantly to the Mesolithic diet (Holden *et al.* 1995; Perry 1999; 2002; Kubiak-Martens 1996; 1999; 2002; 2011a; 2011b; Mason and Hather 2000; Mason *et al.* 2002; Kubiak-Martens and Tobolski 2014; Kubiak-Martens *et al.* 2015). The evidence for the consumption of plant foods in the European Palaeolithic is still sparse, though starchy food remains recovered from the Upper Palaeolithic site of Dolní Věstonice in the Czech Republic (Mason *et al.* 1994; Pryor *et al.* 2013), and Mezhirich in the Ukraine (Adovasio *et al.* 1992) suggest that roots were also consumed at this time.

An overview of the existing archaeobotanical evidence for the use of starchy plants from different Mesolithic sites and different ecological zones is presented here to illustrate our current knowledge of starchy food in hunter-gatherer subsistence in the Mesolithic Europe.

Early to Late Mesolithic

The site of Całowanie in the middle Vistula river valley in Poland provided archaeobotanical evidence for the use of starchy foods from the Early Mesolithic, associated with the early Duvensee/Maglemose tradition (Schild 1989; Kubiak-Martens 1996; Schild *et al.* 1999; Kubiak-Martens and Tobolski 2014). Here, the starchy food was represented by rhizome remains of knotgrass (*Polygonum* sp.), dated to 9510±70 cal BP [9100-8450 cal BC] (CAM-20867) (Fig. 6.2a), tuber remains of arrowhead (*Sagittaria* cf. *sagittifolia*), dated to 9390±70 cal BP [9150-8600 cal BC] (CAM-20870) (Fig. 6.3a), and a possible corm/stem base of a member of the sedge family (Cyperaceae). They were all found in archaeological beds that were embedded in organic sediments formed in the Vistula river valley.

Several *Polygonum* species (e.g., *P. bistorta*, *P. viviparum*, *P. hydropiper*) have edible rhizomes, stems, and leaves (Fig. 6.2b). The rhizomes contain some starch in addition to diverse minerals, including potassium, phosphorus and magnesium (Kuhnlein and Turner 1991). The ethnographic record of recent hunter-gatherers reveals that rhizomes of both *Polygonum bistorta* and *Polygonum viviparum* were used as a plant food in traditional diets of some of the peoples of northern Eurasia, northern Canada and Alaska. The rhizomes were either eaten raw or toasted over a fire (Eidlitz 1969; Kuhnlein and Turner 1991). It is interesting to note that pollen analysis from Całowanie revealed the presence of alpine bistort (*Polygonum viviparum*) in pollen spectra dated to the early Pre-Boreal, suggesting that this *Polygonum* species may have been available for gathering at and/or near the site. Interestingly, the charred rhizome parenchyma of Polygonaceae, possibly *Polygonum*, was also found on another Mesolithic site (Holden *et al.* 1995). There were probably various *Polygonum* species as well as the closely related *Rumex* species in the environment of Mesolithic Europe that may have

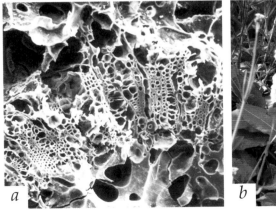

Fig. 6.2. a) SEM micrograph of a charred rhizome fragment of knotgrass (Polygonum *sp.*) *recovered from the Early Mesolithic occupation layer at Całowanie in Poland. Transverse section through the rhizome showing vascular tissue with radially elongated xylem elements and adjacent deteriorated phloem tissue (350×); b) Bistort* (Polygonum bistorta) *in flower (© langenaard.be)*

been collected both for their edible rhizomes and for their greens. The presence of charred rhizomes reflects some method of food processing in order to prepare the rhizomes for food.

In interpreting the remains of charred arrowhead tubers from the Early Mesolithic assemblage at Całowanie, a few aspects must be considered. Arrowhead usually grows in shallow water, stream edges or swampy ground. The egg-shaped tubers are 2–3 cm in diameter (Fig. 6.3b) and they are connected to the mother plant by long stolons and are buried quite deeply in mud. The charred remains found in Całowanie imply that the tubers must have been dug and brought to the site where they were most likely used as food. They must, however, have been exposed to fire as part of their preparation for consumption. There were many places around the Całowanie site during the Early Mesolithic where diverse species of marshy plants (in addition to arrowhead) must have grown and would have been readily available for gathering (Kubiak-Martens and Tobolski 2014).

Ethnographic records indicate that two species of arrowhead (both have high starch content), *S. sagittifolia* and *S. latifolia* (Indian swamp potato), are most commonly reported as having been used as food by various indigenous groups in Eurasia and North America. They were occasionally eaten raw, but were more frequently roasted in hot ashes or underground pits (Maurizio 1926; Arnason *et al.* 1981; Kuhnlein and Turner 1991).

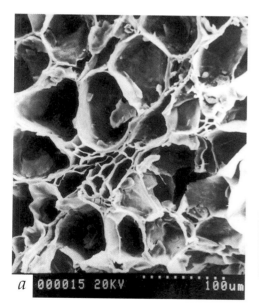

Fig. 6.3. a) SEM micrograph of a charred fragment of parenchyma tissue of arrowhead tuber (Sagittaria cf. sagittifolia) *from the Early Mesolithic site at Całowanie in Poland. Transverse section through the tuber showing vascular tissue composed of phloem elements and surrounded by parenchyma cells) (photo: J. Hather and L. Kubiak-Martens); b) tubers of arrowhead* (Sagittaria sagittifolia) *(photo: W. M. Szymański)*

Another starchy food that may have been appreciated and was easily available in Mesolithic Europe is the horsetail (*Equisetum* spp.). Even though the horsetails are rarely considered as a food resource because of their high silica content, this is based principally on the stems which are rough and silicon-impregnated, while the tubers are highly starchy and have a sweet taste (see Fig. 6.4). The charred archaeological remains of horsetail (*Equisetum* sp.) may therefore relate to the Mesolithic diet. Charred remains of *Equisetum* sp. tubers were recovered from the Early Mesolithic Łajty site in north-east Poland, which was visited over several centuries from 9420±100 cal BP [9150–8350 cal BC] (Gd-8027) to 8870±110 cal BP [8270–7670 cal BC] (Gd-6982) (Sułgostowska 1996), and from the Saltbaek Vig in Denmark that was occupied between *c.* 7000–6000 cal BP [5900–4800 cal BC] (Mason *et al.* 2002). Numerous charred fragments of *Equisetum* parenchyma were found at Saltbaek Vig. Remains of charred parenchyma from *Equisetum* sp. were recovered from the Early–Middle Mesolithic site of Veenkoloniën (Peat District) in the Dutch province of Groningen, where they were interpreted as evidence of either Mesolithic diet, medicine, or technology (Perry 1999). The presence of charred *Equisetum* stems, charred parenchymatous tissue of Pteridophyte and abundant charred vegetative remains of reed (*Phragmites*) have been identified in the Early Mesolithic lake-edge site of Star Carr in northern England (Hather 1998). Even though, at Star Carr, it is suggested that the burning of reed vegetation would have been responsible for occurrences of these charred plant remains in archaeobotanical record (Law 1998), it is essential to note that all of those plants have edible starchy underground organs and could well have been used as food plants.

One of the aims of a recently completed archaeological research project at the Early–Middle Mesolithic site of Yangtze Harbour near Rotterdam, the Netherlands has been

*Fig. 6.4. Tubers of giant horsetail (*Equisetum telmateia*). The white appearance of tuber interior (visible on cut specimens) is due to the starch content (photo: Ł. Łuczaj)*

to provide information on the use of plant foods by the Mesolithic groups visiting the river dune site located in the river landscape created by the Rhine and the Maas. The river dune on which these groups lived in the early Holocene now lies 20 m below the New Amsterdam Water Level (Schiltmans and Vos 2014). Throughout the project, much attention has been devoted to the search for charred remains of edible plants in general and in particular for those of charred parenchyma, which would indicate processing of root foods at the site. There were approximately 25 potentially identifiable fragments recovered of which some were identified with the use of a SEM microscope either to species or taxon level. Both charred vegetative and non-vegetative parenchyma remains were identified. An interesting aspect of the Yangtze Harbour assemblage was the presence of diverse root foods, including lesser celandine (*Ranunculus ficaria*), common club-rush (*Schoenoplectus lacustris*) and possible a member of the sedge family (Cyperaceae) (Kubiak-Martens *et al.* 2015).

The tubers of lesser celandine were preserved as small fragments and identification was based on the anatomy of charred parenchymatous and vascular tissue concentrated in the pith of a tuber (Fig. 6.5a–b). The remains were recovered from the archaeological find layer associated with one or a few early episodes of human presence on the dune. The AMS measurements obtained on charred plant remains recovered from this find layer produced dates of 8135±45 cal BP [7300–7050 cal BC] (GrA-55402) and 7970±45 cal BP [7050–6700 cal BC] (GrA-55483), providing evidence for the Early to Middle Mesolithic use of the site. Even though only few fragments of tuber parenchyma were found, their presence still implies that tubers of lesser celandine were collected and possibly used as food.

Lesser celandine can grow in extended stands (Fig. 6.6) and, through most of the year, can offer a plentiful harvest of tubers from just one plant (Fig. 6.6b). This particular species might therefore have been a well-known root vegetable for Mesolithic groups in temperate Europe.

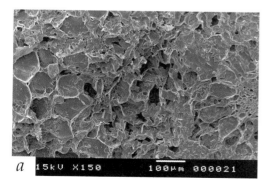 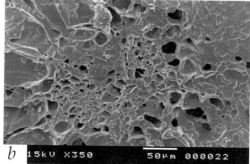

Fig. 6.5. a) SEM micrographs of charred lesser celandine (Ranunculus ficaria) tuber parenchyma from the Middle Mesolithic site at Yangtze Harbour in the Netherlands. Transverse section showing polygonal parenchyma cells, varying in size from 80–100 μm; concentration of vascular tissue (marked by arrow) indicates position of the stele; b) Partially preserved stele with vascular tissue (xylem elements are recognisable), approximately 250 μm across and would been in the centre of the tuber. (photo: L. Kubiak-Martens)

a

b

*Fig. 6.6. a) Lesser celandine (*Ranunculus ficaria*) on moist meadow in riparian woods in Scotland. http://sagebud.com/buttercup-ranunculus (photo: Roger Griffith); b) Tubers of lesser celandine collected in early March just before flowering. Photo: © biax.nl*

The tubers of lesser celandine are also known from other Mesolithic sites both in the Netherlands and Scotland. Charred, almost intact tubers and tuber fragments were found in the nearby river dune site Hardinxveld-Giessendam Polderweg (Bakels and van Beurden 2001) and at Staosnaig on the island of Colonsay in western Scotland (Mason and Hather 2000; Mithen *et al.* 2001). The tubers found at Staosnaig are by far the best evidence for the processing of this root food in Mesolithic Europe. Substantial quantities of tubers were found there together with a massive abundance of hazelnut shells in a feature interpreted as a roasting pit or oven. Consequently, the most reasonable interpretation proposed was intentional and intensive exploitation of *Ranunculus ficaria* (Mithen *et al.* 2001).

It seems that Mesolithic hunter-gatherers were consistently attracted to marsh and water plants as sources of starchy food. At the site Rotterdam Yangtze Harbour, six fragments of charred parenchyma were identified as being derived from the rhizome of common club-rush (*Schoenoplectus lacustris*). For one of the fragments, the AMS date obtained was 8015±45 cal BP [7070–6770 cal BC] (GrA-55481). Of significance is the fact that one of the morphologically intact specimens in cross-section revealed an unusually smooth surface, as if the rhizome was cut prior to charring illustrating clear evidence of human processing (Fig. 6.7a–b). The species was identified using the anatomy of parenchymatous and vascular tissue observed under a SEM microscope (Fig. 6.8 a–b).

Fig. 6.7. Charred fragment of common club-rush (Schoenoplectus lacustris) *rhizome from the Mesolithic site at Yangtze Harbour, Netherlands. The specimen has a conspicuously smooth cross-section, suggesting that the rhizome may have been cut before charring (photos: M. van Waijjen)*

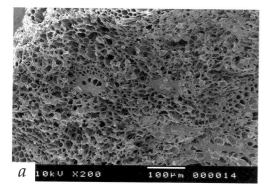
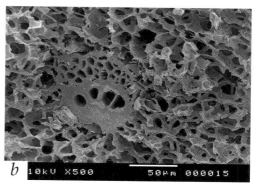

Fig. 6.8. a) SEM micrographs of the internal anatomy of the common club-rush (Schoenoplectus lacustris) *rhizome presented in Fig. 6.7, showing parenchymatous to aerenchymatous tissue with randomly placed vascular bundles; b) The vascular bundles are collateral, having a thick fibre sheath at the xylem side (photos: L. Kubiak-Martens)*

The presence of charred rhizome remains of the common club-rush suggests that people visiting the river dune site at Rotterdam Yangtze Harbour during the late Boreal and/or early Atlantic were making use of the starchy, underground part of this plant. This marsh plant must have grown in abundance near the site (this is clearly indicated by the macro-remains), probably along the riverbanks, where it was easily available for gathering. The simplest way of preparing the rhizomes for consumption would have been to roast them in hot ashes (Mears and Hillman 2007). The interpretation that rhizomes of this species were introduced into the archaeological record as a food plant is supported by archaeobotanical finds from other Mesolithic sites. For example, charred rhizome remains of club-rush (*Schoenoplectus* sp.) were found together with charred parenchymatous tissues of other edible plants at two Early–Middle Mesolithic sites in the Veenkoloniën (Peat District), in the province of Groningen in the Netherlands (Perry 1999).

Late Mesolithic Ertebølle

The evidence for Late Mesolithic settlement in temperate Europe is principally concentrated in coastal zones through large areas of southern Scandinavia. The shallow, brackish lagoons were particularly attractive sites for occupation. The evidence for subsistence activities in these environments is extremely abundant. Two submerged Late Mesolithic Ertebølle sites in Denmark, Tybrind Vig on the west coast of the island of Funen and Halsskov on the west coast of Sjælland, provided good evidence for subsistence activities, including the gathering and processing of root foods among other plant foods (Kubiak-Martens 1999; 2002; Robinson and Harild 2002).

Charred remains of sea beet (*Beta vulgaris* ssp. *maritima*) roots recovered from Tybrind Vig were of special interest because they were both numerous and frequent when compared to most of the charred remains of archaeological parenchyma (Kubiak-Martens 1999). The charred parenchyma of *Beta* root can be identified on the basis of a particular type of anomalous secondary growth. Each bundle of vascular tissue has a cambium associated with it which divides to produce vascular and storage parenchyma. This process is repeated many times, observable in transverse section in the form of concentric rings of vascular tissue between which are the bands of storage parenchyma (Fig. 6.9a).

Sea beet occurs naturally on shingle beaches, tidal drift deposits and the drier areas of salt marshes in temperate Europe. It is still found on the Dutch North Sea coast (Fig. 6.9b). The plant is a biennial dicotyledon with succulent leaves and fleshy roots growing to 30 cm in length and 3-4 cm in thickness. The roots are rich in starch and sugar, both of which are concentrated in the parenchyma tissue. The roots of the sea beet would have been readily available for gathering in many places near the settlement at Tybrind Vig.

It seems that sea beet was a well-known root vegetable to many, if not all, groups living along the coast of the North Sea in early prehistory. Charred remains of root parenchyma were also found in other early sites in Europe, including the Early to Middle Mesolithic site in the aforementioned Veenkoloniën (Peat District) in the Dutch province of Groningen (Perry 1999) and the Middle Neolithic site at Schipluiden Harnaschpolder on the Dutch North Sea coast (Kubiak-Martens 2006). The presence of charred perianths have been identified in the Ertebølle site Møllegabet II in Denmark

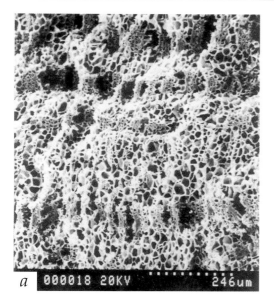 

Fig. 6.9. a) SEM micrograph of a charred fragment of parenchyma derived from the sea beet root (Beta vulgaris ssp. maritima) from the Late Mesolithic Ertebølle site at Tybrind Vig in Denmark, showing concentric rings of xylem and broad bands of storage parenchyma between each ring (500×) (photo: J. Hather and L. Kubiak-Martens); b) Collecting sea beet roots on the Dutch North Sea coast near Bergen in August 2008 (photo: C. Vermeeren)

(Mason 2004), which may also indicate the exploitation of sea beet as a food plant.

Two additional examples of root foods dated to the European Late Mesolithic were found on the site of Halsskov. These included ramsons or wild garlic (*Allium* cf. *ursinum*) bulbs (Fig. 6.10a) and pignut (*Conopodium majus*) tubers (Fig. 6.11a–b) (Kubiak-Martens 2002).

Allium ursinum is a perennial herb with the strong smell and flavour of garlic which often forms extensive patches in shady, damp, deciduous forest. The plant most often grows one bulb but sometimes clusters of elongated bulbs (Fig, 6.10b). Both the leaves and the bulbs are edible. At Halsskov, the bulbs would have been gathered and used as food, or cooked as flavouring with other foods. Cooking would convert the bulb's major carbohydrate, inulin, which is neither easily digestible nor very palatable, into sweet-tasting fructose. What is interesting about the bulbs from Halsskov is their small size, which closely resembles that of specimens of ramsons collected in late March before flowering. For the inhabitants of Halsskov and perhaps other sites, ramsons would have been one of the first root (and green) vegetables to appear in spring.

The evidence for the gathering of *Allium* species is also emerging from archaeological sites associated with the early-agrarian subsistence, for example, two Middle Neolithic sites on the Dutch North Sea coast, Schipluiden and Ypenburg (Kubiak-Martens 2006; 2008). Revealing evidence for use of *Allium* comes also from the Neolithic lakeshore sites in Switzerland, where large amounts of *Allium* pollen were found in the pollen

a

spectra from the occupation layers (Heitz-Weniger 1978) and where *Allium* pollen has been recovered from the pot contents (Hadorn 1994). The presence of *Allium* in both Mesolithic and Neolithic sites is particularly significant, given the abundant ethnographic evidence for its use both as a root and a green vegetable, and also to flavour other foods (Eidlitz, 1969; Turner *et al.* 1990; Kuhnlein and Turner 1991).

Remains of pignut (*Conopodium majus*) tubers, which are another significant root food, were also identified at Halsskov (Fig. 6.11a–b). This perennial plant, which is a member of Apiaceae family, grows in open woodlands and grasslands. Its storage organ is a small irregularly spherical tuber, which grows at the base of the stem (Fig. 6.12). The tubers have a mildly nutty flavour when cooked. The tubers cannot be pulled out of the ground when being harvested as the thin stem breaks away

b

Fig. 6.10. a) *Charred bulb of ramsons (*Allium *cf.* ursinum*) recovered from the Late Mesolithic Ertebølle site at Halsskov in Denmark; the bulb is 6 mm long and 3 mm wide; b) bulbs of ramsons dug out in April before flowering (photos: L. Kubiak-Martens)*

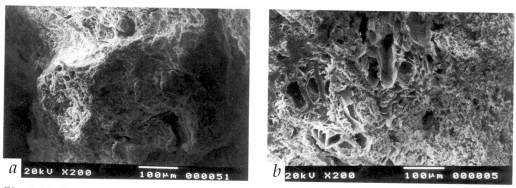

Fig. 6.11. SEM micrographs of charred remains of pignut (Conopodium majus) *tuber from the Late Mesolithic Ertebølle site at Halsskov in Denmark; a) Outer surface of a fragment derived from a small spherical tuber originally 8–10 mm in diameter, showing a root scar; b) Interior anatomy showing diffuse and random organisation of vascular tissue and groups of xylem (photo: L. Kubiak-Martens)*

Fig. 6.12. Pignut (Conopodium majus) *root tubers (photo: © sauvagement-bon.blogspot.com)*

very easily. This suggests that at Halsskov the tubers must have been extracted using some kind of digging tool before being brought to the site, where they were most likely used as food (Kubiak-Martens 2012).

What is remarkable about the charred remains of both the ramsons and pignut from Halsskov is the context in which they were found. Even though the archaeological features at the site consisted of hearths and shallow pits, it was the pits that contributed to the preservation of edible plant remains and not the hearths; the remains of charred parenchyma were mainly recovered from the pits (Kubiak-Martens 2002). The nature of the pits suggests that heating or burning must have occurred within these structures. They were filled up with sandy clay with an admixture of ash, charcoal dust and charcoal fragments (Robinson and Harild 2002). All these characteristics argue for the function of these features as pit-cooking depressions used for cooking root foods and, possibly, other foods. The presence of various root foods in the Halsskov pits is particularly significant, given the abundant ethnographic evidence for pit-cooking as one of the most common methods of preparing root and tuber foods (Kari, 1995; Turner *et al.* 1990).

Archaeological evidence for non-vegetative starchy food from Mesolithic Europe

In the first encounter with the plant assemblages from the European Middle–Late Mesolithic sites, one particular non-vegetative starchy food is usually noticed: oak acorns. Charred acorns are usually preserved either as complete halves or fragmented acorn cotyledons or as small fragments of isolated acorn parenchyma (Kubiak-Martens 1999; Bakels *et al.* 2001; Mason *et al.* 2002; Kubiak-Martens *et al.* 2015).

There are two native oak species in temperate Europe: pedunculate oak (*Quercus robur*) and sessile oak (*Quercus petraea*). Unfortunately, when based solely on morphological or anatomical features of acorn cotyledons, archaeological acorn remains cannot be identified to species level. The species represent different ecological preferences; pedunculate oak prefers rich and wet soils, while sessile oak prefers acidic, sandy soils. The acorns from both species are edible when properly prepared and both are rich in carbohydrates (Mason 1995), in contrast to other nuts, hazelnuts, for example, which are mainly rich in fat. The nutritional content of acorns is comparable to cereals as they are principally a source of carbohydrates (c. 50 g of carbohydrate per 100 g fresh weight) with a small amount of protein and fat (2.8 g and 3.5 g per 100 g fresh weight, respectively; Kuhnlein and Turner 1991).

Acorns from the north-western European species contain tannic acid which gives them a bitter taste and makes them inedible unless the tannin is removed (Mason 1995). There are various methods described in the ethnographic literature for processing acorns; most of these involve the removal of the shell prior to the processing of the acorn (Chestnut 1974; Kuhnlein and Turner 1991). The presence of charred acorn cotyledons (often with the husks removed) in archaeological sites suggests some manner of processing acorns in order to make them edible and palatable. Peeled acorns could have been simply roasted in hot ashes.

Ethnographic evidence for acorn use from various environments, together with emerging archaeobotanical evidence from Mesolithic sites, has led many authors to suggest that acorns might have been an important food, perhaps even a staple food, in the Mesolithic of temperate Europe (e.g., Mason 1995; 2000; Kubiak-Martens 1999; Mason *et al.* 2002; Regnell 2012; Kubiak-Martens *et al.* 2015).

Other non-vegetative starchy plants for which we have good archaeobotanical evidence from various Mesolithic sites in temperate Europe, include water-lilies (*Nymphaea* spp. and *Nuphar lutea*). Both the seeds and the rhizomes of water-lilies are edible and both are rich in starch; the seeds also contain considerable amounts of protein and oil as well as some sugars. Ethnographic records provide considerable evidence for the consumption of both the seeds and the rhizomes (Kuhnlein and Turner 1991). Charred seed remains from Mesolithic sites suggest that they were processed and most likely used as food. On the river dune site in Rotterdam Yangtze Harbour, charred seeds of yellow water-lily (*Nuphar lutea*) were found together with other edible plant remains (Kubiak-Martens *et al.* 2015). At the Late Mesolithic Halsskov site in Denmark, charred seeds of the closely related least water-lily (*Nuphar pumilum*) were found in features considered to be pit-cooking depressions (Kubiak-Martens 2002). The high frequencies of charred seeds of *Nuphar lutea* and another closely related plant, the white water-lily (*Nymhaea alba*), found at the Early Neolithic site at Hoge Vaart in the Netherlands is also notable. Here, the seeds were found in the hearths, suggesting that water-lilies were purposely collected and most likely used as food (Brinkkemper *et al.* 1999).

Conclusions

The identification of starchy foods from European Mesolithic sites has considerable implications for the way in which we view the plant component in late hunter-gatherer diet. It is possible that importance of animal and fish resources has been overestimated, by comparison with starchy foods. Although it is still not possible to estimate relative proportions of plant foods and meat, it is clear that a more balanced view of the Mesolithic diet is emerging from archaeological sites. The study presented here shows a broad diversity with respect to the plant species used and the ecological zones that were explored by Mesolithic groups in the search for vegetative and non-vegetative starchy foods.

Starchy plants were a significant dietary energy source in the European Mesolithic. Starchy roots and tubers, as well as also other non-vegetative starchy foods, were an important part of the diet. However, in order to better understand and assess the amount of starch consumed by hunter-gatherers or, in other words, to balance the complex relationship between animal protein and the starchy plant component in the early human diet, the identification of archaeological parenchyma using SEM has to be applied to networks of hunter-gatherer sites rather than just to individual sites. Even in cases where the specific identification of archaeological parenchyma is difficult or impossible, its presence demonstrates that starchy foods were exploited and processed at the site.

Finally, with regards to the strategies for future archaeobotanical work at hunter-gatherer sites, two points need to be made. Firstly, the advice or, preferably, the presence of an archaeobotanist during the excavation, to supervise both the collection and processing of samples should become standard procedure and secondly, both waterlogged and charred plant remains need to be studied in order to reconstruct the broadest possible spectrum of edible plants. It is only through study of the entire range of plant food remains present on archaeological sites, that far-reaching insights into the role of plant resources in hunter-gatherer diet can be made. Finally, it is only through an increase in the systematic study of well-documented botanical assemblages, that a better understanding of the contribution of wild plant resources, both to Mesolithic diet and to the changing nature of resource use during the transition from the Mesolithic to Neolithic in temperate Europe, can be achieved.

References

Adovasio, J. M., Soffer, D., Dirkmaat, C., Pedler, D., Pedler, D., Thomas, D. and Buyce, R. 1992. *Flotation Samples from Mezhirich, Ukrainian Republic: a micro-view of macro-issues.* Unpublished paper presented at 57th annual meeting of the Society for American Archaeology, Pittsburgh (PA), 8–12 April 1992

Arnason, T., Hebda, R. J. and Jons, T. 1981. Use of plants for food and medicine by native peoples of eastern Canada. *Canadian Journal of Botany* 59, 2189–325

Bakels, C. C. 1988. Hekelingen, a Neolithic site in the swamps of the Maas estuary. In Küster, H. (ed.), *Der prähistorische Mensch und seine Umwelt (Festschrift für Udelgard Körber-Grohne zum 65. Geburtstag)*, 155–61. Stuttgart: Forschungen und Berichte zur Vor- und Frühgeschichte in Baden-Württemberg 31

Bakels, C. C. and Beurden van, L. M. 2001. Archeobotanie. In Louwe Kooijmans, L. P. (ed.), *Hardinxveld-Giessendam – Polderweg. Een mesolithisch jachtkamp in het rivierengebied (5500–5000 v. Chr.)*, 325–378. Amersfoort: Rapportage Archeologische Monumentenzorg 83

Bakels, C. C., Beurden van, L. M. and Vernimmen, T. J. 2001. Archeobotanie. In Louwe Kooijmans, L. P. (ed.), *Hardinxveld-Giessendam – De Bruin. Een kampplaats uit het Laat-Mesolithicum en het begin van de Swifterbant-cultuur (5500-4450 v.Chr.) Polderweg. Een mesolithisch jachtkamp in het rivierengebied (5500–5000 v. Chr.)*, 369–433. Amersfoort: Rapportage Archeologische Monumentenzorg 88

Brinkkemper, O., Hogestijn, W. J., Peeters, H., Visser, D. and Whitton, C. 1999. The Early Neolithic site at Hoge Vaart, Almere, The Netherlands, with particular reference to non-diffusion of crop plants, and the significance of site function and sample location. *Vegetation History & Archaeobotany* 8, 79–86

Brown, W. H. and Poon, T. 2005. *Introduction to Organic Chemistry* 3rd edn. New York: Wiley

Chestnut, V. K. 1974. *Plants Used by the Indians of Mendocino Country, California.* Fort Bragg CA: Contributions from the U.S. National Herbarium VII/Mendocino County Historical Society

Clarke, D. L. 1976. Mesolithic Europe: the economic basis. In Sieveking, G. de G., Longworth, I. H. and Wilson, K. E. (eds), *Problems in Economic and Social Archaeology*, 449–81. London: Duckworth

Eidlitz, K. 1969. Food and emergency food in the circumpolar area. *Studia Ethnographica Upsaliensia* 32, 1–175

Hadorn, P. 1994. Saint-Blaise, Bains des Dames. Palynologie d'un site néolithique et histoire de la végétation des derniers 16,000 ans. *Archéologie Neuchâteloise* 18, 1–121

Hardy, K., Buckley, S., Collins, M. J. C., Estalrrich, A., Brothwell, D., Copeland, L., García-Tabernero, A., García-Vargas, S., de la Rasilla, M., Lalueza-Fox, C., Huguet, R., Bastir, M., Santamaría, D., Madella, M., Huguet, R., Bastir, M., Fernández Cortés, A. and Rosas, A. 2012. Neanderthal medics? Evidence for food, cooking, and medicinal plants entrapped in dental calculus. *Naturwissenschaften* DOI 10.1007/s00114-012-0942-0

Hather, J. G. 1998. Identification of macroscopic charcoal assemblages. In Mellars, P. and Dark, P. (eds), *Star Carr in Context, New Archaeological and Palaeoecological Investigations at the Early Mesolithic site of Star Carr in North Yorkshire*, 183–96 Cambridge: McDonald Institute

Hather, J. G. 1991. The identification of charred archaeological remains of vegetative parenchymatous tissue. *Journal of Archaeological Science* 18, 661–75

Hather, J. G. 1993. *An Archaeobotanical Guide to Root and Tuber Identification. 1: Europe and South Asia*, Oxford: Oxbow Monograph 28

Hather, J. G. 2000. *Archaeological Parenchyma*, London: Archetype

Heitz-Weniger, A. 1978. Pollenanalytische Untersuchungen an den neolithischen und spätbronzezeitlichen Seerandsiedlungen Kleiner Hafner, Grosser Hafner und Alpenquai im untersten Zürichsee (Schweiz). *Botanische Jahrbücher für Systematik, Pflanzengeschichte und Pflanzengeographie* 99(1), 48–107

Henry, A. G., Brooks, A. S. and Piperno, D. R. 2011. Microfossils in calculus demonstrate consumption of plants and cooked foods in Neanderthal diets (Shanidar III, Iraq; Spy I and II, Belgium). *Proceedings of the National Academy of Sciences (USA)* 108, 486–91

Holden, T. G., Hather, J. G. and Watson, J. P. N. 1995. Mesolithic plant exploitation at the Roc del Migdia, Catalonia. *Journal of Archaeological Science* 22, 769–78

Holst, D. 2010. Hazelnut economy of early Holocene hunter-gatherers: a case study from Mesolithic Duvensee, northern Germany. *Journal of Archaeological Science* 37, 2871–80

Kari, P. R. 1995. *Tanaina Plantore Denaina Ketuna. An Ethnobotany of the Denaina Indians of Southcentral Alaska.* Fairbanks AK: Alaska Native Language Center, University of Alaska

Kubiak-Martens, L. 1996. Evidence for possible use of plant foods in Palaeolithic and Mesolithic diet from the site of Całowanie in the central part of the Polish Plain. *Vegetation History & Archaeobotany* 5, 33–8

Kubiak-Martens, L. 1999. The plant food component of the diet at the Late Mesolithic (Ertebølle) settlement at Tybrind Vig, Denmark. *Vegetation History & Archaeobotany* 8, 117–27

Kubiak-Martens, L. 2002. New evidence for the use of root foods in pre-agrarian subsistence recovered from the Late Mesolithic site at Halsskov, Denmark. *Vegetation History & Archaeobotany* 11, 23–31

Kubiak-Martens, L. 2006. Roots, tubers and processed plant food in the local diet. In Louwe Kooijmans, L. P. and Jongste, P. F. B. (eds), *Schipluiden, a Neolithic settlement on the Dutch North Sea Coast c. 3500 cal BC*, 339–352. Leiden: *Analecta Praehistorica Leidensia* 37/38

Kubiak-Martens, L. 2008. Wortels, knollen en bereid plantaardig voedsel. In Koot, H., Bruning, L. and Houkes, E. R. A. (eds), *Ypenburg-Locatie 4. Een nederzetting met grafveld uit het Midden-Neolithicum in het West-Nederlandse kustgebied*, 325–35. Leiden: Hazenberg Archeologie

Kubiak-Martens, L. 2011a, Voedseleconomie: parenchym en andere plantaardige macroresten. In Lohof, E., Hamburg, T. and Flamman, J. (eds), *Steentijd opgespoord. Archeologisch onderzoek in het tracé van de Hanzelijn-Oude Land*, 465–81. Leiden/Amersfoort: Archol Rapport 138/ ADC Rapport 2576

Kubiak-Martens, L. 2011b. Het macrorestenonderzoek. In Noens, G. (ed.), *Een afgedekt Mesolithisch nedderzettingsterrein the Hempens/N31 (Gemeente Leeuwarden, Provincie Friesland, NL.), Algemeen kader voor de studie van een lithische vindplaats*, 231–41. Gent: Academia Press

Kubiak-Martens, L. and Tobolski, K. 2014. Late Pleistocene and Early Holocene Vegetation History and Use of Plant Foods in the Middle Vistula River Valley at Całowanie (Poland). In Schild, R. (ed.) *Całowanie a Final Paleolithic and Early Mesolithic site on an island in the ancient Vistula Chanel*, 333–348. Warsaw: Institute of Archaeology & Ethnology, Polish Academy of Sciences

Kubiak-Martens, L., Kooistra, L. I. and Verbruggen, F. 2015. Archaeobotany: landscape reconstruction and plant food subsistence economy on a meso and microscale. In Moree, J. M. and Sier, M. M. (eds), *Interdisciplinary Archaeological Research Programme Maasvlakte 2, Rotterdam, , Part 1, Twenty metres deep! The Mesolithic period at the Yangtze Harbour site – Rotterdam Maasvlakte, the Netherlands. Early Holocene landscape development and habitation*, 223–286. Rotterdam: BOORrapporten 566

Kuhnlein, H. V. and Turner, N. J. 1991. *Traditional Plant Foods of Canadian Indigenous Peoples: nutrition, botany and use*. Philadelphia PA: Food and Nutrition in History and Anthropology 8

Law, C. 1998. The use and fire-ecology of reedswamps vegetation. In Mellars, P. and Dark, P. (eds), *Star Carr in Context, New Archaeological and Palaeoecological Investigations at the Early Mesolithic site of Star Carr in North Yorkshire*, 197–206. Cambridge: McDonald Institute

Mason, S. L. 1995. Acornutopia? Determining the role of acorns in past human subsistence. In Wilkins, J., Harvey, D. and Dodson, M. (eds), *Food in Antiquity*, 12–24. Exeter: University of Exeter Press

Mason, S. L. 2004. Archaeobotanical analysis-Møllegabet II. In Skaarup, J. and Grøn, O. (eds), *Møllegabet II. A submerged Mesolithic settlement in southern Denmark*, 122–143. Oxford: British Archaeological Report S1328

Mason, S. and Hather, J. G. 2000. Parenchymatous plant remains from Staosnaig. In Mithen, S. (ed.), *Hunter-gatherer Landscape Archaeology. The Southern Hebrides Mesolithic Project 1988–1998*, 415–25. Cambridge: McDonald Institute

Mason, S. L., Hather, J. G. and Hillman, G. C. 1994. Preliminary investigation of the plant macro-remains from Dolní Véstonice II, and its implications for the role of plant foods in Palaeolithic and Mesolithic Europe. *Antiquity* 68, 48–57

Mason, S. L., Hather, J. G. and Hillman, G. C. 2002. The archaebotany of European hunter-gatherers: some preliminary investigations. In Mason, S. L. and Hather, J. G. (eds), *Hunter-gatherer Archaeobotany. Perspectives from the Northern Temperate Zone*, 188–96. London: Institute of Archaeology

Maurizio, A. 1926. *Die Geschichte unserer Pflanzennährug*. Warsaw: Kasa Mianowskiego

Mears, R. and Hillman, G. C. 2007. *Wild Food*. London: Hodder & Stoughton

Mithen, S., Finlay, N., Carruthers, W., Carter, S. and Ashmore, P. 2001. Plant use in the Mesolithic: evidence from Staosnaig, Isle of Colonsay, Scotland. *Journal of Archaeological Science* 28, 223–34

Perry, D. 1999. Vegetative tissues from Mesolithic sites in the northern Netherlands. *Current Anthropology* 40, 231–7

Perry, D. 2002. Preliminary results of an archaeobotanical analysis of Mesolithic sites in the Veenkoloniën, Province of Groningen, the Netherlands. In Mason, S. L. and Hather, J. G. (eds), *Hunter-Gatherer Archaeobotany. Perspectives from the Northern Temperate Zone*, 108–16. London: Institute of Archaeology

Price, T. D. 1989. The reconstruction of Mesolithic diets. In Bonsall, C. (ed.), *The Mesolithic in Europe*, 48–59. Edinburgh: John Donald

Price, T. D. 1991. The Mesolithic of Northern Europe. *Annual Review of Anthropology* 20, 211–33

Pryor, A. J. E., Steele, M., Jones, M. K., Svoboda, J. and Beresford-Jones, D. G. 2013. Plant foods in the Upper Palaeolithic at Dolní Vestonice? *Parenchyma redux. Antiquity* 87, 971–84

Regnell, M. 2012. Plant subsistence and environment at the Mesolithic site Tågerup, southern Sweden: new insights on the 'Nut Age'. *Vegetation History & Archaeobotany* 21, 1–16

Regnell, M., Gaillard, M. J., Bartholin, T. S. and Karsten, P. 1995. Reconstruction of environment and history of plant use during the late Mesolithic (Ertebølle culture) at the inland settlement of Bökeberg III, southern Sweden. *Vegetation History & Archaeobotany* 4, 67–91

Robinson, D. E. and Harild, J. A. 2002. The archaeobotany of an Early Ertebølle (Late Mesolithic) site at Halsskov, Zealand, Denmark. In Mason, S. L. and Hather, J. G. (eds), *Hunter-Gatherer Archaeobotany. Perspectives from the Northern Temperate Zone*, 84–95. London: Institute of Archaeology

Saul, H., Wilson, J., Heron, C. P., Glykou, A., Hartz, S. and Craig, O. 2012. A systematic approach to the recovery and identification of starch from carbonised deposits on ceramic vessels. *Journal of Archaeological Science* 39, 3483–92

Schild, R. 1989. The formation of homogeneous occupation units ('Kshemenitsas') in open air sandy sites and its significance for the interpretation of Mesolithic in Europe. In Bonsall, C. (ed.), *The Mesolithic in Europe*, 89–98. Edinburgh: John Donald

Schild, R., Tobolski, K., Kubiak-Martens, L., Pazdur, M. F., Pazdur, A., Vogel, J. C. and Staffford, T. W. Jr. 1999. Stratigraphy, palaeoecology and radiochronology of the site of Całowanie. *Folia Quaternaria* 70, 239–68

Schiltmans, D. E. A. and Vos, P. C. 2014. Inleiding. In Moree, J. M. and Sier, M. M. (eds), *Twintig Meter diep! Mesolithicum in de Yangtzehaven-Maasvlakte te Rotterdam. Landschapsontwikkeling en bewoning in het Vroeg Holoceen*, 11–24. Rotterdam: BOORrapporten 523

Smith, C. 1989. British antler mattocks. In Bonsall, C. (ed.), *The Mesolithic in Europe*, 272–83. Edinburgh: John Donald

Sułgostowska, Z. 1996. The Early Mesolithic settlement of north-eastern Poland. In Larsson, L. (ed.), *The Earliest Settlement of Scandinavia and its Relationships with Neighboring Areas*, 297–304. Lund: Acta Archaeologica Lundensia Series 8, 24.

Turner, N. J., Thompson, L. C., Thompson, M. T. and York, A. Z. 1990. *Thompson Ethnobotany. Knowledge and usage of plants by the Thompson Indians of British Columbia.* Victoria BC: Royal British Columbia Museum Memoir 3

Zvelebil, M. 1994. Plant use in the Mesolithic and its role in the transition to farming. *Proceedings of the Prehistoric Society* 60, 35–74

7. Tools, use wear and experimentation: extracting plants from stone and bone

Annelou van Gijn and Aimée Little

In this paper we will examine how microwear analysis of ground and chipped stone assemblages, as well as objects of bone and antler, can contribute to our understanding of the role of plants in pre-agricultural societies. With only exceptional cases of direct archaeological evidence of plant craftwork and archaeobotanical food remains surviving, studying the wear traces of plant-working on tools provides important, albeit indirect, evidence for otherwise invisible aspects of hunter-gatherer daily life. Effectively, by studying the microscopic evidence left on pieces of flaked stone, we now know a great deal more about the role of plants. For researchers studying early prehistory – a period where so much of our archaeological data and therefore material understandings of the hunter-gatherer world is based on stone tools – this is significant. It is not uncommon that microwear polishes that form on stone tools as a result of contact with various plants are the only remaining anthropogenic evidence for plant-working by the inhabitants of a site.

In many archaeological contexts organic materials will not have preserved. Only rarely do we see – when the right preservation conditions come together – direct evidence of hunter-gatherer plant-working from within the archaeological record. Most typically these plant eco- and artefacts are found in wetland environments where the preservation is more favourable for organic remains. Some exceptionally well-preserved examples of hunter-gatherer wetland plant craftwork include a number of finely woven fish traps, as revealed at two different commercial excavations in Ireland (FitzGerald 2007; Mossop 2009; McQuade and O'Donnell 2007), intricately woven cord, fish traps, textiles, and net fragments from the Ertebølle excavations at Tybrind Vig, Denmark (Andersen 2013) and the Late Mesolithic sites of Hardinxveld Polderweg and De Bruin (Louwe Kooijmans *et al.* 2001a; 2001b; for Russia and the Baltic region see Chapter 13).

However, this type of preservation is rare, and more commonly we have to rely on indirect methods to obtain information about subsistence and craft activities (Hurcombe 2008a; 2008b). One such method is microwear analysis. This method is based on the fact that using a tool causes wear traces to develop. These traces vary according to the contact material that is worked and the direction in which the tool is used. On flint, but also on bone and antler implements, we can distinguish the following wear phenomena: edge removals (often referred to as use retouch), edge rounding, striations, and polish. It is the combination of features that allows us to distinguish between different contact materials

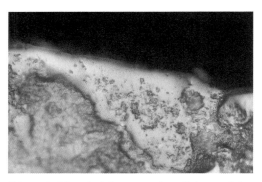

Fig. 7.1. *Very bright, smooth polish resulting from scraping fresh reeds for 300 minutes (200×) (courtesy of Leiden Laboratory of Artefact Studies)*

and to infer the motion of use. For example, polish from contact with siliceous plant materials is very bright and smooth (Fig. 7.1), whereas contact with hide causes a very rough, matt polish with small pits and heavy edge rounding. On ground stone tools we also see attrition and smoothing of grains as features of wear (Adams 2013). Most of these features are seen as resulting from mechanical processes. However, there continues to be a debate as to whether the polish from contact with siliceous plants is due to mechanical polishing of the surface or whether it is an additive (Anderson *et al.* 1996; Vargiolu *et al.* 2003; Fullagar 1991; Unger-Hamilton 1984).

In this paper we will first describe what microwear analysis is, and what sort of functional and ideological information it can reveal about plant-working tools in pre-agricultural society. We will then focus our discussion on examples drawn from Mesolithic sites in north-west Europe, reviewing evidence for various tools involved in subsistence and craft activities, including archaeological and experimentally used examples.

Microwear analysis

History and recent developments

Modern microwear studies began with the pioneering work of Sergei Semenov (1964) who recognised that wear traces develop on tools as a result of use. He distinguished polish, striations, edge rounding and micro-scarring and was the first to systematically use microscopes to study these phenomena. Semenov also stressed the importance of experimentation for providing a reference collection with which to compare wear traces visible on archaeological objects. Other highly important contributions by Semenov include the use of ethnographic information and the fact that he studied objects made of a variety of raw materials. In the decades that followed, most microwear studies were directed at flint tools, with tools made of other raw materials only coming into view from the 1990s onwards (Maigrot 1997; Hamon 2004; Kelly and Van Gijn 2008; Adams 2013). The debate between so-called low and high power proponents – those advocating the use of stereo-microscopes – versus those favouring metallographic ones (e.g., Keeley 1974; Odell and Odel-Vereecken 1980) has long been resolved, with both approaches now considered complementary and best practice when used together (Knutsson *et al.* 1990; Van Gijn 1990).

As already advocated by Semenov, ethnographic and anthropological case studies play an important role in informing microwear analysts about the type and range of plant-related tasks undertaken by past communities. Such information is crucial to the design of our experiments, which should also be closely linked to the information we have on the specific site(s) we are studying. Experimental work is not only important as a reference for archaeological comparisons, but also ensures you become more aware of the limitations and practicalities of a given task (Bamforth 2010). Even seemingly simple modifications, such as a change in the working angle of the tool or an additive such as water, fat or a mineral, can cause a very different type of polish appearance. However, sometimes, despite our best efforts to replicate plant polish types seen recurrently within assemblages, some types of polish remain enigmatic (see discussion below).

Microwear analysis has come a long way since the pioneering work of Semenov, and is now an accepted part of modern artefact studies. The collegial scrutiny, the small interactive international practical microscope sessions, the exchange of experimental results and the publication of blind tests can be regarded as comparatively well-established within the field of microwear. Increasingly we are incorporating other modern technological equipment, in addition to the traditional method of light microscopy, into our analytical 'toolkit'. The diversity of papers presented at the most recent microwear and residue conference in Faro, Portugal in 2012 is testament to the broad variety of analytical methods being developed as a means of investigating wear traces. This conference led to the establishment of AWRANA (The Association of Archaeological Wear and Residue Analysts: http://www.awrana.com/). For a comprehensive overview of recent developments in microscopy methods for identifying wear traces on archaeological tools see Evans *et al.* (2014).

However, despite the addition of new analytical techniques, the primary method for identifying use wear traces on tools remains the same. Most microwear specialists employ a stereo-microscope and a metallographic one, relying on a visual comparison between experimentally obtained traces and archaeological ones. It is important to realise that the inferences microwear analysts make should be considered interpretations, despite the high-tech equipment with which they work (Van Gijn 2014). Objects are used in a multitude of ways; they can also be re-sharpened and re-cycled sometimes thousands of years later, and obviously tools can be used for more than just one activity. So far, analytical techniques have not been able to cope with these complexities and the subjective, yet empirical assessment of the traces by the experienced microwear analyst is still crucial. Wear traces develop and are distinctive, and experienced wear trace analysts usually agree on their inferences. Yet, the subjectivity of these inferences, the lack of quantification and the long apprenticeship required continue to present challenges for microwear analysis.

Yet, despite these shortcomings, this method constitutes a great improvement on the functional inferences based on morphological similarities to present-day tools or objects which occurred before the development of artefact microwear studies. It is not uncommon for use-related polish to be identified that 'contradicts' the artefact's typological classification. In this respect, microwear can provide valuable and often contradictory (based on morphological form alone) insights about the function and use-life of a tool. A good example of this are the so-called 'sickles' found in the northern and western Netherlands that date to the Late Bronze and Early Iron Age. These are bifacially retouched implements with a crescent-shape, displaying a well-developed polish across most of their surface. Their shape and the gloss caused these implements to be not only classified, but even actually regarded as sickles, associated with harvesting cereals. However, the microwear traces on these tools do not resemble the traces on the experimental harvesting knives. Moreover, the tool edges are so blunt as to be completely ineffective. Further experimentation showed the tools to be very suitable for sod cutting, with the resulting microwear traces closely resembling those on the archaeological implements (Van Gijn 1999).

Microwear analysis is pivotal in reconstructing the cultural biography of objects and can offer a major contribution to the understanding of the technologies used in past societies and the materials employed. It can reveal the functional connections between different types of tools, showing how one type of tool is used to make another, or how different types of tools are linked in one specific task. Tools may be very function-specific, or they may, just like Swiss Army Knives, also be used for a variety of purposes, sometimes in a very opportunistic fashion. Microwear study can thus identify toolkits such as sets of tools used for plant working (Van Gijn 2008a; Little and Van Gijn in prep.).

Microwear in practice: how is it done?

The first step in microwear analysis of a tool is to avoid the temptation to look at it under a high power metallographic microscope! A lot of information about tool use can be obtained by studying it macroscopically. By using a simple stereoscope with magnifications typically ranging from 10–100×, the analyst is able to gain a comprehensive overview of the tool including the identification of areas of micro-chipping, edge rounding, micro-cracks, striations and noting areas of polish which leads to the identification of the used edge. This initial assessment, identifying key areas of interest, guides the analyst in the next stage of analysis. The tool is then placed under an incident light or metallurgical microscope and looked at under a range of increasing magnifications – typically 100–500×, with 200 or 300× most commonly used to characterise polish attributes and micro-striations (Fig. 7.2). A record is made of how invasive a polish is, its degree of linkage, on which face(s) it occurs (i.e., more ventral than dorsal), how it relates to use-retouch, its directionality, relative smoothness, and other topographical features such as pits and craters (Vaughan 1985; Van Gijn 1990; Fig. 7.3).

As Semenov discovered, depending on what material a stone tool has been used for (e.g., cutting reeds, scraping fresh hide, peeling tubers, incising pottery), the surface of the tool develops characteristic signatures of that use. Researchers at the Laboratory of Artefact Studies at Leiden University have conducted experimental work over the past couple of decades on a huge variety of materials, producing a corpus of several thousand replica tools with use-wear traces. This extensive reference collection is referred to and developed when trying to interpret the material(s) and action(s) seen microscopically on archaeological specimens. However, as discussed below, we are

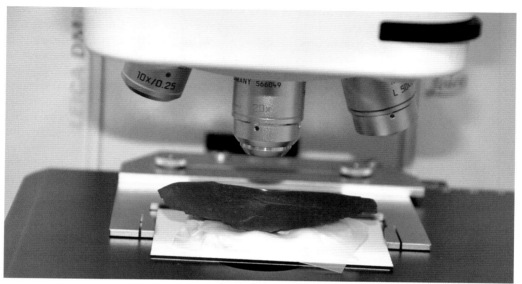

Fig. 7.2. Analysis of a chert tool using a metallurgical microscope (courtesy of Leiden Laboratory of Artefact Studies)

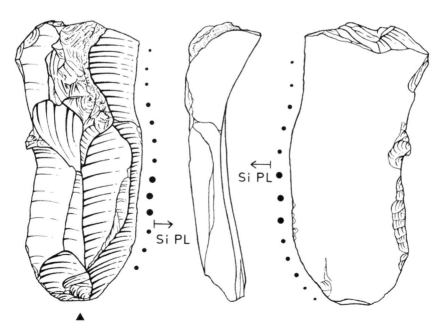

Fig. 7.3. An example of a use-wear analysts' record of siliceous plant traces on a flint blade (original length 4.5 cm) (courtesy of Leiden Laboratory of Artefact Studies)

yet to replicate all the archaeological polish and use wear traces that we encounter under the microscope and nor should we necessarily expect to; our ideas about tool use may not always be congruous with those expressed in the past. Fortunately, the general 'databank' of knowledge is increasing all the time with ever more detailed experimentation. Unfortunately, however, many experiments are not published, and when they are, quality micrographs are sometimes lacking; the dissemination of much of this knowledge could be greatly improved.

What do the results of microwear analysis tell us? What are the inferential limits of this method? First of all, it cannot be stressed enough that all inferences about tool function are essentially archaeological interpretations. The inferences are far more empirically based than the simple analogies often used in the past, but they are always based on comparison or similarities between experimentally obtained traces and archaeological ones. Even then, it can never be excluded that the same traces could result from a different task, one with which we had not yet experimented and is not reflected in the reference collection. Moreover, polishes from different contact materials frequently overlap in terms of their constituent attributes and this can limit our level of interpretive detail; for example, sometimes our statements will be limited to the hardness of the worked material (soft, medium, hard). The fact that people used their tools in a variety of ways – e.g., multiple use, rejuvenation, storage – further complicates the interpretation. On top of this, there are the contextual taphonomic modifications that invariably affect the surfaces of tools in various ways and excavation and post-excavation procedures can also have significant, detrimental effects. Lastly, there is of course the expertise of the individual microwear analyst. Every novice in this field takes a considerable time to learn to recognise and interpret the traces, and an excellent visual memory is essential.

Finally, absence of certain types of wear traces cannot be seen as absence of the associated activities, unless of course there is lot of corroborative evidence. Many activities do not lead to fast-developing wear traces, with some softer contact materials causing hardly any wear at all (Van den Dries and Van Gijn 1997). At complex wetland sites, such as Star Carr, plant-working microwear evidence is curiously absent (Dumont 1988), despite being situated in a plant-rich environment. This lack of evidence in itself does not mean plant-working tasks were not carried out as the samples examined are often relatively small in relation to total assemblage size. Additionally, a number of potentially biasing factors, including the issues of obscuring and overlapping of polishes discussed above, could contribute to this 'absence'. There is also the probability that tasks were spatially structured for practical and ideological reasons, as is known from modern hunter-gatherer communities elsewhere (Jordan 2003). As such, it is conceivable that at sites like Star Carr, plant-working activities took place, but fall outside the limit of modern excavation boundaries. The presence/absence of wear traces can hence be regarded as a seriously complicating factor that should be reviewed alongside other forms of archaeological evidence before conclusions are drawn on what activities are represented at a site.

Plants as subsistence as seen from the tool's perspective

Introduction

Microwear research has shown that flaked and grindstone tools played an important role in acquiring and processing plants for food in prehistory, especially during the Neolithic. Sickle blades, which display a highly reflective gloss acquired during use as cereal harvesting tools, were recognised as early as the late 19th century (Spurrell 1892). Such tools have been well-documented from the Neolithic onwards (Kaminska-Szymczak 2001; Ibáñez *et al.* 2013; Goodale *et al.* 2012; Clemente Conte and Gibaja 1998; Curwen 1930). Due to the distinctive and highly visible sheen that forms on the surface of flint and other stone types as a result of harvesting of cereals, sickle blades are relatively easy to identify, even without the aid of a microscope. However, not all tools displaying a glossy sheen are sickles (Van Gijn 1997). For example, some of them are threshing sledge inserts (Anderson *et al.* 2004) and some were involved in craft activities (see below). The processing of cereals has been well documented on archaeological querns from the European Neolithic (Hamon 2010; Verbaas and Van Gijn 2007), but less is known about Mesolithic grindstones, even though they are not uncommon either as single or multiple finds within assemblages dating to this period (Bos *et al.* 2006; Warren *et al.* 2009; Roda Gilabert *et al.* 2012) and even earlier (Revedin *et al.* 2010; Piperno *et al.* 2004). Moreover, ethnographic, anthropological and archaeological research has long since established that grindstones were a fundamental part of the plant food processing toolkit for hunter-gatherer communities across the globe (Hardy 2010, Field and Fullagar 1998, Fullagar *et al.* 2008)

Experimentation

Wear traces from harvesting and processing domestic cereals develop relatively fast and are distinctive. This is not the case with many other kinds of plant species, notably wild plants that would have been consumed by hunter-gatherers. Experiments have shown that cutting green plants only slowly results in traces of wear on the flint knives involved in procurement (Van den Dries and Van Gijn 1997). In many cases it is far quicker and more practical to hand pick the stems; this is the case with *Rumex*, very young *Urtica*, and various other leafy plants, all of which would provide greens for consumption. Similarly, many other plant foods like berries, tubers, and nuts would require no tool for collection and would require very minimal or no processing (Hardy 2007; Regnell 2012). Indeed, there are many species present in Mesolithic plant macro-remain assemblages that would not require a tool for collection or processing (Fig. 7.4). Other wild plants, on the other hand, are more effectively collected with the aid of a tool, usually a flint blade or flake, and do result in distinct traces. This is the case with most siliceous wild plants such as reeds and rushes like *Typha, Phragmites, Scirpus* and *Juncus*, plants actually more involved in craft activities. The resulting polish is very bright, smooth, highly linked and distributed in a band along the edge. In archaeological contexts, the differences between the traces resulting from contact with these species

Fig. 7.4. The makings of a delicious wild salad – like many plant foods known from European Mesolithic sites, these plants did not require collection or processing with any form of tool (courtesy of Leiden Laboratory of Artefact Studies)

of wild siliceous plants are subtle and are probably not discernable as we also have to deal with taphonomical processes affecting the surfaces of the tools (Plisson and Mauger 1988; Sala 1986).

Our experimental programme incorporated various plant processing techniques including attempts to replicate the transverse plant polishes that we regularly observe on Mesolithic and Early Neolithic blades (Fig. 7.5a–c). Our experiments comprised the peeling of various starchy tubers including yellow and white water lily (*Nuphar lutea* and *Nymphaea alba*), marsh marigold (*Caltha palustris*), and bulrush (*Typha*). The experiments were carried out with small unretouched blades with a slightly curved lateral side that made it ideal to slice along the surface of the root, removing the outer skin. Generally the roots, coming from water, were devoid of any adhering dirt so very few striations occurred. A narrow band of bright polish developed as a result, with very little edge damage occurring. However, by actually consuming these various tubers, we established that they were far tastier if roasted on a fire, after which the charred skin could simply be rubbed off before consumption (D. Pomstra, pers. comm.). Thus it is unlikely that the roots were peeled, although it is conceivable that this may have been done while roots were boiling in water; however, the experimental traces from this activity have not yet been encountered in an archaeological context. Another experiment was wild grass dehusking, using a flint blade held between thumb and forefinger and drawing the blade across the husk in a transverse motion (Fig. 7.6). This resulted in a narrow band of bright, smooth polish (Fig. 7.7) which differed from the transversally oriented polish from more robust siliceous plant species in the sense that it was much less invasive. Plant processing experiments with ground stone tools included the crushing of hazelnuts, acorns and peas. These activities resulted in impact traces on the surface, reflecting the pounding. Though their exact appearance varied somewhat according to the type of stone used, we have been unable to confidently detect any polish or striations that could allow us to differentiate between the various kinds of plants processed.

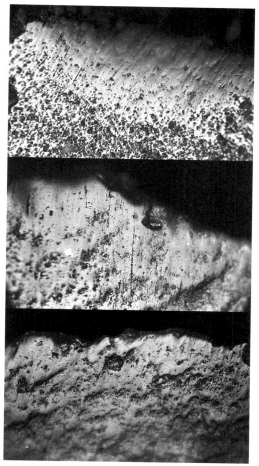

Fig. 7.5. *Variation in transverse siliceous plant polish on unretouched blades from Hardinxveld Polderweg (200×) (courtesy of Leiden Laboratory of Artefact Studies)*

Archaeological examples

While this may appear counter-intuitive, microwear evidence for the use of blade and flake blanks and other formal tool types to obtain or process food are surprisingly rare within northern European Mesolithic assemblages. From microwear evidence it is clear that some flint tools were used to cut plant material, but this does necessarily represent food collection; these microwear traces could equally reflect the procurement of raw materials for plant based crafts. Grinding stones are also rare, but important exceptions exist. The limited evidence we do have for food processing and gathering tools is unsurprising considering that many plant collecting activities can be done by hand. In fact, in many instances, tools can be more cumbersome than useful. Opening nuts can be done by any two unmodified stones, or two pieces of hard wood. Such naturefacts (Oswalt 1976) will rarely have been recognised by excavators as an intentionally used tool.

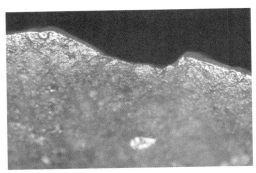

Fig. 7.6. Using a flint blade to dehusk wild grass seeds (courtesy of Leiden Laboratory of Artefact Studies)

Fig. 7.7. Polish from replicating the task of dehusking wild grass seeds (100×) (courtesy of Leiden Laboratory of Artefact Studies)

The traditional division between subsistence and craft is actually very artificial, as many craft activities are geared towards subsistence tasks. This pertains especially to fishing for which many different tools and contraptions are necessary, and often specifically designed for different fish species (Stewart 1977; see also Chapter 13). The same applies to the construction of traps for animals, for which a surprising variety is known, each requiring wood working and often fibres for ropes and attachments (Oswalt 1976). This means that many of the craft activities which will be discussed in the next section actually may have been part and parcel of the technological system behind many subsistence activities.

Plant-based craft activities as seen from the tools' perspective

Introduction

From our microwear data, craft activities are more commonly represented than those from food procurement. This is unsurprising as most craft activities such as wood working, hide scraping, and so forth need a range of tools. Microwear research of different categories of material culture from one site has shown the interconnectivity of these categories in terms of the technological processes (Van Gijn 2008b). However, extensive experimentation with obtaining and processing plants for crafts has shown that the number of tools required is actually very limited. For example, to obtain the siliceous plant material for basketry, an axe, chisels of bone, antler, or wood to obtain the tree bark, and some bone or wooden tools for the actual basketry, weaving and cording are needed (Van Gijn 2010).

Experimentation

We used small blades to scrape various siliceous plants (*Typha, Juncus*, and *Scirpus*) (Fig. 7.8) to make the stems pliable for incorporation into various plant based craft items (Fig. 7.9). Scraping the stems creates a series of consecutive breaks which cause them to retain their flexibility even when dry. This prepares the stems for

Fig. 7.8. Scraping fresh reeds with a flint blade (courtesy of Leiden Laboratory of Artefact Studies)

Fig. 7.9. Some of the plant-based craft products made experimentally. Clockwise from left to right: sandal of bulrushes (Scirpus), knotless net of lime bark, coiled basket made of rushes (Juncus), knotless net made of nettles (Urtica) and knotless net made of willow bark rope

Fig. 7.10. Scraping Salix *bark with a flint tool (courtesy of Leiden Laboratory of Artefact Studies)*

Fig. 7.11. Polish resulting from scraping Salix *bark with a flint tool (200×) (courtesy of Leiden Laboratory of Artefact Studies)*

use in basketry or matting. All of these experiments resulted in a smooth polish that was highly linked, very bright and smooth, with a clear directionality. As mentioned before, the slight differences in wear traces are too subtle to be detected in archaeological contexts and could be due to various other factors such as force exerted, the exact moisture content of the plant stems, gesture and so forth. We also conducted an experiment scraping *Urtica*; this was not done with a blade but with a *quartier d'orange*.

We also scraped willow and hazel bark with the aid of a flint tool (Fig. 7.10). This was effective, especially if using an obtuse angled edge. The polish is less linked than that from siliceous plants, but equally smooth and bright (Fig. 7.11) and the activity produces good quality fibres (Fig. 7.12). Lime bark processing does not require a tool although very soft fibres can be obtained by pounding for a long time with a wooden billet (D. Pomstra pers. comm.). Ground stone tools featured only minimally in our experiments with plant based crafts. We used a hammerstone to crush the *Urtica* stems prior to the removal of fibres but this was not essential and, in general, we preferred a wooden hammer.

Bone tools used in craft activities are principally awls used for making openings in coiled basketry (Fig. 7.13). The polish from different plant species does differ, with some plants causing a much more striated surface than others (Fig. 7.14). For example, making a basket from twined nettle rope produces a matt, finely striated polish, whereas piercing birch bark causes a much rougher polish with short deep striations. We also used a bone needle made from a swan ulna to make a carrying basket using the knotless netting technique; this proved to be a highly effective tool (see Fig. 7.15).

Fig. 7.12. *Rope made from* Salix *bark fibre (courtesy of Leiden Laboratory of Artefact Studies)*

Fig. 7.13. *Using a bone awl for basketry making (courtesy of Leiden Laboratory of Artefact Studies)*

Fig. 7.14. *Polish on bone awl used in weaving* Juncus *for basketry (200×) (courtesy of Leiden Laboratory of Artefact Studies)*

Archaeological examples

Microwear research has revealed a number of tools which appear to be involved in craft-based activities. A good example is the transversely oriented siliceous plant polish on flint tools that is well-known from Mesolithic and early Neolithic contexts in north-west Europe. This polish is very bright, highly linked, and smooth and is normally associated with a somewhat rounded edge oriented perpendicular or slightly oblique to the edge (see Figs 7.5a–c). Although this polish seems ubiquitous across this period at this time, key differences exist – both in terms of the appearance of the polish and the tool form on which it is found. As yet it is not fully understood what these differences represent. For example, in southern Scandinavia this polish is associated with a rough, hide-like polish on the contact side and is found on denticulate and micro-denticulate blade forms dated to the Late Mesolithic Ertebølle culture. It is also known from French and Belgian Mesolithic sites and has been identified within French Chasséen culture assemblages, commonly occurring on notched blades (Gassin *et al.* 2013; Crombé and Beugnier 2013; Gassin *et al.* 2014; Perdaen *et al.* 2008; Guéret 2013). The blades with transverse siliceous plant polish identified from Late Mesolithic and the Early Neolithic Swifterbant culture assemblages in the Netherlands differ morphologically to the denticulates and notched forms in that they are almost always unmodified and the rough, hide-like aspect is missing (Van Gijn 2010; Van Gijn *et al.* 2001a; 2001b; Little and Van Gijn in prep.). Trying to understand the plant type and actions that led to the formation of this polish has been a focus of study for numerous microwear specialists over the past 25 years and despite extensive experimentation, an exact match has not yet been produced. The closest (but not perfect) parallel to the traces on these unretouched blades from the Netherlands and Belgium is the scraping of fresh (green) summer reeds. It should be noted that these tools, ubiquitous in the Dutch coastal sites during the late Mesolithic and the Early Neolithic Swifterbant culture, disappeared by the Middle Neolithic, as seen at the coastal site of Schipluiden (Van Gijn *et al.* 2006). This is noteworthy as crop growing is unequivocally confirmed in the Dutch coastal wetlands at this time. It is

Fig. 7.15. Perforated bone needle made from an ulna of a swan with geometric design from the Mesolithic site of Hardinxveld de Bruin (length of needle = 9.0 cm) (courtesy of Leiden Laboratory of Artefact Studies)

Fig. 7.16. Experimental work: replicating the possible function of the Hardinxveld De Bruin bone needle (courtesy of Archeon, Alphen aan de Rijn)

thus likely that the craft activity these blades were involved in, may have been closely tied to a subsistence task crucial for the broad spectrum economy practiced by hunter-gatherers living in these wetlands, before crop growing took a firm hold (Van Gijn 2010, 66; Little and van Gijn in prep.).

Evidence for plant based craft activities has also been identified from microwear analysis of the bone tools from the Late Mesolithic and Early Neolithic wetland sites in the Hardinxveld region. Specifically, some of the awls had evidence for use on plant materials. The most striking example is the perforated bone needle from Hardinxveld De Bruin, which is decorated with geometric patterns (Fig. 7.15). This tool displayed a highly polished surface which bore a close resemblance to the experimental one described above, suggesting that netting was practiced on site (Fig. 7.16). This is further supported by the discovery of woven fragments made in the knotless netting technique at the nearby site of Hardinxveld Polderweg.

Conclusion

There is no doubt that tools retain traces from plant working. Analysis of these traces provides us, indirectly, with information on the use of plants by past communities. The traces can be seen on a variety of tools, including flint implements, ground stone tools and objects of bone and antler. They can be studied by means of microwear analysis and experimentation. In circumstances where organic preservation is good, such studies enable us to reveal the interconnectivity of different *chaînes opératoires*, that help to understand links between different categories of materials in terms of manufacture and function. For example, at Hardinxveld-Giessendam Polderweg and De Bruin, the bone needle may have been used to produce the fragments of nets that were recovered. In archaeological sites, where preservation is poor and we are left with only stone tools, the wear traces provide indirect information on plant-working activities, which would otherwise remain invisible in the archaeological record. For this reason, microwear research is a valuable method: a lens that enables us to focus on the multitude of utilitarian and non-utilitarian uses of wild plants by past hunter-gatherer communities.

Acknowledgements

Van Gijn would like to acknowledge the financial and above all practical support of the plant based craft experiments by the Lejre Forsogcentre (now called Sagnlandet Lejre) in Denmark. She especially thanks Marianne Rasmussen for her support over the years. Annemieke Verbaas, Jeroen de Groot and Dorothee Olthoff carried out several of the experiments and their enthusiasm and dedication is greatly appreciated. Little was funded by a Marie Curie Intra European Fellowship (2011–13), which provided the opportunity to train as a microwear apprentice with Annelou van Gijn and her laboratory staff. In particular, as well as Annelou van Gijn, she is grateful to Annemieke Verbaas, Eric Mulder, Virginia Garcia-Diaz and Geeske Langejans. From Archaeobotany she would like to extend a warm thank you for 'fruitful' discussions to Erica van Hees and Wim Kuijper.

References

Adams, J. L. 2013. *Ground Stone Analysis; A Technological Approach.* Salt Lake City UT: University of Utah Press

Andersen, S. H. 2013. *Tybrind Vig: Submerged Mesolithic Settlements in Denmark.* Højbjerg: Jutland Archaeological Society

Anderson, P., Astruc, L., Vargiolu, R. and Zahouani, H. 1996. Contribution of quantitative analysis of surface states to a multimethod approach for characterising plant processing traces on flint tools with gloss. *XIII International congress of Prehistoric and Protohistoric Sciences (UISPP)*, 1151–60. Forli: ABACO

Anderson, P. C., Chabot, J. and Van Gijn, A. L. 2004. The functional riddle of 'glossy' Canaanean blades and the Near Eastern threshing sledge. *Journal of Mediterranean Archaeology* 17, 13–87

Bamforth, D. B. 2010. Conducting experimental research as a basis for microwear analysis. In Ferguson, J. R. (ed.), *Designing Experimental Research in Archaeology: examining technology through production and use*, 93–109. Boulder CO: University Press of Colorado

Bos, J. A., Geel, B., Groenewoudt, B. and Lauwerier, R. G. M. 2006. Early Holocene environmental change, the presence and disappearance of early Mesolithic habitation near Zutphen (The Netherlands). *Vegetation History & Archaeobotany* 15, 27–43

Clemente Conte, I. and Gibaja, J. F. 1998. Working processes on cereals: An approach through microwear analysis. *Journal of Archaeological Science* 25, 457–64

Crombé, P. and Beugnier, V. 2013. La fonction des industries en silex et les modalités d'occupation des territoires au Mésolithique. Le cas des zones sableuses du nord-ouest de la Belgique et des Pays-Bas (8700–5400 cal. BC). *L'Anthropologie* 117, 172–94

Curwen, E. C. 1930. Prehistoric flint sickles. *Antiquity* 9, 62–6

Dumont, J. V. 1988. *A Microwear Analysis of Selected Artefact Types from the Mesolithic Sites of Star Carr and Mount Sandel.* Oxford: British Archaeological Report 187

Evans, A. A., Lerner, H., MacDonald, D. A., Stemp, W. J. and Anderson, P. C. 2014. Standardization, calibration and innovation: a special issue on lithic microwear method. *Journal of Archaeological Science* 48, 1–4

Field, J. and Fullagar, R. 1998. Grinding and pounding stones from Cuddie Springs and Jinmium. In Fullagar, R. (ed.), *A Closer Look: Recent Studies of Australian Stone Tools,* 95–108. Sydney: University of Sydney

Fitzgerald, M. 2007. Catch of the day at Clowanstown, Co. Meath. *Archaeology Ireland* 21, 12–15

Fullagar, R. 1991. The role of silica in polish formation. *Journal of Archaeological Science* 18, 1–24

Fullagar, R., Field, J. and Kealhofer, L. 2008. Grinding stone and seeds of change: starch and phytoliths as evidence of plant processing. In Rowan, Y. M. and Ebeling, J. R. (eds), *New Approaches to Old Stones: Recent Studies of Ground Stone Artifacts,* 159–172 London: Equinox

Gassin, B., Gibaja, J. F., Allard, P., Boucherat, D., Claud, E., Clemente, I., Guéret, C., Jacquier, J., Khedhaier, R., Marchand, G., Mazzucco, N., Palomo, A., Perales, U., Perrin, T., Philibert, S. and Torchy, L. 2014. Late Mesolithic notched blades from western Europe and North Africa: technological and functional variability. In Marreiros, J. M., Gibaja Bao, J. F. and Bicho, N. F. (eds), *Use-Wear and Residue Analysis in Archaeology,* 224–231. Newcastle upon Tyne: Cambridge Scholars Publishing

Gassin, B., Marchand, G., Claud, E., Guéret, C. and Philibert, S. 2013. Les lames à coches du second Mésolithique :des outils dédiés au travail des plantes? *Bulletin de la Société Préhistorique Française* 110, 25–46

Goodale, N., Andrefsky, W. J., Otis, H., Kuijt, I., Finlayson, B. and Bart, K. 2012. Reaping 'rewards' in sickle use-wear analysis. *Journal of Archaeological Science* 39, 1908–10

Gueret, C. 2013. *L'Outillage du Premier Mésolithique dan le Nord de la France et en Belgique. Eclairages fonctionnels.* Unpublished PhD thesis, L'Universite Paris 1 Panthéon-Sorbonne, Paris

Hamon, C. 2004. Le statut des outils de broyage et d'abrasion dans l'espace domestique au Néolithique ancien en Bassin Parisien. *Notae Praehistoricae* 24, 117–28

Hamon, C. 2010. Use and function of grinding tools in Early Neolithic settlements of Western Europe. *Il Seminario de Tecnologica Prehistorica: Estudio e Interpretavion del Macroutillaje.* Barcelona: Institucion Milá Fontanals CSIC-Barcelona

Hardy, K. 2007. Food for thought: starch in the Mesolithic diet. *Mesolithic Miscellany* 18, 1–27

Hardy, K. 2010. Ethnography of grinding stones. *Il Seminario de Tecnologica Prehistorica: Estudio e Interpretacion del Macroutillaje.* Barcelona: Institucion Milá Fontanals CSIC-Barcelona

Hurcombe, L. 2008a. Looking for prehistoric basketry and cordage using inorganic remains: the evidence from stone tools. In Longo, L. and Skakun, N. (eds), *'Prehistoric Technology' 40 years later: functional studies and the russian legacy,* 205–16. Oxford: British Archaeological Report S1783

Hurcombe, L. 2008b. Organics from inorganics: using experimental archaeology as a research tool for studying perishable material culture. *World Archaeology* 40, 83–115

Ibáñez, J. J., González-Urquijo, J. E. and Gibaja, J. 2013. Discriminating wild vs domestic cereal harvesting micropolish through laser confocal microscopy. *Journal of Archaeological Science* 48, 96–103

Jordan, P. 2003. *Material Culture and Sacred Landscape: the anthropology of the Siberian Khanty*. Oxford: Alta Mira

Kaminska-Szymczak, J. 2001. Cutting Graminae tools and 'Sickle Gloss' formation. *Lithic Technology* 27, 111–51

Keeley, L. H. 1974. Techniques and methodology in micro-wear studies: a critical review. *World Archaeology* 5, 323–36

Kelly, H. and Van Gijn, A. L. 2008. Understanding the function of coral tools from Anse a la Gourde: an experimental approach. In Hofman, C. L., Hoogland, M. and Van Gijn, A. L. (eds), *Crossing Disicplinary Boundaries and National Borders. New Methods and Techniques in the Study of Archaeological Materials from the Caribbean*, 115–124. Tuscaloosa AL: Alabama University Press

Knutsson, H., Knutsson, K. and Taffinder, J. 1990. *The Interpretative Possibilities of Microwear Studies*. Uppsala: Societas Archaeologica Upsaliensis

Little, A. and Van Gijn, A. in prep. *Flint and Bone Plant-Working Tools in the Dutch Mesolithic and Neolithic*

Louwe Kooijmans, L. P., Hanninen, K. and Vermeeren, C. 2001a. Artefacten van hout. In Louwe Kooijmans, L. P. (ed.) *Hardinxveld-Giessendam De Bruin. Een kampplaats uit het Laat-Mesolithicum en het begin van de Swifterbant-cultuur (5500-4450 v.Chr.)*, 435–478. Amersfoort: RAM 83

Louwe Kooijmans, L. P., Vermeeren, C. and Waveren, A. M. I. V. 2001b. Artefacten van hout en vezels. In Louwe Kooijmans, L. P. (ed.), *Hardinxveld-Giessendam Polderweg. Een Mesolithisch Jachtkamp in Het Rivierengebied (5500–5000 v. Chr.)*, 379–418. Amersfoort: RAM 88

Maigrot, Y. 1997. Tracéologie des outils tranchants en os des Ve et IVe millénaire av. J.-C. en Bassin Parisien. Essai méthodologique et application. *Bulletin de la Societé Préhistorique Française* 94, 198–216

MCQuade, M. and O'Donnell, L. 2007. Late Mesolithic fish traps from the Liffey estuary, Dublin, Ireland. *Antiquity* 81, 569–84

Mossop, M. 2009. Lakeside developments in County Meath, Ireland: a late Mesolithic fishing platform and possible mooring at Clowanstown 1. In McCartan, S., Schulting, R., Warren, G. and Woodman, P. C. (eds), *Mesolithic Horizons*, 895–9. Oxford: Oxbow Books

Odell, G. H. and Odel-Vereecken, F. 1980. Verifying the reliability of lithic usewear assessments by blind tests: the low-power approach. *Journal of Field Archaeology* 7, 87–120

Oswalt, W. 1976. *An Anthropological Analysis of Food-getting Technology*. New York: Wiley

Perdaen, Y., Crombé, P. and Sergant, J. 2008. Redefining the Mesolithic: technological research in Sandy Flanders (Belgium) and its implication for North-western Europe. In Sørensen, M. L. S. and Desrosiers, P. (eds), *Technology in Archaeology*, 125–47. Paris: Mouton

Piperno, D. R., Weiss, E., Holst, I. and Nadel, D. 2004. Processing of wild cereal grains in the Upper Paleolithic revealed by starch grain analysis. *Nature* 430, 670–3

Plisson, H. and Mauger, M. 1988. Chemical and mechanical alteration of microwear polishes. An experimental approach. *Helinium* 28, 3–16

Regnell, M. 2012. Plant subsistence and environment at the Mesolithic site Tågerup, southern Sweden: new insights on the 'Nut Age'. *Vegetation History & Archaeobotany* 21, 1–16

Revedin, A., Aranguren, B., Becattini, R., Longo, L., Marconi, E., Lippi, M. M., Skakun, N., Sinitsyn, A., Spiridonova, E. and Svoboda, J. 2010. Thirty thousand-year-old evidence of plant food processing. *Proceedings of the National Academy of Sciences (USA)*107, 18815–19

Roda Gilabert, X., Martínez-Moreno, J. and Mora Torcal, R. 2012. Pitted stone cobbles in the Mesolithic site of Font del Ros (Southeastern Pre-Pyrenees, Spain): some experimental remarks around a controversial tool type. *Journal of Archaeological Science* 39, 1587–98

Sala, I. L. 1986. Use wear and post-depositional surface modification: a word of caution. *Journal of Archaeological Science* 13, 229–44

Semenov, S. A. 1964. *Prehistoric Technology*. London: Moonraker Press

Spurrell, F. C. J. 1892. Notes on early sickles. *Archaeological Journal* 49, 53–69

Stewart, H. 1977. *Indian Fishing. Early Methods on the Northwest Coast,* Vancouver: J. J. Douglas

Unger-Hamilton, R. 1984. The formation of use-wear polish on flint: the deposit versus abrasion controversy. *Journal of Archaeological Science* 11, 91–8

Van den Dries, M. H. and Van Gijn, A. L. 1997. The representativity of experimental usewear traces. In Ramos-Millán, A. and Bustillo, M. A. (eds) *Siliceous Rocks and Culture,* 499–513. Granada: Universidad de Granada

Van Gijn, A. L. 1990. *The Wear and Tear of Flint. Principles of Functional Analysis Applied to Dutch Neolithic Assemblages.* PhD thesis, Leiden University (appeared as *Analecta Praehistorica Leidensia* 22)

Van Gijn, A. L. 1997. Introduction: there is more to life than butchering and harvesting. In Van Gijn, A. L. (ed.) *Enigmatic Wear Traces on Stone Implements. Evidence for Handicraft in Prehistory. Helinium* 34/2

Van Gijn, A. L. 1999. The interpretation of 'sickles'. A cautionary tale. In Anderson, P. C. (ed.) *The Prehistory of Agriculture,* 363–372. Los Angeles CA: Cotsen Institute of Archaeology

Van Gijn, A. L. 2008a. 'Toolkits' en het technologisch systeem in het Midden-Neolithicum van Schipluiden. In Flamman, J. and Besselsen, E. A. (eds), *Het verleden boven water. Archeologische monumentenzorg in het AHR-project,* 141–155. Delft/Amersfoort: RACM

Van Gijn, A. L. 2008b. Toolkits and technological choices at the Middle Neolithic site of Schipluiden, The Netherlands. In Longo, L. and Skakun, N. (eds), *'Prehistoric Technology' 40 years Later,* 217–226. Oxford: British Archaeological Report S1783

Van Gijn, A. L. 2010. *Flint in Focus. Lithic Biographies in the Neolithic and Bronze Age.* Leiden: Sidestone Press

Van Gijn, A. L. 2014. Science and interpretation in microwear studies. *Journal of Archaeological Science* 48, 166–9

Van Gijn, A. L., Beugnier, V. and Lammers, Y. 2001a. Vuursteen. In Louwe Kooijmans, L. P. (ed.), *Hardinxveld-Giessendam Polderweg. Een mesolithisch jachtkamp in het rivierengebied (5500–5000 v. Chr.),* 119–161. Amersfoort: RAM 83

Van Gijn, A. L., Lammers, Y. and Houkes, R. 2001b. Vuursteen. In Louwe Kooijmans, L. P. (ed.), *Hardinxveld-Giessendam De Bruin. Een kampplaats uit het Laat-Mesolithicum en het begin van de Swifterbant-cultuur (5500–4450 v.Chr.),* 153–191. Amersfoort: RAM 88

Van Gijn, A. L., Van Betuw, V., Verbaas, A. and Wentink, K. 2006. Flint. Procurement and use. In Louwe Kooijmans, L. P. and Jongste, P. F. B. (eds), *Schipluiden. A Neolithic site on the Dutch north sea coast (3800–3500 BC),* 129–166. *Analecta Praehistorica Leidensia* 37/38

Vargiolu, R., Zahouani, H. and Anderson, P. C. 2003. Étude tribologique du processus d'usure des lames de silex et fonctionnement du tribulum. In Anderson, P. C., Cummings, L. S., Schippers, T. K and, Simonel, B. (eds) *Le traitement des récoltes: un regard sur la diversité, du néolithique au présent XXIIIe rencontres internationales d'archéologie et d'histoire d'Antibes,* 439–454. Antibes: Éditions APDCA

Vaughan, P. 1985. *Use-Wear Analysis of Flaked Stone Tools.* Tuscon AZ: University of Arizona Press

Verbaas, A. and Van Gijn, A. L. 2007. Querns and other hard stone tools from Geleen-Janskamperveld. In Velde, P. V. D. (ed.) *Excavations at Geleen-Janskamperveld 1990–1991,* 191–204. *Analecta Praehitsorica Leidensia* 39

Warren, G., Little, A. and Stanley, M. 2009. A Late Mesolithic lithic scatter from Corralanna, County Westmeath, and its place in the Mesolithic landscape of the Irish Midlands. *Proceedings of the Royal Irish Academy* C, 1–35

8. Buccal dental microwear as an indicator of diet in modern and ancient human populations

Laura Mónica Martínez, Ferran Estebaranz-Sánchez and Alejandro Pérez-Pérez

Buccal microwear patterns on enamel surfaces of primate teeth have shown a significantly higher interspecific than intraspecific variability. Striation density and length by orientation categories on enamel buccal surfaces vary depending on ingested foods and the abrasive potential of chewed food particles, including phytoliths from plant foods and silica dust incorporated into food during procurement and processing. Both Cercopithecoidea and Hominoidea primates have shown distinct patterns of buccal microwear that can be related to dietary habits and ecological conditions at species and even population levels. The diversity of hominin and human buccal microwear patterns is sufficient to be used to test current hypotheses about the evolution of dietary habits in association with climatic and ecological conditions throughout the Plio-Pleistocene in Africa. We report on the most significant buccal microwear patterns, and we compare these to evidence derived from both occlusal microwear analyses and the most recent stable isotopic data.

Diet is the most significant factor affecting skeletal morphology, body size, behaviour, and ecology of a species (Grant and Grant 1985; Fleagle 1999). Animals spend much of their daily routine searching, manipulating, and ingesting food. Seasonal food availability also affects their feeding behaviour patterns, population size, and social structure. Moreover, mechanical properties of food influence both cranial and postcranial morphology, in particular those structures related to mastication and ingestion (Grine *et al.* 2006b).

For these reasons, the analysis of dietary habits within the context of the specific environment that species evolve in, contributes to a better understanding of overall feeding adaptations. Shifts in diet and resource exploitation are factors of directional natural selection in relation to those characteristics responsible for food acquisition and breakdown (Reznick and Ghalambor 2001) that can result in speciation processes. Thus, their study can provide clues to reconstructing the ecology of Hominini species.

Buccal microwear formation

Dental wear occurs through repeated movements of the lower maxilla during the chewing cycle. Food chewing movements include two consecutive phases: an initial *puncture and crushing* phase followed by the actual *chewing* phase when foodstuffs are ground.

Dental microwear can be studied on both the surface facing the cheeks, called buccal, and on the tooth-to-tooth contact surface, called occlusal. Buccal enamel microwear is a microscopic abrasion process that affects dental vestibular surfaces during food chewing. Chewing cycles and the action of cheek compression during biting loads causes clear non-occlusal enamel imprints called striations on buccal enamel surfaces (Ungar and Teaford 1996; Pérez-Pérez *et al.* 2003; Lucas 2004; Romero *et al.* 2012). The density, length, and orientation patterns of such buccal microwear features depend on the amount and composition of silica and quartzite particles present in ingested foodstuff. These particles, among which phytoliths are a significant portion, are harder than enamel hydroxyapatite crystals. Thus, they can damage tooth enamel surfaces by scratching them while food moves around the mouth during chewing. Phytoliths are particularly abundant in plant foods such as leaves, culms, shoots and some type of fruits. Striations can also be produced by quartzite particles that are present in dust and ashes that might be incorporated into food during its processing.

Experimental *in vitro* analyses have shown a direct relationship between phytoliths and grit particles with the formation of microwear features on buccal molar surfaces (Puech *et al.* 1985; King *et al.* 1999; Romero *et al.* 2012). In addition, indirect evidence of the abrasive effects of such particles impacted on buccal enamel surfaces or included in dental calculus has been reported, frequently in close association with scratches (Lalueza *et al.* 1996; Gügel *et al.* 2001; Rücker 2006). On the other hand, dental enamel damage produced during tooth-to-tooth contact or tooth-food-tooth interaction is determined by the relative hardness and physical properties of both ingested foods and tooth enamel surfaces (Maas and Dumont 1999; Lucas 2004). Therefore, dental microwear is presumably formed at different rates on occlusal and buccal surfaces. Microscopic examination of dental crown working surfaces shows a distinct and more complex enamel damage pattern on occlusal than buccal surfaces (Krueger *et al.* 2008). Occlusal microwear patterns consist of both pits and scratches of various shapes and sizes, while buccal microwear patterns lack pits and only show striations of variable lengths (Pérez-Pérez *et al.* 2003; Pérez-Pérez 2004; Romero *et al.* 2012). Consistent results from several analyses have emphasized the absence of pit formation on well-preserved buccal surfaces (i.e., those surfaces not affected by *post-mortem* damage) (Fine and Craig 1981; Puech and Albertini 1984; Ungar and Teaford 1996; Galbany *et al.* 2009), as well as on occlusal rims (Ungar and Spencer 1999) and on occlusal shearing facets lacking crushing phases (Goswami *et al.* 2005; Schubert and Ungar 2005; Williams *et al.* 2009).

Dental casts and SEM analysis

Dental microwear analysis is a two-stage technique: firstly, moulds of the original dental samples must be obtained and cast; secondly, buccal enamel surfaces are studied using scanning electron microscopy (SEM) and the microwear pattern is evaluated. Only fossil specimens with well-preserved buccal enamel surfaces can be considered for numerical analyses. Therefore, all teeth that are suspected of having *post-mortem* damage are discarded in order to exclude non-dietary enamel alterations (Ungar and Teaford 1996; King *et al.* 1999; Galbany and Pérez-Pérez 2004; Martínez and Pérez-Pérez 2004).

High-resolution negative molds are generally obtained using President MicroSystem Regular Body (Coltène®) impression material, a polyvinylsiloxane resin widely used in dental microwear studies due to its stable physical properties and high-detail resolution (Beynon 1987; Teaford and Oyen 1989; Galbany *et al.* 2004). Positive casts are then made with a high-resolution epoxy polymer (Epotek 301, Epoxy Technologies, Inc., Billerica, MA) or, alternatively, with a two-component polyurethane Feropur PR-55 (Feroca™). The resulting casts are mounted on aluminum stubs with thermo-fusible gum before SEM observation. Traditionally, in order to prevent electron saturation in the SEM chamber, a colloidal silver layer is applied to the gum ring (Electrodag 1415 M-Acheson Colloiden) and the dental cast is coated with a 400 Å gold layer (Galbany *et al.* 2004). However, environmental microscopes (ESEM) can operate at low pressure with a water vapor atmosphere, so gold-coating is not needed. SEM micrographs (100× magnification) of the buccal enamel surface are obtained in the secondary electron detection mode, at a 15 kV acceleration voltage and at 20–30 mm working distance (Lalueza *et al.* 1996; Pérez-Pérez *et al.* 1999; Galbany *et al.* 2004; Galbany and Pérez-Pérez 2004). SEM images are always obtained from the middle third of the vestibular enamel surfaces, avoiding both occlusal wear (at the occlusal rim) and the cervical rim (Pérez-Pérez *et al.* 1994; 1999; 2003).

The size of all SEM images is standardised to cover exactly 0.56 mm² of the enamel surface using Adobe Photoshop™. A high-pass filter (50 pixels) and an automatic level adjustment are used to normalise the grey level frequency histogram and to increase contrast in order to facilitate scratch characterisation and measurement (Galbany *et al.* 2004). Finally, microwear striations are measured (length and slope or orientation of each observed striation) using a statistical package, such as SigmaScan Pro5 (SPSS) or ImageJ. Scratch slopes or orientations need to be measured in relation to the cement-enamel junction plane, which is used to correctly position the tooth inside the SEM chamber. If scratch slopes are obtained instead of angular measurements, they can be easily transformed into degrees with respect to the horizontal plane. The angular orientation values are then transformed into four 45º discrete orientation categories around the vertical orientation range: vertical (V, 67.5º–112.5º), horizontal (H, 0º–22.5º and 112.5º–180º), mesio-occlusal to disto-cervical (MD), and disto-occlusal to mesio-cervical (DM). The angular values of the MD and DM scratches vary between 22.5º–67.5º and 112.5º–157.5º but their attribution depends on the position of the tooth in the mouth (upper, lower, left, right). Once all the striations have been measured, summary statistics of striation densities (N), average striation lengths (X), and standard deviations of the lengths (S) are derived by orientation categories (V, H, MD, DM), as well as for all striations (T), a procedure that provides 15 quantitative variables (NT, XT, ST, NV, XV, SV, NH, XH, SH, NMD, XMD, SMD, NDM, XDM, SDM). These constitute the buccal microwear pattern of each specimen studied.

Buccal microwear variability of *in vivo* populations

Preliminary analyses of *in vivo* microwear variability have shown that occlusal microwear pattern is dependent on diet abrasiveness (Teaford and Tylenda 1991; Teaford and Lyte 1996). Since then, many occlusal microwear studies have been published, based on the assumption that occlusal microwear patterns are strongly correlated with the amount of abrasives that a particular diet contains. Regarding buccal enamel surfaces, the assumption that cheek loadings against teeth were sufficiently strong to cause the formation of scratches has been a subject that has been discussed for decades. Initially, a direct cause-effect relationship was assumed between chewing cycles and buccal microstriations through the interaction of abrasive particles with enamel surfaces in the absence of tooth-to-tooth contact (Puech *et al.* 1980). Several studies subsequently demonstrated a direct relationship between buccal microwear formation and long-term diet – related abrasion during the life span of an individual (Pérez-Pérez *et al.* 1994; Romero and De Juan 2007). Though many dietary reconstructions based on buccal microwear patterns have been published in the last two decades, several authors have remained sceptical about the reliability of buccal microwear research (Grine *et al.* 2012). The principal criticisms have focused on the belief that buccal microwear is not evidence of diet, but rather a result of post-mortem damage in fossilised teeth. These critics persisted even after Galbany *et al.* (2005; 2009) confirmed that *in vivo* buccal microwear patterns were highly informative about dietary habits in wild extant primates. Galbany's results showed that buccal microwear was a natural process observed in wild primate populations; thus, its formation could no longer be linked to *post-mortem* damage. Nevertheless, controversy was not definitively over until 2012, when Romero (Romero and De Juan 2007; Romero *et al.* 2012) published the results of a 5 year long-term *in vivo* experimental analysis with induced and non-induced diets in humans. These experimental analyses showed that highly abrasive particles included in the diet accelerate the formation of new microwear features on buccal enamel surfaces compared to those formed on enamel surfaces of a subject consuming a normal basal diet (Romero *et al.* 2012). Although these long-term analyses demonstrated scratch turnover values are higher than previously expected (Pérez-Pérez *et al.* 1994; 1999; 2003), they were still inferior to occlusal microwear turnover. Weekly percent rates of formation of new features were much slower than those found in induced or *ad-libitum* diets in occlusal enamel facets: *c.* 2% versus 3.1–3.7%, respectively (Teaford and Lytle 1996; Romero *et al.* 2012). No significant long-term trends in microwear patterns, either for striation density or length, were observed because microwear formation was compensated by scratch disappearance (Romero *et al.* 2012). Consequently, Romero's analyses suggested that an increase in diet abrasiveness had similar effects for both occlusal and buccal microwear patterns. These results confirm that buccal microwear patterns are a dynamic process that are not dependent on tooth-to-tooth contact and are directly related to the chewing of abrasive particles included with ingested food.

Buccal microwear variability of modern hunter-gatherer populations

An analysis of buccal microwear variability of modern human population with well-known diets has recently been published by Romero *et al.* (2013) who analysed the buccal microwear signaling of Pygmy hunter-gatherers and non-Pygmy Bantu pastoralist and farmer populations from Central Africa, who show distinct and well-characterised diets. Differences in buccal microwear pattern were clearly related to actual differences in dietary habits, as well as to food processing techniques, such as cooking. Results showed that Pygmy hunter-gatherers and non-Pygmy Bantu pastoralists shared a microwear pattern showing relatively low striation densities, in contrast to the denser pattern observed in Bantu farmers. Such difference was linked to significant differences in dietary preferences. Both Pygmy hunter-gatherers and Bantu pastoralists share a similar soft dietary pattern based on meat and foraged plant foods, i.e., diets mainly based on nonabrasive foodstuffs, whereas the diet of Bantu farmers incorporated large amounts of exogenous gritty particles from stoneground foods. Thus, Romero's buccal microwear analyses on modern African societies brought new evidence that buccal microwear patterns reflect dietary habits that are independent of ecological conditions. Moreover, as also shown by Galbany *et al.* (2009), among extant human populations, the buccal microwear in African Pygmy and Bantu populations was clearly unrelated to *post-mortem* damage and constitutes, thus, an alternative and fully informative technique to reconstruct present and past dietary adaptations.

Buccal microwear variability of ancient hunter-gatherer populations

Intra-specific patterns of buccal dental microwear variability among Pleistocene and Holocene hunter-gatherers and agro-pastoral populations (Pérez-Pérez *et al.* 1994; 1999; 2003; Lalueza *et al.* 1996; Romero and De Juan 2007) have shown significant variability of striation patterns in relation to dietary habits, ecological, and climatic adaptations, as well as economic strategies related to food availability and food processing techniques. Buccal microwear analyses in mainly carnivorous hunter-gatherers, such as the Inuit or Fuegians, have shown microwear patterns characterised by lower striation densities than the Khoe-San or Australian Aboriginal hunter-gatherers from arid or mesothermal environments, which is indicative of a diet low in abrasives. Moreover, the carnivorous hunter-gatherer groups were characterised by high frequencies of vertical striations (Fig. 8.1) while the agriculturalists, such as the Hindu group, showed high densities of horizontal striations (Lalueza *et al.* 1996); this may be related to higher amounts of phytoliths in their diets (Baker *et al.* 1959; Ciochon *et al.* 1990).

Paleoecology and buccal microwear of Plio-Pleistocene hominins

During the Pliocene (5.3–2.6 myr) climatic conditions in East Africa became progressively drier and cooler (Ravelo *et al.* 2004), with increasing seasonality (Cerling *et al.* 1997). In this temporal span, *A. anamensis* fossils appeared in a wide range of environments, including open woodlands and bushlands with abundant grasses (Andrews and Humphrey 1999) and gallery forests (Leakey *et al.* 1995). By 3.5 myr, *A. afarensis* paleoecological reconstructions included grassland savannahs, wooded bushlands, and woodlands (White *et al.* 1993; Reed, 1997; Bonnefille *et al.* 2004) in a wetter clime (Denys 1999).

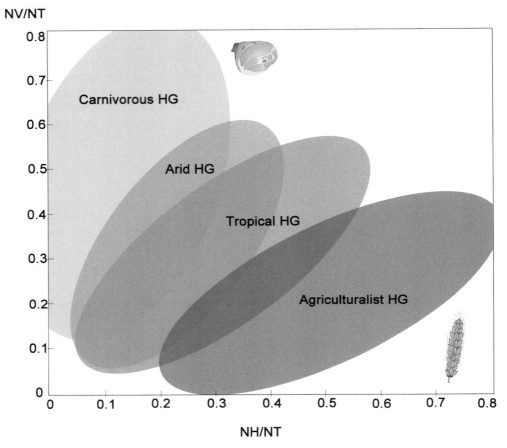

Fig. 8.1. *Plot NH/NT vs NV/NT of modern dietary groups (adapted from Lalueza* et al. *1996). Ellipses represent 95% dispersion*

Between 2.8 and 1.7 myr three climatic shifts towards more arid conditions have been recorded (Bobé and Behrensmeyer 2004; Quinn *et al.* 2007). This climatic trend led to an ecological change to more open environments (Vrba 1985; Potts 1998) and the appearance in the fossil record of two novel lineages: *Paranthropus* and *Homo*. However, this climatic change was not homogeneous and contrasting year-round seasons have been described (De Menocal 1995) with different habitats coexisting at the same time, both in east and South Africa (Behrensmeyer *et al.* 1997; Reed and Russak 2009; Bobe and Leakey 2009).

At around 1.4 myr, open grasslands were predominant in East Africa, where hominin species were forced into an intense exploitation of savannah plant foods and the generalised use of lithic tools that likely offered *Early Homo* new opportunities to consume different food resources in these highly variable environments (Kimbel *et al.* 1996; Kimbel 2009).

In this scenario, diet is likely to have played a significant role in human evolution and adaptation as an important selection force. Habitat shift may have contributed to changes in the ecological niches occupied by hominins, with a progressive increase in the intensity of exploitation of savannah resources (including underground storage organs, USOs) through time, both in east and South Africa (Vrba 1985; Cerling 1992; Behrensmeyer *et al.* 1997; Potts 1998; Bamford *et al.* 2008). Isotopic evidence has shown a temporal trend toward greater consumption of C_4 resources among early hominins over time (Bamford *et al.* 2008; Cerling *et al.* 2010; Sponheimer *et al.* 2013) in relation to changing climatic conditions, food availability and the ways it was obtained and consumed, which can also be linked to the emergence of bipedalism, the increase in encephalisation and new social and cultural behaviours of hominins.

Isotopic evidence, molar structure and occlusal microwear analyses have suggested that *Ardipithecus ramidus* may have been a frugivorous-omnivorous species with a large intake of soft fruits (Suwa *et al.* 2009a; White *et al.* 2009). This dietary scenario for *Ardipithecus* matches with paleoecological reconstructions suggested for this species (tropical forest, closed woods, and riparian forests nearby rivers). *Ardipithecus* is considered to be the ancestral morphotype of the East African *Australopithecus*. It is now considered that all Pliocene East African hominins (*Ardipithecus, Australopithecus anamensis,* and *Australopithecus afarensis*) represent a unique phyletic speciation process; in which case, these species are all different chronospecies of the same anagenetic evolving lineage. Along this anagenetic process, a clear tendency towards an increase in size of the masticatory apparatus is observed, with a directional increase in both postcanine dental size and enamel thickness. *Ardipithecus ramidus* lacked adaptations to enable the breakdown of abrasive and hard foodstuffs. Thus, a directional increase in the abrasiveness of diet through the evolving lineage (*Ardipithecus-Au. anamensis-Au. afarensis*) is expected. *Au. anamensis* had thicker post-canine tooth enamel and displayed a relative megadontia compared to *Ardipithecus* (Wood and Richmond 2000; Suwa *et al.* 2009b), which suggests that *Au. anamensis* underwent a dietary shift towards a more abrasive and harder diet (Teaford and Ungar 2000; Wood and Richmond 2000; Suwa *et al.* 2009b).

Though several microwear studies have addressed the question of the diet of *Au. anamensis* (Grine *et al.* 2006a; Estebaranz *et al.* 2012), no isotopic analyses have been published as yet for *Au. anamensis*. Traditionally, the diet of *Au. afarensis* has been reconstructed based on its dental and masticatory morphology. Based on its broad molars, thick enamel and low bulbous cusps, as well as its thick mandibular corpus, *Au. afarensis* was assumed to have had a very abrasive diet, predominantly based on nuts, seeds and hard fruits (Wood and Richmond 2000; White *et al.* 2000; Ungar 2004). This interpretation of the diet of *Au. afarensis* predominated for nearly three decades, until the early 2000s, when dental topography analyses suggested that these food items would not represent the actual diet of *Au. afarensis*, but rather fallback resources (Ungar 2004). More recently, research on *Au. afarensis'* diet has focused new evidence based on both occlusal (Grine *et al.* 2006b; Delezene *et al.* 2013) and buccal (Estebaranz *et al.* 2009) microwear analyses, as well as isotopic evidence (Wynn *et al.* 2013). In contrast to previous interpretations, occlusal microwear analyses now identify *Au. afarensis* as a frugivorous species that would have relied seasonally on hard fallback foods (Grine *et al.* 2006b). Alternatively, the *Au. afarensis* isotopic signal from Hadar has shown a significant C_4/CAM food intake (with a mean of 22% and highly variable range), and no temporal trend in the amount consumed (Wynn *et al.* 2013); These values are similar to those shown for *Au. africanus* from South Africa (Sponheimer *et al.* 2013).

Buccal dental microwear analyses do not support the hypothesis of an increase in the abrasiveness of diet from *Au. anamensis* to *Au. afarensis*, that is suggested by dental morphology (Fig. 8.2). Thus, the hypothetical directional shift towards a progressively tougher diet in *Australopithecus* from Pliocene East Africa is not supported by microwear

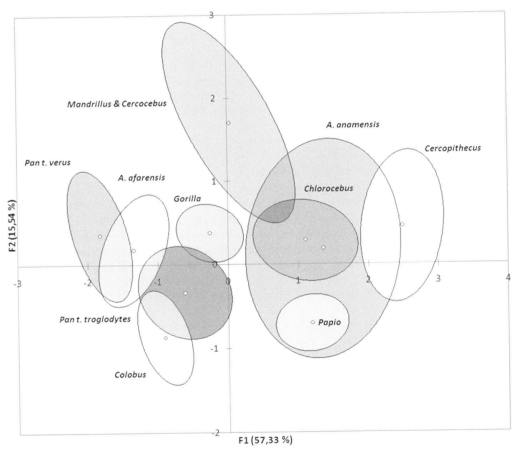

Fig. 8.2. Descriptive linear discriminant analysis of ten groups considered, including Au. anamensis, Au. afarensis *and extant primates. DF1 was mainly correlated with NV and NT; while DF2 was correlated with NFM and NT (Estebaranz et al. 2012)* Au. anamensis *clusters with Cecopithecoidea, whereas* Au. afarensis *overlaps with Pan. Ellipses included 95% confidence intervals of the sample means*

evidence (Estebaranz *et al.* 2009; 2012). By contrast, buccal microwear analyses suggest that both *Australopithecus* species (*Au. anamensis* and *Au. afarensis*) underwent a unique and distinct specialisation process. *Au. anamensis* demonstrates a very abrasive buccal microwear pattern (Fig. 8.3), with clearly higher striation densities (NT) than its putative descendant, *Au. afarensis* (Fig. 8.2), as well as all the extant African non-human hominids (i.e., *Gorilla gorilla* and *Pan troglodytes*). Surprisingly, *Au. anamensis'* buccal dental microwear pattern resembles more closely those of extant African *Cercopithecoidea* species, such as *Cercocebus* sp, *Cercopithecus aethiops*, and *Mandrillus sphinx*. These species have a common dietary pattern based on fruit and seed intake. In this way based on shared microwear characteristics, *Au. anamensis* is likely to have been a granivorous-frugivorous species that consumed seeds, tubercles, nuts and roots, as present-day baboons do. This scenario is directly in opposition to previous dietary reconstructions based on occlusal microwear. However, a granivorous diet is consistent both with the masticatory pattern observed in *Au. anamensis* as well as the palaeoenvironments it inhabited.

In contrast, *Au. afarensis'* buccal dental microwear pattern showed high similarities to those of extant Cameroon gorillas and chimpanzees (Fig. 8.3). Thus, *Au. afarensis* might have been a highly frugivorous species during good seasons, as are extant African *Gorilla gorilla* and *Pan troglodytes*. However, during the leaner seasons, *Au. afarensis* would have depended on fallback foods. These resources, such as such as barks or leaves, are abundant all year-round though energetically poor, and are mainly consumed during the dry season. Therefore, *Au. afarensis* would have been a frugivorous species, in contrast to its ancestor morphotype, *Au. anamensis*. Nevertheless, one of the most highlighted aspects of *Au. afarensis* buccal microwear is the lack of any temporal or paleoecological trend, i.e., no variation in buccal microwear patterns is observed neither throughout the period of time *Au. afarensis* inhabited East Africa nor for the different paleoenvironments it occupied (Grine *et al.* 2006b; Estebaranz *et al.* 2009). These results are fairly unexpected, since it implies that *Au. afarensis'* dietary preferences were not affected by the overall trend towards more arid climates and more open environments. As Grine *et al.* (2006b) have pointed out, *Au. afarensis* could be an example of *dietary stasis*, with no significant dietary change over the course of almost one million years. *Au. afarensis* might have been an opportunistic and active pursuer species, as modern gorillas and chimpanzees are. This would mean that *Au. afarensis* might have actively searched preferred foods items in a wide range of environments or, at least, with similar chemical and physical properties. Nevertheless, it seems feasible that different palaeoecological environments would afford differences in food and, particularly, fruit availability (*Au. afarensis* is thought to have been an active frugivorous species). Thus, presumably, the energy investment required of *Au. afarensis* to obtain food might had differed in different environments. Finally, the inability of *Au. afarensis* to change its dietary preferences or habits during this climatic cooling process could explain its ultimate extinction as the energetic investment required in its food search would have increased as aridity spread. Thus, only some East African regions, which retained heavily forested and wet woodlands, could have harboured *Au. afarensis* populations. These would have become progressively smaller and more isolated, surviving in reducing patches of favourable environmental conditions. In this scenario, genetic drift and eventually extinction would become more probable, as was the case for *Au.*

afarensis at the end of the Pliocene. This occurred at the same time as the onset of the North Hemisphere Glaciation and the global increase in aridity, which resulted in the appearance of different hominin lineages in East and South Africa, with completely different masticatory adaptations – notably *Paranthropus* sp and *Early Homo.*

The C_4 isotopic signal for *P. boisei* shows a C_4/CAM specialisation (80% in OH 5, Van der Merwe *et al.* 2008). This isotopic signal is related to the consumption of C_4 sedges, grasses or USOs, for which *Paranthopus* would have little competition (Cerling *et al.* 2011; Sponheimer *et al.* 2013). However, this isotopic signal could also be indicative of consumption of animals that consumed these plants (Fontes-Villalba *et al.* 2013). The massive cranio-dental development in *Paranthropus* supports the view of it as a C_4 plant consumer as this is consistent with the thick mandibular body, large postcanine tooth crown areas, extended muscle insertion areas in the skull, and thick dental enamel with a microstructure that would support high biting stresses and would prevent dentine exposure and dental cracking. Post-canine tooth chipping has documented peak bite forces that correlate with the consumption of hard foodstuffs (Constantino *et al.* 2010; 2011). *Paranthropus boisei* C_4 signal does not overlap with that described for *P. robustus* from South Africa (Sponheimer *et al.* 2013), the former having a more variable diet that would imply different niche exploitation for both species.

All these morphological and isotopic reconstructions are linked with heavy stress loads and repetitive chewing of hard, tough or abrasive grasses (Jolly 1970; Grine 1981; Strait *et al.* 2009). However, occlusal and our preliminary buccal microwear signals suggest a different scenario. Both low complexity values in occlusal microwear and reduced buccal microwear densities demonstrate a low abrasive diet for this species (Fig. 8.3). Occlusal surfaces are heavily worn and dentine exposure obscures occlusal working facets that are normally analysed as part of occlusal microwear analyses. In contrast, preserved buccal surfaces have shown the lowest striation densities yet observed in hominins. Some studies have postulated that the durophagy model could be applied to *Paranthropus* (Shabel 2010), perhaps with seasonal consumption of hard objects as fallback foods, when preferred foods were not available. This view implies that *Paranthropus boisei* could be a generalist (Wood and Collard 1999) that could have included C_4 sedge USOs in its diet, perhaps also corms with fatty acids and rich in starches (Macho 2014) or small C_4 invertebrates (Shabel 2010).

In contrast, the isotopic signal of Early *Homo* shows a mixed C_3/C_4 diet, with a marked tendency favouring C_3 consumption (Van der Merwe *et al.* 2008). Increase in habitat aridity would have resulted in savannahs and open woodlands rich in succulent food resources, as well as xeric plant foods (O'Connell *et al.* 1999; Wrangham *et al.* 1999). The increase in brain size, the reduced denthognathic proportions and the use of lithic industries for food processing have been interpreted as highly indicative of meat consumption as a major food resource in the *Homo* genus, mainly through scavenging strategies (Blumenschine 1995; Domínguez Rodrigo 1997). The *'expensive tissue hypothesis'* has suggested a concurrent reduction in hominin gut size and metabolic activity with an increase in brain size (Aiello and Wheeler 1995). This process, starting about 2.5 myr ago, would have caused an increase in consumption of a high quality diet including animal proteins and structural fatty acids that are a requirement for the development of brain tissues (Broadhurst *et al.* 2002). At the same

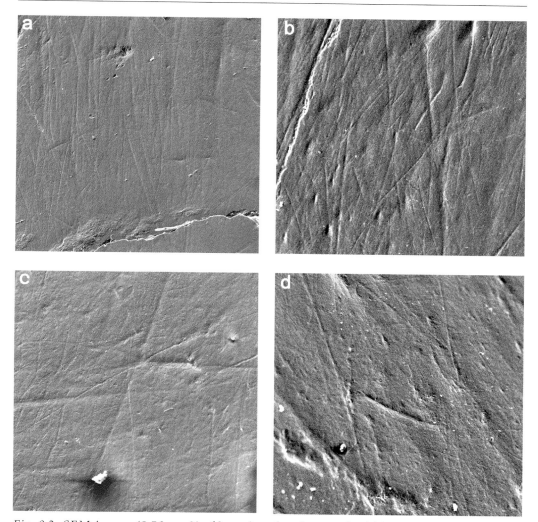

Fig. 8.3. SEM images (0.56 mm2) of buccal surface from teeth of (a) Au. anamensis *(KNM-ER_35231 RM2) (b)* Au. afarensis *(LH4 LM2) (c)* P. boisei *(KNM-WT-18600 LP3) (d)* Early Homo *(OH 41, LM1)*

time, the reduction of tooth size would have limited the types of foods available to *Homo ergaster* (Wood and Collard 1999; Bailey 2004) in increasingly opened grassland ecosystems. *Homo ergaster* would have faced environment changes with more versatile subsistence strategies, with increased complexity in social interactions that meat procurement would have required. Our preliminary buccal dental microwear analyses of this species have shown a significantly high density of buccal striations, suggestive that Early *Homo* individuals were chewing hard and/or tough items, perhaps also including grit particles incorporated to food during its preparation (Fig. 8.3).

Buccal microwear analyses of hominin specimens are still revealing new insights into dietary adaptations of our ancestors. Traditional hypotheses are continuously put into question and no methodological approach seems to have the last word regarding the complex interaction between ecology and dietary adaptations. However, morphological traits alone have proved to be insufficient to tell the full story. Buccal microwear studies have raised some questions regarding the so called final interpretation provided by occlusal microwear analyses and can be used to test the sometimes contradictory results that have emerged from both isotopic and occlusal evidences.

References

Aiello, L., Wheeler, P., 1995. The expensive-tissue hypothesis: the brain and the digestive system in human and primate evolution. *Current Anthropology* 36(2), 199–221

Andrews, P. and Humphrey, L. 1999. African Miocene environments and the transition to early hominines. In Bromage, T. G. and Schrenk, F. (eds), *African Biogeography, Climate Change and Human Evolution*, 282–300. New York: Oxford University Press

Bailey, S. 2004. A morphometric analysis of maxillary molar crowns of Middle–Late Pleistocene hominins. *Journal of Human Evolution* 47, 183–8

Baker, V., Jones, H. P. and Wardrop, I. D. 1959. Causes of wear in sheep's teeth. *Nature* 184, 1583–4

Bamford, M. K., Stanistreet, I. G., Stollhofen, H. and Albert, R. M. 2008. Late Pliocene grassland from Olduvai, Tanzania. *Palaeogeography, Palaeoclimatology, Palaeoecology* 257, 280–93

Behrensmeyer, A. K., Todd, N. E., Potts, R. and McBrinn, G. E. 1997. Faunal turnover in the Turkana Basin, Kenya and Ethiopia. *Science* 278, 1589–94

Beynon, A. D. 1987. Replication technique for studying microstructure in fossil enamel. *Scanning Microscopy* 1, 663–9

Blumenschine, R. J. 1995. Percussion marks, tooth marks and experimental determinations of the timing of hominid and carnivore access to long bones at FLK *Zinjanthropus*, Olduvai Gorge, Tanzania. *Journal of Human Evolution* 29, 21–51

Bobé, R. and Behrensmeyer, K. 2004. The expansion of grassland ecosystems in Africa in relation to mammalian evolution and the origin of the genus *Homo*. *Palaeogeography, Palaeoclimatology, Palaeoecology* 207(3–4), 389–420

Bobé, R. and Leakey, M. G. 2009. Ecology of Plio-Pleistocene mammals in the Omo-Turkana Basin and the emergence of *Homo*. In Grine, F. E., Fleagle, F. J. and Leakey, R. E. (eds), *The First Humans: origin and early evolution of the genus Homo*, 173–84. Dordrecht: Springer

Bonnefille, R., Potts, R., Chalie, F., Jolly, D. and Peyron, O. 2004. High-resolution vegetation and climate change associated with Pliocene *Australopithecus afarensis*. *Proceedings of the National Academy of Science (USA)* 101, 12125–9

Broadhurst, C. L., Wang, Y., Crawford, M. A., Cunnane, S. C., Parkington, J. E. and Schmidt, W. F. 2002. Brain-specific lipids from marine, lacustrine or terrestrial food resources: potential impact on early *Homo sapiens*. *Comparative Biochemistry & Physiology* (B) 131, 653–73

Cerling, T. E. 1992. Development of grasslands and savannas in East Africa during the Neogene. *Palaeogeography, Palaeoclimatology, Palaeoecology* 97, 241–7

Cerling, T. E., Harris, J. M., MacFadden, B. J., Leakey, M. G., Quade, J., Eisenmann, V. and Ehleringer, J. R. 1997. Global vegetation change through the Miocene/Pliocene boundary. *Nature* 389, 153–8

Cerling, T. E., Levin, N. E., Quade, J., Wynn, J. G., Fox, D. L., Kingston, J. D., Klein, R. G. and Brown, F. H. 2010. Comment on the Paleoenvironment of *Ardipithecus ramidus*. *Science* 328(5982), 1105/ *DOI*: 10.1126/science.1185274

Cerling, T. E., Mbua, E., Kirera, F. M., Manthi, F. K., Grine, F. E., Leakey, M. G., Sponheimer, M. and Uno, K. T. 2011. Diet of *Paranthropis boisei* in the early Pleistocene of East Africa. *Proceedings of the National Academy of Science (USA)* 108(23), 9337–41

Ciochon, R. L., Piperno, D. and Thompson, R. G. 1990. Opal phytoliths found on the teeth of the extinct ape *Gigantopithecus blacki*: Implications for paleodietary studies. *Proceedings of the National Academy of Science (USA)* 87, 8120–4

Constantino, P. J., Lee, J. J-W., Morris, D., Lucas, P. W., Hartstone-Rose, A., Lee, W-K., Dominy, N. J., Cunningham, A., Wagner, M. and Lawn, B. R. 2011. Adaptation to hard-object feeding in sea otters and hominins. *Journal of Human Evolution* 61(1), 89–96

Constantino, P. J., Lee, J. J-W., Vchai, H., Zipfel, B., Ziscovici, C., Lawn, B. R. and Lucas, P. W. 2010. Tooth chipping can reveal the diet and bite forces of fossil hominins. *Biology Letters* 6(6), 826–9

De Menocal, P. B. 1995. Plio-Pleistocene African climate. *Science* 270, 53–9

Delezene, L. K., Zolnierz, M. S., Teaford, M. F., Kimbel, W. H., Grine, F. E. and Ungar, P. S. 2013 Premolar microwear and tooth use in *Australopithecus afarensis*. *Journal of Human Evolution* 65(3), 282–93

Denys, C., 1999. Of mice and men: evolution in east and south Africa Pliocene and Pleistocene hominids. In Bromage, T. G. and Schrenk, F. (eds), *African Biogeography, Climate Change and Human Evolution*, 226–52. New York: Oxford University Press

Domínguez-Rodrigo, M. 1997. Meat eating by early hominids at FLK22 *Zinjanthropus* site, Olduvai Gorge (Tanzania): an experimental approach using cut-mark data. *Journal of Human Evolution* 33(6), 669–90

Estebaranz, F., Galbany, J., Martínez, L. M., Turbón, D., Pérez-Pérez, A. 2012. Buccal dental microwear analyses support greater specialization in consumption of hard foodstuffs for *Australopithecus anamensis*. *Journal of Anthropological Sciences* 90, 1–24.

Estebaranz, F., Martínez, L. M., Galbany, J., Turbón, D. and Pérez-Pérez, A. 2009. Testing hypotheses of dietary reconstruction from buccal dental microwear in *Australopithecus afarensis*. *Journal of Human Evolution* 57, 739–50

Fine, D. and Craig, G. T. 1981. Buccal surface wear of human premolar and molar teeth: a potential indicator of dietary and social differentiation. *Journal of Human Evolution* 10, 335–44

Fleagle, J. G. 1999. *Primate Adaptation and Evolution* 2nd edn. San Diego: Academic Press

Fontes-Villalba, M., Carrera-Bastos, P. and Cordain, L. 2013. African hominin stable isotopic data do not necessarily indicate grass consumption. *Proceedings of the National Academy of Science (USA)* 110 (43): E4055 /doi: 10.1073/pnas.1311461110

Galbany, J. and Pérez-Pérez, A. 2004. Buccal enamel microwear variability in Cercopithecoidea primates as a reflection of dietary habits in forested and open savannah environments. *Anthropologie* 42, 13–19

Galbany, J., Martínez, L. M. and Pérez-Pérez, A. 2004. Tooth replication techniques, SEM imaging, and microwear analysis in primates: methodological obstacles. *Anthropologie* 42 (1), 5–12

Galbany, J., Moyà-Solà, S. and Pérez-Pérez, A. 2005. Dental microwear variability on buccal tooth enamel surfaces of extant Catarrhini and the Miocene fossil *Dryopithecus laietanus* (Hominoidea). *Folia Primatologica* 76, 325–41

Galbany, J., Estebaranz, F., Martínez, L. M. and Pérez-Pérez, A. 2009. Buccal dental microwear variability in extant African Hominoidea: taxonomy versus ecology. *Primates* 50, 221–30

Goswami, A., Flynn, J., Ranivoharimanana, L. and Wyss, A. R. 2005. Dental microwear in Triassic Amniotes: implications for paleoecology and masticatory mechanics. *Journal of Vertebrate Paleontology* 25, 320–9

Grant, P. R. and Grant, B. R. 1985. Predicting microevolutionary responses to directional selection on heritable variation. *Evolution* 49, 241–2

Grine, F. E. 1981. Trophic differences between 'gracile' and 'robust' australopithecines: a scanning microscope analysis of occlusal events. *South African Journal of Science* 77, 203–30

Gügel, I. L., Grupe, G. and Kunzelmann, K. H. 2001. Simulation of dental microwear: characteristic traces by opal phytoliths give clue to ancient human dietary behaviour. *American Journal of Physical Anthropology* 114, 124–38

Grine, F., Ungar, P. S. and Teaford, M. F. 2006a. Was the Early Pliocene hominin '*Australopithecus*' *anamensis* a hard object feeder? *South African Journal of Science* 102, 301–10

Grine, F. E., Ungar, P. S., Teaford, M. F. and El-Zaatari, S. 2006b. Molar microwear in *Praeanthropus afarensis*: evidence for dietary stasis through time and under diverse paleoecological conditions. *Journal of Human Evolution* 51, 297–319

Grine, F. E., Sponheimer, M., Ungar, P. S., Lee-Thorp, J. and Teaford, M. K. 2012. Dental microwear and stable isotopes inform the paleoecology of extinct hominins. *American Journal Physical Anthropology* 148(2), 285–317

Jolly, C. J. 1970. The seed-eaters: a new model of hominid differentiation based on a baboon analogy. *Man* 5, 1–26

Kimbel, W. H. 2009. The origin of *Homo*. In Grine, F. E., Fleagle, J. G. and Leakey, R. E. (eds), *The First Humans: origin and early evolution of the genus* Homo, 31–7. Dordrecht: Springer

Kimbel, W. H., Walter, R. C., Johanson, D. C., Reed, K. E., Aronson, J. L., Assefa, Z., Marean, C. W., Eck, G. C., Bobe, R., Hovers, E., Rak, Y., Vondra, C., Yemane, T., York, D., Chen, Y., Evensen, N. M. and Smith, P. E. 1996. Late Pliocene *Homo* and Oldowan tools from the Hadar Formation (Kada Hadar Member, Ethiopia). *Journal of Human Evolution* 31, 549–61

King, T., Aiello, L. C. and Andrews, P. 1999. Dental microwear of *Griphopithecus alpani*. *Journal of Human Evolution* 36, 3–31

Krueger, K. L., Scott, J. R., Kay, R. F. and Ungar, P. S. 2008. Technical note: dental microwear textures of 'Phase I' and 'Phase II' facets. *American Journal of Physical Anthropology* 137, 485–90

Lalueza, C., Pérez-Pérez, A. and Turbón, D. 1996. Dietary inferences through buccal microwear analysis of Middle and Upper Pleistocene human fossils. *American Journal of Physical Anthropology* 100, 367–87

Leakey, M. G., Feibel, C. S., McDougall, I. and Walker, A. 1995. New four-million-year-old hominid species from Kanapoi and Allia Bay, Kenya. *Nature* 376, 565–71

Lucas, P. W. 2004. *Dental Functional Morphology: how teeth work*. Cambridge: Cambridge University Press

Maas, M. C. and Dumont, E. R. 1999. Built to last: the structure, function, and evolution of primate dental enamel. *Evolutionary Anthropology* 8, 133–52

Macho, G. 2014. Baboon feeding ecology informs the dietary niche of *Paranthropus boisei*. *PLOS ONE* 91(1): e84942. doi:10.1371/journal.pone.0084942

Martínez, L. M. and Pérez-Pérez, A. 2004. Post-mortem wear as indicator of taphonomic processes affecting enamel surfaces of hominin teeth from Laetoli and Olduvai (Tanzania): implications to dietary Interpretations. *Anthropologie* 42, 37–42

O'Connell, J. F., Hawkes, K. and Blurton Jones, N. G. 1999. Grandmothering and the evolution of *Homo erectus*. *Journal of Human Evolution* 36, 461–85

Pérez-Pérez, A. 2004. Why buccal microwear? *Anthropologie* 42(1), 1–3

Pérez-Pérez, A., Bermúdez de Castro, J. M. and Arsuaga, J. L. 1999. Non-occlusal dental microwear analysis of a 300,000 year-old *Homo heidelbergensis* teeth from Sima de los Huesos (Sierra de Atapuerca, Spain): implications of intrapopulation variability for dietary analysis of hominid fossil remains. *American Journal of Physical Anthropology* 108, 433–57

Pérez-Pérez, A., Espurz, V., Bermúdez de Castro, J. M., de Lumley, M. A. and Turbón, D. 2003. Non-occlusal dental microwear variability in a sample of Middle and Upper Pleistocene human populations from Europe and the Near East. *Journal of Human Evolution* 44, 497–513

Pérez-Pérez, A., Lalueza, C., Turbon, D. 1994. Intraindividual and intragroup variability of buccal tooth striation pattern. *American Journal of Physical Anthropology* 94, 175–87

Potts, R. 1998. Environmental hypotheses of hominin evolution. *Yearbook of Physical Anthropology* 41, 93–136

Puech, P. F. 1985. Precisions sur la méthode dite des 'répliques' pour l'éstude des surfaces dentaires et osseusses. *Bulletin de la Société Préhistorique Française* 82, 72

Puech P. F. and Albertini, H. 1984. Dental microwear and mechanisms in early hominids from Laetoli and Hadar. *American Journal of Physical Anthropology* 65, 87–91

Puech, P. F., Albertini, H. and Mills, N. T. W. 1980. Dental reconstruction in Broken Hill Man. *Journal of Human Evolution* 9, 33–9

Puech, P. F., Prone, A., Roth, H. and Cianfarani F. 1985. *Reproduction Expérimentale de Processus d'Usure des Surfaces Dentaires des Hominidés Fossils: Consequences Morphoscopiques et Exoscopiques avec Application à l'Hominidé I de Garusi*, 59–64. Comptes Rendus de l'Academie des Sciences Paris 301(1)

Quinn, R. L., Lepre, C. J., Wright, J. D. and Feibel, C. S. 2007. Paleogeographic variations of pedogenic carbonate $\delta^{13}C$ values from Koobi Fora, Kenya: implications for floral composition of Plio-Pleistocene hominin environments. *Journal of Human Evolution* 53, 560–73

Ravelo, A. C., Andreasen, D. H., Lyle, M., Olivarez Lyle, A. and Wara, M. W. 2004. Regional climate shifts caused by gradual global cooling in the Pliocene epoch. *Nature* 426, 263–7

Reed, K. E. 1997. Early hominid evolution and ecological change through the African Plio-Pleistocene. *Journal of Human Evolution* 32, 289–322

Reed, K. E. and Russak, S. M. 2009. Tracking ecological change in relation to the emergence of *Homo* near the Plio-Pleistocene boundary. In Grine, F. E., Fleagle, J. G. and Leakey, R. E. (eds), *The First Humans: origin and early evolution of the genus* Homo, 159–83. Dordrecht: Springer

Reznick, D. N. and Ghalambor, C. K. 2001. The population ecology of contemporary adaptations: what empirical studies reveal about the conditions that promote adaptive evolution. *Genetics* 112–113, 183–98

Romero, A. and De Juan, J. 2007. Intra- and interpopulation human buccal tooth surface microwear analysis: inferences about diet and formation processes. *Anthropologie* 45, 61–70

Romero, A., Galbany, J., De Juan, J. and Pérez-Pérez, A. 2012. Short and long-term in vivo human buccal dental-microwear turnover. *American Journal of Physical Anthropology* 148, 467–72

Romero, A., Ramírez-Rozzi, F. V., De Juan, J. and Pérez-Pérez, A. 2013. Diet-Related buccal microwear patterns in central African Pygmy foragers and Bantu-speaking farmer and pastoralists populations. *PLOS ONE* 8 (12), e84804

Rücker, C. 2006. Les micros-striations dentaires, alternative à la distinction de groupes humains. In Buchet L., Dauphin, C. and Séguy, I. (eds), *La Paléodémographie. Mémoire d'Os, Mémoire d'Hommes*, 49–56. Antibes: APDCA

Schubert, B. W. and Ungar, P. S. 2005. Wear facets and enamel spalling in tyrannosaurid dinosaurs. *Acta Palaeontologica Polonica* 50, 93–9

Shabel, A. B. 2010. Brain size in carnivoran mammals that forage at the land -water ecotone with implications for robust Australopithecine paleobiology. In Cunnane, S. C. and Stewart, K. M. (eds), *Human Brain Evolution: the influence of freshwater and marine food resources*, 173–87. New Jersey: Wiley-Blackwell

Sponheimer, M., Alemseged, Z., Cerling, T. E., Grine, F. E., Kimbel, W. H., Leakey, M. G., Lee-Thorp, J. A., Manthi, F. K., Reed, K. E., Wood, B. A. and Wynn, J. G. 2013. Isotopic evidence of early hominin diet. *Proceedings of the National Academy of Science (USA)* 110(26), 10513–10518. /pnas.1222579110

Strait, D. S., Weber, G. W., Neubauer, S., Chalk, J., Richmond, B. G., Lucas, P. W., Spencer, M. A., Schrein, C., Dechow, P. C., Ross, C. F., Grosse, I. R., Wright, B. W., Constantinod, P., Wood, B. A., Lawn, B., Hylander, W. L., Wang, Q., Byron, C., Sliceb, D. E. and Smith, A. L. 2009.

The feeding biomechanics and dietary ecology of *Australopithecus africanus*. *Proceedings of the National Academy of Science (USA)* 106, 2124–9

Suwa, G., Asfaw, B., Kono, R. T., Kubo, D., Lovejoy, C. O. and White, T. D. 2009a. The *Ardipithecus ramidus* skull and its implications for hominid origins. *Science* 326, 68e1–7

Suwa, G., Kono, R. T., Simpson, S. W., Asfaw, B., Lovejoy, C. O. and White, T. D. 2009b. Paleobiological implications of the *Ardipithecus ramidus* dentition. *Science* 326, 94–9

Teaford, M. F. and Lytle, J. D. 1996. Brief communication: diet-induced changes in rates of human tooth microwear: a case study involving stone-ground maize. *American Journal of Physical Anthropology* 100, 143–7

Teaford, M. F. and Oyen, O. J. 1989. Differences in the rate of molar wear between monkeys raised on different diets. *Journal of Dentistry Research* 68, 1513–18

Teaford, M. F. and Tylenda, C. A. 1991. A new approach to the study of tooth microwear. *Journal of Dentistry Research* 70, 204–7

Teaford, M. F. and Ungar, P. S. 2000. Diet and the evolution of the earliest human ancestors. *Proceedings of the National Academy of Science (USA)* 97, 13506–11

Ungar, P. S. 2004. Dental topography and diets of *Australopithecus afarensis* and early *Homo*. *Journal of Human Evolution* 42, 605–22

Ungar, P. S. and Spencer, M. A. 1999. Incisor microwear, diet, and tooth use in three Amerindian populations. *American Journal of Physical Anthropology* 109, 387–396

Ungar, P. S. and Teaford, M. F. 1996. A preliminary examination of non-occlusal dental microwear in Anthropoids: implications for the study of fossils. *American Journal of Physical Anthropology* 100, 101–13

Van der Merwe, Masao, F. T. and Bamford, M. K. 2008. Isotopic evidence for contrasting diets of early hominins *Homo habilis* and *Paranthropus boisei* of Tanzania. *South African Journal Science* 104, 153–5

Vrba, E. S. 1985. Environment and evolution: alternative causes of the temporal distribution of evolutionary events. *South African Journal Science* 81, 229–36

White, T. D., Ambrose, S. H., Suwa, G., Su, D. F., DeGusta, D., Bernor, R. L., Boisserie, J. R., Brunet, M., Delson, E., Frost, S., García, N., Giaourtsakis, I. X., Haile-Selassie, Y., Howell, F. C., Lehman, T., Likius, A., Pehlevan, C., Saegusa, H., Semprebon, G., Teaford, M. and Vrba, E. 2009. Macrovertebrate paleontology and the Pliocene habitat of *Ardipithecus ramidus*. *Science* 326, 87–93

White, T. D., Suwa, G., Simpson, S. and Asfaw, B. 2000. Jaws and teeth of *Australopithecus afarensis* from Maka, Middle Awash, Ethiopia. *American Journal of Physical Anthropology* 111, 45–68

White, T. D., Suwa, G., Hart, W. K., Walter, R. C., WoldeGabriel, G., de Heinxelin, K., Clark, J. D., Asfaw, B. and Vrba, E. 1993. New discoveries of *Australopithecus* at Maka in Ethiopia. *Nature* 366, 261–5

Williams, V. S., Barrett, P. M. and Purnell, M. A. 2009. Quantitative analysis of dental microwear in hadrosaurid dinosaurs, and the implications for hypotheses of jaw mechanics and feeding. *Proceedings of the National Academy of Science (USA)* 106, 11194–9

Wood, B. and Collard, M. 1999. The human genus. *Science* 284 (65), 65–71

Wood, B. and Richmond, B. G. 2000. Human evolution: taxononmy and paleobiology. *Journal of Anatomy* 196, 9–60

Wrangham, R. W. Conklin-Brittain, N. and Hunt, K. D. 1999. Dietary responses of chimpanzee and cercopithecines to seasonal variation in fruit abundance. I. Antifeedants. *International Journal of Primatology* 19, 949–70

Wynn, J. G., Sponheimer, M., Kimbel, W. H., Alemseged, Z., Reed, K., Bedaso, Z. K. and Wilson, J. N. 2013. (Early Edition). Diet of *Australopithecus afarensis* from the Pliocene Hadar Formation, Ethiopia. *Proceedings of the National Academy of Science (USA)*. 110(26), 10495–500

9. What early human populations ate: the use of phytoliths for identifying plant remains in the archaeological record at Olduvai

Rosa María Albert and Irene Esteban

While the incorporation of meat in the diet has been considered as a key factor in the development of the cognitive, social, and physical capabilities of early humans, less attention has been paid to the influence of plant consumption, even though plants served as a primary source of energy in the diet of our human ancestors. Long before farming developed, populations were completely dependent on wild resources for survival. In order to understand what these early hominins were eating, it is essential to reconstruct the vegetation at the time in order to evaluate the resources that were available in relation to the costs, in terms of energy, predation risk, and nutritional value. These results can then be compared with the available archaeological record.

Using the available data, which comprise phytoliths, macro-plant remains, charcoal, and techniques such as FTIR, we attempt to gain a deeper understanding of environmental influence on human evolution and the mode of acquisition of plant resources by early human populations for different purposes, including diet. This ongoing research focuses on two key chronological periods in human evolution: i) hominin populations during the Plio-Pleistocene in eastern Africa, with Olduvai Gorge as a study site; ii) emergence of early Homo sapiens during the Middle Stone Age period in South Africa, based on the studies conducted at Pinnacle Point archaeological complex.

Phytoliths provide a record of past vegetation and can therefore be used in the reconstruction of palaeoenvironmental conditions. In archaeological contexts, they can also be used to infer what humans were collecting and eating. Reconstruction of the palaeovegetation is based on the comparative study of modern analogous landscapes; this includes a detailed description of the vegetation, and compilation of a modern reference collection of plants and soils which are analysed for phytoliths. Results are later compared to archaeological and palaeoanthropological samples.

The phytolith record at Olduvai Gorge, together with other related studies on plants, bones and sediments, have depicted a more humid landscape than initially thought for Bed I and Bed II which represent the times in which Homo habilis and Paranthropus boisei were present in the area. For these time periods, phytoliths identify a range of plants usually related to fresh water sources, including grasses, sedges, and a significant number of palm species. Many of these would provide important edible resources during the rainy season.

At Mossel Bay, our research in caves PP13B and 5/6 has documented the presence of dicotyledonous-leaf phytoliths related to some of the hearths. The use of these leaves as fuel, food or other purposes is still an open question. Current study on modern vegetation and food resources available to those modern H. sapiens will try to shed more light on this issue.

Introduction

Traditionally, the incorporation of meat in the diet has been signalled as a key factor in the development of early human cognitive, social and physical capacities. Less attention has been paid to the role of plants despite the fact that plant foods clearly provided a major source of energy (Milton 2000) as humans need large amounts of carbohydrates (Wrangham 2009). All modern hunting-gathering societies consume plants to a certain extent, in some cases this is as much as 60–70% of the diet (Zihlman and Tanner 1978). One example of this is the !Kung from southern Africa who have a diet consisting of both animal and plant foods with plants representing around 67% of the total dietary intake, and animal-based products around 33% (Milton 2000). Similarly, the diet of the Hadza hunter-gatherers of Tanzania comprises over 60% wild plants even though they refer to themselves as hunters (Milton 2000).

Our plant diet has transformed through the years, and has been strongly related to climatic changes and vegetation shifts, as we have adapted to the available plant resources. We can distinguish three key developments in the evolution of plant consumption: i) the incorporation of gathering; ii) the use of fire for cooking purposes; iii) the first agricultural processes. In this chapter we focus on the first two.

When and why did gathering evolve?

Gathering depends on the ability to carry bundles of food, memory to return to the locations where wild staples are found and special tools for digging, when collecting underground storage organs (USOs) such as geophytes, etc. At the same time, gathering implies a certain social organisation since it presupposes the existence of 'base camps' where the collected food is brought and shared. For modern hunter-gatherers, gathering is critical since hunters often return empty-handed, in which case the group must rely entirely on gathered foods. According to Milton (2000), gathering for these societies frequently depends on one or a very small number of dependable wild staples which provide much of their energy needs. Did early hominins have the cognitive capacity for gathering? What breakthroughs in technology or changes in social context enabled gathering to develop (Wrangham 2009)?

The control of fire represents one of the most significant technological advances with important implications for hominin physical, social, and cognitive evolution (Oakley 1970; Perlès 1977; Wrangham 2009). Possible controlled use of fire has recently been documented in Wonderwerk Cave (South Africa), aged around 1 million years (Berna *et al.* 2012) and in Gesher-Benot Ya`akov (Israel), around 760,000 years ago (Goren

Inbar *et al.* 2004). It has been consistently recorded since the final stages of the Lower Palaeolithic from around 400–500,000 years ago (Rolland 2004; Karkanas *et al.* 2007). Hearths may have been used for different purposes including cooking. Wrangham (2009) proposes that the anatomy of *H. erectus* with its small jaws and teeth was poorly adapted for eating tough, raw meat and suggests they might have developed the ability to use fire for cooking. Furthermore, Wrangham *et al.* (1999) note that cooking increases the digestibility of underground storage organs (USOs), which are thought to have been an important part of their diet.

Having identified the key issues, the next challenge is to recognise plant consumption, gathering activities, and cooking processes, if present, in the archaeological record. This is not always easy since organic remains usually disappear soon after deposition. Nevertheless, inorganic plant remains, such as phytoliths, can be recovered from the archaeological record and these can provide evidence for the presence of plants. Direct evidence of plant consumption can be found in teeth through the identification of phytoliths and/or starches preserved in dental calculus (Buchet *et al.* 2001; Henry *et al.* 2011; Hardy *et al.* 2012). Other methods used to infer diet directly from the fossil record are based on biological adaptations (tooth size and morphology), or the effects of food on individuals during their lifetimes (dental microwear, stable isotopes ratios, and trace elements) (Sponheimer *et al.* 2005; Cerling *et al.* 2011; Grine *et al.* 2012). Indirect remains of plant consumption include stone tools, and the imprints left by the uses they were put to (cutting wood, leather, bones, etc.) or through the preservation of other micro-remains such as phytoliths attached to their surface (Shafer and Holloway 1979; Barton *et al.* 1998; Hardy *et al.* 2001). Preserved plant remains can also be recovered from sediments; these include phytoliths, carbonised plant remains, charcoal, burnt seeds and fruits, pollen, and starch granules. In this case, their most likely uses are inferred in relation to their context (food, fuel, bedding, etc.).

A different approach to understanding dietary plants comes from the development of ecological models and the reconstruction of the past vegetation. Early human populations were completely dependent on available food resources, potable water, safe shelter and refuge from predators. Thus if we want to know what these early hominins were eating, it is essential to reconstruct the vegetation and the availability of resources in terms of energy, predation risk and nutritional values. The results can then be compared with the available archaeological record.

Our research aims are to reconstruct the vegetation during two key periods in human evolution: i) hominin populations during the Plio-Pleistocene in eastern Africa, with Olduvai Gorge as a study site (Fig. 9.1); ii) the emergence of early *H. sapiens* in South Africa, based on the studies conducted at Pinnacle Point archaeological complex (Fig. 9.2). The palaeovegetation reconstruction takes places through the study of phytoliths. Phytoliths, which are siliceous micro-remains that reproduce the cellular structure of some plants, reflect the vegetation within an area and are used to reconstruct the environment. In archaeological contexts, they are also used to infer what humans were collecting and eating (Piperno 2006). Because phytoliths are inorganic and thus resistant to most postdepositional processes, they survive in well-preserved conditions over long periods of time and can be found in contexts up to few million years old (Wolde Gabriel *et al.* 2009; McInerney *et al.* 2011).

Olduvai Gorge has a rich vertebrate fossil record from Plio-Pleistocene deposits (Leakey 1971; Hay 1976; Blumenschine *et al.* 2003). The site became world famous in 1959 with the discovery, by Mary Leakey, of the *Paranthropus boisei* cranium (OH5), and in 1960 of the *H. habilis* mandible (OH7) by Jonathan Leakey. Also remarkable was Mary Leakey's detailed excavation and interpretation of the stone artefacts at FLK North (FLK N) and HWK East (HWK E) (Leakey 1971) where most of our studies are focused.

Pinnacle Point in Mossel Bay is located on the south coast of South Africa. Detailed palaeoenvironmental and archaeological records suggest that during the glacial MIS marine isotope stage (MIS) 6 this area retained diverse terrestrial plant and animal resources, as well as coastal resources, which would have supported hunter-gatherers at a time when other areas in Africa were uninhabitable (Marean 2008). Pinnacle Point includes a complex of around 50 caves and rockshelters with MSA deposits, and has one of the longest and most detailed palaeoclimatic and human evolution records in Africa, from MIS 6 to MIS 3 (190–30 kya approximately); it also has many hearth structures in all the studied levels (Marean 2010).

Using a combined approach, which links the results of phytolith analysis with other related disciplines including macroplant remains, charcoal and FTIR, our aim is to gain a deeper understanding of environmental influence on human evolution and the mode of acquisition of plant resources by early human populations, for use as raw materials and as part of the diet.

Materials and methods

The palaeovegetation reconstruction is based on actualistic studies. Actualistics is the study of modern analogous landscapes, these are then compared to archaeological and palaeoanthropological samples. Our research includes a detailed description of the vegetation, and compilation of modern plant and soil reference collections which are then analysed in terms of phytoliths. The results are integrated into a database and detailed vegetation maps are completed and incorporated into Geographic Information System (GIS) reconstructions.

The research at Olduvai Gorge is framed within the *Olduvai Landscape and Paleoanthropology Project* (OLAPP) (Peters and Blumenschine 1995; Blumenschine and Peters 1998). The palaeovegetation actualistic studies conducted to date include, Ngorongoro crater, Ol'Balbal, Lake Manyara, Serengeti Plain, Lakes Masek and Ndutu, and Lake Eyasi (Fig. 9.1). These locations, situated in the proximity of excavated sites, reflect environmental conditions that are most likely to have been dominant at Olduvai approximately 1.8 million years ago and include the presence of high-salinity lakes, fresh water sources and volcanic sediments. One of the main goals of this study is to recognise the real distribution of C_3/C_4 grasses and USOs in the Olduvai area in order to better understand climatic conditions and the availability of resources for hominin populations. The results of the palaeoanthropological samples studied are integrated into the new compartment system developed by Stanistreet (2012). The intention is to derive as much palaeoenvironmental information as possible when studying a particular stratigraphic level by combining the full range of perspectives. These include phytoliths,

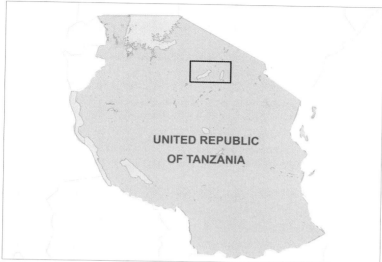

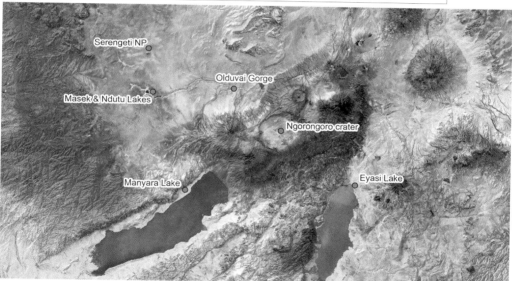

Fig. 9.1. a) Location map of Olduvai Gorge; b) Olduvai Gorge and the modern analogous landscapes studied to date: Ngorongoro crater, Lake Manyara, Lake Eyasi, Lakes Masek and Ndutu and Serengeti National Park

plant macrofossils, vertebrate and invertebrate remains, lithic debris, diatoms, sponge spicules as well as termiteria and roots in relation to site formation processes. Results are also compared to previous palynological results (Bonnefille 1984).

The *South Africa Coastal Paleoclimate, Paleoenvironment, Paleoecology, and Paleoanthropology* project (SACP4) is currently working in the area of Pinnacle Point in Mossel Bay. The phytolith research conducted to date has focused on the caves PP13B, 9b–9c and 5/6 (Fig. 9.2). In parallel with the archaeological research project, a study of modern landscapes from the Cape Floral Region (CFR), included in the Fynbos biome, is also taking place following a similar protocol to that of Olduvai. The CFR is the smallest of all known plant biomes and is geographically restricted to South Africa (Dallman 1998). The CFR has a Mediterranean vegetation with nearly 2000 species of geophytes, mostly from the monocotyledonous group, although dicotyledonous plants such as *Pelargonium* (*Geraniaceae*) have also been reported (Proches *et al.* 2006).

Regarding the archaeological study, we attempt to recognise plants that could have been used as combustibles in the hearths and to identify edible plants that might have been part of the diet, with a particular emphasis on geophytes. For that purpose,

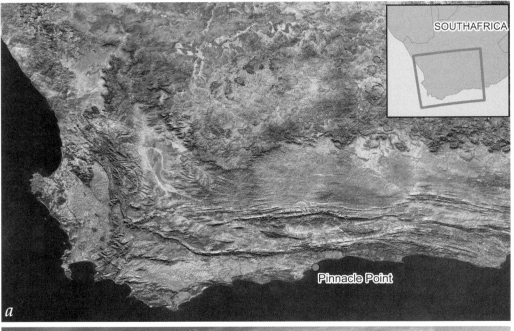

Fig. 9.2. a) Map showing the south coast of South Africa and the location of Pinnacle Point; b) Panoramic photograph of sites PP13B and PP5/6

deposits derived from anthropogenic processes, including hearths and occupation levels are compared to the phytolith records to ascertain the relationship between various human activities and phytolith distribution. Influencing factors such as dripping water, proximity to the cave mouth, and sediment type are also compared to the phytolith spatial distribution. Phytolith results are then contrasted with other palaeobotanical evidences such as charcoal whenever possible (Albert and Marean 2012).

Laboratory work

The chemical extraction of siliceous microremains (phytoliths, diatoms, sponge spicules) from both modern and fossil samples, takes place at the Department of Prehistory, Ancient History and Archaeology of the Universitat de Barcelona (http://www.archeoscience.com).

All the sediment samples (modern and fossils) are first analysed using FTIR (Fourier Transformed Infrared Spectrometry). FTIR is widely used in archaeology to identify both crystalline and amorphous minerals (such as the opal that form the phytoliths; Weiner 2010). Knowing what the mineral composition of sediments is prior to conducting the phytolith analyses, permits us to evaluate the preservation state of the sediment which enables us to evaluate the possibilities of detecting phytolith rich levels. FTIR can also determine if the sediment has been thermally altered (Berna *et al.* 2007).

The chemical extraction of phytoliths from modern plants follows the dry ashing method (Parr *et al.* 2001). Samples are heated to 500ºC then phytoliths are extracted from the ash using a simple acid extraction method.

For the extraction of siliceous microremains from both modern and archaeological soils, we initially used the fast method developed by Katz *et al.* (2010). This method uses a minute amount of the sample (20–40 mg) and reduces sampling preparation time from days to hours. This method is less effective for soils with high clay content and in these cases we follow the method described in Albert *et al.* (1999).

The morphological identification of phytoliths takes place using a petrographic microscope at 400× magnifications and follows our catalogue database (Albert *et al.* 2011) as well as standard literature (Twiss *et al.* 1969; Runge 1999; Mercader *et al.* 2000; Piperno 2006). Phytoliths are classified according to the anatomical origin of the plant in which they form. When this is not possible, geometric criteria are followed (Albert and Weiner 2001). The terminology used to describe the phytoliths follows the International Code for Phytolith Nomenclature (ICPN) wherever possible (Madella *et al.* 2005). Morphometric analyses, which are based on the identification and measurement of phytoliths using digital images, are also conducted (Ball *et al.* 1996; Portillo *et al.* 2006).

To determine whether phytoliths have undergone combustion processes, we use the Refractive Index, and FTIR. Elbaum *et al.* (2003) stated that the Refractive Index of phytoliths increases with temperature and fixed the limit between burned and unburned phytoliths at R.I. of 1.440. Nevertheless, later studies showed that this limit may vary under certain conditions such as the use of fresh or naturally-dried wood (Albert and Cabanes 2009). The FTIR is used to detect those sediments that have been heated to over 500º C (Berna *et al.* 2007) by analysing changes in the mineralogical composition ratio.

Summary and interpretation of results

At Olduvai Gorge the differential phytolith preservation that we noted in the fossil samples (Fig. 9.3a–b), led us to develop a model of postdepositional effects for phytoliths. This comprised their formation in plants, their deposition in the soils and their preservation in contexts 1.8 million years old (Albert *et al.* 2006). Diagenetic processes affecting phytoliths need to be understood in order to minimise the effect of taphonomic biases in palaeoecological interpretation.

The actualistics showed that phytoliths can be altered by several factors including their formation process in the plants. The anatomical origin or the type of cell mineralised will influence the degree of silicification and therefore the quality of preservation conditions (Albert *et al.* 2006). For example in sedges, hat-shaped phytoliths, although abundant in modern plants (Fig. 9.3c–e), are rarely preserved in soils, regardless of the mineralogical conditions (Albert *et al.* 2006). However, once deposited in sediments, phytoliths are vulnerable to any postdepositional process that may affect the preservation of silica (Fig. 9.3f–k). Volcanic soils and lake margins with more acidic conditions preserve phytoliths better than pyroclastic and saline-alkaline soils (Albert *et al.* 2006).

In the fossil samples, phytolith preservation during the deposition of Uppermost Bed I (UMBI) and Lowermost Bed II (LMBII), varied both among different localities (HWKE/EE, HWKW, FLKN, KK, VEK, and MCK) and between stratigraphic levels in the same locality and in relation to the ancient palaeolake (regression or transgression periods, or calcareous sedimentation) (Albert *et al.* 2006; Bamford *et al.* 2006; Albert and Bamford 2012).

Regarding the vegetation reconstruction, the noteworthy results based on analysis of phytoliths and macroplants obtained at the FLK peninsula and the adjacent channel, wetland edge and wetland interior during *P. Boisei* times (1.84 myr) (Blumenschine *et al.* 2012), are as follows: i) the channel produced a predominance of phytoliths indicative of monocotyledons and dicotyledons as well as sedge rhizomes; ii) despite, the fact that the samples from the FLK peninsula produced very few phytoliths, the identification

Fig. 9.3. (facing page) Microphotographs of modern and fossil phytoliths from Olduvai Gorge area. Pictures taken at 400 magnifications. a–b) Comparison of phytoliths from fossil samples with signs of dissolution (a) parallelepiped morphotype identified in HWKEE locality (T-107.5) and (b) well preserved phytoliths (Spheroid (globular) echinate morphotype characteristic of palms and diatoms from HWKE locality (T-104.2); c–e) hat-shape phytoliths from modern Killynga sedge collected at the Olduvai Gorge area; f–k) comparison of phytoliths from modern plants and soils which show the good preservation of certain morphotypes after their deposition in soils; (f) bulliform cell from the C$_4$ Sporobolus consimilis grass from Lake Masek; (g–h) bulliform cell phytoliths from grasses identified in modern soil reference samples from Lake Manyara; (i) short cell phytolith from the C$_4$ Sporobolus consimilis grass collected at Lake Masek; (j) short cell rondel characteristic of C$_3$ grasses identified in modern soil reference samples from Lake Manyara; (k) short cell saddle diagnostic of C$_4$ grasses recovered from modern soil samples at Lake Manyara; l) bulliform phytolith from grass from HWKE locality (T-151); m) prickle phytolith from KK locality (T-136); n–o) short cell phytoliths (rondel and saddle types) associated with diatoms from HWKE locality (T-104.2); p–q); Spheroid (globular) echinate phytoliths characteristic of palm trees from T-151; r) diatoms from HWKE locality (T-104.2).

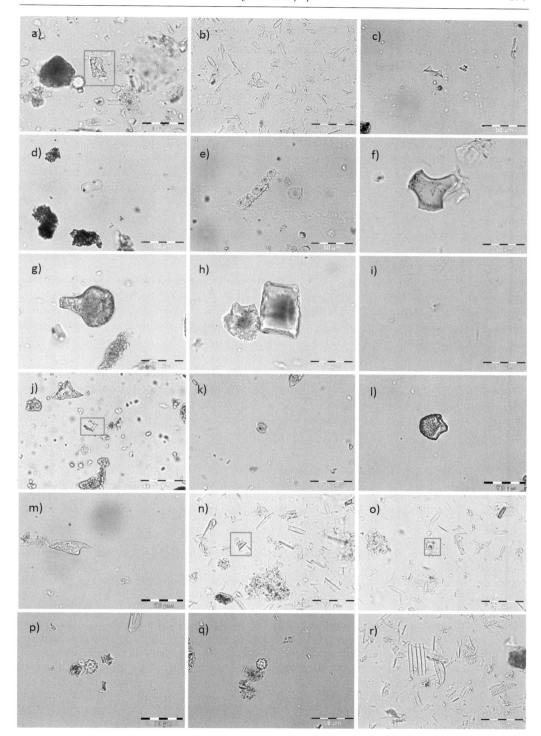

on the wetland area of woody branches in an aligned position suggested that these branches might have been transported from the peninsula, due to erosion processes that had removed the aerial plant parts, and thus the phytoliths; iii) the wetland edge had phytoliths from sedges and woody dicotyledons, but mostly from grasses; iv) the wetland interior identified by the phytoliths, corresponded to grasses, along with monocotyledons, dicotyledons and fern or moss spores. The macroplant record also identified the presence of *Typha* spp. (Blumenschine *et al.* 2012).

Phytoliths from grasses and sedges were also identified in different localities such as FLKN, VEK, HWKE, MCK and KK during the deposition periods of LMBII, Tuff IF and UMBI (Fig 9.3, l–m) (Bamford *et al.* 2008; Albert and Bamford 2012).

In addition to grasses and sedges, at HWKE and HWKEE, during LMBII, diatoms and palm phytoliths were also present (Fig. 9.3n–r). The presence of palms HWKEE (T-107) in some samples was over 80% of the total phytolith record (Albert *et al.* 2006; 2009; Bamford *et al.* 2006). Palms were also identified by Barboni *et al.* (2010) during Upper Bed times at FLK. Our research continues today in other localities from older periods where palms have also been recognised in abundance, which suggests that these plants were an important vegetational component.

At Mossel Bay, the results obtained at PP13B, were consistent with the micromorphological results (Karkanas and Goldberg 2010) and showed differential preservation of phytoliths in relation to their location in the cave, the material was better preserved at the back of the cave, and poorly preserved at the entrance of the cave (Albert and Marean 2012). In addition, dicotyledonous leaf phytoliths were recovered from DB Sand 4c level at the back of the cave, and dated to MIS 6 (Fig. 9.4a–d) (Jacobs 2010; Marean *et al.* 2010). During this cold period when the coast was tens of kilometres away, occupation of the cave is considered to be low (Fisher *et al.* 2010). Shellfish are extremely rare in this layer and occupants may have focused on other activities such as large mammal and plant food foraging (Albert and Marean 2012). No phytoliths were recovered from PP9b and 9c.

PP5/6 is currently under study. As in PP13B, preliminary observations show that phytoliths preserve differently depending on the stratigraphic levels as well as within the levels themselves, in relation to the different locations. Dicotyledonous leaves were identified in combustion features from the oldest occupation moments, during the MIS 5 (around 100,000 years ago) in Light Brown Sand and Roofspall stratigraphic aggregate (LBSR) (Fig. 9.4e–h). Ongoing actualistic studies on modern vegetation and plant foods available to those modern *H. sapiens* will try to shed more light on this issue (Esteban *et al.* in press).

Fig. 9.4. (facing page) Microphotographs of phytoliths from sites PP13B (a–d) and PP5/6 (e–h). a–b) multicellular structure with polyhedral shapes from sample 32329 (W3CY1; 200× and 400×, respectively); c–d) multicellular structures of polyhedral shapes from sample 32328 (W3DJ2; 400×); e) multicellular structure polyhedral shape from sample 162782 (400×); f) multicellular structure polyhedral shape from sample 162778 (400×); g) fruit phytolith from sample 162778 (400×); h) multicellular structure polyhedral shape from sample 162781 (400×).

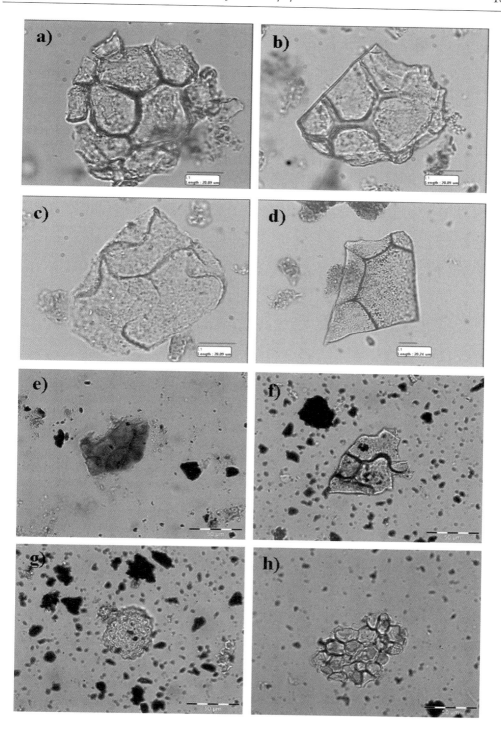

Discussion

The landscape at Olduvai during the time *H. habilis* and *P. boisei* lived in the area, would have been composed of a fluctuating saline-alkaline lake, fresh water courses, marshlands, and edaphic grasslands (Hay 1976). Grasses, sedges, palms and dicotyledons, shrubs and trees were part of the vegetation. In certain locations from the eastern lake margin, such as FLK, MCK, or HWK E, in this new more varied environment, vegetation which had been affected by pyroclastic events such as the Olmoti eruption responsible for the deposition of Tuff 1F which divided Bed I from Bed II would dominate (Bamford 2012; Stollhofen and Stanistreet 2012). The archaeological evidence shows that while small groups of hominins and animals visited the eastern palaeolake margin during UMBI times, there is no direct evidence of hominins visiting during Tuff IF times while there is increased activity during LMBII times (Blumenschine *et al.* 2007).

The phytolith results indicate that palms were widespread during LMBII times at HWKEE (Albert *et al.* 2006; 2009; Bamford *et al.* 2006) while ongoing studies suggest that palms were also extensive during earlier times in the HWK area (Albert *et al.* in prep.).

Just as modern primates are capable of eating a broad range of foods, hominins were surely equally capable. In the light of our palaeoenvironmental reconstruction, we are now in a position to explore the plant diet of both *H. habilis* and *P. boisei* when they visited this area.

The identification of phytoliths, as well as macro plant remains, from grasses and sedges during LMBII, Tuff 1F and UMBI times (Bamford *et al.* 2008; Albert and Bamford 2012; Blumenschine *et al.* 2012), confirm that the roots, corms, tubers, and bulbs from these plants would have been an available food resource for these populations. Isotopic analyses indicate that *P. boisei* had a diet that was almost 70% C_4 (Ungar and Sponheimer 2011; Cerling *et al.* 2011). This high C_4 diet might be explained by the consumption of starchy sedge and grass underground storage organs (USOs), perhaps particularly in the dry season (Wrangham 2009), although they were probably available all year round. The acquisition of USOs would not be a technical problem for these hominins, since chimpanzees, which have smaller brains, dig for tubers (Hernandez-Aguilar *et al.* 2007).

The phytolith results have also shown that in addition to the availability of dicotyledonous fruits and seeds, for example at FLK, the HWKE area would be rich in a range of edible plants including palm fruits, which are related to fresh water areas (Hay 1976; Albert *et al.* 2009). One modern palm analogue is wild date palm (*Phoenix reclinata*) which is common in current savannah environments usually related to fresh-water areas such as riverbanks or swamps (Fig. 9.5a–c). This palm also produces high amounts of characteristic phytoliths which closely resemble those identified in the fossil samples (Fig. 9.5d–f). *P. reclinata* is today a commonly exploited palm. Both modern hunter-gatherer populations such as those occupying the Tana River district of north-eastern Kenya as well as the Tana River crested mangabey (*Cercocebus galeritus galeritus*), exploit it mostly for its fruits and seeds (Kinnaird 1995). Other primates such as Mchelelo baboons eat the fruits green, ripe or dry and this provides them with 42% of their food consumption (Bentley-Condit 2009). The fruits are available between April-May and September-October (Fig. 9.5c).

Fig. 9.5. *a–b)* Phoenix reclinata *palms, on the riverbank of River Bonar, Serengeti National Park; c) closer view of the fruits of* Phoenix reclinata *palm, River Bonar, Serengeti National Park; d–e) microphotograph of Spheroid (Globular) echinate phytoliths from modern* Phoenix reclinata *palm from Lake Manyara; f) Spheroid (Globular) echinate phytolith identified in HWKE locality (T-151).*

Since *P. reclinata* is not a C$_4$ plant, isotopic results do not seem to support the idea that *P. boisei* may have eaten this palm's fruit, at least not habitually. With regards to *H. habilis* however, the isotopic and dental microwear results indicate that they had a softer diet than later *H. ergaster*, this may well have included the consumption of fruits and seeds (Martinez *et al.* 2006). Studies of 23 individuals of *H. habilis* from Olduvai indicated that about 27.3% of these were affected by hypoplasia with a periodicity of approximately half a year which might have coincided with the presence of two different seasons in the area and may indicate a seasonal cycle of stress (Martinez *et al.* 2006). Our results suggest that together with the consumption of fruits and seeds from dicotyledonous trees and shrubs, present for example at FLK, date palms were also available in HWKE and thus a food source during the rainy season when these fruits are available (Albert *et al.* 2009; Albert *et al.* 2015). This hypothesis, though, needs to be tested in relation to new experimental and dental microwear analyses.

At Mossel Bay, the use of dicotyledonous leaves in hearth contexts from the back of PP13B was documented during periods of low occupation when the sea was further away. These results may have implications regarding different activities conducted during these periods. Nevertheless the ongoing study at PP5/6 also showed the presence of dicotyledonous leaves, with very similar morphological traits, in relation to combustion features around 100,000 years ago. The important difference in respect to PP13B is that the dicotyledonous-leaf phytoliths here were recovered when the sea was around 2 km away from the cave signifying that these plants were collected from the Fynbos biome and not from the exposed new lands. Additionally, the hearths from where these dicotyledonous-leaf phytoliths were recovered, presented some differences in relation to other hearths from the same areas, mostly in relation to the amount of archaeological material and the type of burned residues recovered which suggest a differential use for these hearths. In this sense the identification of these phytolith morphotypes from different periods implies that these plants were continuously present and abundant and that they were used either for diet, as fuel, or other purposes, such as bedding.

If these leaves represent evidence of diet, the question then becomes - which plants might have been collected as food and of these, which plants could have left these types of phytoliths? One possible answer is geophytes (Fig. 9.6). They are extremely abundant in the Fynbos biome and have been found in Holocene sites (Parkington 1976; 1981) suggesting that they were a key food resource for hunter-gatherers during MIS 6 (Marean 2010). Furthermore, it has been suggested that most of these plants needed to be cooked in order to be edible (Wrangham *et al.* 1991; Carmody and Wrangham 2009). It is plausible that early modern humans would have consumed geophytes on a regular basis since they were available throughout the year. The difficulty arises when trying to identify uncooked geophytes in the archaeological record. Phytoliths are usually produced in the above-ground organs of plants (leaves, inflorescences, stems, wood/bark) and very few are formed in their underground structures. In the future we plan to analyse dicotyledonous plant geophytes in order to improve detection on archaeological sites.

The possible uses of the dicotyledonous-leaf phytoliths in the hearths of PP13B and PP5/6 remains an opening question which can only be answered through more detailed studies and refinement of the actualistics.

Fig. 9.6. Photographs of geophytes collected from the Fynbos biome during the autumn season 2010 near Mossel Bay. a) Satyrium membranaceum *from* Orchidaceae *family; b)* Cyanella hyacinthoides *from* Iridaceae *family; c)* Moraea fugax *from* Iridaceae *family; d)* Tritonia croata *from* Iridaceae *family.*

Conclusions

The phytolith results at Olduvai Gorge, in combination with other related studies on plants, bones and sediments, have depicted a more humid landscape than initially thought for the times *H. habilis* and *P. boisei* were occupying the area. Phytoliths indicate that palms usually associated with fresh water sources would be abundant in Beds I and II, and would provide important edible sources, mostly for *H. habilis*, during the rainy season.

At Mossel Bay, the results show that early *Homo sapiens* was collecting plants for different purposes. These are linked to the evidence for fire related activities and it is possible that some of them would have been part of the diet.

The results show that phytolith analysis is an efficient tool which contributes to the identification of plant remains as dietary products during very early periods, when other evidence for plants is no longer available.

References

Albert, R. M. and Bamford, M. K. 2012. Vegetation during uppermost Bed I and deposition of Tuff IF at Olduvai Gorge, Tanzania, based on phytoliths and plant remains. *Journal of Human Evolution* 63(2), 342–50

Albert, R. M. and Cabanes, D. 2009. Fire in prehistory. Combustion processes and phytolith remains: an experimental approach. *Israel Journal of Earth Sciences* 56, 175–89

Albert, R. M. and Marean, C. 2012. Early *Homo sapiens* pyrotechnology: exploitation of plant resources and taphonomical implications. *Geoarchaeology Journal* 27, 363–84

Albert, R. M. and Weiner, S. 2001. Study of phytoliths in prehistoric ash layers using a quantitative approach. In Meunier, J. D. and Coline, F. (eds), *Phytoliths: Applications in Earth Sciences and Human History*, 251–66. Lisse: A.A. Balkema

Albert, R. M., Bamford, M. K. and Cabanes, D. 2006. Taphonomy of phytoliths and macroplants in different soils from Olduvai Gorge, Tanzania: application to Plio-Pleistocene palaeoanthropological samples. *Quaternary International* 148, 78–94

Albert, R. M., Bamford, M. K. and Cabanes, D. 2009. Palaeoecological significance of palms at Olduvai Gorge, Tanzania based on phytolith remains. *Quaternary International* 193, 41–8

Albert, R. M., Bamford, M. K. and Esteban, I. 2015. Reconstruction of ancient palm vegetation landscapes using a phytolith approach. *Quaternary International* 369, 51–66

Albert, R. M., Esteve, X., Portillo, M., Rodríguez-Cintas, A., Cabanes, D., Esteban, I. and Hernández, F. 2011. *PhytolithCoRe, Phytolith Reference Collection*. Retrieved [November, 2012], from http://comunidad.archeoscience.com/page/bd-fitolitos.

Albert, R. M., Tsatskin, A., Ronen, A., Lavi, O., Estroff, L., Lev-Yadun, S. and Weiner, S. 1999. Mode of occupation of Tabun Cave, Mt. Carmel Israel during the Mousterian period: A study of the sediments and the phytoliths. *Journal of Archaeological Science* 26, 1249–60

Ball, T. B., Brotherson, J. D. and Gardner, J. S. 1996. Identifying phytoliths produced by the inflorescence bracts of three species of wheat (*Triticum monoccocum* L., *T. dicoccon* Schrank, and *T. aestivum* L.) using computer-assisted image and statistical analyses. *Journal of Archaeological Science* 23, 619–32

Bamford, M. K. 2012. Fossil sedges, macroplants and roots from Olduvai Gorge, Tanzania. *Journal of Human Evolution* 63 (2), 351–63

Bamford, M. K., Albert, R. M. and Cabanes, D. 2006. Assessment of the Lowermost Bed II Plio-Pleistocene vegetation in the eastern palaeolake margin of Olduvai Gorge (Tanzania) and preliminary results from fossil macroplant and phytolith remains. *Quaternary International* 148, 95–112

Bamford, M. K., Stanistreet, I. G., Stollhofen, H. and Albert, R. M. 2008. Late Pliocene fossil grass from Olduvai Gorge, Tanzania. *Palaeogeography, Palaeoclimatology, Palaeoecology* 257, 280–93

Barboni, D., Ashley, G., Dominguez-Rodrigo, M., Bunn, H. T., Mabulla, A. Z. P. and Baquedano, E. 2010. Phytoliths infer locally dense and heterogeneous paleovegetation at FLK North and surrounding localities during upper Bed I time, Olduvai Gorge, Tanzania. *Quaternary Research* 74, 344–54

Barton, H., Torrence, R. and Fullagar, R. 1998. Clues to stone tool function re-examined: comparing starch grain frequencies on used and unused obsidian artifacts. *Journal of Archaeological Science* 25, 1231–8

Bentley-Condit, V. K. 2009. Food choices and habitat use by the Tana River Yellow Baboons (*Papio cynocephalus*): a preliminary report on five years of data. *American Journal of Primatology* 71, 432–6

Berna, F., Behar, A., Shahack-Gross, R., Berg, J., Boaretto, E., Gilboa, A., Sharon, I., Shalev, S., Shilshtein, S., Yahalom-Mack, N., Zorn, J. R. and Weiner, S. 2007. Sediments exposed to high temperatures: reconstructing pyrotechnological processes in Late Bronze and Iron Age Strata at Tel-Dor (Israel). *Journal of Archaeological Science* 34, 358–73

Berna, F., Goldberg, P., Kolska Horwitz, L., Brink, J., Holt, Sh., Bamford, M. and Chazan, M. 2012. Microstratigraphic evidence of in situ fire in the Acheulean strata of Wonderwerk Cave, Northern Cape province, South Africa. *Proceedings of the National Academy of Sciences (USA)* 109(20), 1215–20

Blumenschine, R. J. and Peters, C. R. 1998. Archaeological predictions for hominid land use in the paleo-Olduvai Basin, Tanzania, during lowermost Bed II times. *Journal of Human Evolution* 34, 565–607

Blumenschine, R. J., Peters, C. R., Capaldo, S. D., Andrews, P., Njau, J. K. and Pobiner, B. L. 2007. Vertebrate taphonomic perspectives on Oldowan hominin land use in the Plio-Pleistocene Olduvai basin, Tanzania. In Pickering, T., Schick, K. and Toth, N. (eds), *African Taphonomy: a tribute to the career of C.K. 'Bob' Brain*, 161–79. Bloomington, IN: CRAFT Press

Blumenschine, R. J., Peters, C. R., Masao, F. T., Clarke, R. J., Deino, A. L., Hay, R. L., Swisher, C. C., Stanistreet, I. G., Ashley, G. M., McHenry, L. J., Sikes, N. E., van der Merwe, N. J., Tactikos, J. C., Cushing, A. M., Deocampo, D. M., Njau, J. K. and Ebert, J. I. 2003. Late Pliocene *Homo* and hominid land use from western Olduvai Gorge, Tanzania. *Science* 299, 1217–21

Blumenschine, R. J., Stanistreet, I. G., Njau, J. K., Bamford, M. K., Masao, F. T., Albert, R. M., Stollhofen, H., Andrews, P., Fernanadez-Jalvo, Y., Prassack, K. A., McHenry, L. J., Camilli, E. L. and Ebert, J. I. 2012. Environments and activity traces of hominins across the FLK Peninsula during *Zinjanthropus* times (1.84 MA), Olduvai Gorge Tanzania. *Journal of Human Evolution* 63(2), 364–83

Bonnefille, R. 1984. Palynological research at Olduvai Gorge. *National Geographic Society Research Reports* 17, 227–43

Buchet, L., Cremoni, N., Rucker, C. and Verdin, P. 2001. Comparison between the distribution of dental micro-striations and plant material included in the calculus of human teeth. In Meunier J. D. and Coline, F. (eds), *Phytoliths: applications in earth sciences and human history*, 117–18. Lisse: A. A. Balkema

Carmody, R. N. and Wrangham, R. W. 2009. The energetic significance of cooking. *Journal of Human Evolution* 57, 379–91

Cerling, T. E., Mbua, E., Kirera, F. M., Manthi, F. K., Grines, F. E., Leakey, M. G., Sponheimer, M. and Uno, K. T. 2011. Diet of *Paranthropus boisei* in the early Pleistocene of East Africa. *Proceedings of the National Academy of Sciences (USA)* 108(23), 9337–41

Dallman, P. R. 1998. *Plant Life in the World's Mediterranean Climates. Oxford:* Oxford University Press

Elbaum, R., Albert, R. M., Elbaum, M. and Weiner, S. 2003. Detection of burning of plant materials in the archaeological record by a change in the refractive indices of siliceous phytoliths. *Journal of Archaeological Science* 30(1), 217–26

Esteban, I., de Vynck, J. C., Singels, E., Vlok, J., Marean, C. W., Cowling, R. M., Fisher, E. C., Cabanes, D. and Albert, R. M. (in press) Modern Soil Phytolith Assemblages used as proxies for paleoscape reconstruction on the South Coast of South Africa. *Quaternary International*

Fisher, E. C., Herries, A. I. R., Brown, K. S., Williams, H. M., Bernatchez, J., Ayalon, A. and Nilssen, P. J. 2010. A high resolution and continuous isotopic speleothem record of paleoclimate and paleoenvironment from 90 to 53 ka from Pinnacle Point on the south coast of South Africa. *Quaternary Science Reviews* 29, 2131–45

Goren-Inbar., N., Alperson, N., Kislev, M. E., Simchoni, O., Melamed, Y., Ben-Nun, A. and Werker, E. 2004. Evidence of Hominin control of fire at Gesher Benot Ya'aqov, Israel. *Science* 304, 725–7

Grine, F. E., Sponheimer, M., Ungar, P. S., Lee-Thorp, J. and Teaford, M. 2012. Dental microwear and stable isotopes inform the paleoecology of extinct hominins. *American Journal of Physical Anthropology* 148, 285–317

Hardy, B. L., Kay, M., Marks, A. E. and Monigal, K. 2001. Stone tool function at the Paleolithic sites of Starosele and Buran Kaya III, Crimea: behavioral implications. *Proceedings of the National Academy of Sciences* (USA) 98, 10972–7

Hardy, K., Buckley, S., Collins, M. J., Estalrrich, A., Brothwell, D., Copeland, L., García-Tabernero, A., García-Vargas, S., de la Rasilla, M., Lalueza-Fox, C., Huguet, R., Bastir, M., Santamaría, D., Madella, M., Wilson, J., Fernández-Cortés, Á. and Rosas, A. 2012. Neanderthal medics? Evidence for food, cooking, and medicinal plants entrapped in dental calculus. *Naturwissenschaften* 99, 617–26

Hay, R. L. 1976. *Geology of the Olduvai Gorge.* Berkeley CA: University of California Press

Hernández-Aguilar, A., Moore, J. and Pickering, T. R. 2007. Savanna chimpanzees use tools to harvest the underground storage organs of plants. *Proceedings of the National Academy of Sciences (USA)* 104(49), 19210–13

Henry, A. G., Brooks, A. S. and Piperno, D. R. 2011. Microfossils in calculus demonstrate consumption of plants and cooked foods in Neanderthal diets (Shanidar III, Iraq; Spy I and II, Belgium). *Proceedings of the National Academy of Sciences (USA)* 108(2), 486–91

Jacobs, Z. 2010. An OSL chronology for the sedimentary deposits from Pinnacle Point Cave 13B – a punctuated presence. *Journal of Human Evolution* 59(3–4), 289–305

Karkanas, P. and Goldberg, P. 2010. Site formation processes at Pinnacle Point Cave 13B (Mossel Bay, Western Cape Province, South Africa): resolving stratigraphic and depositional complexities with micromorphology. *Journal of Human Evolution* 59 (3–4), 256–73

Karkanas, P., Shahack-Gross, R., Ayalon, A., Bar-Matthews, M., Barkai, R., Frumkin, A., Gopher, A. and Stiner, M. C. 2007. Evidence for habitual use of fire at the end of the Lower Paleolithic: site-formation processes at Qesem Cave, Israel. *Journal of Human Evolution* 53(2), 197–212

Katz, O., Cabanes, D., Weiner, S., Maeir, A. M., Boaretto, E. and Shahack-Gross, R. 2010. Rapid phytolith extraction for analysis of phytolith concentrations and assemblages during an excavation: an application at Tell es-Safi/Gath, Israel. *Journal of Archaeological Science* 37, 1557–63

Kinnaird, M. F. 1995. Competition for a forest palm: use of *Phoenix reclinata* by human and nonhuman primates. *Conservation Biology* 6, 101–7

Leakey, M. D. 1971. *Olduvai Gorge, Volume 3: Excavations in Beds I and II, 1960–1963.* Cambridge: Cambridge University Press

Madella, M., Alexandre, A. and Ball, T. B. 2005. International code for phytolith nomenclature 1.0. *Annals of Botany* 96, 253–60

Marean, C. W. 2008. *The African Origins of Modern Human Behavior*. Nobel Conference 44. Available from: http://gustavus.edu/events/nobelconference/2008

Marean, C. W. 2010. Coastal South Africa and the co-evolution of the modern human lineage and coastal adaptations. In Bicho, N., Haws, J. A. and Davis, L. G. (eds), *Trekking the Shore: changing coastlines and the antiquity of coastal settlement*, 421–40. New York: Springer

Marean, C. W., Bar-Matthews, M., Fisher, E. C., Goldberg, P., Herries, A. I. R., Karkanas, P., Nilssen, P. J. and Thompson, E. 2010. The stratigraphy of the Middle Stone Age sediments at Pinnacle Point Cave 13B (Mossel Bay, Western Cape Province, South Africa). *Journal of Human Evolution* 59 (3–4), 234–55

Martínez, L., Pérez-Pérez, A. N. and Turbón, D. 2006. *Nutrition in Ancient Hominids*, 16–22. Amsterdam: Elsevier International Congress Series 1296

McInerney, F. A., Strömberg, C. A. E. and White, J. W. C. 2011. The Neogene transition from C3 to C4 grasslands in North America: stable carbon isotope ratios of fossil phytoliths. *Paleobiology* 37(1), 23–49

Mercader, J., Runge, F., Vrydaghs, L., Doutrelepont, H., Corneile, E. and Juan-Treserras, J. 2000. Phytoliths from archaeological sites in the tropical forest of Ituri, Democratic Republic of Congo. *Quaternary Research* 54, 102–12

Milton, K. 2000. Hunter-gatherer diets, a different perspective. *American Journal of Clinical Nutrition* 71(3), 665–67

Oakley, K. 1970. On man's use of fire, with comments on tool-making and hunting. *Viking Fund Publications in Anthropology* 31, 176–93

Parkington, J. 1976. Coastal settlement between the mouths of the Berg and Olifants Rivers, Cape Province. *South African Archaeological Bulletin* 31, 127–40

Parkington, J. 1981. The effects of environmental change on the scheduling of visits to the Elands Bay Cave, Cape Province, S. A. In Hodder, I., Isaac, G. and Hammond, N. (eds), *Patterns of the Past*, 341–59. Cambridge: Cambridge University Press

Parr, J. F., Lentfer, C. J. and Boyd, W. E. 2001. A comparative analysis of wet and dry ashing techniques for the extraction of phytoliths from plant material. *Journal of Archaeological Science* 28, 875–86

Perlès, C. 1977. *Préhistoire du Feu*. Paris: Masson

Peters, C. R. and Blumenschine, R. J. 1995. Landscape perspectives on possible land use patterns for early hominids in the Olduvai Basin. *Journal of Human Evolution* 29, 321–62

Piperno, D. R. 2006. *Phytoliths: a comprehensive guide for archaeologists and paleoecologists*. Lanham MD: AltaMira Press

Portillo, M., Ball, T. and Manwaring, J. 2006. Morphometric analysis of inflorescence phytoliths produced by *Avenasativa* L. and *Avenastrigos schreb*. *Economic Botany* 60(2), 121–9

Proches, S., Cowling, R. M., Goldblatt, P., Manning, J. C. and Snijman, D. A. 2006. An overview of the Cape geophytes. *Biological Journal of the Linnean Society* 87, 27–43

Rolland, N. 2004. Was the emergence of home bases and domestic fire a punctuated event? A review of the Middle Pleistocene record in Eurasia. *Asian Perspectives* 43, 248–80

Runge, F. 1999. The opal phytolith inventory of soils in central Africa - quantities, shapes, classification and spectra. *Review of Palaeobotany and Palynology* 107, 23–53

Shafer, H. J. and Holloway, R. G. 1979. Organic residue analysis in determining stone tool function. In Hayden, B. (ed.) *Lithic Use-wear Analysis*, 385–99. New York: Academic Press

Sponheimer, M., Lee-Thorp, J., de Ruiter, D., Codron, D., Codron, J., Baugh, A. T. and Thackeray, F. 2005. Hominins, sedges, and termites: new carbon isotope data from the Sterkfontein valley and Kruger National Park. *Journal of Human Evolution* 48(3), 301–12

Stanistreet, I. G. 2012. Fine resolution of early hominin time, Beds I and II, Olduvai Gorge, Tanzania. *Journal of Human Evolution* 63(2), 300–8

Stollhofen, H. and Stanistreet, I. G. 2012. Plio-Pleistocene synsedimentary fault compartments, foundation for the eastern Olduvai Basin paleoenvironmental mosaic, Tanzania. *Journal of Human Evolution* 63(2), 309–27

Twiss, P. C., Suess, E. and Smith, R. M. 1969. Morphological classification of grass phytoliths. *Soil Science Society of America Journal* 33(1), 109–15

Ungar, P. S. and Sponheimer, M. 2011. The diets of early hominins. *Science* 334(6053), 190–3

Weiner, S. 2010. *Microarchaeology. Beyond the Visible Archaeological Record*. New York: Cambridge University Press

Wolde Gabriel, G., Ambrose, S. H., Barboni, D., Bonnefille, R., Bremond, L., Currie, B., DeGusta, D., Hart, W. K., Murray, A. M., Renne, P. R., Jolly-Saad, M. C., Stewart, K. M. and White, T. D. 2009. The geological, isotopic, botanical, invertebrate and lower vertebrate surroundings of *Ardipithecus ramidus*. *Science* 326, 5949, 65 65e1–65e5

Wrangham, R. W. 2009. *Catching Fire: how cooking made us human*. New York: Basic Books

Wrangham, R. W., Conklin, N. L., Chapman, C. A. and Hunt, K. D. 1991.The significance of fibrous foods for Kibale Forest chimpanzees. *Philosophical Transactions of the Royal Society B: Biological Sciences* 334, 171–8

Wrangham, R. W., Holland Jones, J., Laden, G., Pilbeam, D. and Conklin-Brittain, N. L. 1999. The raw and the stolen: cooking and the ecology of human origins. *Current Anthropology* 40, 567–94

Zihlman, A. L. and Tanner, N. M. 1978. Gathering and hominid adaptation. In Tiger, L. and Fowler, H. T. (eds), *Female Hierarchies*, 163–94. Chicago MI: Beresford Food Services

10. Phytolith evidence of the use of plants as food by Late Natufians at Raqefet Cave

Robert C. Power, Arlene M. Rosen and Dani Nadel

The Natufian culture was a critical junction in human subsistence, marking the first steps towards sedentism and the domestication of cereals and pulses. The Natufians are known to have collected large seeded grasses (wild barley and wheat) and small-seeded grasses prior to the onset of cultivation but in most of the Levant little direct evidence of this survives. A variety of stone mortars and mortars carved into bedrock are common on Natufian sites. These were often interpreted as reflecting intensification of cereal or acorn food preparation, though the proposition was rarely tested. This study uses phytoliths to interpret Late Natufian plant-use at Raqefet Cave in south-eastern Mount Carmel, Israel. Phytolith analysis focused on 35 samples from bedrock mortars, human burials, and cave floor sediments in the cave. We show that the Late Natufians at Raqefet were consuming both small-grained grasses and wheat and barleys. Our findings also suggest that the mortars did not serve primarily as specialised grass-seed processing implements. There is also evidence the occupants of Raqefet Cave continued to utilise forest resources, even as Late Natufians shifted towards more exploitation of wild grasses elsewhere in the southern Levant.

Introduction

Raqefet Cave is a Late Natufian archaeological site on the slopes of Mount Carmel, in present day Israel (Fig. 10.1). Five seasons of excavations exposed a Natufian cemetery in the first chamber of the cave, with 29 human burials. Annexed to these, and in several locations in the first chamber and on the terrace, around 100 bedrock features were exposed, some preserving intact Natufian deposits. The site is unique for its abundance of both burials and varied assemblages of mortar features. Many of these are bedrock mortars whose function is still unknown. Excavations suggest that the cave was occupied by Late Natufians who probably cleared most of the cave floor in the first chamber so they could carve these mortars (Nadel *et al.* 2009a).

Understanding Natufian subsistence is important as it may help to explain the subsequent development of agriculture in the PPN. Natufian diet is often assumed to be the product of climatic and socio-economic forces (Munroe 2004; Rosen 2010). Scholars accept that the Natufians had intensive subsistence strategies that utilised grass seeds and many other Mediterranean forest plants and animals. Bedrock mortars are

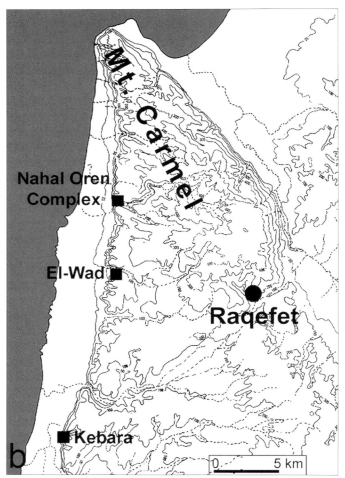

Fig. 10.1. Raqefet Cave and other Natufian sites of the Southern Levant discussed in the text. The Mediterranean Sea is on the left (after Google Earth 2011)

characteristic of the Late Natufian period but there has been no direct evidence of their function up until now. Phytolith-based studies support an increase in grass-seed use during the Early to Late Natufian shift (Rosen 2010), yet this has never been examined in the Mount Carmel region. Since macrobotanical plant remains were not preserved at Raqefet Cave, the study of plant-use is reliant on plant micro-remains phytoliths that were preserved in the site's sediments. The increasing use of these plant micro-remains is helping to refine our knowledge of subsistence in a wide span of archaeological cultures (e.g., Tromp 2012; Power *et al.* 2014a). The aim of this paper is to show how phytoliths can reveal past plant usage relating to Natufian subsistence. Our analysis focused on 35 samples from 16 burials, 12 samples from bedrock mortars, three tufa samples, two samples from a bedrock basin, and two off-site control samples.

Background

The cultural and climatic chronology

The Early Natufian (~15,500/15,000–13,700 cal BP; Stutz *et al.* 2009; Blockley and Pinhasi 2011; Weinstein-Evron *et al.* 2012) is defined by the revolutionary emergence of sizable settlements such as the site of Eynan on the edge of the Hula marsh in northern Israel (Perrot 1966; Valla *et al.* 2004, 2007; Blockley and Pinhasi 2011). These sites are interpreted as hallmarks of reduced group mobility and growing populations (Belfer-Cohen 1991). This commencement of near-settled life was a revolutionary change in human organisation (Bar-Yosef 1998). Unsurprisingly, the Natufian period is known for its expressions of economic reorganization including stone architecture and immobile stone artefacts (Boyd 2002; 2006; Grosman 2003). Ultimately, the Late Natufians (~13,700–11,600 cal BP) would be the last hunter-gatherers in the southern Levant before the arrival of agriculture in the subsequent PPN, which would gradually transform human subsistence across western Eurasia. However, during this period in some of the region people returned once again to smaller occupational sites, hinting at higher mobility strategies as site occupation intensity decreased (Valla 1995; Bar-Yosef 1998; Belfer-Cohen and Bar-Yosef 2000; Bar-Yosef and Belfer-Cohen 2002). Some researchers have proposed an increased reliance on consumption of the wild progenitors of cereals, involving investment in specialisation during the Late Natufian (Bar-Yosef and Meadow 1995). Most researchers accept that climatic events occurring during this period influenced Natufian cultures (Bar-Yosef and Belfer-Cohen 1989; 1991; Rosen 2007a). Foremost among these was the Younger Dryas cold episode, which coincided with the Late Natufian, lasting from ~13,600 cal BP to ~11,400–11,200 cal BP (Bar-Matthews *et al.* 1999; 2003). The oxygen isotope curves from Nahal Soreq Cave, Israel, indicate rapid cooling and drying in the eastern Mediterranean at this time (Bar-Matthews *et al.* 1999; 2003; Blockley and Pinhasi 2011). Moore and colleagues (Moore *et al.* 2000) proposed these changes resulted in an abrupt reduction in edible plants which were available in the Quercus-Pistacia park-woodland which covered much of the Southern Levant.

Natufian subsistence

Natufian use of animal resources is increasingly well-studied (Bar-Oz *et al.* 2004; Munro 2004). Zooarchaeological assemblages suggest Natufian foragers adopted a number of different hunting intensification strategies (Stutz *et al.* 2009). However, data on plant food subsistence is less available. Yet, understanding plant utilisation is critical to explaining the development of cultivation and domestication that would subsequently emerge. Plant use was widespread, as is evident from microwear on grinding implements and knapped sickle tools (which are associated with grass-seed harvesting) (Unger-Hamilton 1989; Dubreuil 2004). However, few Natufian sites in the southern Levant have produced macrobotanical remains that could reveal detailed information. This is due both to the excavation of some key sites prior to the introduction of archaeobotanical flotation and screening, and the destructive local taphonomic processes across the region (Bar-Yosef

Table 10.1: Plant genera and species with evidence of exploitation by Natufians in the southern Levant

Plant species	Fossil type	Site
Amygdalus communis Almond	Shell	Hayonim[1]
Lens sp. Lentil	Seed	Raqefet[3]
cf. *Pisum sp.* Pea	Seed	Hayonim[1], Raqefet[3]
Vicia sp. Vetch	Seed	Nahal Oren[4]
Lupinus pilosus Lupine	Seed	Hayonim[1]
Olea spp. Wild olive	Seed	Nahal Oren[4]
Vitis sp. Grape	Seed	Nahal Oren[4]
Hordeum spontaneum Wild barley	Seed	Hayonim[1]
Hordeum sp. Barley	Phytoliths	Eynan, Hilazon Tachtit[2]
Triticum sp. Wheat	Phytoliths	Eynan, Hilazon Tachtit[2]
Small-seeded grasses	Phytoliths	Eynan, Hilazon Tachtit[2] Natal Oren[4]

Seed total is < 70. [1]Hopf and Bar-Yosef, 1987, [2]Rosen, 2010. [3]Garrard, 1980. [4]Noy *et al*. 1973. Remains from Nahal Oren and Raqefet 1970s excavations were not found in secure archaeological contexts. The new Raqefet results are excluded

1998). Although in the Northern Levant a small number of sites such as Abu Hureyra and Dederiyeh Cave preserve plant remains (Moore *et al.* 2000; Tanno *et al.* 2013), in the southern Levant only tiny quantities of charred seeds have been reported (<70), some which are in questionable contexts, See Table 10.1 (Noy *et al.* 1973; Garrard 1980; Hopf and Bar-Yosef 1987; Zohary *et al.* 2012).

Multi-cell phytoliths from Natufian contexts at Eynan near Lake Hula confirmed grass seeds were collected, including wheat and barley but most collected species were of small-seeded varieties (Rosen 2004; 2007b). Researchers believe that the consumption of grass seeds has a very long history in the Levant (Piperno *et al.* 2004; Weiss *et al.* 2004; Henry *et al.* 2010). However, the dominance of small-seeded grass phytoliths showed Natufians were targeting an array of species without specialising on wheat or barley (Rosen 2010). Rosen (2010) explained grass seed use as an adaptation stimulated by the declining availability of forest staples, (e.g., olive, pistachio, almond, and acorn) during the climatic deterioration of the Younger Dryas. Ethnographic studies have verified that although grass seeds are energy rich, many hunter-gatherers prefer to prioritise other plants foods such as nuts due to lower processing costs (Keeley 1992; Mason 1995). Grass seeds enter hunter-gatherer diets as a 'second choice' to cross over nutritional shortfalls (Keeley 1992).

Plant processing

Deciphering the use and changes in Natufian groundstone tools is a key part of understanding Late Pleistocene – early Holocene Levantine subsistence. Groundstone implements may be animal, plant or non-subsistence processing installations, but their importance in plant processing makes them a valuable proxy for understanding the fluctuating trends in plant exploitation (Dubreuil 2004; Eitam 2009; Wright 1994). Subsistence change from the Early Epipalaeolithic to the Neolithic is increasingly being understood as a changing system of resource specialisation and preparation rather than a shift to new staple foods. Wollstonecroft (2007) has stated that from the Early Epipalaeolithic, hunter-gatherers developed progressively more specialised food procurement and processing technologies. Groundstone technology becomes more common across the Levant from the Early Epipalaeolithic onward (Wright 1994). Increases in the number and frequency of groundstone implements at Levantine sites were also accompanied by the development of deep bedrock features and conical pestles (Eitam 2009). Their creation required an impressive amount of manufacturing effort. The sheer abundance of bedrock mortars at Raqefet shows a high technological investment. In the Early Natufian, groundstone technology appears to have become far more economically important than in preceeding periods, but it also developed substantially during the Natufian. Typologically, the Late Natufian is richer, and also includes bedrock rock-carved mortars which were scarce during the Early Late Natufian. More intense food preparation with groundstone implements would decrease plant food particle-size and thus increase bioavailable energy but with the trade-off of increased cost in processing (Wright 1994).

The Raqefet Cave site

Raqefet Cave is located in a south-eastern projection of Mount Carmel (lat. 32.65 long. 35.08) at 230 m OD (Fig. 10.1). Facing west, the cave mouth is 50 m above a wadi bed. The cave is comprised of five chambers, the most eastern of which has an open chimney (Nadel *et al.* 2008). Early excavations revealed deposits dating to the Early Upper Palaeolithic, the Levantine Aurignacian, the Kebaran, the Geometric Kebaran, and the Natufian (Noy *et al.* 1973). The renewed excavations focused on the Natufian remains in the first chamber and the terrace (Lengyel *et al.* 2005; Nadel *et al.* 2008; 2009; 2012). The excavated portion of the cave included an area containing numerous Natufian burials (Loci 1 and 3; Fig. 10.2), a large bedrock basin with a burial and two boulder mortars (Locus 2), and two areas of cemented archaeological sediment (Loci 4, 5) (Nadel *et al.* 2008). Across the site, features such as shallow and deep mortars, pipe-like forms, basins and cupmarks were carved into the cave's bedrock.

Among the burials in Locus 3 were Homos 18 and 19, which were found in an elongated pit dug into grey Middle Palaeolithic sediment. These skeletons were directly dated by radiocarbon to 13,700–13,000 cal BC (Nadel *et al.* 2013). Although we cannot prove the burials are contemporary with bedrock features, *in situ* lithics in these carved features confirms their Late Natufian origin (Nadel *et al.* 2008). Several other nearby Natufian burials were cut into this layer but were less well-preserved. Flowering plant impressions were identified inside a number of the Natufian burials including Burial 18, 19, 25, 28, and 31. A narrow shelf on the southern wall of Chamber 1 (Locus 5) was also investigated revealing a Kebaran tufa deposit containing lithic and faunal remains (Nadel *et al.* 2009).

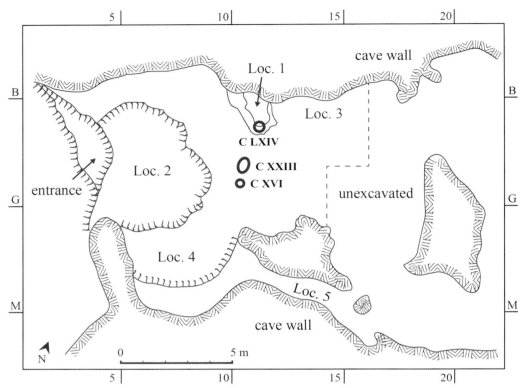

Fig. 10.2. Plan of the first chamber, Raqefet Cave. Note the location of the burial area (Loci 1 and 3) and the three bedrock mortars discussed in the text

Bedrock cut features

Bedrock-cut installations are characteristic of the Natufian in caves, rock shelters and open-air sites. They are present in many Late Natufian sites, usually ranging in number from a handful to several dozen (Nadel and Rosenberg 2010). They range in form and size, from tiny holes 2–5 cm wide and deep, through cupmarks and small bowls to deep narrow specimens (some 50–80 cm deep; Fig. 10.3). The function and use of these features is not understood. Researchers have suggested that some mortars were used in a variety of social or symbolic contexts (Nadel and Lengyel 2009; Nadel and Rosenberg 2010; Nadel *et al.* 2012). More commonly, certain types are seen as implements for plant grinding or pounding; Eitam (2008) has proposed that some of these installations may have been for barley-dehusking rather than grinding. The rounded parabolic (as opposed to flat) bases in several Raqefet and Nahal Oren examples reinforce this view. Parabolic bases typically are used for dehusking as opposed to grinding, including on implements used up to recent times (Hillman 1984). Nonetheless, there are other potential plant staples such as sea club rush tubers

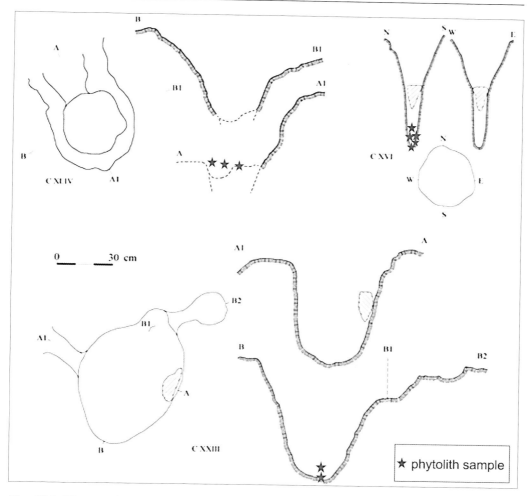

Fig. 10.3. *Plans and sections of the three bedrock mortars discussed in the text. Note the large volume of C XXIII and the deep narrow shaft of C XVI. Bedrock mortar C XLIV had tufa adhering to the base and the bottom contour was not revealed*

(*Bolboschoenus* spp.) which require mortar dehusking. Many other plant foods also require dehusking and grinding e.g., acorns (*Quercus* spp.), water lily (*Nuphar* spp. and *Nymphaea* spp.), or just pulverising for instance bird cherry (*Prunus padus* L.) (Mears and Hillman 2007; Wollstonecroft 2007). Although the number of bedrock features at Raqefet (*c.* 100) is not matched by an equivalent number of pestles (n<5), wooden pestles offer some technological advantages over stone pestles and might explain pestle rarity (Hillman *et al.* 1989).

Methodology

Phytolith extraction and quantification

The samples were prepared using gravity sedimentation, dry-ashing, and heavy liquid flotation (Power *et al.* 2014b). Each sample consisted of 0.5–2 g of pre-sieved <0.25 mm size fraction. These were treated with 15 ml of 10% HCl. HCl was then removed by adding distilled water, centrifuging, and discarding the supernatant. Clays were dispersed by adding 20 ml of a 5% solution of sodium hexametaphosphate and distilled water. Water was then added to a height of 8 cm. The samples were left to sit for 70 minutes to allow settling of the silt and fine-sand fractions. The suspension containing clay particles was then discarded. The beakers were then refilled, left to settle for 60 minutes and poured again. This process was repeated until the suspension was clear. The residue was transferred to crucibles, dried, and then placed in a muffle furnace for two hours at 500°C to remove organic matter. The samples were transferred into 15 ml tubes. Then, 3 ml of sodium polytungstate solution calibrated to 2.3 sp.gr was added to each. The sample was then centrifuged at 800 rpm (115–120 × g) for 10 minutes. The tubes were removed and the suspension containing the phytoliths was poured into clean 15 ml tubes. Distilled water was added to the tubes with the phytoliths to reduce the specific gravity and then centrifuged at 2000 rpm (721–752 × g) for 5 minutes. After two such treatments, the clean phytoliths were transferred into beakers, dried, and weighed. Entellan (Merck) or Permount was used to mount a weighed aliquot of 2–3 mg of residue from each sample. Phytoliths were counted at 400× magnification. Single-cell phytolith forms were counted to 300 or more individuals. Multi-cells forms were counted as up to 100 individuals where possible. During mounting 24 × 24 mm cover slides were used. This created 2304 fields of view under 400× magnification. Quantification necessitated the documentation of the number of fields used in phytolith counting. Any mounting agent with phytoliths which had spilled beyond the cover slide was allowed for in this calculation. The number of phytoliths on the slide was calculated using the following algorithm:

n phytoliths per slide = n counted/n fields counted x total n fields on slide

This value was used to derive the number of phytoliths per gram sediment, which served as a comparable unit of quantification across all samples used in this study. This value was calculated with the following formula:

n phytoliths per gram = n phytoliths per slide/total amount of sediment mounted (mg) × total phytolith amount (mg)/total initial sediment (mg) × 1000

This formula calculated the number of each phytolith type on each slide and the number of each type per gram of dry <0.25 mm sediment. This was carried out separately for single-cell and multi-cell phytoliths.

Phytolith identification

In monocotyledons (grasses, sedges, palms) phytoliths form as silica bodies within the cells of the living plants and result in microfossils that share cell morphology. As such, they are distinctive in their shapes and similar morphotypes regularly recur. The monocot phytoliths can be identified according to the anatomical part of the plant in which they formed (Rosen 1992; 2010; Power *et al.* 2014b). Phytolith morphotypes such as 'dendritic long-cells' are found only in the florets of grasses, and are accepted as proxies for grass grains. Smooth 'psilate long-cells' form in the culms of grasses and sedges and therefore represent the stems of these plants. In contrast, silica bodies from dicotyledons (woody herbaceous or arboreal plants) usually have irregular forms often in the shape of irregular platelets when formed in wood. Dicot leaves produce phytoliths within the epidermal cells and they can be identified from their multi-cell polyhedral or jigsaw puzzle-like configurations.

In addition to plant parts, we can identify some genera of grasses, notably large seeded grasses, from the multi-cell configuration of floral epidermal cells, allowing us to distinguish between wheat (*Triticum* sp.) and barley (*Hordeum* sp.) and the broader category of small-seeded grasses (Rosen 1992) as well as common reed (*Phragmites* sp.). However, some of those barley phytoliths may in fact come from weedy small-seeded varieties of barley, such as *Hordeum bulbosum* L. We can also identify plants in the sedge family (Cyperaceae) by the characteristic cone-shaped single-cell morphotypes and the diagnostic suite of cells manifested in the multi-cell forms.

Results

Following our analysis of the 35 samples (Table 10.2), we found that although the absolute quantities of phytoliths varied, non-archaeological control samples showed lower densities of phytoliths. The non-archaeological phytoliths were mostly eroded phytoliths from grass stems. A high-level of phytolith turnover may explain phytolith density outside the cave. Phytoliths in archaeological samples were far more numerous than these controls (Figure 10.4), with one context (Mortar C-XVI) containing a particularly high concentration. This shows that an archaeological phytolith signal is present.

Economic uses of plants

In all of the samples, grasses were the dominant plant category. The per cent of 'dendritic' long-cell phytoliths from grass husks was low in each sample (Fig. 10.5), especially in the bedrock features. Our Figure 10.6 box plot shows that the bedrock features generally had no distinct association with dendritic phytoliths and only track sample size of classified category. Although the multi-cell forms generally mirror this trend, two samples from clay mortar LXXIV did not (RQ-11-47 and RQ-11-48) as multi-cell floral types exceeded 50 per cent (Fig. 10.4). Likely, this is not a bedrock mortar feature but a sediment-negative of a mortar or a bowl that is no longer present. The floral multi-cells in clay mortar C-LXXIV may represent residues of plant processing. This includes small-seeded grasses and a small amount of barley (Fig. 10.3).

Table 10.2. *The provenance of sediment samples studied for phytoliths*

	Lab no.	Grid	Locus	Depth	Feature	Description
Burials	RQ-11-26	C14c	3	−2.06/−2.15	Homo 11	Infant burial
	RQ-11-4	B12a	1	−2.49	Homo 15	Near tibia
	RQ-11-10	B12a	1	−2.49	Homo 15	Near tibias
	RQ-11-9	C12	1	−2.87/−2.88	Homo 17	Under thorax
	RQ-11-12	C12a–d	1	−2.85	Home 17	Under slab B4
	RQ-11-13	C12	1	−2.80/−2.82	Homo 17	Left humerus/ radius
	RQ-11-20	D16	3	−2.38	Homo 18	Under stone 15 – skull & hands
	RQ-11-21	D16	3	−2.25	Homo 18	Under sacrum
	RQ-11-22	D16	3	−2.27	Homo 18	Under proximal ulna
	RQ-11-17	D16a	3	−2.20	Homo 19	Under stone 47 at abdomen
	RQ-11-18	D16c	3	−2.27/−2.28	Homo 19	Under stone 64 – proximal ulna
	RQ-11-19	D16c	3	−2.20	Homo 19	Under stone 37 at upper chest
	RQ-11-23	D14d/D15c	3	−2.06/−2.15	Homo 20	Under skull
	RQ-11-24	E15c	3	−2.16	Homo 22	Under thorax
	RQ-12-56	E14b	3	−254/−259	Homo 31	Chest area
	RQ-12-57	E14b	3	−243/−248	Homo 31	Above/near hand
Bedrock features/ mortars	RQ-11-27	F11d	–	46–50cm above base	Mortar C-XVI	
	RQ-11-28	F11d	–	50–53cm above base	Mortar C-XVI	
	RQ-12-49	F11d	–	53–55cm from top	Mortar C-XVI	

	Lab no.	Grid	Locus	Depth	Feature	Description
	RQ-12-50	F11d	–	55–56cm from top	Mortar C-XVI	
	RA-07-01	F11d	–	Base	Mortar C-XVI	Sealed and 50cm deep
	RQ-11-29	D12a	1	–2.73	Mortar C-XLIV	North wall, under Homo 9
Bedrock	RQ-11-30	D12b	1	–2.69	Mortar C-XLIV	North rim
features/	RQ-11-32	D12b	1	–2.65/–2.70	Mortar C-XLIV	Hard internal sediment
mortars	RQ-11-1	F12	–	–5cm above base	Mortar C-XXIII	5 cm above base
	RA-07-02	F12	–	Base of feature	Mortar C-XXIII	Base of 80 cm feature
	RQ-12-47	E14b	–	–2.50	Clay mortar LXXIV	Mortar negative fragment 4
	RQ-12-48	E14b	–	–2.50	Clay mortar LXXIV	Mortar negative fragment 6
	RA-07-04	K7	4	–2.15/–2.27	Tufa	Late Natufian matrix
Tufa,	RA-07-05	K8	4	–	Tufa	Late Natufian matrix
controls	RA-07-06	L14a	5	–	Kebaran tufa	Archaeological matrix
and cave	RQ-12-52	D14c	3	–2.48–2.55	Basin sediment	Natufian natural basin
basin	RQ-12-53	E14a	3	–2.70–2.75	Basin sediment	Natufian natural basin
	RQ-12-45	–	–	–	Control	40 m NW
	RQ-12-46	–	–	–	Control	100 m NW of cave

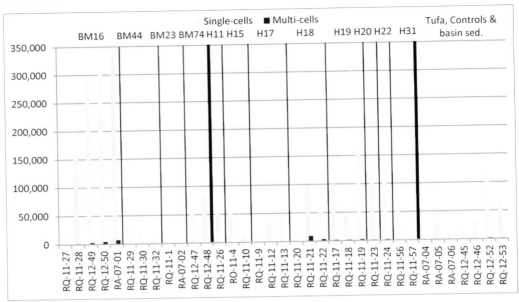

Fig. 10.4. *Single and multi-cell phytoliths per gramme of sediment of the four contexts groups (Power* et al. *2014)*

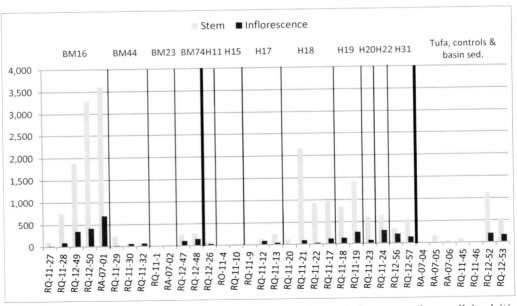

Fig. 10.5. *Multi-cell stem (long-cell psilate phytoliths) and grass inflorescence (long-cell dendritic phytoliths) per gramme of sediments (single-cells only) (Power* et al. *2014)*

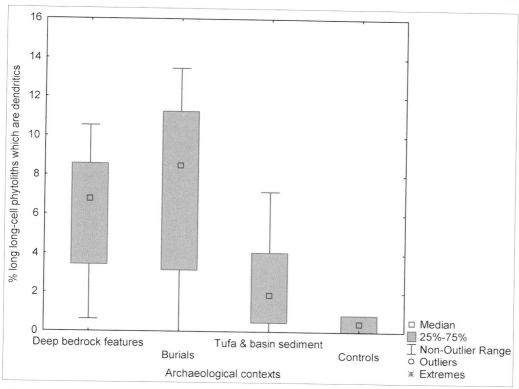

Fig. 10.6. *Dendritic phytoliths as a per cent of long-cell grass phytoliths as a box plot indicate grass husks occur across the site (single-cell only). Dendritic phytoliths probably relate to samples size. Mortars n=12, Burials n=16, Tufa and basin sediment n= 5, Controls n=2 (Power et al. 2014)*

Across the site, dendritic morphotypes from grass florets and seed-husks are more commonly associated with burials than deep bedrock features. Thus, it is plausible that these husks relate to the Burial of Homo 9 on the north margin of C-XLIV rather than with the bedrock feature itself. Accordingly, C-XLIV is inconclusive of *in situ* plant processing. This trend is most clear in C-XXIII.

Bedrock mortar C-XVI is of particular interest. It is a deep, narrow specimen, funnel-like in shape and thus unique at the site (Fig. 10.3). It is 62 cm deep, 33 cm wide at the top and only 7 cm wide at the bottom. Mid-way down, there was an isolated stone in the shaft which matched the contour of the mortar (it contained no other stones). Evidently, it was placed there intentionally, probably as a seal, as was the case in several other Natufian sites (Nadel *et al.* 2009b). The sediment in the lower half was typical of the local Natufian sediment in terms of matrix. Furthermore, it contained Natufian lunates and animal bones. Thus, the five phytolith samples from here are some of the few to provide data from within a sealed deep, narrow Natufian mortar (Terradas *et al.* 2013).

Table 10.3: Multi-cell phytoliths of wheat, barley and small-seeded grasses, which were identified (Power et al. 2014)

Sample	Wheat	Barley	Small-seeded grasses
RQ-11-27			6
RA-07-01	277	60	277
RQ-11-30		6	24
RQ-11-32		9	37
RQ-12-47	18	6	
RQ-12-48		17	8
RQ-11-26		4	8
RQ-11-12	12	18	6
RQ-11-13		5	5
RQ-11-18		26	53
RQ-11-17	17	17	
RQ-11-19			48
RQ-11-24	18	35	89
RQ-12-56			8
RQ-12-57			9

Table excludes phytoliths that could not be distinguished between wheat and barley. Small-seeded includes species of Rye, *Aegilops* and weedy species of barley eg, *Hordeum bulbosum* L.

Homo 19 and Homo 22 contain abundant well-preserved husk phytoliths that have little similarity with the other burial assemblages. This is most clear in RQ-11-19 and RQ-11-24, which came from the abdomens. This sample contained many dendritics (Fig. 10.5, Table 10.3). When identification was possible, multi-cell husks were shown to be mostly small-seeded grasses but barley *(Hordeum* sp.) also occurred. This could indicate visceral contents suggesting a grass seed meal before death or an offering of grains included with the deceased.

We can better understand the use of grass stems and seed-husks by considering the samples from non-bedrock features. Furthest from the bedrock features (8 m south) are the samples from the hard tufa sediment matrix of the cave floor at Loci 4 (Natufian) and 5 (Kebaran). This part of the cave yielded moderate to low levels of grass stem morphotypes. These samples indicate a tailing-off of the localities indicating the cultural use of grasses near the southern cave wall. The observed spatial pattern probably represents occasional subsistence-based activities near the northern cave wall, the use of grass as a floor covering in that area, or the use of these materials in graves (see below).

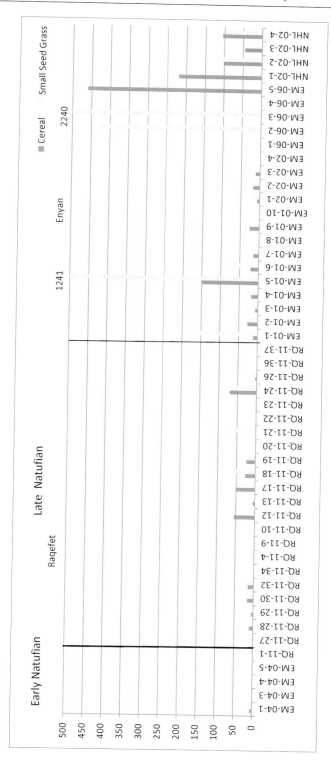

Fig. 10.7. Large-seeded grasses (wheat and barley) and small-seeded grass husk abundance per gramme of sediment at Early Natufian Eynan, Late Natufian Raqefet and Eynan

Across the site only 29% of multi-cell phytoliths were identifiable but when genera could be characterised, small-seeded grasses were similar (n=578) to wheat and barley combined (n=545) (Fig. 10.3, Table 10.3). Although the data are tentative due to the low abundance of multi-cell phytoliths, they agree with the broader picture of Late Natufian data on grass seed subsistence (Fig. 10.7; Rosen 2010; Rosen and Rivera-Collazo 2012). Of the large-seeded grasses, wheat (n=342) is much more common than barley (n=203).

Discussion

Variations of phytolith presence and density

Site formation processes may have also played a role influencing the phytoliths within the cave. Burning of phytoliths and root activity in archaeological deposits has been correlated with increased level of phytolith breakdown (Cabanes *et al.* 2011). Therefore, burning in the cave may have reduced the number of multi-cell phytoliths in particular. However, dissolution was not sufficient to preclude the widespread preservation of silicified plant hairs, one of the most soluble morphotypes in the Raqefet sediments (Cabanes *et al.* 2011).

The highest phytolith concentrations occur in sediments relating to human activity. The low concentrations in the non-archaeological samples reveal that the phytolith-rich samples are anthropogenic in origin. The patterning suggests that human activities at Raqefet were probably limited, possibly indicating short-lived occupation. The occurrence of grass husks may reflect a spring–summer (March–June) occupation, but this does not rule out occupation in other seasons.

Bedrock mortar use

Grass phytoliths are dominant in the cave sediments but this does not indicate a predominance of grasses within the diet. However, the rarity of husks, even in the sealed bedrock mortar (C-XVI), is incongruent with the notion of a specialised grass seed processing subsistence strategy. This supports Wright's hypothesis that Natufian bedrock features had a broad non-specialised function (Wright, 1994). The diversity of phytolith forms, even within a single site such as Raqefet Cave supports the hypothesis that these bedrock features fulfilled a variety of functions, some in the social or even symbolic realms, without actually being used to process anything (Nadel and Lengyel 2009). One should also take into account the probability of complex behaviour regarding the bedrock mortars. Even if some did function as food preparation devices, they may have been occasionally or repeatedly cleaned, or left uncovered and exposed to the elements, and thus the indicative evidence (grains and husks) would scarcely remain *in situ*. Furthermore, we found Natufian sediments in only a handful of bedrock mortars and cupmarks. Thus, all interpretations and reconstructions are limited due to a non-representative sample of these features. Our results also demonstrate the need to highlight the increasingly troublesome gaps in our knowledge about how Natufians organised plant processing (Wright 1994; Dubreuil 2004; Dubreuil and Rosen 2010).

Furthermore, the patterning in the phytoliths may reflect non-routine treatments of the bedrock features at the end of their use, rather than their original purpose. Many microfossil studies of portable grindstones for plant food processing fail to detect plant residues due to post-depositional factors (Field *et al.* 2009). Panicoid grasses (C_4 warm-moist loving genera), as opposed to pooid grasses (C_3 temperate genera), are abundantly represented in several of the mortars C-XVI, C-XLIV and C-XXIII (Fig. 10.8). It is plausible that the Panicoid rich assemblages revealed by the 'bilobate' phytoliths represent residues of basketry such as use of a coiled vessel for the storage of grain particularly in C-XXIII as reflected in its large size (80 cm at the mouth) (Rosen 2005; Ryan 2011). However, the phytoliths from deep narrow features (C-XVI) (Fig. 10.9) may suggest processing of sedge tubers as well as sedge matting or basketry. At Wadi Kubbaniya in Egypt, large amounts of tubers from wild nut-grass (*Cyperus rotundus*) and sea club rush (*Bolboschoenus maritimus*) were processed and consumed by late Pleistocene hunter-gatherers (Hillman 1989; Wollstonecroft 2007).

Some component of phytoliths in the bedrock features may represent non-archaeological microfossils from the environment. However, there is evidence of tuber and grass-seed collection and processing close to the sampled contexts. This characteristic can be observed by comparing the samples to the cave floor matrix in cave breccia samples RA-07-4, RA-07-05 and RA-07-06. These three samples show far fewer phytoliths and these mainly comprised grass stems. Alternative explanations for the presence of the grass stems in the mortars include cave floor coverings, grave linings, bedding or matting from structural partitions.

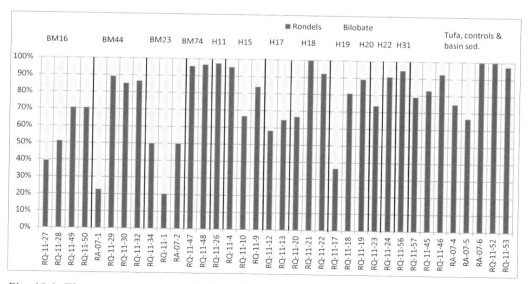

Fig. 10.8. *The per cent of short-cell phytoliths, an indicator of grass sub-family. Bilobates are indicative of C4 panicoid grasses while rondels indicate C3 pooid grasses (Power* et al. 2014)

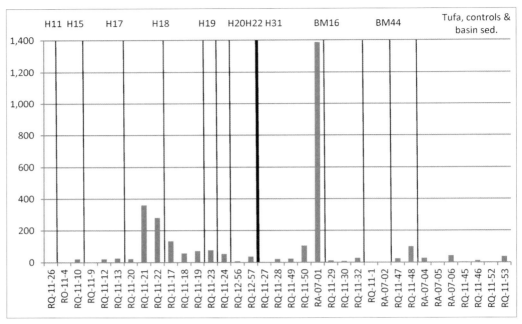

Fig. 10.9. Cyperaceae stem and husk multi-cell phytoliths per gramme of sediment

Natufian diet

We argue that the abundance of dendritic phytoliths adjacent to the abdomen sediment of Homo 19 and Homo 22 could possibly reflect residue from visceral contents (Fox *et al.* 1996). If this is the case, the multi-cell assemblage indicates that small-seeded grasses were Homo 19's final meal. Significantly, this may constitute direct evidence of the consumption of small-seeded grasses in Raqefet Cave.

We note the role of grasses in Natufian diet has frequently been overstated due to presumptions of their importance arising from their role in succeeding periods. Plant consumption in the Natufian Period should be reconstructed without pre-assuming the importance of grasses. Alternative proxies of diet may indicate relative importance of other staples. Some studies show that legumes may have been more important than grasses (Dubreuil 2004). This pattern of intensive utilization of legumes has been identified with macrobotanical remains outside the Levant at the late Epipaleolithic site of Hallan Çemi in eastern Anatolia (Savard *et al.* 2006).

Temporal and regional trends in plant use

Our research has shown that grass-grain became increasingly predominant in diet during the Late Natufian in the Mount Carmel region whilst dicotyledons became proportionately less important (Figs 10.7 and 10.10). This region bears out trends observed elsewhere in the southern Levant (Portillo *et al.* 2010; Rosen 2010). Yet Figure 10.10 shows Raqefet is 'intermediate' in this trend. Our findings confirm that, although wheat and barley were collected by Late Natufians in the southern Mount Carmel region, the small-seeded grasses were just as important (Table 10.3). In other words, Late Natufians at Raqefet practised generalised grass-seed foraging. We support the possibility of a 'lawn-mower' seed collection strategy (Hillman 1989). By gathering species that grew closer to camp than wheat or barley, the energy expense of travelling to patches could be negated (Charnov 1976). This can be compared to the widespread utilisation of small-seeded grasses at the Upper Palaeolithic site of Ohalo II (23,000 cal BP; Weiss *et al.* 2004). This is a strategy which is both widespread and long-lived in the Levant (Rosen 2010; Savard *et al.* 2006). These patterns bolster the concept of continuity through human adaptive cycles in the Levant (Rosen and Rivera-Collazo 2012). Since Raqefet Cave is situated in a habitat where both open valleys and the more forested hills would have been accessible in Natufian times, this locality would have allowed Natufians to exploit resources from a variety of vegetation communities. They would have been well-placed to continue to exploit some woodland resources even as the Younger Dryas was reshaping the woodland/ grassland boundaries in the Levant.

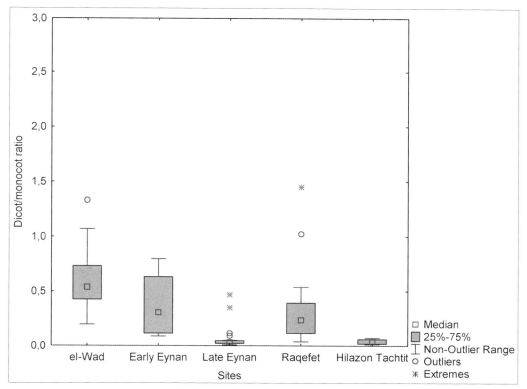

Fig. 10.10. The dicot to monocot ratio shows a trend towards monocots in the Late Natufian. Samples from the Early Natufian include

References

Bar-Matthews, M., Ayalon, A., Kaufman, A. and Wasserburg, G. J. 1999. The eastern Mediterranean paleoclimate as a reflection of regional events: Soreq cave, Israel. *Earth and Planetary Science Letters* 166, 85–95

Bar-Matthews, M., Ayalon, A., Gilmour, M., Matthews, N. and Hawkesworth, C. S. 2003. Sea-land oxygen isotopic relationships from planktonic foraminifera and speleothems in the Eastern Mediterranean region and their implication for paleo rainfall during interglacial intervals. *Geochimica et Cosmochimica Acta* 67(17), 3181–99

Bar-Oz, G., Dayan, T., Kaufman, D. and Weinstein-Evron, M. 2004. The Natufian economy at el-Wad Terrace with special reference to gazelle exploitation patterns. *Journal of Archaeological Science* 31, 217–31

Bar-Yosef, O. 1998. The Natufian culture in the Levant: threshold to the origins of agriculture. *Evolutionary Anthropology* 6(5), 159–77

Bar-Yosef, O. and Belfer-Cohen, A. 1989. The origins of sedentism and farming communities in the Levant. *Journal of World Prehistory* 3(4), 447–98

Bar-Yosef, O. and Belfer-Cohen, A. 1991. From sedentary hunter-gatherers to territorial farmers in the Levant. In Gregg, S. A. (ed.), *Between Bands and States*, 181–202. Carbondale IL: Center for Archaeological Investigations

Bar-Yosef, O. and Belfer-Cohen, A. 2002. Facing environmental crisis: social and cultural changes at the transition from the Younger Dryas to the Holocene in the Levant. In Cappers, R. T. J. and Bottema, S. (eds), *The Dawn of Farming in the Near East*, 55–66. Berlin: Studies in Early Near Eastern Production, Subsistence, and Environment 6

Bar-Yosef, O. and Meadow, R. H. 1995. The origins of agriculture in the Near East. In Price, T. N. and Bebauer, A. B. (eds), *Last Hunters, First Farmers: new perspective on the prehistoric transition to agriculture*, 39–94. Santa Fe NM: School of American Research Press

Belfer-Cohen, A. 1991. The Natufian in the Levant. *Annual Review of Anthropology* 20, 167–86

Belfer-Cohen, A. and Bar-Yosef, O. 2000. Early sedentism in the Near East: A bumpy ride to village life. In Kuijt, I. (ed.), *Life in Neolithic Farming Communities: social organization, identity, and differentiation*, 19–37. New York: Kluwer Academic/Plenum

Blockley, S. P. E. and Pinhasi, R. 2011. A revised chronology for the adoption of agriculture in the Southern Levant and the role of late glacial climatic change. *Quaternary Science Reviews* 30(1–2), 98–108

Boyd, B. 2002. Ways of eating/ways of being in the later Epipaleolithic (Natufian) Levant. In Mailakis, Y., Pluciennik, M. and Tarlow, S. (eds), *Thinking Through the Body: archaeologies of corporeality*, 137–52. New York: Kulwer Academic/Plenum

Boyd, B. 2006. On sedentism in the Later Epipalaeolithic (Natufian) Levant. *World Archaeology* 38(2), 164–78

Cabanes, D., Weiner, S. and Shahack-Gross, R. 2011. Stability of phytoliths in the archaeological record: a dissolution study of modern and fossil phytoliths. *Journal of Archaeological Science* 38(9), 2480–90

Charnov, E. L. 1976. Optimal foraging, the marginal value theorem. *Theoretical Population Biology* 9, 129–36

Dubreuil, L. 2004. Long-term trends in Natufian subsistence: a use-wear analysis of ground stone tools. *Journal of Archaeological Science* 31(11), 1613–1629.

Dubreuil, L., Rosen, A., 2010. Alternative methods for gathering: direct and indirect evidence of plant exploitation during the Natufian. *Eurasian Prehistory* 7(1), 5–7

Eitam, D. 2008. Plant food in the Late Natufian: the oblong conic mortar as a case study, Mitekufat Haeven. *Journal of the Israel Prehistoric Society* 38, 133–51

Eitam, D. 2009. Late Epipalaeolithic rock-cut installations and groundstone tools in the Southern Levant: methodology and classification system. *Paléorient* 35, 77–104

Field, J., Cosgrove, R., Fullagar, R. and Lance, B. 2009. Starch residues on grinding stones in private collections: a study of morahs from the tropical rainforests of NE Queensland. In Haslam, M., Robertson, G., Crowther, A., Nugent, S. and Kirkwood, L. (eds), *Archaeological Science under a Microscope: studies in residue and ancient DNA analysis in honour of Thomas H. Loy*, 228–38. Canberra: Terra Australis Monograph 30

Fox, C. L., Juan, J. and Albert, R. M. 1996. Phytolith analysis on dental calculus, enamel surface, and burial soil: information about diet and paleoenvironment. *American Journal of Physical Anthropology* 101, 1–13

Garrard, A. N. 1980. *Man-Animal-Plant Relationships during the Upper Pleistocene and Early Holocene.* Unpublished PhD thesis, University of Cambridge, Cambridge.

Grosman, L. 2003. Preserving cultural traditions in a period of instability: the Late Natufian of the hilly Mediterranean zone. *Current Anthropology* 44(4), 571–80

Henry, A. G., Brooks, A. S. and Piperno, D. 2010. Microfossils in calculus demonstrate consumption of plants and cooked foods in Neanderthal diets (Shanidar III, Iraq; Spy I and II, Belgium). *Proceedings of the National Academy of the Sciences (USA)* 108(2), 486–91

Hillman, G. C. 1984. Traditional husbandry and processing of archaic cereals in modern times, Part I, the glume-wheats. *Bulletin on Sumerian Agriculture* I, 114–52

Hillman, G. C. 1989. Late Palaeolithic plant foods from Wadi Kubbaniya in Upper Egypt: dietary diversity, infant weaning, and seasonality in a riverine environment. In Harris, D. R. and Hillman, G. C. (eds), *Foraging and Farming: the evolution of plant exploitation*, 207–239. London: One World Archaeology 13

Hillman, G. C., Madeyska, E. and Hather, J. G. 1989. Wild plant foods and diet of Late Palaeolithic Wadi Kubbaniya: the evidence from charred remains. In Wendorf, F., Schild, R. and Close, A. (eds), *The Prehistory of Wadi Kubbaniya 2: stratigraphy, palaeoeconomy and environment*, 162–242. Dallas TX: Southern Methodist University Press

Hopf, M. and Bar-Yosef, O. 1987. Plant remains from Hayonim Cave, Western Galilee. *Paléorient* 13, 117–2

Keeley, L. R. 1992. The use of plant foods among hunter-gatherers: A cross-cultural survey. In Anderson-Gerfaud, P. (ed.), *Préhistoire de l'Agriculture: Nouvelles Approches Expérimentales et Ethnographiques*, 29–38. Paris: CNRS

Lengyel, G., Nadel, D., Tsatskin, A., Bar-Oz, G., Bar-Yosef Mayer, D., Be'eri, R. and Hershkovitz, I. 2005. Back to Raqefet Cave, Mount Carmel, Israel. *Journal of the Israel Prehistoric Society* 35, 245–70

Mason, S. L. R. 1995. Acornutopia? Determining the role of acorns in past human subsistence. In Wilkins, J., Harvey, D. and Dobson, M. (eds), *Food in Antiquity*, 12–24. Exeter: Exeter University Press

Mears, R. and Hillman, G. C. 2007. *Wild Food.* London: Hodder & Stoughton

Moore, A. M. T., Hillman, G. C. and Legge, A. J. 2000. *Village on the Euphrates: from foraging to farming at Abu Hureyra.* Oxford: Oxford University Press

Munro, N. 2004. Zooarchaeological measures of hunting pressure and occupation intensity in the Natufian. *Current Anthropology* 45(S4), S5–34

Nadel, D. and Lengyel, Y. 2009. Human-made bedrock holes (mortars and cupmarks) as a Late Natufian social phenomenon. Archaeology, *Ethnology & Anthropology of Eurasia* 37, 37–48

Nadel, D. and Rosenberg, D. 2010. New insights into late Natufian bedrock features (mortars and cupmarks). *Eurasian Prehistory* 7, 66–87

Nadel, D., Danin, A., Power, R. C., Rosen, A. M., Bocquentin, F., Tsatskin, A., Rosenberg, D., Yeshurun, R., Weissbrod, L., Rebollo, N., Barzilai, O. and Boaretto, E. 2013. Earliest floral grave lining from 13,700–11,700 year old Natufian burials at Raqefet Cave, Mt. Carmel, Israel. *Proceedings of the National Academy of Sciences (USA)* 110(29), 11774–8

Nadel, D., Lambert, A., Bosset, G., Bocquentin, F., Rosenberg, D., Yeshurun, R., Weissbrod, L., Tsatskin, A., Bachrach, N., Bar-Matthews, M., Ayalon, A., Zaidner, Y. Beeri, R. and Greenberg, H. 2012. The 2010 and 2011 seasons of excavation at Raqefet Cave. *Journal of the Israel Prehistoric Society* 42, 35–73

Nadel, D., Lengyel, G., Bocquentin, F., Tsatskin, A., Rosenberg, D., R., Yeshurun, Bar-Oz, G., Bar-Yosef Mayer, D. E., Beeri, R., Conyers, L., Filin, S., Hershkovitz, I., Kurzawska, A. and Weissbrod, L. 2008. The Late Natufian at Raqefet Cave: the 2006 excavation season. *Journal of the Israel Prehistoric Society* 38, 59–131

Nadel, D., Lengyel, G., Panades, T. C., Bocquentin, F., Rosenberg, D., Yeshurun, R., Brown-Goodman, R., Tsatskin, A., Bar-Oz, G. and Filin, S. 2009a. The Raqefet Cave 2008 excavation season. *Journal of the Israel Prehistoric Society* 39, 21–61

Nadel, D., Rosenberg, D. ad Yeshurun, R. 2009b. The deep and the shallow: the role of Natufian bedrock features at Rosh Zin, Central Negev, Israel. *Bulletin of the American Schools of Oriental Research* 355, 1–29

Noy, T., Legge, A. J. and Higgs, E. S. 1973. Recent excavations at Nahal Oren, Israel. *Proceedings of the Prehistory Society* 39, 75–99

Perrot, J. 1966. Le gisement Natoufien de Mallaha (Eynan), Israel. *L'Anthropologie* 70(5–6), 437–84

Piperno, D. R., Weiss, E., Holst, I. and Nadel, D. 2004. Processing of wild cereal grains in the Upper Palaeolithic revealed by starch grain analysis. *Nature* 430(7000), 670–3

Portillo, M., Rosen, A. M. and Weinstein-Evron, M. 2010. Natufian plant uses at el-Wad terrace (Mount Carmel, Israel): the phytolith evidence. *Eurasian Prehistory* 7(1), 99–112

Power, R. C., Salazar-García, D. C., Wittig, R. M. and Henry, A. G. 2014a. Assessing use and suitability of scanning electron microscopy in the analysis of microremains in dental calculus. *Journal of Archaeological Science* 49, 160–9

Power, R. C., Rosen, A. M. and Nadel, D. 2014b. The economic and ritual utilization of plants at the Raqefet Cave Natufian site: the evidence from phytoliths. *Journal of Anthropological Archaeology* 33, 49–65

Rosen, A. M. 1992. Preliminary identification of silica skeletons from Near Eastern archaeological sites: an anatomical approach. In Rapp, G. Jr and Mulholland, S. (eds), *Phytolith Systematics*, 129–47. New York: Plenum

Rosen, A. M. 2004. Phytolith evidence for plant use at Mallaha/Eynan. In Valla, F. *et al.* 2004, 189–201

Rosen, A. M. 2005. Phytolith indicators of plant and land use at Çatalhöyük. In Hodder, I. (ed.), *Inhabiting Çatalhöyük: reports from the 1995–1999 seasons*, 203–12. Cambridge: McDonald Institute Monographs, Çatalhöyük Project 4 (A)

Rosen, A. M. 2007a. Phytolith remains from Final Natufian contexts at Mallaha/Eynan. *Journal of the Israel Prehistoric Society* 37, 340–55

Rosen, A. M. 2007b. *Civilizing Climate. Social Responses to Climate Change in the Ancient Near East.* Lanham MD: Altamira Press

Rosen, A. M. 2010. Natufian plant exploitation: managing risk and stability in an environment of change. *Eurasian Prehistory* 7, 117–31

Rosen, A. M. and Rivera-Collazo, I. 2012. Climate change, adaptive cycles and the persistence of foraging economies during the Late Pleistocene/Holocene transition in the Levant. *Proceedings of the National Academy of Sciences (USA)* 109(10), 3640–5

Ryan, P. 2011. Plants as material culture in the Near Eastern Neolithic: Perspectives from the silica skeleton artifactual remains at Çatalhöyük. *Journal of Anthropological Archaeology* 30(3), 292–305

Savard, M., Nesbitt, M. and Jones, M. K. 2006. The role of wild grasses in subsistence and sedentism: new evidence from the northern fertile crescent. *World Archaeology* 38(2), 179–96

Stutz, A. J., Munro, D. N. and Bar-Oz, G. 2009. Increasing the resolution of the Broad Spectrum Revolution in the Southern Levantine Epipaleolithic (19–12 ka). *Journal of Human Evolution* 56(3), 294–306

Tanno, K., Willcox, G., Muhesen, S. and Nishiaki, Y. 2013. "Preliminary Results from Analyses of Charred Plant Remains from a Burnt Natufian Building at Dederiyeh Cave in Northwest Syria." In *Natufian Foragers in the Levant*, eds Ofer Bar-Yosef and François R. Valla. Ann Arbor: International Monographs in Prehistory, 83–87

Terradas, X., Ibañez, J. J, Braemer, F., Hardy, K., Iriarte, E., Madella, M., Ortega, D., Radini, A. and Teira, L. C., 2013. Natufian bedrock mortars at Qarassa 3: preliminary results from an interdisciplinary methodology. In Borrell, F., Ibáñez, J. J. and Molist, M. (eds), *Stone Tools in Transition: From Hunter-Gatherers to Farming Societies in the Near East*, 441–56. Barcelona: Universitat Autònoma de Barcelona

Tromp, M. 2012. *Large-scale Analysis of Microfossils Extracted from Human Rapanui Dental Calculus: a Dual-Method Approach Using SEM-EDS and Light Microscopy to Address Ancient Dietary Hypotheses*. Unpublished MSc thesis, Idaho State University, Pocatello, Retrieved 2013-9-10 at http://www.isu.edu/anthro/dudgeon/pubs/Tromp_Thesis_2012.pdf

Unger-Hamilton, R. 1989. The Epipalaeolithic southern Levant and the origins of cultivation. *Current Anthropology* 30(1), 88–103

Valla, F. R. 1995. The first settled societies – Natufian (12,500–10,200 BP). In Levi, T. E. (ed.), *The Archaeology of Society in the Holy Land*, 170–87. Leicester: Leicester University Press

Valla, F. R., Khalaily, H., Valladas, H., Tisnérat-LaBorde, N., Samuelian, N., Bocquentin, F., Rabinovich, R., A. Bridault, T. Simmons, G. Le Dosseur, Rosen, A. M., Dubreuil, L., Bar-Yosef Mayer, D. E. 2004. Les Fouilles de Mallaha en 2000 et 2001: 3ème rapport préliminaire. *Journal of the Israel Prehistoric Society* 34, 49–244

Valla, R. F., Khalaily, H., Valladas, H., Kaltnecker, E., Bocquentin, F., Cabellos, T., Bar-Yosef-Mayer, D. E., Le Dosseur, G., Regev, L., Chu, V., Weiner, S., Boaretto, E., Samuelian, N., Valentin, B., Delerue, S., Poupeau, G., Bridault, A., Rabinovich, R., Simmons, T., Zohar, I., Ashkenazi, S., Delgado Huertas, A., Spiro, B., Mienis, H. K., Rosen, A. M., Porat, N. and Belfer-Cohen, A. 2007. Les fouilles de Ain Mallaha (Eynan) de 2003 à 2005: quatrième rapport préliminaire. *Journal of the Israel Prehistoric Society* 37, 135–383

Weinstein-Evron, M., Yeshurun, R., Kaufman, D., Eckmeier, E. and Boaretto, E. 2012. New 14C dates for the Early Natufian of el-Wad Terrace, Mount Carmel, Israel. *Radiocarbon* 54(3–4), 813–22

Weiss, E., Wetterstrom, W., Nadel, D. and Bar-Yosef, O. 2004. The broad spectrum revisited: evidence from plant remains. *Proceedings of the National Academy of Sciences (USA)* 101(26), 9551–5

Wollstonecroft, M. 2007. *Post-harvest Intensification in Late Pleistocene Southwest Asia: plant food processing as a critical variable in Epipalaeolithic subsistence and subsistence change*. Unpublished PhD thesis. Institute of Archaeology, University College London

Wright, K. I. 1994. Ground-stone tools and hunter-gather subsistence in southwest Asia: implications for the transition to farming. *American Antiquity* 59, 238–63

Zohary, D., Hopf, M. and Weiss, E. 2012. *Domestication of Plants in the Old World: the origin and spread of domesticated plants in Southwest Asia, Europe, and the Mediterranean Basin*. Oxford: Oxford University Press

11. Evidence of plant foods obtained from the dental calculus of individuals from a Brazilian shell mound

Célia Helena C. Boyadjian, Sabine Eggers and Rita Scheel-Ybert

The human groups associated with sambaquis (Brazilian shell mounds) lived for thousands of years along most of the Brazilian coastline. They have always been thought of as fishermen and mollusc gatherers whereas the role of plants in their daily activities and subsistence remained underestimated for decades. This is due both to the low preservation of non-carbonised macrobotanical remains and the fact that systematic archaeobotanical studies have only recently begun at these sites. Recent analysis of carbonised plant remains show that, in fact, a range of different plants was used as fuel and possibly also as construction material, in rituals, and as food. However, direct evidence of the plants used as food has only recently become possible through the analysis of micro-remains trapped in the dental calculus from individuals buried at these sites. In this case study, we present microfossil data from samples of dental calculus of 19 individuals from Jabuticabeira II (1214–830 cal BC to 118–413 cal AD), a large sambaqui in southern Brazil. Starch granules, phytoliths, and diatoms were the most abundant micro-particles recovered. The high number and variety of starch granules point to a significant consumption of starchy plants. A range of phytolith types also indicate use of a variety of plants. Based upon analysis of these micro-remains, we suggest that plants from the Arum family, palms, sweet potato, yams, ariá and Poaceae (including maize) were among the plants eaten in Jabuticabeira II. The identification of diatoms indicates that some resources were obtained from the lagoon nearby. The number and distribution of micro-remains varied considerably between individuals, suggesting that some of them had a more diversified plant diet. Nevertheless, the microfossil assemblages from different groups according to sex, age at death, caries frequency, paleopathologies, and burial features were compared, but no difference was found between them. Damaged starches were also found, possibly indicating that plant food could have been processed prior to consumption. Though we cannot confirm the hypothesis that plants were cultivated at Jabuticabeira II, we cannot exclude the possibility that some botanical taxa may have been managed. Despite the preliminary nature of this study, the results presented here provide new insights into the diet and lifestyle of this sambaqui builders group.

Introduction

Use of plants, especially in diet, and food processing in prehistoric Brazil is still little known. Investigation into plant use by early human groups has been hampered until recently by the low macrobotanical preservation in most archaeological contexts and the generalised lack of systematic studies and analyses that could maximise recovered remains (Bianchini *et al.* 2007; Scheel-Ybert *et al.* 2009; Wesolowski *et al.* 2010).

Fortunately, in the last two decades archaeobotanical remains have received more attention and palaeoethnobotanical studies are gradually developing in Brazilian archaeology. Research based on the analyses of charcoal and, whenever possible, non-carbonised wood, fibres, fruits and seeds recovered from different sites, have detected the exploitation of many different *taxa*, including edible species (Scheel-Ybert 2000; 2001; Ceccantini and Gussella 2001; Ceccantini 2003; 2005; Freitas *et al.* 2003; Blanchot and Amenomori 2005; Kamase 2005; Scheel-Ybert and Solari 2005; Bianchini *et al.* 2007; Peixe *et al.* 2007; Bianchini 2008; Beauclair *et al.* 2009). Even so, it does not necessarily mean that those plants were used as food.

There are many different ways to reconstruct diet, nutrition and subsistence patterns from past human groups. However the analysis of coprolites, gut or dental calculus contents are the only ones that provide direct evidence of what was effectively ingested (Ciochon *et al.* 1990; Reinhard and Bryant 1992; 2008; Holt, 1993; Reinhard 1993; Middleton and Rovner 1994; Pearsall 2000; Reinhard *et al.* 2001; Hardy *et al.* 2009; Vinton *et al.* 2009). While the preservation of coprolites is rare in most contexts, dental calculus can be found on individuals from most archaeological sites, regardless of their location and chronology.

Despite being relatively recent, the analysis of dental calculus contents is already recognised as a powerful tool, not only for the reconstruction of past diets (Fox *et al.* 1996; Juan-Tresseras *et al.* 1997; Scott-Cummings and Magennis 1997; Lieverse 1999; Gobetz and Bozarth 2001; Henry and Piperno 2008; Piperno and Dillehay 2008; Hardy *et al.* 2009; Henry *et al.* 2010; Li *et al.* 2010; Wesolowski *et al.* 2010; Asevedo *et al.* 2012; Dudgeon and Tromp 2012; Hardy *et al.* 2012; Henry *et al.* 2012; Mickleburgh and Pagán-Jiménez 2012; Scott and Poulson 2012; Power *et al.* 2014), but also for other features of past life including identification of medicinal plants (Hardy *et al.* 2012) and investigation of infectious diseases (De La Fuente *et al.* 2012; Warinner *et al.* 2014). This kind of analysis is particularly important in archaeological sites where information about plant use would otherwise be absent.

As our aim is to investigate diet and plant use by human groups associated with *sambaquis*, for this study we analysed micro-remains retrieved from samples of dental calculus from Jabuticabeira II, in a case study.

Sambaquis

Sambaquis (Fig. 11.1) are 2–30 m high shell mounds distributed along almost the entire Brazilian coast that date to around 8000–1000 years ago (Prous 1992; Gaspar 1998; Lima 2000; Lima *et al.* 2002). Usually, they are composed of mollusc shells and sediments, but may contain other faunal remains, especially fish bones and charcoal (Prous and

Piazza 1977; Prous 1992; DeBlasis *et al.* 2007). Most *sambaquis* contain hearths, post-holes and shell, bone, and lithic artefacts (DeBlasis *et al.* 1998; Gaspar 1998; Tenório 2004). However, human burials, which are extremely numerous in some sites (Fish *et al.* 2000), are not always present (Gaspar 1998).

There is regional variation in internal structure and material culture of these sites (Lima 2000; DeBlasis *et al.* 2007; Wagner *et al.* 2011). For example, zooliths, which are highly elaborate zoomorphic sculptures considered as having a ritual purpose (Prous 1976; Schmitz 1987; Lima 2000), have only been found in the *sambaquis* in the south-south-eastern regions (Prous 1976; 1992). Another example is the monumentality of the mounds found in southern Brazil but not in other regions. Most specifically, in Santa Catarina State, there are sites that reach as much as 70 m high and 500 m long (DeBlasis *et al.* 2007). This regional variation has fuelled a long-standing debate to whether it represents culturally distinct groups or not (Gaspar 1998; Lima 2000).

Diet, subsistence and use of plants in *Sambaquis*

The enormous amounts of shells that comprise most of the *sambaquis* were traditionally considered to be a result of food leftover accumulation and the humans associated with those sites were seen as small nomadic, shellfish-gathering bands (Kneip 1980; Heredia *et al.* 1989; Lima 2000). However, zooarchaeological and stable isotope studies have shown that those groups subsisted mainly on fish while shellfish played a secondary role in diet and shells were probably used primarily as construction material (Bandeira 1992; Figuti 1992; Afonso and DeBlasis 1994; De Masi 1999; Fish *et al.* 2000; De Masi 2001; Klökler 2001; Klökler 2008). Nowadays, archaeological reinterpretations picture those individuals as sedentary groups of fisher-gatherers living in higher demographic densities and more complex social structures than imagined before (DeBlasis *et al.* 1998; 2007; Gaspar 1998; Lima 2000).

The *sambaquis* were strategically located in very resource-rich areas, associated with a variety of ecosystems (sea, lagoons, or rivers, mangroves, tropical rain forest), where both faunal and botanical resources were abundant and available throughout the year, supporting sedentary settlements and population growth (Lima 2000; Scheel-Ybert 2001). Lima (2000) proposed that the increasing complexity of *sambaqui* groups was favoured by the abundance of resources from this kind of environment, in the absence of food production.

However, some southern *sambaquis* contained pottery remains that were associated with the later phases of occupation; this was traditionally interpreted as evidence of the introduction of horticulture (Beck 1972); however these remains were not linked to any other observable technological changes (Gaspar 1998; Lima 2000). It was later proposed therefore, that they might represent either the use of abandoned *sambaquis* by ceramic groups or the use of pottery by *sambaqui* people without any substantial changes in their subsistence (Lima 2000).

In fact, evidence of the adoption of domesticated plants and food production by pre-ceramic groups has already been found in other contexts (Piperno *et al.* 2000; Iriarte *et al.* 2004; Perry *et al.* 2006; Dickau *et al.* 2007). Nevertheless, we still know very little about

Fig. 11.1. Features of Brazilian shell mounds (sambaquis): a) Garopaba do Sul, a sambaqui from Santa Catarina State; b) detail of an excavated sambaqui showing the post-holes and the layers of shells intercalated with sediments; c) a zoolith and d) a lithic artefact recovered from different sambaquis (image b) by José Filippini)

the consumption, cultivation, and domestication of plants by the *sambaqui* groups. The few reports that have discussed plant use in *sambaquis* were based on indirect evidence such as lithic artefacts (interpreted as plant processing tools), and on the presence of a few seeds and charred palm nuts (Kneip 1977; 1994; Oliveira 1991; Bryan 1993; Gaspar 1998). Though the remains of wooden or plant-fibre artefacts were also found (Afonso and DeBlasis 1994; Peixe *et al.* 2007), these remains are very rare.

More recently, charcoal studies have shed some light on this issue. They have demonstrated that a wide range of plants was exploited as fuel and may also have been used in construction and in ritual contexts (Scheel-Ybert 2000; 2001; Scheel-Ybert *et al.* 2009). Scheel-Ybert *et al.* (2003) also suggest that the management or even the incipient cultivation of some species might have been practiced at sites from south-south-east Brazil. Many of the charred seeds and wood fragments identified through charcoal analysis correspond to taxonomic groups that produce edible fruits (Scheel-Ybert 2001; Bianchini 2008), while charred tubers of some monocotyledons, including yams (*Dioscorea* sp.) were also found (Scheel-Ybert 2001). It is therefore important to investigate whether these plants may have been used as food sources.

Recovering plant micro-remains from dental calculus samples is currently the most appropriate way to reconstruct the use of plants as food for most *sambaquis* excavated in the past, particularly when the osteological collections are already located in museums as the plant remains were not recovered and new excavations are no longer possible.

This technique has only been conducted on material from a few sites in Brazil (Reinhard *et al.* 2001; Boyadjian *et al.* 2007; Wesolowski *et al.* 2010; Kucera *et al.* 2011; Boyadjian and Eggers 2014). One of these is Jabuticabeira II, a *sambaqui* located on the southern coast.

The site

Jabuticabeira II (2880±75–1805±65 BP; Lab ID: Az9880 and Az9884; DeBlasis *et al.* 2007; cal BP: 3163–2779 cal BP to 1832–1537 cal BP; cal BC: 1214–830 cal BC to 118–413 cal AD) is a *sambaqui* that has been interpreted as the result of incremental funerary rituals (DeBlasis *et al.* 1998; Fish *et al.* 2000). It is one of the more than 60 shell mounds surrounding the Camacho lagoon, in Santa Catarina State (Fig. 11.2), and currently lies between the lagoon and the shoreline (3 km and 6 km away, respectively).

Indirect evidence suggests that plant resources were very important for the Jabuticabeira II group and artefacts interpreted as plant processing tools (mortars and pestles) were found (Scheel-Ybert *et al.* 2009). Additionally, charcoal analysis revealed that the site was located in a typical Brazilian coastal ecosystem rich in fruits, seeds, and starchy storage organs (Scheel-Ybert 2001; Scheel-Ybert *et al.* 2009): the *Restinga*.[1] The people who lived in that region could have exploited tubers from the *Poaceae* family, as well as yams (*Dioscorea* spp.), both of which are commonly found in the *Restinga*. Data from charcoal analyses demonstrated that the Jabuticabeira II group also explored the

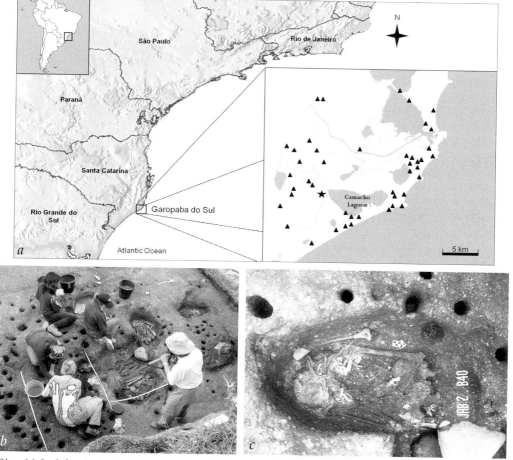

Fig. 11.2. Jabuticabeira II Shell Mound (Santa Catarina State, Brazil): a) map showing the many sambaquis from Camacho region, with Jabuticabeira II highlighted (star); b) excavation of the studied site; and c) detail of a burial (image c) by Maria Dulce Gaspa)

tropical rainforest from the nearby hills to obtain wood (Scheel-Ybert 2000; 2001; Scheel-Ybert *et al.* 2009; Bianchini 2008), and it has been suggested that they actively managed some species (Scheel-Ybert *et al.* 2009).

Seeds from *Annonaceae, Arecaceae, Cucurbitaceae,* and *Myrtaceae,* all plant families that produce edible fruits, were identified in the funerary layers of the site and their use seems to be related to funerary rituals, as food offerings or for feasting (Bianchini and Scheel-Ybert 2012). However, until recently, there was no direct evidence that these plants were consumed by the individuals buried at Jabuticabeira II.

Aims

The main objectives of this research were:

a) To analyse and identify micro-remains retrieved from the dental calculus of individuals buried in Jabuticabeira II in order to conduct preliminary investigations into certain aspects of plant consumption and increase the knowledge of plant use among the *sambaqui* builders. We focused particularly on starch granules, as they are intimately linked to human nutrition. However, in order not to miss any possible additional information on diet, we also analysed phytoliths and diatoms.

b) To compare starch granule assemblages to verify intragroup variation in diet.

Material and methods

We obtained dental calculus samples from 19 adult individuals (seven females, nine males, and three unidentified). Before sampling, the chosen teeth were brushed with a soft sterilised toothbrush (a different one for each individual) and distilled water. This procedure was commonly applied by different researchers in order to wash soil residues off the surface of the calculus deposits, thus avoiding contamination (Fox *et al.* 1996; Piperno and Dillehay 2008; Wesolowski *et al.* 2010; Mickleburgh and Pagán-Jiménez 2012). A new protocol (Warinner *et al.* 2014), more appropriate for decontamination, was proposed after the current work was undertaken. However, in a previous study (Boyadjian *et al.* 2007), control samples were prepared with material from Jabuticabeira II in order to test whether the micro-remains found in the calculus samples could have come from the *sambaqui* sediments. No micro-remains were found in the control samples, indicating that the remains found in the experimental samples were not the result of contamination (Boyadjian *et al.* 2007).

After the cleaning, calculus fragments were mechanically detached from one tooth per individual (Table 11.1) with a sterilised dental curette (Fox *et al.* 1996; Reinhard *et al.* 2001; Henry and Piperno 2008; Hardy *et al.* 2009; Henry *et al.* 2010; Wesolowski *et al.* 2010; Mickleburgh and Pagán-Jiménez 2012). This procedure does not harm the teeth, as demonstrated by Henry and Piperno (2008).

Immediately after sampling, the fragments were transferred to sterilised 2.0 ml microtubes. They were then treated with 10% hydrochloric acid and centrifuged to dissolve the calculus matrix and release the micro-remains. The samples were centrifuged and

Table 11.1. Number and distribution of the three main types of micro-remains found in the dental calculus samples from Jabuticabeira II

Sample number	Burial	Tooth*	Calculus location**	Starch granules	Phytoliths	Diatoms
105	12A L1.25	M	Lab. sub.	5	2	–
106	12C L1.25	UL PM²	Lab. supra	1	1	–
107	17A L1.05	LR M¹	Lin. supra	2	8	1
108	34 L2.05	LR M¹	Lab. supra	3	–	–
109	36A L2.05	LL M²	Lin. supra	1	–	–
110	40 L2.05	UL M¹	Lin. sub	3	–	–
111	41A L2.05	LL M¹	Lin. supra	5	–	–
112	43 L1.77	UL M²	Lab. supra	31	1	1
113	3B L6B3(E3)	M³	Lab./Lin. (mixed)supra	–	–	–
114	17A L2.05	LL M¹	Lab. sub	1	–	–
115	25A L2.65	UL M¹	Lab. supra	1	–	–
116	27A L2T15	LR I²	Lin. sub	4	1	2
118	28A L2T15	I	Lab. sub	2	5	4
119	42A L1.76	M²	Lab. supra	6	–	1
120	102 L1.75	LR M³	Lab. sub	3	1	–
121	108 L2.05	UR M²	Lab. supra	4	1	2
122	107 T18	UR M²	Lab. supra	2	3	2
123	110 L2	UL M²	Lab. supra	13	1	1
125	115B L6	LR M¹	Lin. supra	3	5	–
Total				90	29	14

* UR= upper right, UL= upper left, LR= lower right, LL= Lower left; PM= pre-molar, M= molar, I=incisive; ** Lab.= labial, Lin.= lingual; supra= supra gingival, sub= sub gingival

the supernatant removed. Distilled water was then added to the micro-tubes, which were agitated and centrifuged again. The supernatant was again decanted and the rinsing process repeated once more. Finally, the water was substituted with a solution of 95% ethanol and the samples were transferred to labelled vials (for more details on this protocol see Wesolowski *et al.* 2010 and Boyadjian 2012). The final solutions containing the micro-remains were used to prepare microscope slides that were scanned under an optical compound microscope with polarised light at 400× magnification. For each individual, three slides were examined, each one mounted with 10 μL of solution and a drop of glycerol.

The micro-remains were counted, measured, described, and recorded as digital images and drawings. At first, the micro-remains were classified according to morphotype by size, shape, and surface characteristics. For the starch granules, additional aspects related to the hilum, extinction cross, lamellae, fissures, and pressure facets were also taken into account for the classification (Lentfer *et al.* 2002; Torrence 2006; Crowther 2009). In order to identify the taxa, where possible, all micro-remains were then compared to our developing reference collection, and to images and descriptions available in the literature (Reichert 1913; Ugent *et al.* 1986; Piperno and Holst 1998; Pearsall 2000; Perry 2002; 2004; Piperno *et al.* 2000; 2004; Iriarte 2003; Pearsall *et al.* 2004; Horrocks *et al.* 2004; Chandler-Ezzel *et al.* 2006; Fullagar *et al.* 2006; Piperno 2006; Torrence and Barton 2006; Piperno and Dillehay 2008; Zarrillo *et al.* 2008; Henry *et al.* 2009; Revedin *et al.* 2010; Messner 2011). Additionally, modified starches were compared to reported patterns of damage (Babot 2003; 2006; Piperno *et al.* 2004; Samuel 2006; Henry *et al.* 2009) in order to investigate food preparation processes, as has been carried out in other dental calculus studies (Henry and Piperno 2008; Piperno and Dillehay 2008; Hardy *et al.* 2009; Henry *et al.* 2010; Mickleburgh and Pagán-Jiménez 2012).

Finally, to verify intragroup dietary variation, the starch granule assemblages of the individuals were compared according to sex, age at death, caries frequency, paleopathologies, and burial characteristics (such as burial offerings suggestive of status differences).

Results and discussion

At the present time there is no standardised method for the processing of dental calculus and the analysis and quantification of its contents. Different research groups use different methods and each one is adapted to the nature of the calculus deposits used. For example, some calculus deposits are so faint that it is only possible to obtain a fine powder that permits the mounting of the sample directly on only one slide. In other collections, the deposits are very abundant, thick and dense, which allows the elimination of the most superficial layers of the calculus fragments in order to promote decontamination, and also permits subsampling. Aside from this, despite the fast advances in the field in recent years, there are still many gaps in the knowledge about calculus formation processes and the preservation of micro-remains in its matrix.

In the current study, three slides were prepared from each sample, accounting for just a fraction of the micro-remains that could have been recovered. We adopted this strategy so we could save valuable material for future analysis in the case that more accurate methods are developed. Furthermore, since it is still not possible to estimate how much of the sample needs to be analysed in order to have a satisfactory representation of the dental record of an individual's diet, we decided that this conservative strategy would be more adequate for this case study. It should, therefore, be kept in mind that the results obtained in this project are still preliminary and the interpretation of the data in some of the topics discussed in this session was performed in a more exploratory way.

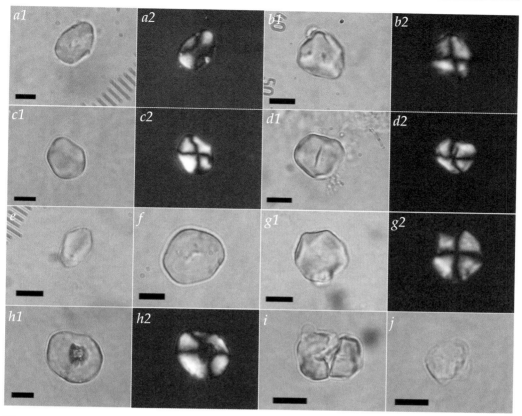

Fig. 11.3. Starch granules recovered from calculus samples from Jabuticabeira II: a) consistent with Eugenia sp. *(Myrtaceae); b)* Ipomoea batatas *(sweet potato); c) most likely from Araceae; d) possibly Poaceae (similar to Zea mays); e–j) unidentified starch granules; h) unidentified starch granule with a central cavity; i) very damaged starch granule; j) partially gelatinised starch granule. Scale bars are 10 μm. The dark pictures were taken under polarised light*

Variety of plants exploited

An abundance and variety of microfossils were obtained, as reported in similar studies (Henry and Piperno 2008; Piperno and Dillehay 2008; Henry *et al.* 2010; Wesolowski *et al.* 2010; Mickleburgh and Págan-Jiménez 2012; Dudgeon and Tromp 2012). Whereas phytoliths and diatoms were found only in some samples (Table 11.1), starch granules were ubiquitous and numerous, as reported also in other contexts (Scott-Cummings and Magennis 1997; Henry and Piperno 2008; Piperno and Dillehay 2008; Henry *et al.* 2010; Wesolowski *et al.* 2010; Hardy *et al.* 2012; Mickleburgh and Págan-Jiménez 2012).

Most of the starch granules found have not yet been identified. Therefore, in this paper we decided to focus on the variation and distribution of starch morphotypes (Lentfer *et al.* 2002; Barton 2006).

The high number and the high diversity of starch morphotypes found (Tables 11.1 and 11.2 and Fig. 11.3), indicate that a great variety of starchy plants may have been ingested by the people of Jabuticabeira II. Of course, care needs to be taken as the relationship between the diversity of starch types found in dental calculus, or any archaeological context, and the variety of plants exploited is anything but simple and straightforward (Mercader *et al.* 2008; Mickleburgh and Pagán-Jiménez 2012). Not every plant produces diagnostic starch granules and there is a 'redundancy' of starch morphologies in different taxa (Mercarder *et al.* 2008; Henry *et al.* 2010; Mickleburgh and Pagán-Jiménez 2012). Furthermore, some species are heteromorphic (Piperno and Holst 1998; ICSN 2011), producing different types of starch granules (size or shape), in some cases, even in the same organ, such as *Dioscorea bulbifera* (Fullagar *et al.* 2006, 600) and *Triticum aestivum* (Gott *et al.* 2006, 41 and pl. 14).

However, considering the original settlement area, people from Jabuticabeira II must have had access to an abundance and variety of resources, including many plant *taxa* that could be used as food. The high diversity of starch types found could therefore correspond to the exploitation of a wide variety of starchy plants as food staples.

On the other hand, even if a human group inhabits a region that is rich in resources, it does not necessarily mean that all the available reserves could be exploited by that group (Piperno and Pearsall 1998). Many criteria guide human food choices. Revedin and colleagues (2010), for example, affirm that 'the reasons for choosing a plant for food are dictated by its dominance in the local vegetation, as well as by its size and appearance' (Serban *et al.* 2008; Revedin *et al.* 2010). Accordingly, food choice is also based on energetic considerations (Piperno and Pearsall 1998). Additionally, there are also cultural criteria, such as food taboos, playing an important role in food choices. Thus, plants that offer potential as food sources are not necessarily consumed today, or may not have been consumed in the past (Piperno and Pearsall 1998).

Nevertheless, besides the starch granules, we also found a wide variety of phytoliths; this contributes to the hypothesis that a wide range of plants were indeed exploited for food.

Additionally, we cannot exclude the possibility that some of those remains arose from non-food use of plants. The environment would provide many botanical resources that could have been used as medicine or for artefact production. In the latter, the micro-remains would enter the dental calculus accidentally, following the processing of raw material, or directly by the use of the teeth as tools (Hardy 2008) while holding, chewing, scraping, and scratching plant fibres and other plant structures during the production of material items such as ropes, baskets and fishing nets (Boyadjian and Eggers 2014). However, even if the mechanism that causes the remains to become entrapped in the calculus matrix is still not completely understood, it seems more likely that the majority of micro-remains detected derive from the mastication of food.

Last, but not least, it is worth remembering that many plants could have been ingested without leaving any trace in the dental calculus. Not all plants consumed can be identified through the analysis of starch granules from dental calculus, simply because not all ingested plant parts contain starch. Some species may produce high concentrations of oils and fats, or may have other forms of carbohydrate reserves, such as fructans (Messner 2011), which are not visible in the dental calculus.

In order to tentatively reconstruct the interaction between plants and people, studies of macrobotanical remains, starch, pollen grains, phytoliths, and other microbotanical remains need to be integrated, whenever possible, as each one offers a different line of evidence (Messner 2011). Also, alternative techniques for investigating calculus contents, based on molecular data, complementary to the morphological analysis of micro-remains have emerged (Hardy *et al.* 2012; Warinner *et al.* 2014), contributing to the types of evidence that can be obtained.

Intragroup dietary variation

The number and distribution of starch granules varied considerably between individuals (Tables 11.1 and 11.2) suggesting that, even if the group as a whole exploited a wide

Table 11.2 Number and distribution of starch granule morphotypes found in the dental calculus from Jabuticabeira II

Sample number	Burial	Sex	Age class	No. of morphotypes	#morphotypes*
105	12A L1.25	♀	Old adult	3	1, 2, 3
106	12C L1.25	♂	Young adult	1	4
107	17A L1.05	♂	Middle adult	2	5, 6
108	34 L2.05	♀	Yong adult	2	1, 8c
109	36A L2.05	♂	Young adult	0	–
110	40 L2.05	♀	Adult	2	23, 30
111	41A L2.05	♂	Adult	4	10, 11, 27, 31
112	43 L1.77	♂	Middle adult	9	8b(18), 8c, 10, 11, 13, 15, 19, 25, 28
113	3B L6B3(E3)	♀	Adult	0	–
114	17A L2.05	♂	Adult	0	–
115	25A L2.65	♀	Middle adult	0	–
116	27A L2T15	Indet.	Adult	3	8d, 10, 11
118	28A L2T15	Indet.	Adult	2	10, 13
119	42A L1.76	Indet.	Middle adult	4	8b(2), 8d(2), 8e, 14
120	102 L1.75	♀	Adult	2	8a, 15
121	108 L2.05	♀	Middle adult	4	8a, 8f, 17, 18
122	107 T18	♂	Old adult	2	19, 20
123	110 L2	♂	Young adult	7	8a, 8b, 8c(2), 8d, 21, 23, 24
125	115B L6	♂	Old adult	3	4, 8g, 29

* The number in parenthesis next to some morphotypes indicates how many granules of that specific type were found in the sample. When there was only one granule found, there is no parenthesis

variety of starchy plant foods, the evidence from each individual comprises a small and variable portion of what was available.

However, similar studies also reported (sometimes considerable) differences in the amount and the types of starch granules among teeth from individuals from the same archaeological site, or even from the same individual (Boyadjian *et al.* 2007; Henry and Piperno 2008; Henry *et al.* 2010; Boyadjian and Eggers 2014). This may be connected to the formation rates of the dental calculus itself, which is influenced by many variables including diet, oral pH, salivary flux, genetic and hormonal factors, and hygiene practices, among others (Middleton and Rovner 1994; Hazen 1995; Lieverse 1999; Roberts-Harry *et al.* 2000; Henry and Piperno 2008; Hardy *et al.* 2009; Mickleburgh and Pagán-Jiménez 2012). Although it is known that the variations in the rates of calculus deposition exist, even between individuals of the same group (Roberts-Harry *et al.* 2000; Hardy *et al.* 2009), many questions regarding the mechanisms of calculus formation remain unanswered (Middleton and Rovner 1994; Hazen 1995; Lieverse 1999; Henry and Piperno 2008).

Besides that, Mickleburg and Pagán-Jiménez (2012) explain that 'starch grains trapped in dental calculus do not simply reflect the range of plant foods consumed by an individual, as consumption of a plant does not guarantee that its starches are preserved in the calculus' but 'presumably frequent consumption raises the chances of starches becoming trapped'. Also, 'there are differences in starch production in different plant *taxa* and different plant organs'. Thus it is still difficult (and even speculative) to estimate the frequency of consumption of a specific plant based on its record in the dental calculus of an individual (Mickleburgh and Pagán-Jiménez 2012).

Consequently, despite the difficulties in inferring the amount and variety of plant foods consumed per individual through the analysis of micro-botanical remains

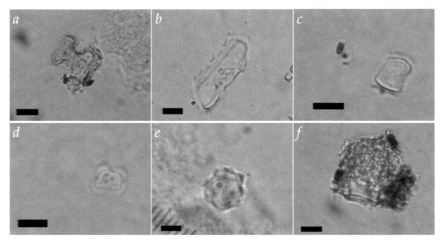

Fig. 11.4. Phytoliths recovered from calculus samples from Jabuticabeira II: a–d) phytoliths from the grass family (Poaceae), a) bilobate from Panicoideae subfamily; b) long wavy trapezoid from Pooideae subfamily; c) saddle-shaped from Chloridoideae subfamily; d) cross-shaped phytolith; e) spherical spinulose phytolith from Araceae (Palms); f) phytolith similar to the type found in the rhizome of Calathea *sp. (Marantaceae). Scale bars are 10 μm*

from dental calculus, by evaluating all the samples together it is possible to offer an approximation of the range of plants consumed for the group as a whole. In the same way, it is also possible to model paleodietary characteristics of population subgroups from the same community, as carried out herein.

The micro-remains assemblages from individuals were compared according to sex, age at death, oral pathologies, paleopathologies, and burial features; however, no differences were found, suggesting there was apparently no difference in the access of botanical resources among these subgroups from Jabuticabeira II. Notwithstanding, there is only one outranging individual among the 19 analysed. This middle adult from burial 43 showed the highest number and variety of starch morphotypes, while also being unique regarding his bilateral auditory exostosis, his great robusticity, and his beautiful mollusc shell necklace. Do those features suggest that this individual enjoyed a special status and a better access to plant food resources?

Initially, it has been suggested that the existence of a social system characterised by inequality and based on hierarchy, such as in a system of chiefdoms, could have triggered strategies mainly associated with the construction of funerary mounds (DeBlasis *et al.* 2007; DeBlasis and Gaspar 2009). Regarding this idea, differences in the mounds dimensions associated with the regional distribution of the sites could reflect not only demographic asymmetry and timespan of occupation, but also characteristics

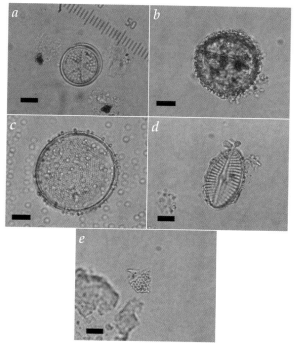

Fig. 11.5. Diatoms found in the calculus samples from Jabuticabeira II: a) Actinoptychus senarius; *b)* Paralia sulcata; *c)* Coscinodiscus *sp.; d)* Diploneis ovalis; *e) unidentified fragment. Scale bar is 10 μm. Identification based on Amaral (2008)*

related to different degrees of socio-political hierarchy (DeBlasis and Gaspar 2009, DeBlasis *et al.* 2007). Also, in spite of being rare, more elaborate and differentiated burials were found in some *sambaquis* (Hurt 1974; Prous 1992), indicating that some people could have been more important than others (DeBlasis *et al.* 2007).

However, since there is no clear evidence of social inequality, and also because no marked variations in terms of sex or age status were observed in the Jabuticabeira II burials, as described in previous studies, DeBlasis and colleagues (2007) propose a homogeneous social system in *sambaquis* from southern Brazil. Furthermore, these authors interpret the 60 *sambaquis* registered around the Camacho Lagoon (including Jabuticabeira II) as a result of non-hierarchical communal systems.

At the moment, the results obtained here do not point to dietary inequality of plant resources in individuals buried in Jabuticabeira II. We suggest therefore, that there were no subgroups regulated by social, political or economic issues; or, if these regulations existed, they were not identifiable by differences in the evidence for dietary starch. As our results are still preliminary, this hypothesis needs further testing.

Plants used

The evidence based on the tentative identification of starch granules and phytoliths from comparative reference collections (Table 11.3, Figs 11.3 and 11.4) indicates that underground storage organs (USOs) were a constituent part of the diet. We observed starch granules that may be from the Arum family (Araceae), from yams (genus *Dioscorea*), and consistent with sweet potato (*Ipomoea batatas*), as well as phytoliths similar to the type found in the rhizome of *Calathea* sp. (ariá), a genus from the Maranthaceae family.

We also found evidence of *Arecaceae* (palms), *Myrtaceae, Lauraceae* and *Poaceae* (the grass family, including starch granules most likely from *Zea mays*).

There is not yet enough evidence to know whether or not the Jabuticabeira II group cultivated maize (*Zea mays*). However, there is evidence that maize was used by highland groups from southern Brazil during the late Holocene (Miller 1971; De Masi 1999; Behling *et al.* 2005). It is therefore possible that the people from Jabuticabeira II obtained maize through contact with those inlanders. Indeed, starch granules suggestive of maize were also found in calculus samples from two other *sambaquis* contemporaneous with Jabuticabeira II from the southern coast of Brazil (Wesolowski *et al.* 2010). Starch and phytoliths from Brazilian pine seeds (*Araucaria angustifolia*), which occur only in the southern highlands, and during winter time, were also found in dental calculus from the same sites which further supports seasonal contact between *sambaqui* and inland populations (Wesolowski *et al.* 2010).

Some of the Poaceae phytoliths that were found are typical of leaves and their presence in the dental calculus may be related to non-food use, as they are not considered very palatable for humans. They may have been used as medicine or they may have been accidentally ingested with food prepared directly in hot ashes. Burnt wood and large amounts of phytoliths from Poaceae leaves identified through micromorphlological analyses of sediments at Jabuticabeira II, might indicate that dry grass was used as fuel for hearths (Villagrán 2008), supporting the hypothesis of accidental ingestion.

Table 11.3 Botanical taxa identified from the dental calculus samples from the Sambaqui Jabuticabeira II

Taxonomic identification*	Micro-remain	Samples	Burials
Araceae	Starch	110, 112, 118	40, 43, 28a
Arecaceae	Phytolith	107	17aL1.05
Convolvulaceae – *Ipomoea batatas*	Starch	112, 125	43, 115b
Dioscoreaceae – *Dioscorea* sp	Starch	111	41a
aff. Lauraceae	Phytolith	107	17aL1.05
Myrtaceae – cf. *Eugenia uniflora*	Starch	125	115b
Marantaceae – cf. *Calathea* sp.	Starch	107, 112, 121	17aL1.05, 43, 108
Poaceae	Starch/phytolith	107, 112, 121,122	17aL1.05, 43, 108,107
Poaceae – cf. *Zea mays*	Starch	110, 112, 120,123	40, 43, 102, 110
Araceae, Marantaceae, Fabaceae (?) *Manihot* sp .(?)	Starch	105, 108	12a, 34
Underground storage organ (Unidentified)	Starch	107	17aL1.05

* The question mark next to a *taxon* means that there is doubt about the identification.

Diatoms

Diatom analyses can be extremely useful for paleoenvironmental reconstructions. These microscopic unicellular algae are found in a range of different environments, they have a wide geographic distribution, and many species are ecologically sensitive, having different levels of tolerance to environmental factors such as temperature, salinity, pH, among others and occupying specific niches (Mann and Droop 1996; Stoermer and Smol 2001; Amaral 2008).

The siliceous carapace of diatoms, the frustule, is used for taxonomic identification (Collins and Kalinsky 1977; Stoermer and Smol 2001) and preserves very well in sediments (Ybert *et al.* 2003), as well as inside the dental calculus matrix (Gobetz and Bozarth 2001; Boyadjian *et al.* 2007; Dudgeon and Tromp 2012). Diatoms can therefore render interesting complementary information about the resources used by the groups studied, as demonstrated by Dudgeon and Tromp (2012). We found diatoms in samples from eight individuals (42% from the total samples), yielding five securely identified different taxonomic groups (Table 11.4 and Fig. 11.5).

Contrary to Dudgeon and Tromp (2012), who found freshwater diatoms in dental calculus samples from the Rapa Nui suggesting different sources of drinking water, we have found diatoms that inhabit the marine environment but which are principally found in brackish water. This means that they are common in the estuarine environment that characterised the region (Amaral, P., pers. comm.) since the mid-Holocene (Kneip 2004; DeBlasis *et al.* 2007) and in which Jabuticabeira II is located. Assuming that

Table 11.4 *Types of diatoms found in the calculus samples from the site Jabuticabeira II*

Sample no.	Burial	Actinoptychus senarius (Type 1)	Cyclotella striata (Type 5)	Coscinodiscus sp. (Type 3)	Diploneis ovalis (Type 4)	Paralia sulcata (Type 2)	Unident. fragments	Total
107	17A/L1.05	1						1
112	43/L1.77		1					1
116	27A/L2T15					1	1	2
118	28A/L2T15			1	2	1		4
119	42A/L1.76					1		1
121	108/L2.05					1	1	2
122	107/T18	1		1				2
123	110/L2					1		1
	TOTAL	2	1	2	2	5	2	14

diatoms were ingested accidentally together with the food, which seems reasonable, this data corroborates the zooarchaeological findings, which indicate that the aquatic food resources were mainly obtained from the nearby paleolagoon (Klökler 2008), not from the sea.

Food preparation

Around 20% of the starch granules recovered was severely damaged. They were obtained from calculus samples from almost 60 % of the individuals observed.

Lithic artefacts that were found at the site (Scheel-Ybert *et al.* 2009), and patterns of starch damage consistent with pressure (cracking around the surface of the granule), support the hypothesis that starchy food may have been mechanically processed either through grinding or pounding, though dry heat (consistent with roasting), or even chewing, would result in similar patterns. Recovery and analysis of micro-remains from the surface of these artefacts, as well as experimental reproduction, are the next steps required to explore the nature of food processing in *sambaquis*.

Some of the damaged granules displayed a central cavity with irregular borders in the area of the hilum (Fig. 11.3e). Although this remains to be confirmed, this characteristic may suggest that the granules were damaged through heating in absence of water (for example, through roasting the starchy food directly in hot ashes), possibly followed by mechanical attrition (milling), such as has been suggested for the traditional Andean production of maize flour (Babot 2006).

Sand and dark fragments found in dental calculus from other *sambaquis* of Southern Brazil were interpreted as evidence of food preparation in direct contact with charcoal and soil (Wesolowski 2007), such as might occur in underground ovens that have been reported in ethnographic studies (Flowers 1983; Beck 2006). Accordingly, the patterns of some starch granules, the presence of grass phytoliths (as described earlier in this chapter), and the presence of dark particles in the dental calculus samples from an earlier study (Boyadjian and Eggers 2014), suggest that the group from Jabuticabeira II may have also prepared their food in underground ovens. However, tubers, roots and other underground storage organs could have been roasted directly in hot ashes, in the open air. Unfortunately, we cannot identify the plants processed this way because the starch granules have lost most of their diagnostic characteristics.

Other possible food processing methods include cooking in the presence of water, which gelatinises starch granules (Babot 2006; Hardy *et al.* 2009). However, in the calculus samples from Jabuticabeira II we found very few partially gelatinised starch granules, which might indicate that this was not a common practice for the group.

Although the analysis of the damage patterns found in many granules suggest these may be related to processing food prior to consumption, the possibility that some of them are the result of diagenetic modification, as highlighted by Collins and Copeland (2011), cannot be discarded. There are many factors that can provoke the degradation of starch granules in soil and archaeological sediments (Haslam 2004). However, the starch granules analysed herein were embedded in the calculus matrix, and this may have protected them to some extent against diagenetic changes (Hardy *et al.* 2009). Nonetheless, 'despite the growing interest in the analysis of ancient starch granules, studies of diagenesis are rare' (Collins and Copeland 2011), and in particular in respect to the preservation of starch from ancient dental calculus.

Cultivation

The gathering, management, and cultivation of some species of plants by *sambaqui* people has been suggested before, based on indirect evidence (Dias and Carvalho 1983; Oliveira 1991; Wesolowski 2000; Scheel-Ybert 2001; Scheel-Ybert *et al.* 2009).

The very low caries frequency and the intense dental wear of the individuals from Jabuticabeira II (Storto *et al.* 1999; Okumura and Eggers 2005) are similar to hunter-gatherer populations from other studies (Turner 1979; Larsen 1997). However, low frequencies of carious lesions were noticed not only in hunter-gatherers, but also in agricultural populations with carbohydrate rich diets. Indeed, many factors are involved in caries formation and manifestation, with diet being only one of them (Tayles *et al.* 2000; Lanfranco and Eggers 2010).

Judging by the variety of micro-remains obtained from the Jabuticabeira II calculus samples, we suggest that opportunistic gathering of a wide variety of plant foods, complementary to the protein resources that would have been obtained from the paleo-lagoon, seems more likely than intensive cultivation of a few species. However, since the preliminary data presented here show that the diet of the group included both wild and domesticated plants, it is possible that they were practicing horticulture[2] which could also have contributed to the variety of starch granules and phytoliths found. These hypotheses need further testing.

Conclusions

Because of the exploratory nature of this study, and due to the fact that it is still not possible to identify many of the starch granules and phytoliths from our samples, most of the results presented here are still preliminary. Consequently, many questions remain unanswered. However, we have only recently begun to develop a starch reference collection of native plants from the studied region. Once this expands, we will be able to identify more of the morphotypes observed.

For the present, based on the variety of starch and phytolith types found, and the rich environment surrounding the site (Scheel-Ybert 2001; Scheel-Ybert *et al.* 2009), which contains an abundance of edible plants available throughout the year, we suggest that a wide range of plant resources were used as food by the Jabuticabeira II group. The variation we detected in the micro-particle assemblages of individuals from this site could not be explained through differences of sex, age, presence or absence of pathologies or burial features. This suggests equality of access to plant food resources; however, as our sample size was small, statistical analyses were uninformative and thus not presented.

Based on the evidence from starch and phytoliths, we suggest that plants from the *Araceae, Arecaceae, Convolvulaceae, Dioscoreaceae, Maranthaceae*, and *Poaceae* families were among those eaten in Jabuticabeira II. The presence of *Poaceae* could also be related to non-dietary use. The diatom *taxa* found indicate that the aquatic resources exploited came from the estuarine environment nearby, not from the sea.

The analysis of damaged starch granules indicate that food was most likely processed mechanically, using lithic artefacts, and/or roasted directly in hot ashes or underground ovens. The cooking of plant food in water was probably rare. The possible effect of diagenesis on starch granules also needs to be taken into consideration.

We propose that even though we cannot confirm cultivation, this group appears to have lived in a system with a mixed economy, where fishing, gathering, hunting, and possibly horticulture provided most of the daily nutrition. Additionally, some resources such as maize (if it was not cultivated by the group itself) may have been exchanged with highland groups.

Despite the limited results, we are starting to shed some light into diet and lifestyle of the studied human groups. In the near future, with the development of our methods and advances in the knowledge about dental calculus contents we will expand this investigation.

Acknowledgements

We would like to thank Karen Hardy for this opportunity and the anonymous reviewer for the valuable comments. We would specially like to thank Prof. Karl Reinhard, from University of Nebraska-Lincoln (USA) who started this all and generously gave access to his laboratory and equipment, essential to perform this research. We also thank Veronica Wesolowski, for helping initially with the dental calculus protocol. We are extremely grateful to Prof. Paulo DeBlasis and his team (MAE-USP), responsible for the archaeological projects that permitted the burials to be analysed. We would also like to thank Profs Karol Chandler-Ezell (Stephen F. Austin State University), Linda Perry (Foundation for Archaeobotanical Research in Microfossils), Paula Amaral (IO-USP), Jorge Casallas (Museu Nacional – UFRJ), and Gilson Laone (PPGH-PUC Porto Alegre), for helping with the classification and identification of some of the micro-remains. We extend our thanks to Rodrigo Elias (LEEH-IBUSP) and the team from Laboratório de Arqueobotânica e Paisagem (Museu Nacional – UFRJ) for the helping hands with the plants from our reference collection and Luis Pezo, MaDu Gaspar and Leonardo Waisman for the figures. This research was supported by FAPESP (Grant 2008/53351-7), CAPES (BEX 1279/09-2) and CNPq.

Notes

1   The *Restinga* ecosystem, a mosaic of vegetation types with diverse physiognomies, occupies the coastal sandy beach ridges. It varies from sparse open plant communities, such as herbaceous and scrub formations ('open *restinga*') to dense evergreen forests ('*restinga* forest').

2   According to Piperno and Pearsall (1998), horticulture is the small-scale cultivation of wild or domesticated species close to the settlement area (in gardens, for example).

References

Afonso, M. C. and DeBlasis, P. A. D. 1994. Aspectos da formação de um grande sambaqui: alguns indicadores em Espinheiros II, Joinville, SC. *Revista do Museu de Arqueologia e Etnologia* 4, 21–30

Amaral, P. G. C. D. 2008. *Evolução da sedimentação lagunar holocênica da região de Jaguaruna, estado de Santa Catarina: uma abordagem sedimentológica-micropaleontológica integrada.* PhD Thesis, Universidade de São Paulo (USP), Brazil

Asevedo, L., Winck, G. R., Mothé, D. and Avila, L. S. 2012. Ancient diet of the Pleistocene gomphothere Notiomastodon platensis (Mammalia, Proboscidea, Gomphotheriidae) from lowland mid-latitudes of South America: Stereomicrowear and tooth calculus analyses combined. *Quaternary International* 255, 42–52

Babot, M. D. P. 2003. Starch grain damage as an indicator of food processing. In Hart, D. M. and Wallis, L. A. (eds), *Phytolith and Starch Research in the Australian-Pacific-Asian regions: The State of the Art.* Pandanus Books, Canberra, 69–81.

Babot, M. D. P. 2006. Damage on starch from processing Andean food plants. In Torrence, R. and Barton (eds) 2006, 66–7

Bandeira, D. D. R., 1992. *Mudança na estratégia de subsistência do sítio arqueológico Enseada I – Um estudo de caso.* MSc Dissertation, Universidade Federal de Santa Catarina (UFSC), Santa Catarina, Brazil

Barton, H., 2006. Starch granules from Niah Cave sediments In Torrence, R. and Barton (eds) 2006, 132–4

Beauclair, M., Scheel-Ybert, R., Bianchini, G. F. and Buarque, A. 2009. Fire and ritual: bark hearths in South-American Tupiguarani mortuary rites. *Journal of Archaeological Science* 36, 1409–1

Beck, A. 1972. *A variação do conteúdo cultural dos sambaquis, litoral de Santa Catarina.* PhD Thesis, FFLCH, Universidade de São Paulo, Brazil

Beck, W. 2006. Processing Cheeky Yam in northern Australia. In Torrence, R. and Barton (eds) 2006, 58–60

Behling, H., Pillar, V. D. and Bauermann, S. G. 2005. Late Quaternary grassland (Campos), gallery forest, fire, and climate dynamics, studied by pollen, charcoal and multivariate annalysis of the São Francisco de Assis core in Western Rio Grande do Sul (Southern Brazil). *Review of Palaeobotany & Palynology*, 133(3–4), 235–48

Bianchini, G. F. 2008. *Fogo e Paisagem: evidências de práticas rituais e construção do ambiente a partir da análise antracológica de um sambaqui no litoral sul de Santa Catarina.* MSc Dissertation, Museu Nacional, Universidade Federal do Rio de Janeiro (UFRJ), Brazil

Bianchini, G. F. and Scheel-Ybert, R. 2012. Plants in a funerary context at the Jabuticabeira-II shellmound (Santa Catarina, Brazil) – feasting or ritual offerings? In Badal, E., Carrion, Y., Macias, M. and Antinou, M. (Coord.) *Wood and Charcoal Evidence for Human and Natural History. Sagvntvm: papeles del laboratorio de arqueologia de Valencia, Extra-13*, 253–8. Valencia: Universitat de Valencia, Departament de Prehistòria i Arqueologia de la Facultad de Geografía i Història

Bianchini, G. F., Scheel-Ybert, R. and Gaspar, M. D. 2007. Estaca de Lauraceae em contexto funerário (sítio Jaboticabeira II, Santa Catarina, Brasil). *Revista do Museu de Arqueologia e Etnologia* 17, 223–9

Blanchot, H. and Amenomori, S. N. 2005. Levantamento dos vestígios vegetais do Abrigo Rupestre de Santa Elina. In Vilhena-Vialou, A. and Vialou, D. (eds), *Pre-história do Mato Grosso: Uma Pesquisa Brasileira-Francesa Pluridisciplinar*, 211–14. São Paulo: Edusp

Boyadjian, C. H. C. 2012. *Análise e identificação de microvestígios vegetais de cálculo dentário para a reconstrução de dieta sambaquieira: estudo de caso de Jabuticabeira II, SC.* PhD Thesis, Universidade de São Paulo (USP), Brazil

Boyadjian, C. H. C. and Eggers, S. 2014. micro-remains trapped in dental calculus reveal plants consumed by Brazilian shell mound builders. In Roksandic, M., Burchel, M., Eggers, S., Klökler, D. and Mendonça de Souza, S. (eds), *The Cultural Dynamics of Shell-Matrix Sites*, 279–288. Albuquerque NM: University of New Mexico Press.

Boyadjian, C. H. C., Eggers, S. and Reinhard, K. 2007. Dental wash: a problematic method for extracting microfossils from teeth. *Journal of Archaeological Science* 34, 1622–8

Bryan, A. L. 1993. *The Sambaqui at Forte Marechal Luz, State of Santa Catarina, Brazil.* Corvallis OR: Center for the Study of the First Americans, Oregon State University

Ceccantini, G. C. T. 2003. Anatomy of wood stakes from the Santa Elina archeological site, Mato Grosso, Brazil: paleoenvironmental and ethnobotanical interpretations. *IAWA Journal* 24, 313

Ceccantini, G. C. T. 2005. Anatomia da madeira e identificação de estacas arqueológicas do Abrigo Rupestre de Santa Elina. In Vilhena-Vialou, A. and Vialou, D. (eds), *Pre-história do Mato Grosso: Uma Pesquisa Brasileira-Francesa Pluridisciplinar*, 189–200. São Paulo: Edusp

Ceccantini, G. C. T. and Gussella, L. W. 2001. Os Novelos de Fibras do Abrigo Rupestre Santa Elina (Jangada, MT, Brasil): Anatomia vegetal e paleoetnobotânica. *Revista do Museu de Arqueologia e Etnologia* 11, 189–200

Chandler-Ezell, A. K., Pearsall, D. M., Zeidler, J. A. and Chandler-ezell, K. 2006. Root and tuber phytoliths and starch grains document manioc (*Manihot Esculenta*), Arrowroot (*Maranta Arundinacea*), and Llerén (*Calathea* sp.) at the Real Alto Site, Ecuador. *Economic Botany* 60, 103–20

Ciochon, R. L., Piperno, D. R. and Thompson, R. G. 1990. Opal phytoliths found on the teeth of the extinct ape Gigantopithecus blacki: implications for paleodietary studies. *Proceedings of the National Academy of Sciences (USA)* 87, 8120–4

Collins, M. J. and Copeland, L. 2011. Ancient starch: Cooked or just old? *Proceedings of the National Academy of Sciences (USA)* 108, E145, author reply E146

Collins, G. B. and Kalinsky, R. G. 1977. Studies on Ohio diatoms: I. Diatoms of the Scioto river basin, II. Referenced checklist of diatoms from Ohio, exclusive of lake Erie and the Ohio river. *Bulletin of the Ohio Biology Survey* 5(3), 1–76

Crowther, A. 2009. Starch Identification. In Charles, M., Crowther, A., Ertug, F., Herbig, C., Jones, G., Kutterer, J., Longford, C., Madella, M., Maier, U., Out, W., Pessin, H. and Zurro, D., *Archaeobotanical Online Tutorial* http://archaeobotany.dept.shef.ac.uk/wiki/index.php/Main_Page

DeBlasis, P. A. and Gaspar, M. D. 2009. Os Sambaquis do Sul Catarinense: Retrospectiva e Perspectivas de Dez Anos de Pesquisa. *Caderno de Ciências Humanas* 11/12(20/21), 83–125

DeBlasis, P., Fish, S. K., Gaspar, M. D. and Fish, P. R. 1998. Some references for the discussion of complexity among the Sambaqui moundbuilders from the southern shores of Brazil. *Revista de Arqueologia Americana* 15, 75–105

DeBlasis, P., Kneip, A., Scheel-Ybert, R., Giannini, P. C. and Gaspar, M. D. 2007. Sambaquis e Paisagem: Dinâmica natural e arqueologia regional no litoral do sul do Brasil. *Arqueología Suramericana/ Arqueologia Sul-Americana* 3, 29–61

De La Fuente, C., Flores, S. and Moraga, M. 2012. DNA from ancient human bacteria: a novel source of genetic evidence from archaeological dental calculus. *Archaeometry* 55(4), 767–78

De Masi, M. A. N. 1999. *Prehistoric Hunter-gatherer Mobility on Southern Brazilian Coast: Santa Catarina Island*. PhD Thesis, Stanford University, Califórnia

De Masi, M. A. N. 2001. Pescadores coletores da costa sul do Brasil. *Pesquisas Antropologia* 57, 1–136

Dias, O. F. and Carvalho, E. T. 1983. Um possível foco de domesticação de plantas no Estado do Rio de Janeiro: RJ-JC-64 (Sítio Corondó). *Boletim do Instituto de Arqueologia Brasileira* 1, 4–18

Dickau, R., Ranere, A. J. and Cooke, R. G. 2007. Starch grain evidence for the preceramic dispersals of maize and root crops into tropical dry and humid forests of Panama. *Proceedings of the National Academy of Sciences (USA)* 104(9), 3651–6

Dudgeon, J. V. and, Tromp, M. 2012. Diet, geography and drinking water in Polynesia: microfossil research from archaeological human dental calculus, Rapa Nui (Easter Island). *International Journal of Osteoarchaeology* doi, 10.1002/oa.2249

Figuti, L. 1992. *Les Sambaquis COSIPA (4200 à 1200 Ans BP): Étude De La Subsistance Chez Les Peuples Préhistoriques De Pêcheurs-ramasseurs De Bivalves De La Cote Centrale De L'état De São Paulo, Brésil*. Paris: Museum National d'Histoire Naturelle

Fish, S. K., DeBlasis, P. A. D., Gaspar, M. D. and Fish, P. R. 2000. Eventos Incrementais na Construção de Sambaquis, Litoral Sul do Estado de Santa Catarina. *Revista do Museu de Arqueologia e Etnologia* 10, 69–87

Flowers, N. M. 1983. Seasonal factors in subsistence, nutrition and child growth in a central Brazilian Indian community. In Hames, R. B. and Vickers, W. T. (eds), *Adaptive Responses of Native Amazonians*, 357–89. New York: Academic

Fox, C. L., Juan, J. and Albert, R. M. 1996. Phytolith analysis on dental calculus, enamel surface, and burial soil: information about diet and paleoenvironment. *American Journal of Physical Anthropology* 101, 101–13

Freitas, F. O., Bendel, G., Allaby, R. G. and Brown, T. A. 2003. DNA from primitive maize landscapes and archaeological remains: implications for the domestication of maize and its expansion into South America. *Journal of Archaeological Science* 30, 901–8

Fullagar, R., Field, J., Denham, T. and Lentfer, C. 2006. Early and mid-Holocene tool-use and processing of taro (*Colocasia esculenta*), yam (*Dioscorea* sp.) and other plants at Kuk Swamp in the highlands of Papua New Guinea. *Journal of Archaeological Science* 33, 595–614

Gaspar, M. D. 1998. Considerations of the sambaquis of the Brazilian coast. *Antiquity* 72, 592–615

Gobetz, K. E. and Bozarth, S. R. 2001. Implications for Late Pleistocene Mastodon diet from opal phytoliths in tooth calculus. *Quaternary Research* 55, 115–22

Gott, B., Barton, H., Samuel, D. and Torrence, R. 2006. Biology of starch. In Torrence, R. and Barton (eds) 2006, 35–45

Hardy, K., Blakeney, T., Copeland, L., Kirkham, J., Wrangham, R. and Collins, M. 2009. Starch granules, dental calculus and new perspectives on ancient diet. *Journal of Archaeological Science* 36, 248–55

Hardy, K., Buckley, S., Collins, M. J., Estalrrich, A., Brothwell, D., Copeland, L., García-Tabernero, A., García-Vargas, S., De la Rasilla, M., Lalueza-Fox, C., Huguet, R., Bastir, M., Santamaría, D., Madella, M., Wilson, J., Cortés, A. F. and Rosas, A. 2012. Neanderthal medics? Evidence for food, cooking, and medicinal plants entrapped in dental calculus. *Naturwissenschaften* 99, 617–26

Haslam, M. 2004. The decomposition of starch grains in soils: implications for archaeological residue analyses. *Journal of Archaeological Science* 31, 1715–34

Hazen, S. P. 1995. Supragingival dental calculus. *Periodontology* 2000(8), 125–36

Henry, A. G., Brooks, A. S. and Piperno, D. R. 2010. Microfossils in calculus demonstrate consumption of plants and cooked foods in Neanderthal diets (Shanidar III, Iraq; Spy I and II, Belgium). *Proceedings of the National Academy of Sciences (USA)* 108, 486–91

Henry, A. G., Hudson, H. F. and Piperno, D. R. 2009. Changes in starch grain morphologies from cooking. *Journal of Archaeological Science* 36, 915–22

Henry, A. G. and Piperno, D. R. 2008. Using plant microfossils from dental calculus to recover human diet: a case study from Tell al-Raqā'i, Syria. *Journal of Archaeological Science* 35, 1943–1950.

Henry, A. G., Ungar, P. S., Passey, B. H., Sponheimer, M., Rossouw, L., Bamford, M., Sandberg, P., De Ruiter, D. J. and Berger, L. 2012. The diet of *Australopithecus sediba*. *Nature* 487, 90–3

Heredia, O. R. Tenorio, M. C. Gaspar, M. D. and Buarque, A. M. G. 1989. Environment exploitation by prehistorical population of Brazil. In Neves, C. (ed.), *Coastlines of Brazil*, 230–9. New York: American Society of Civil Engineers

Holt, B. 1993. Phytoliths from dental calculus: direct evidence on prehistoric diet. *Phytolitherien Newsletter* 7, 8

Horrocks, M., Irwin, G., Jones, M. and Sutton, D. 2004. Starch grains and xylem cells of sweet potato (*Ipomoea batatas*) and bracken (*Pteridium esculentum*) in archaeological deposits from northern New Zealand. *Journal of Archaeological Science* 31, 251–8

Hurt, W. R. 1974. *The interrelationship between the natural environment and four sambaquis, coast of Santa Catarina, Brasil*. Bloomington IN: Indiana University Museum Occasional Papers and Monographs 1

ICSN. 2011. *The International Code for Starch Nomenclature*, www.fossilfarm.org/ICSN/Code.html. Accessed in March 2012.

Iriarte, J. 2003. Assessing the feasibility of identifying maize through the analysis of cross-shaped size and three-dimensional morphology of phytoliths in the grasslands of southeastern South America. *Journal of Archaeological Science* 30, 1085–94

Iriarte, J., Holst, I., Marozzi, O., Listopad, C., Alonso, E., Rinderknecht, A. and Montaña, J. 2004. Evidence for cultivar adoption and emerging complexity during the mid- Holocene in the La Plata basin. *Nature* 432, 614–17

Juan-Tresserras, J., Lalueza, C., Albert, R. M. and Calvo, M. 1997. Identification of phytoliths from prehistoric human dental remains from the Iberian Peninsula and the Balearic Islands.

In Pinilla, A., Juan-Tresserras, J. and Machado, M. J. (eds), *Primer encuentro Europeo sobre el estudio de fitolitos*, 197–203. Madrid: Gráficas Fersán

Kamase, L. M. 2005. As estacas de madeira. In Vilhena-Vialou, A. and Vialou, D. (eds), *Pré-história do Mato Grosso: uma pesquisa brasileira-francesa pluridisciplinar*, 201–9. São Paulo: Edusp

Klökler, D. M. 2001. *Construindo ou Deixando um Sambaqui? Análise de Sedimentos de um Sambaqui do Litoral Meridional Brasileiro: Processos Formativos*. Região de Laguna, SC. MSc Dissertation, Universidade de São Paulo (USP), Brasil

Klökler, D. M. 2008. Food *for Body and Soul: mortuary ritual in shell mounds (Laguna-Brazil)*. PhD Thesis, University of Arizona, USA.

Kneip, L. M. 1977. Pescadores e coletores pré-históricos do litoral de Cabo Frio, RJ. Coleção Museu Paulista, sér. *Arqueologia* 5, 7–169

Kneip, L. M. 1980. A sequêcia cultural do sambaqui do Forte, Cabo Frio, Rio de Janeiro. *Pesquisas* série *Antropológica* 31, 87–100

Kneip, L. M. 1994. Cultura material e subsistência das populações pré-históricas de Saquarema, RJ. Documento de Trabalho, sér. *Arqueologia* 2, 1–120

Kneip, A. 2004. *O povo da Lagoa: uso do SIG para modelamento e simulação na área arqueológica do Camacho*. PhD Thesis, Universidade de São Paulo, Brasil.

Kucera, M., Pany-Kucera, D., Boyadjian, C. H., Reinhard, K. and Eggers, S. 2011. Efficient but destructive: a test of the dental wash technique using secondary electron microscopy. *Journal of Archaeological Science* 38, 129–35

Lanfranco, L. and Eggers, S. 2010. The usefulness of caries frequency, depth, and location in determining cariogenicity and past subsistence: A test on early and later agriculturalists from the Peruvian coast. *American Journal of Physical Anthropology* 143(1), 75–91

Larsen, C. S. 1997. *Bioarchaeology: interpreting behavior from the human skeleton*. Cambridge: Cambridge University Press

Lentfer, C., Therin, M. and Torrence, R. 2002. Starch grains and environmental reconstruction: a modern test case from West New Britain, Papua New Guinea. *Journal of Archaeological Science* 29, 687–98

Li, M., Yang, X., Wang, H., Wang, Q., Jia, X. and Ge, Q. 2010. Starch grains from dental calculus reveal ancient plant foodstuffs at Chenqimogou site, Gansu Province. *Science China Earth Sciences* 53, 694–9

Lieverse, A. 1999. Diet and the aetiology of dental calculus. *International Journal of Osteoarchaeology* 232, 219–32

Lima, T. A. 2000. Em busca dos frutos do mar: os pescadores-coletores do litoral centro-sul do Brasil. *Revista USP* 44, 270–327

Lima, T. A., Macario, K. D., Anjos, R. M., Gomes, P. R. S., Coimbra, M. M. and Elmore, D. 2002. The antiquity of the prehistoric settlement of the central-south Brazilian coast. *Radiocarbon* 44(3), 733–8

Mann, D. G. and Droop, J. M. 1996. Biodiversity, biogeography and conservation of diatoms. *Hydrobiologia* 336, 19–32

Mercader, J., Bennett, T. and Raja, M. 2008. Middle Stone Age starch acquisition in the Niassa Rift, Mozambique. *Quaternary Research* 70, 283–300

Messner, T. C. 2011. *Acorns and Bitter Roots: starch grain research in prehistoric eastern woodlands*. The Tuscaloosa AL: University of Alabama Press

Mickleburgh, H. L. and Pagán-Jiménez, J. R. 2012. New insights into the consumption of maize and other food plants in the pre-Columbian Caribbean from starch grains trapped in human dental calculus. *Journal of Archaeological Science* 39, 2468–78

Middleton, W. D. and Rovner, I. 1994. Extraction of opal phytoliths from herbivore dental calculus. *Journal of Archaeological Science* 21, 469–73

Miller, E. T. 1971. *Pesquisas Arqueológicas Efectuadas no Planalto Meridional, Rio Grande do Sul (rios Uruguai, Pelotas, e das Antas).* Publicações Avulsas do Museu Paraense Emílio Goeldi, (PRONAPA 4) 15, 37–60

Okumura, M. M. M. and Eggers, S. 2005. The people of Jabuticabeira II: reconstruction of the way of life in a Brazilian shellmound. HOMO – *Journal of Comparative Human Biology* 55, 263–81

Oliveira, M. C. T. 1991. *A importância da coleta no advento da agricultura.* MSc Dissertation, Universidade Federal do Rio de Janeiro (UFRJ), Brazil

Pearsall, D. M. 2000. *Paleoethnobotany: a handbook of procedures* 2nd edn. New York: Academic

Pearsall, D. M., Chandler-Ezell, K. and Zeidler, J. 2004. Maize in ancient Ecuador: results of residue analysis of stone tools from the Real Alto site. *Journal of Archaeological Science* 31, 423–42

Peixe, S. P., Melo, J. C. F. Jr and Bandeira, D. da R. 2007. Paleoetnobotânica dos macrorestos vegetais do tipo trançados de fibras encontrados no sambaqui Cubatão I, Joinville-SC. *Revista do Museu de Arqueologia e Etnologia* 17, 211–22

Perry, L. 2002. Starch granule size and the domestication of manioc (*Manihot esculenta*) and sweet potato (*Ipomoea batatas*). *Economic Botany* 56(4), 335–49

Perry, L. 2004. Starch analyses reveal the relationship between tool type and function: an example from the Orinoco valley of Venezuela. *Journal of Archaeological Science* 31, 1069–81

Perry, L., Sandweiss, D. H., Piperno, D. R., Rademaker, K., Malpass, M. A., Umire, A. N. and De Le Vera, P., 2006. Early maize agriculture and interzonal interaction in southern Peru. *Nature* 440(2), 76–9

Piperno, D. R. 2006. *Phytoliths: a comprehensive guide for archaeologists and paleoecologists.* Lanham MD: AltaMira Press

Piperno, D. R. and Dillehay, T. D. 2008. Starch grains on human teeth reveal early broad crop diet in northern Peru. *Proceedings of the National Academy of Sciences (USA)* 105, 19622–7

Piperno, D. R. and Holst, I. 1998. The presence of starch grains on prehistoric stone tools from the humid neotropics: indications of early tuber use and agriculture in Panama. *Journal of Archaeological Science* 25, 765–76

Piperno, D. R. and Pearsall, D. M. 1998. The neotropical ecosystem in the present and the past. In Piperno, D. R. and Pearsall, D. M. *The Origins of Agriculture in the Lowland Neotropics,* 39–107. San Diego: Academic Press

Piperno, D. R., Ranere, a J., Holst, I. and Hansell, P. 2000. Starch grains reveal early root crop horticulture in the Panamanian tropical forest. *Nature* 407, 894–7

Piperno, D. R., Weiss, E., Holst, I. and Nadel, D. 2004. Processing of wild cereal grains in the Upper Paleolithic revealed by starch grain analysis. *Nature* 430, 670–3

Power, R. C., Salazar-García, D. C., Wittig, R. M. and Henry, A. G. 2014. Assessing use and suitability of scanning electron microscopy in the analysis of micro remains in dental calculus. *Journal of Archaeological Science* 49,160–169

Prous, A. 1976. *Lês sculptures zoomorfes du Sud Brésilien et de l'Uruguay.* Paris: Cahiers d'Archéologie d'Amérique du Sud 5.

Prous, A. 1992. *A Arqueologia Brasileira.* Brasília: Editora da Universidade de Brasília

Prous, A. and Piazza, W. 1977. *Documents pour La préhistoire du Brésil Meridional 2: l'etat de Santa Catarina.* Paris: Cahiers d'Archéologie d'Amérique du Sud 4.

Reichert, E. T. 1913. *The Differentiation and Specificity of Starches in Relation to Genera, Species, etc.; stereochemistry applied to protoplasmic processes and products, and as a strictly scientific basis for the classification of plants and animals.* Washington DC: Carnegie Institution of Washington

Reinhard, K. J. 1993. The utility of pollen concentration on coprolite analysis: Expanding upon Dean's comments. *Journal of Ethnobiology* 13, 114–28

Reinhard, K. J. and Bryant, V. M. Jr. 1992. Coprolite analysis: a biological perspective on archaeology. In Schiffer, M. B. (ed.) *Advances in Archaeological Method and Theory* 14, 245–88. Tucson AZ: University of Arizona Press

Reinhard, K. J. and Bryant V. M. Jr. 2008. Pathoecology and the future of coprolite studies. In Stodder, A. W. M. (ed.), *Reanalysis and Reinterpretation in Southwestern Bioarchaeology*, 199–216. Tempe AZ: Arizona State University Press

Reinhard, K. J., Souza, S. F. M., Rodriguez, C., Kimmerle, E. and Dorsey-Vinton, S. 2001. Microfossils in dental calculus: a new perspective on diet and dental disease. In Williams, E. (ed), *Human Remains: conservation, retrieval and analysis*, 113–18. Oxford: British Archaeological Report S934

Revedin, A., Aranguren, B., Becattini, R., Longo, L., Marconi, E., Lippi, M. M., Skakun, N., Sinitsyn, A., Spiridonova, E. and Svoboda, J. 2010. Thirty thousand-year-old evidence of plant food processing. Proceedings *of the National Academy of Sciences (USA)* 107, 18815–19

Roberts-Harry, E. A., Clerehugh, V., Shore, R. C., Kirkham, J. and Robinson, C. 2000. Morphology and elemental composition of subgingival calculus in two ethnic groups. *Journal of Periodontology* 71(9), 1401–11

Samuel, D. 2006. Modified starch. In Torrence and Barton (eds) 2006, 205–16

Serban, P., Wilson, J. R. U., Vamosi, J. C. and Richardson, D. M. 2008. Plant diversity in the human diet: Weak phylogenetic signal indicates breadth. *BioScience* 58, 151–9

Scheel-Ybert, R. 2000. Vegetation stability in the southeastern Brazilian coastal area from 5500 to 1400 14C yr BP deduced from charcoal analysis. *Review of Palaeobotany and Palynology* 110, 111–38

Scheel-Ybert, R. 2001. Man and vegetation in southeastern Brazil during the Late Holocene. *Journal of Archaeological Science* 28, 471–80

Scheel-Ybert, R. and Solari, M. E. 2005. Macro-restos vegetais do Abrigo Santa Elina: Antracologia e Carpologia. In Vilhena-Vialou, A. and Vialou, D. (eds), *Pré-história do Mato Grosso: uma pesquisa brasileira-francesa pluridisciplinar*, 139–47. São Paulo: Edusp

Scheel-Ybert, R., Eggers, S., Wesolowski, V., Petronilho, C. C., Boyadjian, C. H., DeBlasis, P. A. D., Barbosa-Guimarães, M. and Gaspar, M. D. 2003. Novas perspectivas na reconstituição do modo de vida dos sambaquieiros: uma abordagem multidisciplinar. *Revista Arqueologia* 16, 109–37

Scheel-Ybert, R., Eggers, S., Wesolowski, V., Petronilho, C. C., Boyadjian, C. H., Gaspar, M. D., Tenório, M. C. and DeBlasis, P. 2009. Subsistence and lifeway of coastal Brazilian mound builders. In Capparelli. A., Chevalier, A. and Piqué, R. (eds), *La alimentación en la América precolombina y colonial: un aproximación interdisciplinary*, 37–53. Barcelona: Treballs d'Etnoarqueologia 7

Scott Cummings, L. and Magennis, A. 1997. Phytolith and starch record of food and grit in Mayan human tooth tartar. In Pinilla, A., Juan-Tresserras, J. and Machado, M. J. (eds), *Primer encuentro Europeo sobre el estudio de fitolitos*, 211–18. Madrid: Gráficas Fersán

Scott, G. R. and Poulson, S. R. 2012. Stable carbon and nitrogen isotopes of human dental calculus: a potentially new non-destructive proxy for paleodietary analysis. *Journal of Archaeological Science* 39, 1338–93

Schmitz, P. I. 1987. Prehistoric hunters and gatherers of Brazil. *Journal of World Prehistory* 1, 53–26

Stoermer, E. F. and Smol, J. P. 2001. Applications and uses of diatoms: prologue. In Stoermer, E. F. and Smol, J. P. (eds), *The Diatoms: applications for the environmental and earth sciences*. Cambridge: Cambridge University Press

Storto, C., Eggers, S. and Lahr, M. M. 1999. Estudo preliminar das paleopatologias da população do sambaqui Jaboticabeira II, Jaguaruna, SC. *Revista do Museu de Arqueologia e Etnologia* 9, 61–71

Tayles, N., Domett, K. and Nelsen, K. 2000. Agriculture and dental caries? The case of rice in prehistoric southeast Asia. *World Archaeology* 32(1), 68–83

Tenório, M. C. 2004. Identidade Cultural e Origem Dos Sambaquis. *Revista do Museu de Arqueologia e Etnologia* 14, 169–78

Torrence, R. 2006. Starch and archaeology. In Torrence and Barton (eds) 2006, 17–34

Torrence, R. and Barton, H. (eds) 2006. *Ancient Starch Research.* Walnut Creek CA: Left Coast Press

Turner, C. G. II. 1979. Dental anthropological indications of agriculture among the Jomon people of central Japan. *American Journal of Physical Anthropology* 51(4), 619–36

Ugent, D., Pozorski, S. and Pozorski, T. 1986. Archaeological manioc (*Manihot*) from coastal Peru. *Economic Botany* 40(1), 78–102

Villagrán, X. 2008. *Análise de arqueofácies na camada preta do sambaqui Jabuticabeira II.* MSc Dissertation, Universidade de São Paulo (USP), Brazil

Vinton, S. D., Perry, L., Reinhard, K .J., Santoro, C. M. and Teixeira-Santos, I. 2009. Impact of empire expansion on household diet: the Inka in Northern Chile's Atacama Desert. *PLOS ONE* 4, e8069

Wagner, G., Hilbert, K., Bandeira, D., Tenório, M. C. and Okumura, M. M. 2011. Sambaquis (shell mounds) of the Brazilian coast. *Quaternary International* 239, 51–60

Warinner, C., Rodrigues, J. F. M., Vyas, R., Trachsel, C., Shved, N., Grossmann, J., Radini, A., Hancock, Y., Tito, R. Y., Fiddyment, S., Speller, C., Hendy, J., Charlton, S., Luder, H. U., Salazar-García, D. C., Eppler, E., Seiler, R., Hansen, L. H., Castruita, J. A. S., Barkow-Oesterreicher, S., Teoh, K. Y., Kelstrup, C. D., Olsen, J. V., Nanni, P., Kawai, T., Willerslev, E., von Mering, C., Lewis C. M. Jr, Collins, C. M., Gilbert, M. T. P., Rühli, F. and Cappellini, E. 2014. Pathogens and host immunity in the ancient human oral cavity. *Nature Genetics* 46, 336–44

Wesolowski, V. 2000. *A prática da horticultura entre os construtores de sambaquis a acampamentos litorâneos da região da Baía de São Francisco, Santa Catarina: uma abordagem bioantropológica.* MSc Dissertation, Universidade de São Paulo (USP), Brazil

Wesolowski, V. 2007. *Cáries, desgaste, cálculos dentários e micro-resíduos da dieta entre grupos pré-históricos do litoral norte de Santa-Catarina: É possível comer amido e não ter cárie?* PhD Thesis, Escola Nacional de Saúde Pública, Fundação Oswaldo Cruz (FIOCRUZ), Rio de Janeiro, Brazil

Wesolowski, V., Ferraz Mendonça de Souza, S. M., Reinhard, K. J. and Ceccantini, G. 2010. Evaluating microfossil content of dental calculus from Brazilian sambaquis. *Journal of Archaeological Science* 37, 1326–38

Ybert, J. P., Bissa, W. M., Catharino, E. L. M. and Kutner, M. 2003. Environmental and sea-level variations on the southeastern Brazilian coast during the Late Holocene with comments on prehistoric human occupation. *Palaeogeography, Palaeoclimatology, Palaeoecology* 189, 11–24

Zarrillo, S., Pearsall, D. M., Raymond, J. S., Tisdale, M. A. and Quon, D. J. 2008. Directly dated starch residues document early formative maize (*Zea mays* L.) in tropical Ecuador. *Proceedings of the National Academy of Sciences (USA)* 105, 5006–11

12. Stable isotopes and mass spectrometry

Karen Hardy and Stephen Buckley

Stable isotope analysis is the most commonly used analytical technique in early human and hominin dietary reconstruction today. This chapter begins with a background to stable isotope analysis and describes how it works in terms of detecting and differentiating between C$_3$ and C$_4$ plants, and the use of carbon and nitrogen stable isotopes. This is followed by a summary of the research that has been conducted into early hominin diet using the different carbon isotopes. Finally, the use of the dual methods of thermal desorption-gas chromatography-mass spectrometry (TD-GC-MS) and pyrolysis-gas chromatography-mass spectrometry (Py-GC-MS) which detect a wide variety of biomolecules in residues is discussed.

Introduction

Stable isotope analysis is the most common analytical technique in use today in the investigation of early human and hominin diet. There are two main fields of stable isotope analysis used in paleodietary reconstruction; variation between different carbon isotopes (C$_3$, C$_4$, CAM), and the analysis and interpretation of carbon and nitrogen isotopes (^{13}C/^{15}N). Carbon isotopes alone are used to differentiate between major plants groups based on variation in their methods of CO$_2$ entrapment during photosynthesis that results in retention of variable quantities of ^{13}C which can be measured. This has proved to be extremely useful in reconstructing major changes in early hominin diet, notably the adoption of savannah resources, and in documenting the spread of certain domesticated C$_4$ plants (sorghum, millet, maize); however, because virtually all plants in temperate regions are C$_3$, this method is of little value in dietary reconstruction outside these contexts. C/N isotope analysis is focused principally on identifying the level of protein to reconstruct trophic level of the consumers and also to differentiate between terrestrial and marine dietary components. It is a reliable way to extract information about marine or aquatic animal food consumption, though it is less reliable in determining relative proportions of animal food in the diet because it relies heavily on nitrogen isotope values which can change with the environment and soils (Lee-Thorp 2008). There is still little understanding of all the influences on nitrogen balance in humans; for example, using hair samples, variation in ^{15}N has been detected in relation to physiological factors unrelated to diet (D'Ortenzio *et al.* 2015). Likewise, macronutrient scrambling, which is based on the fact that other macronutrient components, such as carbohydrates and lipids may also contribute carbon to collagen, potentially affects the linear interpolations between theoretical endpoints which are used to infer terrestrial or marine diets (Craig *et al.* 2013).

The plant component of the diet, which leaves a very low nitrogen signature, is more difficult to detect using this method, though a complete lack of animal products in the diet (i.e. vegan) was identified in a modern study using hair samples that compared these to vegetarians eating dairy products (O'Connell and Hedges 1999). The way stable isotope analysis works is that food that is eaten is reflected in proxies in bones and teeth which can be detected and therefore provide a quantitative way of comparing between individuals and populations. However, care is needed in the interpretation of the results, and the need to take into consideration the potential of competing influences in the isotopic composition.

In this chapter, we begin with a description of what stable isotope analysis is and how it works. We then provide a summary of the research that has been conducted in the diets of Pliocene hominins using the variable rates of ^{13}C. Stable isotope analysis is now a very large topic, and this chapter is not exhaustive; it focuses on perspectives that are relevant to the presence of plants in the diet. We also summarise the use of GC-MS techniques in detection and identification of residues.

The principles of stable isotopes

There are a number of good descriptions of how stable isotope analysis works. The description below which is taken from Lee-Thorp (2008), Schoeninger (2010), and Brown and Brown (2011), provides a summary of the key points. For a more detailed explanation of this and of the use of stable isotopes in archaeology, please refer to these publications and others in the reference list.

Isotopes of a single element share the same chemical properties and have the same number of protons and electrons, but they have different numbers of neutrons in their nuclei (mass). Some isotopes are radioactive and decay, for example ^{14}C. Others such as ^{12}C and ^{13}C are stable and do not decay. Lighter isotopes tend to be more abundant than heavier ones, so for example ^{12}C comprises 98.93% of all carbon atoms. The different masses of the different isotopes result in differences during physical and chemical processes as lighter isotopes react faster and their bonds break more easily than those with heavier isotopes, so for example ^{12}C reacts more quickly than ^{13}C. This results in isotope fractionation in which the relative proportions of the isotopes changes. This effect is apparent among the light elements, notably hydrogen (H), carbon (C), nitrogen (N) and oxygen (O), because there is a relatively greater mass difference. These changes are measured in relation to acknowledged worldwide standards that are used as the basic measurement (‰ parts per thousand or per mil) of the different isotope proportions present. The standards are represented by the δ symbol which always precedes an isotope measurement. A good example of how isotope fractionation works is in the way plants take up their carbon.

Plants take their carbon from the atmosphere using three different photosynthetic pathways, C_3, C_4 and CAM (Crassulacean Acid Metabolism). The main difference between C_3 and C_4 plants is that different enzymes are used to fix the CO_2 during the photosynthetic process. C_3 plants, so called because the absorbed CO_2 is first incorporated into 3-carbon compound, use the enzyme Rubisco in photosynthesis which works particularly well in cool, moist conditions. C_4 plants use PEP carboxylase for initial CO_2 uptake which allows CO_2 absorption to be rapid thereby reducing the amount of water lost in transpiration. The CO_2 is then transferred to Rubisco in the plant, for photosynthesis. This makes C_4 plants better adapted to hot, dry, places with intense sunlight. CAM plants fix their CO_2 at night and are adapted to very arid conditions. The abundance of the ^{13}C isotope differs between C_3 and C_4 plants, due to the differences in their photosynthetic pathways; this leads to a different ratio between the heavy and light isotopes in C_3 and C_4 plants. Average ^{13}C values for C_3 plants are between -26‰ – -28‰ while for C_4 plants it is around -12 ‰. The relative proportion of isotopes from the plants eaten is retained in animal tissue and can be measured using tissue samples. In this way, the different plant groups (C_3 and C_4) in the diet can be reconstructed. Around 95% of plants worldwide use the C_3 pathway; these include herbaceous plants and trees while C_4 plants include most grasses and some sedges. The early domesticates, sorghum and millet, are C_4. As most plants eaten are C_3, the domestication centres and spread of these C_4 plants can be measured in part using stable isotopes.

The other main use of stable isotopes in dietary reconstruction of early human/ hominin diet is based on carbon and nitrogen isotope analysis. This is used as an indicator of the main sources of dietary protein and is focused largely on terrestrial animal and marine products (Richards and Trinkaus 2009); it does not distinguish direct consumption of plants, but rather the consumption of the animals which ate the plants. It is therefore of limited use in detecting the consumption of plant materials. The proportions of ^{13}C in marine environments differ from that in terrestrial environments due to the use of other photosynthetic pathways in marine plants. This results in a different fractionation of ^{13}C in plants from marine environments at approximately -20‰. As fish, shellfish, and other marine foodstuffs eat these plants, the proportion of ^{13}C in their tissues is different to terrestrial animals (around -16‰) and this is passed on to the human populations that consume these marine-based products (average in humans eating a marine-based diet is -12‰).

Nitrogen has two stable isotopes, ^{14}N and ^{15}N. As with the carbon isotopes, the quantity of ^{15}N is different in terrestrial and marine plants; the more complex ecosystems in marine environments can lead to higher concentrations of ^{15}N (Richards and Trinkaus 2009). This means that by measuring both the ratios of $^{13}C/^{12}C$ and $^{15}N/^{14}N$ it is possible to determine whether the individual had a diet that was based predominantly on marine or terrestrial resources. Both carbon and nitrogen are needed to differentiate between the types of food consumed. For example, if only carbon is used not only do the values for C_3 plants and C_3-based terrestrial fauna overlap to some degree, but they also overlap with freshwater fish. Likewise, if only nitrogen is used C_3 and C_4 plants overlap to some degree, and C_3-based terrestrial fauna, reef shell fish and tropical reef fish overlap significantly with each other (Tykot 2006). However, when both carbon stable isotope and nitrogen stable isotope values are plotted two dimensionally for a sample, all these, and other food sources, plot with a good degree of separation (Tykot 2006).

Dietary interpretations based on C/N stable isotope analysis are based on comparative data obtained from contemporary animals, including both herbivores and carnivores. ^{15}N concentrates by an average of 3‰ at each trophic level (Schoeninger 2014). Stable isotope values can differ according to climate and the values for Pleistocene Europe are not necessarily the same as Holocene Europe (Richards and Trinkaus 2009). In Holocene Europe, plants have the lowest nitrogen values (0–2‰); this is followed by herbivores (3–7‰), then carnivores (6–12‰). The reason Neanderthals are considered to be highly carnivorous is that their trophic levels are consistently equal to or above that of contemporary or near contemporary carnivores such as hyenas (Bocherens 2009) and for this reason they are thought to have included very few plants in their diet (Richards and Schmitz 2008; Richards and Trinkaus 2009).

Samples of bone or tooth enamel are the most common sources of stable isotopic data used in reconstruction of prehistoric human diet. Collagen is a very robust molecule which is the main protein in bone and dentine and the stable isotopes are extracted from this. Tooth enamel and dentine, which are formed early in life and largely not replaced, provide information on diet in childhood, while bone continues to be replaced during much of adult life and therefore can provide information on a broader timescale in the life of an individual; bone collagen reflects the most recent 10–25 years of a person's life (Hedges *et al.* 2009). Bone and tooth enamel carbonate (apatite) has also been used in addition to collagen for some studies (Harrison and Kaztenburg 2003). Bone collagen is produced in higher amounts from protein in the diet, while bone carbonate and tooth enamel carbonate is produced from all of the components of the diet. This means that bone carbonate reflects the whole diet rather than just the protein part of it (Harrison and Katzenburg 2003). While this has enabled more detailed information on plants to be obtained (e.g. Harrison and Katzenburg 2003), the apatite is subject to diagenetic processes that can cause alterations in the carbonate over time and this can complicate its use (Lee-Thorp 2008). Samples are extracted for analysis using chemicals which demineralise the bone or enamel. The samples are then analysed using IRMS (isotope ratio mass spectrometry) for bulk isotopic determination or GC-C-IRMS (gas chromatography-combustion-isotope ratio mass spectrometry) for compound specific isotopic determination, which analyses the relative quantities of the different isotopes which are then matched against comparative data.

Stable isotope analysis is based on interpretation of results in the light of environmental, geographical and archaeological contexts and it provides information on broad scale dietary trends. For example, a mid-continental sample would not be considered for marine-based dietary products even if the measurement could indicate this. Rather, the results would be examined in relation to the local conditions.

Organic residue analysis of pot sherds is a relatively new method that is being used today to detect the remains of food that has become embedded in the pot matrix. Multiple analyses, including bulk isotope measurements of carbon and nitrogen, lipid biomarker analysis, and compound-specific carbon isotope determinations are used though this approach is focused largely at present on detection of lipids (Heron and Evershed 1993; Heron and Craig 2015; Heron *et al.* 2015).

The use of carbon isotopes in reconstruction of early hominin diets

The analysis of differences in carbon isotopes has played a leading role in reconstruction of the different characteristics of Pliocene hominin diet. The aim of this section is to provide a brief summary of the key points. Further reading, using the key references provided here, is recommended for a fuller understanding of the subject and its complexities.

The data are extracted from tooth enamel which can survive well in these extended time periods. The estimates of diet for this time period in Africa are based on averages that comprise +2‰ for a 'pure-C_4 grazing' and −12‰ for a 'pure-C_3 browsing' diet (van der Merwe *et al.* 2008). Carbon isotope data now exists for 11 hominin taxa that are known to have existed in Africa between 4.4 and 1.3 million years ago (Sponheimer *et al.* 2013); this includes taxa that eventually became extinct and those that are ancestral to the later Pleistocene hominins and ultimately *Homo* (Sponheimer *et al.* 2013). As with C/N stable isotopes, the results are interpretive therefore meat from animals that ate the plant types cannot be excluded. Small animal items such as termites (Sponheimer *et al.* 2005) and invertebrates, reptiles, birds and small mammals (Peters and Vogel 2005) may also have been eaten, but in general it is thought that predominant food resources are likely to have been the plants themselves. This interpretation is also based on dental morphology as well as evidence from modern African human hunter gatherer groups (Lee-Thorp *et al.* 2012).

Early hominin species are found principally in South and East Africa, and there appears to be a general increase in the use of C_4 resources beginning sometime between 4 and 3 million years ago (Cerling *et al.* 2013). Before 4 million years ago, hominin species ate a principally C_3 -based diet. By around 3.5 million years ago, C_4 plants were becoming more common in the diet of the hominin species. *Ardipithecus ramidus* (~4.4 million years) and *Australopithecus anamensis* (~4 million years) have relatively low $\delta^{13}C$ markers suggesting a focus on C_3 plants. *Paranthropus robustus* and *Australopithecus afarensis* (~3.5 million years) ate a predominantly C_3 diet but also included some C_4 plants (Lee-Thorp *et al.* 2012). *Paranthropus boisei* (1.9 million years) had a diet that was heavily focused on C_4 or possibly even CAM plants (Schoeninger 2014) with an average $\delta^{13}C$ value of -1.3±0.9 ‰ (Cerling *et al.* 2011). Its diet could have been based principally on grasses or sedge storage organs; neither of these foods is easy to understand as both have challenges linked to their collection or retrieval (Lee-Thorp 2011). However, three specimens of *Australopithecus bahrelghazali* from Chad, dating to 3.5 million years ago, were recently found to have a $\delta^{13}C$ range similar to *Paranthropus boisei*, with a predominant focus on C_4 plants, and a range of between −0.8 to −4.4‰ (Lee-Thorp *et al.* 2012). This is 1.5 million years earlier than had previously been understood for the exploitation of C_4 plants. Based on the carbon isotope ratios, it appears that some species such as *P. boisei* were specialised feeders, while others including early *Homo* consumed a mixed C_3/C_4 diet. *A.afarensis* and *Kenyanthropus platyops* dating to around ~2.0 million years ago had diets that appear to have been broader, as they overlapped with both the C_3 and C_4 plant eaters, though *A. sediba*, ~2 million years ago had a diet focused strongly on C_3 plant resources (Sponheimer *et al.* 2013; Schoeninger 2012).

These results have enabled a range of questions to be addressed such as the difference between niche specialists and generalists, how environmental change affected dietary trends, and investigation into the food within these broad groups, that might have been eaten (Lee Thorp *et al.* 2010; Sponheimer *et al.* 2013). Though the sources of C_4 in the diet remains uncertain, they are thought to include sedge roots, grasses insects and small mammals (Schoeninger 2014). The focus on C_4 plants by *A. bahrelghazali* suggests a capacity to survive on savannah resources. This implies that by 3.5 million years ago some hominin species had succeeded in becoming entirely separate from the ancestral forest environments and their C_3 resources (Lee-Thorp *et al.* 2012). Other studies using different methods including laser ablation profiles of strontium/calcium, barium/calcium, and strontium (Balter *et al.* 2012) have challenged some of these interpretations, which suggests that further developments in the understanding of the diet of Pliocene hominins, is likely.

The stable isotope studies of the Plio-Pleistocene diet provide perhaps the best insight into early hominin plant use, in the broadest sense, that has been conducted to the present. It is particularly useful in that the C_3 and C_4 plants come largely from different environments, and these studies of Pliocene hominin diet have therefore contributed directly to the timing of expansion into different environments, notably savannah.

Thermal desorption-gas chromatography-mass spectrometry (TD-GC-MS) and pyrolysis-gas chromatography-mass spectrometry (Py-GC-MS)

TD-GC-MS and Py-GC-MS are used extensively in organic geochemistry to characterise petroleum source rocks in petroleum exploration as they can give clues to the oil potential of the source rocks under study. They can, however, also be used in a range of analytical contexts to characterise a wide range of organic materials, particularly when combined sequentially. More specifically, TD-GC-MS uses moderate heat energies to volatilise – 'thermally extract' – a wide range of organic chemical compounds without breaking carbon-carbon bonds; these vapourised compounds are then swept onto a specially coated coiled column held within the oven of a gas chromatograph (GC) where the temperature can be programmed to heat the column at a rate which maximises the separation of the volatilised organic components extracted on the basis of size (smaller molecules move through the column more quickly) and relative chemical affinity for the column coating. As each organic compound emerges from the GC column they are partially ionised by a stream of electrons in a mass spectrometer source where the resulting ions, characteristic of each chemical compound, are separated according to mass to give a 'chemical fingerprint' of each organic molecule present. The main advantages of thermal desorption over more conventional GC-MS are that very small sample sizes are needed, and little sample preparation is necessary, reducing the potential for contamination and sample loss.

Py-GC-MS uses high heat energies to break the molecular bonds in large molecules resulting in smaller volatile compounds which can then be analysed by gas chromatography combined with mass spectrometry, allowing the separation and characterisation of monomeric units of the original polymers. Since the chemical bonds in each molecule break in a predictable way, depending on their energy, a pattern of small molecules results and these can be used to identify the original macro-molecular material. Biopolymers are common not only in organic geochemical contexts, but also in many archaeological materials, making this a highly appropriate technique for such research, particularly in cases where there are no convenient alternatives.

In studies of dental calculus for example, the mineral component of the dental calculus traps organic compounds associated with it (e.g., from food consumed) and provides a protective environment for the more resilient organic molecules present. This permits identification of surviving biological marker compounds ('biomarkers') characteristic of the original inputs, such as the foods consumed. Recent studies have identified a range of (unidentified) plant foods including the presence of lignin suggestive of 'green 'plants as well as two medicinal plants, yarrow and camomile, in Neanderthal individuals from the site of El Sidrón, Spain (Hardy *et al.* 2012), evidence of ingestion of the plant, *Cyperus rotundus*, throughout the sequence, at the multi-period site of Al Khiday, in Sudan (Buckley *et al.* 2014), and evidence of ingestion of the essential plant-based polyunsaturated fatty linoleic and linolenic acids by Lower Palaeolithic individuals from Qesem Cave, Israel (Hardy *et al.* 2015). A recent GC-MS study of faecal biomarkers from Neanderthal coprolites demonstrated that Neanderthals ate meat and plants (Sistiaga *et al.* 2014).

Discussion

TD-GC-MS and Py-GC-MS when used together on samples of dental calculus, offer a powerful way for direct access to highly specific details on some plant foods ingested in the past. This method has already provided new insight into several plant species that were used at different times in the past and it can add a significant new perspective on human use of plants from contexts in which actual remains may not survive through the detection of specific compounds that identify plants or specific compounds in plants.

Stable isotope analysis is a very well recognised method for prehistoric dietary reconstruction. However, stable isotope ratios taken from collagen samples, largely reflect the protein component of the diet, while the carbohydrates ingested are invisible (Richards and Hedges 1999) and have not therefore been a focus of stable isotope dietary reconstruction (Barras 2012). Apart from the identification of variation in C_3 and C_4 plants in the diet of early hominins, stable isotope analysis has yet to make a contribution to a better understanding of plants more broadly in human pre-agrarian diet. The lack of stable isotope evidence for plants has led to a tendency for broad dietary interpretations to be focused exclusively on protein consumption in a wide range of contexts which may offer only a partial perspective on pre-agrarian diet. Humans need to eat plants (see Introduction and Chapters 1 and 2) and as much as 25% of the diet can be based on plants without being visible in the isotopic signal (Jones 2009). The dominance of C/N stable isotopes in pre-agrarian dietary reconstruction has therefore contributed to a perception of diet that may be physiologically unsustainable in the long term. Recently, direct evidence for plant consumption among Neanderthals

(Henry *et al.* 2011, Hardy *et al.* 2012), which appears to contradict the stable isotope evidence for an exclusively carnivore diet, suggests a need for stable isotope methods to be developed that can focus more specifically on detection of the plants themselves, rather than the animals that consumed the plants, which were then eaten by humans or hominins.

The need to understand the input of other macronutrients such as carbohydrates and lipids to the carbon in collagen is likely to generate more research as the need to unscramble this is crucial for the integrity of the dietary interpretations that are made. This may ultimately contribute to a clearer picture of carbohydrate input into the diet; for example, based principally on variation in nitrogen isotope values, Lelli *et al.* (2012) have detected variable carbohydrate consumption in an early agricultural context in Italy. Organic residue analysis on early pottery, though currently focused largely on detection of lipids (Heron and Craig 2015; Heron *et al.* 2015) may eventually develop towards detection of identifiable plant components. A combined study using GC-MS and stable isotopes, detected specific, identifiable plant oils in Egyptian pot residues (Copley *et al.* 2005).

One way forward for stable isotope research may be in the use of complex statistics to tease out the nuances of the C/N readings (Froehle *et al.* 2012). It is hoped that if this is developed, it could eventually contribute to a broader perspective on the dietary use of plants in pre-agrarian diet that will enhance and complement the many other techniques that are currently used. New methods are always in development, and the use of other stable isotopes, as highlighted in Balter *et al.* (2012), may also offer alternative ways to detect plant signals extracted from bone or tooth enamel.

References

Balter, V., Braga, J., Télouk, P. and Thackeray, J. F. 2012. Evidence for dietary change but not landscape use in South African early hominins. *Nature* 489(7417), 558–60

Barras, C. 2012. Neanderthal dental tartar reveals evidence of medicine. *New Scientist*, 18 July 2012. Available at: http://www.newscientist.com/ article/dn22075-neanderthal-dental-tartar-revealsevidence- of-medicine.html (accessed 26 April 2013).

Bocherens, H. 2009. Neanderthal dietary habits: review of the isotopic evidence. In Hublin, J. J. and Richards, M. P. (eds), *The Evolution of Hominin Diets: integrating approaches to the study of palaeolithic subsistence*, 241–50. Dordrecht: Springer

Brown, R. and Brown, K. 2011. *Biomolecular Archaeology: an introduction.* Chichester: Wiley Blackwell

Buckley, S., Usai, D., Jakob, T., Radini, A. and Hardy, K. 2014. Dental calculus reveals evidence for food, medicine, cooking and plant processing in prehistoric Central Sudan. *PLOS ONE* 9(7), e100808. doi:10.1371/journal.pone.0100808

Cerling, T. E., Mbua, E., Kirera, F. M., Manthi, F. K., Grine, F. E., Leakey, M. G., Sponheimer, M. and Uno, K. T. 2011. Diet of Paranthropus boisei in the early Pleistocene of East Africa. *Proceedings of the National Academy of Sciences (USA)* 108(23), 9337–41

Cerling, T. E., Manthi, F. K., Mbua, E. N., Leakey, L. N., Leakey, M. G., Leakey, R. E., Brown, F. H., Grine, F. E., Hart, J. A., Kaleme, P., Roche, H., Uno, K. T. and Wood, B. A. 2013. Stable isotope-based diet reconstructions of Turkana Basin hominins. *Proceedings of the National Academy of Sciences (USA)* 110(26), 10501–6

Copley, M. S., Bland, H. A., Rose, P., Horton, M. and Evershed, R. P. 2005. Gas chromatographic, mass spectrometric and stable carbon isotopic investigations of organic residues of plant oils and animal fats employed as illuminants in archaeological lamps from Egypt. *Analyst*, 130(6), 860–871

Craig, O. E., Bondioli, L., Fattore, L., Higham, T. and Hedges, R. (2013). Evaluating marine diets through radiocarbon dating and stable isotope analysis of victims of the AD79 eruption of Vesuvius. *American journal of physical anthropology*, 152(3), 345–352

D'Ortenzio, L., Brickley, M., Schwarcz, H. and Prowse, T. 2015. You are not what you eat during physiological stress: Isotopic evaluation of human hair.*American Journal of Physical Anthropology*

Froehle, A. W., Kellner, C. M. and Schoeninger, M. J. 2012. Multivariate carbon and nitrogen stable isotope model for the reconstruction of prehistoric human diet. *American Journal of Physical Anthropology* 147(3), 352–69

Hardy, K., Buckley, S., Collins, M. J., Estalrrich, A., Brothwell, D., Copeland, L., García-Tabernero, A., García-Vargas, S., de la Rasilla, M., Lalueza-Fox, C., Huguet, R., Bastir, M., Santamaría, D., Madella, M., Fernández Cortés, A. and Rosas, A. 2012. Neanderthal medics? Evidence for food, cooking and medicinal plants entrapped in dental calculus. *Naturwissenschaften* 99(8)617–626. DOI: 10.1007/s00114-012-0942-0

Hardy, K., Radini, A., Buckley, S., Sarig, R., Copeland, L., Gopher, A. and Barkai, R. 2015. Dental calculus reveals inhaled environmental contamination and ingestion of essential plant-based nutrients at Lower Palaeolithic Qesem Cave Israel. Quaternary International DOI 10.1016/j.quaint.2015.04.033

Harrison, R. G. and Katzenberg, M. A. 2003. Paleodiet studies using stable carbon isotopes from bone apatite and collagen: examples from Southern Ontario and San Nicolas Island, California. *Journal of Anthropological Archaeology*, 22(3), 227–44

Hedges, R., Rush, E. and Aalbersberg, W. 2009. Correspondence between human diet, body composition and stable isotopic composition of hair and breath in Fijian villages. Isotopes Environmental Health Studies 45:1–17

Henry, A. G., Brooks, A. S. and Piperno, D. R. 2011. Microfossils in calculus demonstrate consumption of plants and cooked foods in Neanderthal diets (Shanidar III, Iraq; Spy I and II, Belgium). *Proceedings of the National Academy of Sciences (USA)*108, 486–91

Heron, C. and Craig, O. E. 2015. Aquatic Resources in Foodcrusts: Identification and Implication. *Radiocarbon*, 57(4), 707–719

Heron, C., Craig, O. E., Luquin, A., Steele, V. J., Thompson, A. and Piličiauskas, G. 2015. Cooking fish and drinking milk? Patterns in pottery use in the southeastern Baltic, 3300–2400 cal BC. *Journal of Archaeological Science* 63, 33–43)

Heron, C. and Evershed, R. P. 1993. The analysis of organic residues and the study of pottery use. *Archaeological Method and Theory*, 247–284

Jones, M., 2009. Moving north: archaeobotanical evidence for plant diet in Middle and Upper Paleolithic Europe. In Hublin, J. J. and Richards, M. P. (eds), *The Evolution of Hominin Diets: Integrating Approaches to the Study of Palaeolithic Subsistence*, 171–80. Dordrecht: Springer

Lee-Thorp, J. A. 2008. On isotopes and old bones, *Archaeometry* 50, 925–50

Lee-Thorp, J. A., Sponheimer, M., Passey, B. H., De Ruiter, D. and Cerling, T. E. 2010. Stable isotopes in fossil hominin tooth enamel suggest a fundamental dietary shift in the Pliocene. *Philosophical Transactions of the Royal Society B* 365, 3389–96

Lee-Thorp, J. A. 2011. The demise of 'Nutcracker man'. *Proceedings of the National Academy of Sciences (USA)* 108 (23), 9319–20

Lee-Thorp, J. A., Likius, A., Mackaye, T. S., Vignaud, P., Sponheimer, M. and Brunet, M. 2012. Isotopic evidence for an early shift to C$_4$ resources by Pliocene hominids in Chad, *Proceedings of the National Academy of Sciences (USA)* 109 (50), 20369–72

Lelli, R., Allen, R., Biondi, G., Calattini, M., Barbaro, C. C., Gorgoglione, M. A. and Craig, O. E. (2012). Examining dietary variability of the earliest farmers of South-Eastern Italy. *American Journal of Physical Anthropology*, 149(3), 380–390

O'Connell, T. C. and Hedges, R. E. 1999. Investigations into the effect of diet on modern human hair isotopic values. *American Journal of Physical Anthropology*, 108(4), 409–425

Peters, C. R. and Vogel, J. C. 2005. Africa's wild C_4 plant foods and possible early hominid diets. *Journal of Human Evolution* 48(3), 219–36

Richards, M. P. and Hedges, R. E. 1999. Stable isotope evidence for similarities in the types of marine foods used by Late Mesolithic humans at sites along the Atlantic coast of Europe. *Journal of Archaeological Science* 26(6), 717–22

Richards, M. P. and Schmitz, R. W. 2008. Isotope evidence for the diet of the Neanderthal type specimen *Antiquity* 82, 553–9

Richards, M. P. and Trinkaus, E. 2009. Isotopic evidence for the diets of European Neanderthals and early modern humans. *Proceedings of the National Academy of Sciences (USA)* 106, 16034–9

Schoeninger, M. J. 2010. Diet reconstruction and ecology using stable isotope ratios. In Larson, C. S. (ed.), *A Companion to Biological Anthropology*, 445–64. Chichester Wiley Blackwell

Schoeninger, M. J. 2012. Palaeoanthropology: the ancestral dinner table. *Nature* 487(7405), 42–3

Schoeninger, M. J. 2014. Stable isotope analyses and the evolution of human diets. *Annual Review of Anthropology* 43(1), 413–430

Sistiaga, A., Mallol, C., Galván, B. and Summons, R. E. 2014. The Neanderthal Meal: A New Perspective Using Faecal Biomarkers. PLoS ONE 9(6): e101045. doi:10.1371/journal.pone.0101045

Sponheimer, M., Alemseged, Z., Cerling, T., Grine, F. E., Kimbel, W. H., Leakey, M. G., Lee-Thorp, J. A., Manthi, F. K., Reed, K. E. and Wood, B. A. 2013. Isotopic evidence of early hominin diets. *Proceedings of the National Academy of Sciences (USA)* 10 (26), 10513–18

Sponheimer, M., Lee-Thorp, J., de Ruiter, D., Codron, D., Codron, J., Baugh, A. T. and Thackeray, F. 2005. Hominins, sedges, and termites: new carbon isotope data from the Sterkfontein valley and Kruger National Park. *Journal of Human Evolution* 48(3), 301–12

Tykot, R. H. 2006. Isotope analyses and the histories of maize. In Staller, J. E., Tykot, R. H. and Benz, B. F. (eds), *Histories of Maize: multidisciplinary approaches to the prehistory, linguistics, biogeography, domestication, and evolution of maize*, 131–42. New York: Academic Press

van der Merwe, N. J., Masao, F. T. and Bamford, M. K. 2008. Isotopic evidence for contrasting diets of early hominins *Homo habilis* and *Australopithecus boisei* of Tanzania. *South African Journal of Science* 104, 153–5

PART 3
PROVIDING A CONTEXT:
ETHNOGRAPHY, ETHNOBOTANY,
ETHNOHISTORY, ETHNOARCHAEOLOGY

Ethnobotany is the study and collection of information on traditional, applied knowledge of plants from living populations. Ethnoarchaeology also studies living people, though here the focus is on addressing questions that can be used to help in understanding aspects of the past, while ethnohistory uses historical documents with the same objectives. The 'ethnos' provide a broad context for the interpretation of archaeological remains, and for the reconstruction of prehistoric subsistence strategies and diet, technology and seasonality. Traditional knowledge and use of plants is a huge subject as non-industrialised people depend and depended heavily on the plants they could obtain in their local environment for food, medicine and raw materials.

This part has five chapters which cover various climatic zones; from circumpolar to tropical regions. The first chapter (13) describes the wooden and wicker technology of fish traps and fishing equipment from the Late Mesolithic site of Zamostje 2 in Russia and offers a perspective on the enduring legacy of Mesolithic fishing technology across the Russian European Plain and the eastern Baltic region, in some places, into recent historical periods. The second chapter (14) examines current use of plants by Aboriginal people to explore both palaeoenvironment and past people's lives (prehistoric subsistence patterns and technology in particular) in northern Australia. It also demonstrates the importance of macro-botanical remains in the interpretation of Australian archaeological sites, and highlights the potential for future archaeobotanical research as well as covering some of the issues related to methodology. The third chapter (15) combines ethnoarchaeology with ethnohistory to reconstruct social perspectives on variable plant use among recent hunter-gatherers of Tierra del Fuego. The authors of this chapter argue that even though the plant biodiversity in this circumpolar region is low and availability of plant food resources is highly seasonal, still plants contributed a significant amount of nutrients to the Fuegian diet. Chapter 16 focuses on the use of plants by a group of modern Hadza hunter-gatherers and explores the evolutionary implications of this. The Hadza diet includes a wide variety of plant foods (with a high contribution of tubers), game, birds, larvae, and honey. Even though the diet is well balanced, plant food resources make up the majority of the diet. Many plant species are also used as medicine and as raw materials. The final chapter (Chapter 17) is a summary of the results of an ethnobotanical study from

Senegal, specifically on use of wild plant resources in a modern farming population, with a particular focus on fallback or emergency foods. This final chapter examines the many different ways plants are used in the diet and offers a context for the way plants need to be evaluated in relation to their qualities and potential value in the diet.

13. Prehistoric fish traps and fishing structures from Zamostje 2, Russian European Plain: archaeological and ethnographical contexts

Ignacio Clemente Conte, Vladimir M. Lozovski, Ermengol Gassiot Ballbè, Andrey N. Mazurkevich, and Olga V. Lozovskaya

Fish traps and nets documented ethnographically on the Russian European Plain and in Siberia are described and compared with similar archaeological finds from the north and north-west of Russia and the Baltic States of Latvia and Lithuania. We analyse the features they have in common in order to interpret the fishing structures from Zamostje 2. The site has yielded occupation levels covering the transition from the Mesolithic to the Early and Middle Neolithic. A comparative study of the fish remains from Zamostje 2 and modern species in the Volga and Oka river basins suggests that the greatest diversity of species (11) is found in the earliest Mesolithic level, with only six species recorded for the Late Mesolithic and Early Neolithic. The number of individuals caught also varies over time, with a steady increase in the capture of cyprinids, which is particularly striking in the last phases of the occupation. Wooden artefacts connected with fishing demonstrate a high level of technological expertise while the finds of several possible paddle fragments and one whole paddle confirm the use of boats.

Thirteen sites have been identified along a 2 km stretch of the Dubna River. The sites lie on what is today, the banks of the river which runs through a broad alluvial plain of lacustrine origin. The sites are located in the village of Zamostje, 50 km north of the city of Sergiev Posad and 110 km from Moscow (Fig. 13.1). Archaeological excavations at Zamostje 2 (1989–1991; 1995–2000; 2010–2013) extended along a 160 m² stretch of the west river bank and 190 m² in an area that is today, underwater (Lozovski *et al.* 2013). Altogether the excavations covered an area of 40 × 4–5 m.

The archaeological occupations are located in a sedimentary sequence that is directly associated with the lakeside context and evidence for fluctuations in water levels. The two oldest layers, 'LL' and 'UL', are Late Mesolithic and date to 7900–7600 uncal BP and 7400–7100 uncal BP (7000–6600 cal BC and 6400–6000 cal BC). They are separated by a level of peat that has a significantly lower density of archaeological remains. Above this, two layers, 'EN' (Upper Volga culture = UVC) and 'MN' (Lialovo culture), have been interpreted as Neolithic due to the presence of pottery. The first of these occupation horizons (UVC) coincides with a warm period; it is dated to 6800–6200 uncal BP (5800–5200 cal BC) and is considered to be Early Neolithic. The second occupation horizon dates to approximately 5700 uncal BP (*c.* 4900–4300 cal BC) and took place

Fig. 13.1. Location map of Zamostje 2

during a period of lakeside soil formation amid receding water levels (Lozovskaya *et al.* 2013; Lozovski *et al.* 2014). There are few archaeological remains dating to the Late Neolithic, possibly because they were destroyed by the large transgression which occurred towards the end of the 4th millennium BC, which signaled the final abandonment of the site (Lozovski 1996; 1997).

Following analysis of the fishbone assemblage (Radu, Desse-Berset 2012; 2013; Lozovski *et al.* 2013), a comparative study was conducted between the modern fish species from the Volga and Oka rivers with those exploited during the Mesolithic and Neolithic. The most numerous range of species identified occurred during the Early Mesolithic (11 species); in the Late Mesolithic and Early Neolithic only six species were identified, while seven species were identified towards the end of the site's occupation. The following species were present in all of the excavated layers: *Carassius carassius* (crucian carp), *Exos lucius* (pike), *Leuciscus idus* (ide), *Perca fluviatilis* (perch), and *Rutilus rutilus (roach)*. The quantity of fish caught also changes to a certain extent over time; most notable is the increasing evidence for cyprinids (carps), in particular towards the end of the occupation. This may be related to an increased use of nets (either with or without boats), which would permit greater numbers of fish to be caught in less time. Analysis of bone size suggests that pike were caught in the year following their birth, and mostly in the spring, which also coincides with the spawning season.

There is a small amount of evidence for human action on the bones, though little evidence for contact with fire; this occurs on only around 0.5% of all identified bones. Most of the evidence for human activity was found on pike bones, but even here there are no cut marks. There was proportionally less evidence for certain bones, in particular vertebrates of large individuals. It is possible that these were preserved whole for consumption at a later date. Tools for skinning and cleaning fish were identified, which suggest activities related to preservation, either through drying or smoking. They include knives manufactured from elk rib bones; these all have use wear traces that are identical to those on the experimental tools (Clemente *et al.* 2002; Clemente and Gyria 2003; García and Clemente 2011) and which also correspond with ethnographic tools documented across Siberia (Fig. 13.2).

We have also investigated the possibility that specific tool types or activities were linked to fish species and we have begun a series of experiments to determine whether the harpoons and hooks found at Zamostje 2 were in fact used or not (Clemente *et al.* 2013; Gyria *et al.* 2013; Maigrot *et al.* 2014a; 2014b). Other material linked to fishing found at Zamostje 2 include fish-screens, fish-fences, wooden fish traps, and needles, which were used for mending nets, several small fibre knots which may have formed part of a net, and pine bark floats. Several possible fragments of paddle, as well as one whole paddle were also recovered, confirming the use of the river, presumably for transport as well as fishing (Figs 13.2–13.6).

In this chapter, we will describe the artefacts used in construction of the stationary fish traps at Zamostje 2. First, however, we will contextualize these structures by comparing them to other archaeological traps found in the region. We will describe these, and we will also outline the ethnographic record for fish traps in the European part of Russia and Siberia (Lozovski 1997; 1999).

Ethnographic sources for the use of fences and enclosed traps for fishing in the Russian Plain and Siberia

Ethnographically, the use of fish traps, manufactured both in stone and a wide variety of wood, is common in the Russian Plain and Siberia. Zelenin writes that fish traps were

> made from thin twigs or nets. They were constructed of two cones, a smaller one which is inserted into a larger one though there are many variations both on materials used to construct them and on the structures themselves. In some cases they were made from willow slats … In the north of Russia, the fishermen would cover the inside surface with bait … but the most common way was to construct a fence in the river with an 'opening' in the centre where they would place the bait … (Zelenin, 1991, 103)

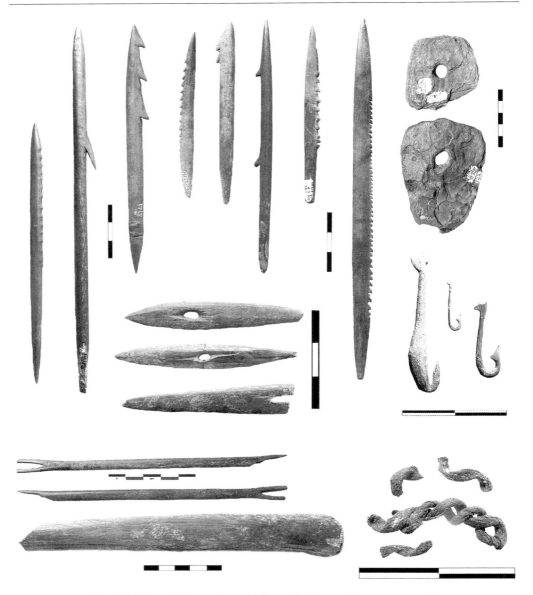

Fig. 13.2. Tools from Zamostje 2 for fishing and processing fish

The exterior opening was either square or oval. Ethnographic accounts suggest that thin branches (willow, elm, etc) were used in their construction. In some cases, slats were used, for example among the Korjaki of Siberia (Tolstov 1956). At the end of the 19th and early 20th centuries, the Comí (a Finno-Ugrian population), used pieces of pine 0.4–0.7 m long (Kondakov 1983) to construct their fish traps. All the ethnographic

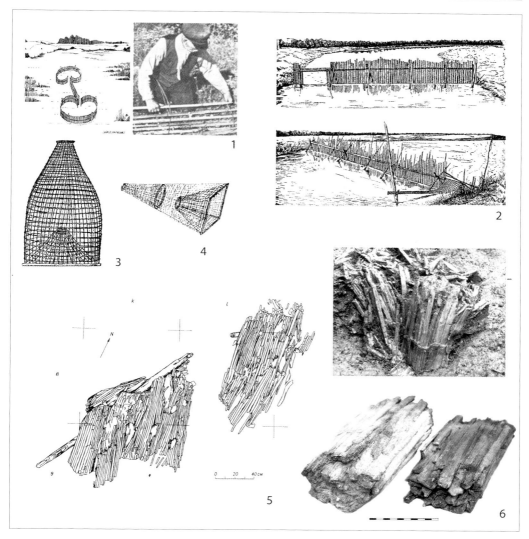

Fig. 13.3. 1) Ethnographic evidence for the use and placement of fences and traps in Latvia (Berzins 2008); 2–3) fish traps from Siberia (Tolstov 1956); 4) fish trap from northern Siberia (Zelenin 1991); 5) remains of fish traps from the Terminal Neolithic at Abora (Loze 1979); 6) archaeological remains related to fishing from the Neolithic site of Sarnate (Berzins 2008)

sources that describe the manufacture and use of fish traps, explain that it is necessary to build a fence, (or dam) using logs, posts and stakes that stretches across the width of the river, or, in the case of lakes, near the shore (Fig. 13.3).

Zelenin (1991) distinguishes two structures that were used across Siberia; 1) light structures made using thin reeds or sticks; 2) larger, stronger structures that were made using heavy posts. These larger structures could be left in place and were able to survive the spring melts. The Tartars from western Siberia built small fences across rivers using thick branches tied together with string or cords. At one edge, near the bank, they would leave a hole into which they inserted the traps (Tolstov 1956). Stakes were hammered into the river bed to hold the fences and traps in place. Fish traps, made from small branches, were inserted into the openings that had been left in the fences for the purpose. These fish traps were used even in winter (Tolstov 1956). Even more common were structures made out of two parallel lines of stakes, hammered into the river bed and filled with sticks and small branches. These were widely used by the Comí, the Uragiri and the Korjaki people (Tolstov 1956) and also in the Ivanovo region, Central Russia (Krajnov 1991).

In some cases, these fences or dams were built using different materials. The Tuva used stones to fill the gaps between the lines of posts (Tolstov 1956). In other cases, different fishing structures were constructed. For example the Ob Ugrians built barriers made solely out of small sticks and twigs; as the fish tried to swim through, they became caught amongst the branches (Vasil'ev 1962). These were exceptions though; most Siberian people, including western Siberian Komes, Shrotses, Selkups, Kets, Evenks, Tartars, Nentser, Ukagirs, Koryals etc., used the more usual method of damming the river and placing fish traps in the openings they had left in the dams. The placing of these fish traps was not always the same; sometimes they were in the middle of the structure (for example among the Russian and Koriak populations) in other cases, they were placed near the shore (Khants, West Siberian Tartars). However, all the traps were secured by stakes hammered into the river beds or were held in place by being weighted down with stones.

Archaeological fish traps from Russia and the Baltic regions

Latvia

Zvidze is a Neolithic site on the edge of Lake Lubans, in east Latvia. This site is located on the edge of a rolling plain that gradually drops down to the lowlands on the western side of the ancient Lake Lubans. The site is dated to between 7565–7170 cal BP (5876–4810 cal BC) which corresponds with the Early Neolithic. The remains of the fish fences were found at a depth of 1.2 m in a stratum of decomposing organic sediment; these had been dug into an Earlier Neolithic layer (post-hole with charcoal) to 60 cm depth. The fish fence was 6 m from the ancient edge of the marshy Lake Zvidze. Loze (1986) describes the remains of the fish traps and fence found here and suggests that the structure stretched along tens of metres following the shoreline of the lake; he suggests it was used in spring during the pike spawning season. The fence was constructed using vertical and horizontal posts. Fish traps, made from pine and willow, were attached to both sides of the fence. The frame of the fence comprised three rows of posts vertically hammered into the lake bed. The diameter of the posts was 5–8 cm.

Loze believes that the two lines of posts were the walls of the fence and that the third line supported the completed construction. Between the lines of posts, four layers of large posts, made from elm and willow were inserted and slats and bundles of twigs were placed at a slight angle. The remains of nets and hooks were also found in amongst the posts. The preservation of the fish traps was uneven and appeared in the archaeological deposits as concentrations of wooden slats. Four traps were identified though only one was in good condition. This fish trap was in two pieces, possibly representing the area around the wide opening of the trap. The large fragment consisted of a main body, made of slats with an inner funnel made of small branches, 1–2 cm in diameter.

The fish trap was formed with ten rows of slats and an inner row constructed of sticks. The outer layer of slats overlay the perpendicular line of sticks. The sticks were 1–2 cm in diameter, and with a rectangular or square cross-section. A large number of pike vertebrae and mandibles were found between the rows of sticks. In one trap, the bones of up to 16 pike were recorded (Loze 1986).

Another site is Sarnate, which is situated in western Latvia, around 2.5 km from the Baltic Sea. This site was first excavated between 1938 and 1940 by E. Shturms. After a temporary interruption during WWII, the History Museum of Lithuania and the Institute of History of the Lithuanian Academy of Sciences, took over the investigations which were directed by L. V. Vankina in 1949 and between 1953–1959. Forty dwellings were identified, 25 of these classified as *sarnatski'* type and dated to the 3rd millennium BC. A large number of well-crafted wooden items were recovered including construction materials and six fish traps. The traps were made of slats 1–2 cm wide and 0.5 cm deep. Only two fish traps were well preserved. One of these, with 2.5 m long slats, was located near one of the houses. The slats were attached together with fibres in three different places. The fish traps were conical; the narrow end containing straps while the opening was located at the wider end. The second trap, which was found beside the first one, was broken in two. On the larger fragment, a thin band of fibres joined the slats together in two places. The trap appears to have had two walls – an outer wall with larger slats and an inner layer made out of thinner sticks – these walls were bound together with cords (Vankina 1970).

Fish traps have also been found in Abora, a site in Eastern Latvia which was excavated by Loze between 1964–1971. This site was located in the right bank of the Abora River in the lowlands of Lubans. Radiocarbon dates place the site in the Late Neolithic (4510–4010 cal BP and 4530–3980 cal BP)[2] (3870±70 BP, LE-671; 3860±100 BP, LE-749) (2561–2141 cal BC; 2581–2017 cal BC) (Loze 1979). The remains of three fish traps were found in peaty, decaying, organic deposits at a depth of 1 m. Three bundles of sticks were tied together. The first measured 0.80–1.15 m. long, and consisted of three layers of slats and a post, 1.88 m long, which was attached. This trap was found beside fragments of other traps. Each trap was held together by cords which were between 13–18 cm apart. The remains of the other two traps were found to the east of the first trap and measured 1.5 × 0.75 m and 0.6 × 0.5 m respectively. The traps also had various attached slats and pieces of wood. It has been suggested that these three traps formed part of a large structure and that, at some point after abandonment of the site they fell off (Loze 1979).

Lithuania

The fishing-related finds from Šventoji (Shvjantoji) were mainly found in sites excavated by R. K. Rimantene near Shvjantoji in the Baltic coast where Neolithic and Early Bronze Age sites were located on the shores of a lake or sea lagoon (Rimantene 1979; 1992). The remains of the fish traps were divided into two groups: 1) Fish traps made of 2–3 cm pine slats tied with cord. These objects were made with a narrow end which closed with a lid; 2) fish traps made from nets attached to a wooden frame. The first type is found in the Early Neolithic of Shvjantoji 2[b], in this case the narrow end of the trap had a lid that was 10 cm in diameter. In Shvjantoji 1[a] (Late Neolithic– Early Bronze Age) a fence had been constructed within the site and the remains of three fish traps were found near this. One trap measured 80 cm long and 20 cm wide; it was built in two layers within which were flat pebbles, 5cm in diameter that may have acted as a means for closing the ends of the basket. The other two objects were not well preserved but appeared similar to this (Rimantene 1991). A further fishing fence was identified in a stream that connected the lake to the sea at Shvjantoji 9 (Late Neolithic–Early Bronze Age). This fence measured about 40 m, and was constructed in the form of an arc. It began on the north shore of the old channel and stretched into the centre of the stream. Several holes were located in the centre of the fence, which is where the baskets would have been placed. The fence was constructed out of two rows of stakes driven into the stream bed. The stakes were near each other and were more abundant in the middle than at the edge of the structure. The area between the rows of stakes was filled with large pieces of bark. The stakes, which were 8–10 cm in diameter, measured 1.20–1.47 m in length; however, the tops of the stakes were all at exactly the same level. To improve the secure fixing of the stakes, these were attached with cross-pieces held in place by planks which measured 2.60 m. Near the central openings of the system, several bundles of sticks were found with some transverse small branches attached, which may also have been part of the fish traps. Two examples of a second type of trap were recovered from the site of Shvjantoji 2B. The frames that held the mouth of the traps were constructed using small branches that were 1.5 cm thick. The net was made of fibres tied with a 'fisherman's knot' to the wooden frame (Rimantene 1980).

The north part of European Russia

Vis 2 is situated in the Vichegda river basin in a peat bog in the lowlands surrounding Lake Sindor. The remains of a fishing fence constructed in an ancient lake was found here (Burov 1968). The fence consisted of posts hammered vertically into the lake bed and supported by transverse planks which were fixed between the upright large posts. The materials from the excavation site suggest the site is Early Iron Age and dates to the 2nd milennium BC (Burov 1969).

Marmugino is located on the Ug River near the city of Veliky Ust'ug. Here, Burov excavated sites in two peat bogs where the remains of fish traps were located in deposits of clay and decomposed organic matter, at a depth of 2.8–3.5 m beneath the surface. Both constructions were made with slats measuring 2.20 m long, arranged in three rows, and joined together (Burov 1969). The radiocarbon dates for this site are 4510±50 uncal BP(Le-703), 4700±60 uncal BP(Le-711) (Sementsov *et al.* 1969; Burov, 1988).

Fixed fishing structures have also been recorded in the north-west part of the Russian Plain. Serteya I was discovered by A. M. Miklyaev on the banks of the Serteika river near Smolensk in the Velizh region. Since 2010, underwater archaeological excavations, led by A. Mazurkevich and E. Dolbunova, have investigated 50 m of the river bed and three accumulations of materials have been identified, one of which was *in situ*. One fishing structure has been dated to the Middle Neolithic through its association with pottery types, and by radiocarbon dating of some of the wooden stakes. Several wooden slats with a rectangular cross section, had been hammered vertically into the river bed and were attached together by fibre cords to create a fence. One end of the slats was pointed while the other ends had been broken and around 70 cm of the length had survived. The fence crossed a stream that, at the time, joined two lakes. Near the western end of the fence, ten net-sinkers and two large stones (*c.* 20 cm wide) were piled together; string and cord fragments were found amongst them. The net-sinkers consisted of pebbles, *c.* 7 × 3 cm in size, wrapped in birch bark. A fragment of fibre net is preserved on one of these pebbles. One metre downstream from here is the remains of another structure, though it is less well preserved. Weights and slat fragments were located in other places along the river, and are thought to have originally been part of the fishing structures.

Other similar Early Neolithic structures have been located nearby, at Rudnya Serteïskaya and Seteya X and XIV, though none was well preserved (Dolukhanov *et al.* 1989; Mazurkevich and Miklyaev 1998). The most striking find comes from Rudnya Serteïskaya where two rows of stakes hammered into the river bed held a concentration of hazel branches and pine sticks that were arranged in two levels forming a 90 ° angle. No lithic or ceramic finds were found here.

Central Russia

Lugovskoje is a site located in a peat bog in the upper region of the Svijaga River, near the city of Uljanovsk. Here, a line of posts aligned in a north-west to south-east direction and made from elm and alder were found in a peaty sediment at a depth of 2.45–2.50 m. Most posts were in a horizontal position, though one post leaned slightly (Burov 1972).

Sakhtysh is located in the Upper Volga region, near Lake Sakhtish and has been in excavation for many years by D. A. Krajnov. A large dwelling, dating to between the Middle and Late Neolithic was found with the entrance facing the river. Fish traps with a Late Neolithic date were located both near and and inside the house. One fish trap was found just inside the entrance to the house, at a depth of 1.30–1.40 m. The trap was rectangular and constructed out of wooden slats. It measured 1.50 m long, with a 0.50 m opening and it narrowed to 0.20 m at its thin end. The second trap, which was partially destroyed, was found in the north-west corner of the house, also near the entrance. The surviving fragments measured 0.70 m long and 0.4 m wide. The trap was constructed out of rectangular wooden slats that were still attached together. The third object, which was 2 m long, was recovered in the south-west corner of the house; this may have been a trap, though the excavator thought it could also have been a mat (Krajnov 1991).

Podzorovo is a Late Neolithic site, situated in the upper stretches of the Voronezh River and was excavated by V. O. Levenok. A bundle of pine slats were found, 2.2 m below the surface. The slats measured 2.50 × 3.00 m long, 1 cm wide and 0.5 cm deep and were separated from each other by 0.5–1 cm. Levenkov (1969) suggested that this construction was very similar to modern fish traps.

Finally, wicker (*Salix* sp.) fish traps dated to *c.* 9620 BP (uncal) (9230–8810 cal BC; 9224–8824 cal BC) were also found at Stanovoje 4 and Sakhtysh 2a; these traps are similar to those found at Zamostje 2 (Zhilin 2004: figs 27 and 28).

General features

All of the prehistoric fish traps described, have some common features. First, all the traps were made with wooden slats from coniferous trees (mainly pine). Though at times the traps recovered had disintegrated, it was possible in certain cases to distinguish a conical shape and at Zvidze, Šventoji and Sakhtysh the slats were attached either by ropes and/or crossed posts. Only the three objects recovered from Šventoji had lids attached to the trap openings. One remarkable find was at Abora where a series of additional stakes were found driven in to the ground beneath the actual structure. These can only be explained as having served to attach the trap to the fence. The location of the traps, which is always connected to lake or river sediments and the presence of large numbers of fish remains inside some of the traps, such as occured at Zvidze, provides direct evidence for their use in fishing. This is particulary clear in the cases of Zvidze and Šventoji 9 where the structure consisted of fences with interrelated traps. In each case, the gap between 2–3 rows of vertical posts had been filled with horizontal tree trunks, branches and bark. It is very likely that the structures from Vis 1 and Marmugino were the same. The fish traps inside and near the houses at Sarnate and Sakhtysh 1 may have been incomplete and /or in a state of repair.

The fish traps from Zamostje 2

Two traps were recovered during the 1989 excavation season (Lozovski 1996; 1997). Of these, only one was in a good state of preservation. It was almost intact and was only missing the narrow end of the cone. Only parts of the frame and some wooden posts were recovered from the second trap. However, a large fragment of bast fibre was recovered on one part of the frame; this measured 2–2.5 cm in diameter and 1 m long, The bast fibre tethering was 2 mm thick × 3 cm wide and *c.* 12 cm long (Figs 13.4, 13.6 and 13.7).

The upper part of the best preserved trap had a series of planks and branches which presumably were originally part of a fence that had been linked to the trap. The trap was constructed out of pine slats with a rectangular cross section and measured 1 × 0.5 cm

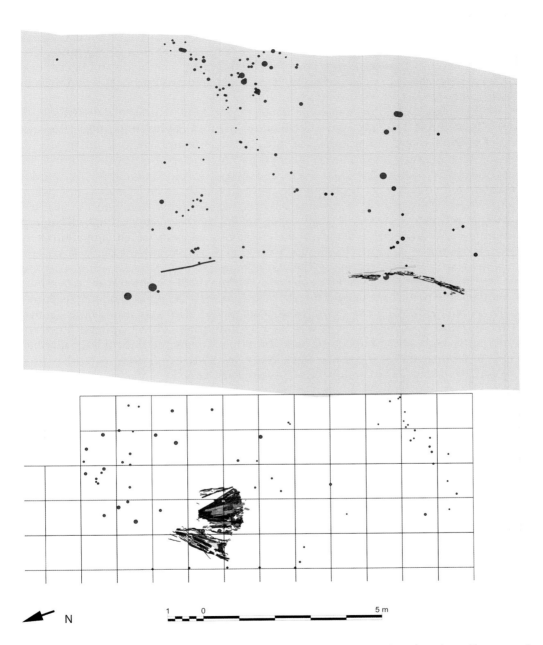

Fig. 13.4. Southern part of the survey and excavation area at Zamostje 2: location of hammered in stakes, and fish trap

around 1.5–2 m long. The slats were linked together every 28–30 cm by a fine fibre cord made of common reed (*Phragmites australis*; Lozovskaya *et al.* 2012; Lozovski *et al.* 2013). Three rows of this cord were found preserved beneath the planks and branches. A slat fragment, with a piece of cord attached was recovered in 2010 and has been radiocarbon dated to 6450±45 BP, 5482–5337 cal BC (CNA 1081). This places it in the Early Neolithic in the region of the Russian Plain.

Usually, the slats lie on south facing slope, with an angle of 25°–30°. The distance between the uppermost part of the slats and the lowest part of the ground surface is 40 cm; this suggests they were lying on the ancient river bank which inclined upwards towards the shore. The best preserved trap was excavated in a block in 2011 and was taken to the Hermitage laboratories for study and conservation, this is still in process. During the excavation of summer 2011, another conical shaped object was found to the west of the other traps. This object was *c.* 2 m long. The orientation of this structure is consistent with the others suggesting that they most likely came from the same fishing complex; however, this trap is thinner and had no surviving fibre attachments. It is located at a higher level, most likely again reflecting the river bank which appears to drop around 10–15cm to 50 cm in the southernmost part, which is similar to the previous examples. An almost complete paddle with an asymmetric blade was found the upper part of this structure. (Figs 13.4, 13.6 and 13.7). The radiocarbon dates for this trap are similar to the previous example, 6539±43 BP, 5613–5588 cal BC (CNA1341) and the paddle, which was found inside the trap was dated to 6676±47 BP, 5636–5602 cal BC (CNA1342). All three traps were found in the same sedimentary layer. This layer comprises a yellowish clayey sediment within a grey–brown sapropel (putrefied mud), with some patches containing small branches, aquatic snail shells and small fish skeletons all in anatomical position (Lozovski *et al.* 2013).

Several structures, also made with pine slats with associated shells and posts, were located 7–8 m further south-south-east, beyond the artificial dam constructed to carry out the excavation and below the current Dubna River bed sediments. One of these structures measures *c.* 4 m and is made from several layers of semi-parallel slats and one long straight slat positioned in a north–south direction. The slats are thin and narrow and some are slightly twisted. Most have rectangular cross sections though some are square or trapezoidal. A second object was also located underwater though somewhat deeper. So far it has not been fully excavated though over 2 m that has been exposed, here, six lines of stitching, 25 cm apart, using bulrush (*Scirpus lacustris*) fibres has been identified (Lozovskaya *et al.* 2012; Lozovski *et al.* 2013). The slats are laid flat and in a parallel line. To the north, the slats are bent and enter the sediment almost vertically, suggesting a relatively large structure with a width of 40 cm. One of these traps has been radiocarbon dated to 6216–6047 cal BC (7248±35 BP; Lozovski *et al.* 2013, 74–5). The remainder of the underwater objects that have been recorded lie north-west to south-east, which is almost perpendicular to the current flow of the river. Although there is no doubt that they are manufactured, it is unclear whether they are traps or something else, possibly mats, or parts of a fence. However they are all manufactured from long wooden slats of 2–4 m in length.

Stakes

Analysis of the stakes that are hammered in along the full 154 m² length of the excavated area on the left bank of the current (artificial) course of the Dubna river, shows that the vast majority of these were located in the southern part of the site, in the same area as the fish traps. Two main groups of stakes have been identified; one, comprising 29 stakes, was found to the north of the traps; this forms a structure facing north-west to south-east. The second group, which is linked to an area of large branches and long trunks, consists of 18–20 stakes. This group, which was found in 1990, lay 4–5 m south of the fish traps and also lay in a north-west to south-east facing direction. Other than these, only 22 individual posts with no apparent link between them, have been recorded altogether on this site.

Additional stakes have been found hammered into the river bed. Approximately 70 m² east of the traps, 116 stakes have been recorded, each with a diameter of 4–10 cm, though only 12 are over 8 cm. Half of these artefacts have removal scars linked to the point at the distal end, while there are three pieces that comprise only distal pointed fragments; 12 of the stakes retain their bark. The lack of evidence for removal scars in the rest of the stakes, may be because they were dug in deeper than the level studied. As the river was opened into a canal in the 1980s, some stakes were destroyed together with archaeological levels, and their correct archaeological context cannot be reconstructed. However, the trees used to make the stakes has been identified; these comprise: alder (*Alnus* sp.), European bird cherry (*Padus racemosa* Gilib), elm (*Ulmus* sp.), poplar (*Populus* sp.), hornbeam (*Carpinus betulus* L.), maple (*Acer* sp.), pine (*Pinus sylvestris* L.), birch (*Betula*), ash (*Fraxinus* sp.) and willow (*Salix*) (Lozovski *et al.* 2013).

The stakes can be divided into three groups. The southernmost group, which consists of 21 posts, stretches for 7 m and crosses the whole river. It is possible that it is connected to the line of stakes that was discovered in 1990. A structure made of long wooden posts was also found, lying at right angles to this group. Three stakes were found lying next to the posts, and one actually crosses this line of stakes transversally.

The second group is around 4–5 m further north, and apart from two stakes that were found close to the dam that was constructed for the excavation of 2010–11, the remainder are found in an area measuring 4 × 4 m. Seven stakes were found in the north part of this area, and aligned in a straight line. The stakes were all of a similar diameter and were made of wood from the same trees.

Finally, the third group was located on the east bank of the river. This comprises a dense cluster of stakes, forming a right angle. This large structure was discovered below the ground on the right bank of the river and was not fully excavated, therefore a full analysis will need to wait for future excavations. Most stakes are small (5–6 cm in diameter) though three are 10 cm in diameter and four are 8–8.5 cm. One of the stakes (no. 34) was radiocarbon dated to 5580±40 uncal BP (Beta-283034), 4490–4340 cal BC, which places the structure in the Middle Neolithic period.

Interpretation and discussion

The recent excavations at Zamostje 2 have permitted the confirmation and verification of the structures as moveable fish traps in association with other structures including static fish traps first identified by V. M. Lozovski in the 1989 excavation. The construction of this fishing equipment suggests a deep technological understanding of wood and woodworking. The people of Zamostje 2 had an elaborate array of tools including axes, adzes, wedges, gouges, and chisels made on flint and other rocks, but mainly on raw materials from animals (Clemente *et al.* 2002; Clemente and Lozovskaya 2011). The use of fish traps represents a breakthrough in fishing technology, since the fish are captured effortlessly while the population can engage in other productive or leisure activities (Morales 2010). Fifteen radiocarbon dates were obtained from materials excavated in 2010 and 2011. These include both fish traps, and mats made from pine, recovered both on land and in the current Dubna river, as well as stakes and the oar found in one of the fish traps (Figs 13.6 and 13.7).

The dates cover a wide period between 6327–6320 cal BC and 4447–4418 cal BC but they fall into three groups. The oldest dates for these structures (6327–6023 cal BC), is linked to a stake near the fish trap that contains the oar, and the traps and other underwater structures manufactured using pine posts, all found in the southern part

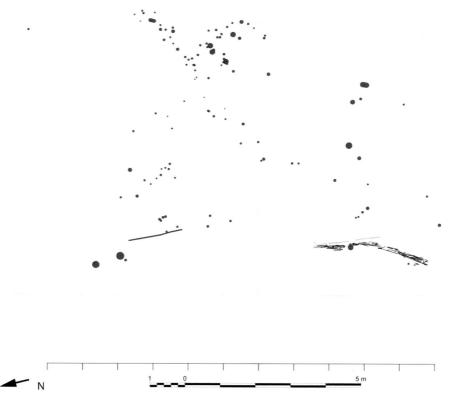

Fig. 13.5. *Detail of posts and fishing structures located in the modern Dubna River*

Fig. 13.6. Detail of the fish traps recovered on land in the most recent excavation season. The fish trap at the most easterly point was identified by V. M. Lozovski in 1989

Fig. 13.7. Photograph of the two fish traps recovered on the terrestrial excavations at Zamostje 2. Image taken from the south profile. A paddle is preserved inside the easternmost trap

of the excavation (Figs 13.4 and 13.5). On the other hand, the time period 5636–5462 cal BC contains the oar, the two traps excavated in 1989/2010 and 2011 and two stakes that were near the last trap (Figs 13.4 and 13.6). Finally, stakes nos 12, 35, 38, 88, 101, and 115, all date to between 4992 and 4418 cal BC. This final group was located in the central and northernmost part of the area excavated underwater. It remains to be determined whether all the stakes and posts recovered, are in fact related to fish traps structures.

Even so, we suggest that the fishing structures are likely to have required modifications and replacements over the course of 2000 years. Fluctuations in water level may have meant that the static fish structures had to be moved. An example of this is found in one of the oldest traps (N114), dated to 6070–6023 cal BC, which is crossed by a stake (no. 115), that has a date of 4992–4898 cal BC. It is clear that part of the site could correspond to the lakeshore where a complex structured composed of walls made of branches, planks and bark, and at least three traps, made from thin, long pine slats. Future work further south of the excavated area, in the actual area of the river could provide new records of these traps and perhaps older chronologies.

The work undertaken so far at Zamostje 2 has enabled us to obtain sufficient data to suggest the site was home to a stable population. There is evidence for hunting of animals, especially moose and beavers in summer and winter, migratory birds normally present here only in autumn, as well as evidence for wild fruits present in late summer and autumn. The evidence for fishing suggests it was carried out in spring-summer (Chaix 2003; Radu and Desse-Berset 2012; 2013). Fishing is likely to have had an important role in these societies from the end of the Mesolithic and through the Neolithic in Russian European Plain. Fish are a predictable resource, and with the technological expertise that was evidently available, fish could be captured in large quantities to be stored and consumed later though this requires methods for drying or smoking the fish for storage. Zamostje 2 should be considered as a site that still has great potential for the study of fishing practices in past societies. The potential comprises not only the presence and excellent preservation of fish traps and timber structures related to the capture of fish, but also for the study of other wooden artefacts and the outstanding bone tool collection, which provides insights into the technological capabilities of working with hard materials. Together, the assemblage comprises a wide array of items in bone, teeth and antler of animals all of which are related to woodworking. Taken together this site offers a perspective of impressive expertise in woodworking, and in particular in the use of pine, in the manufacture of all these fishing structures (Lozovskaya 2011a).

Acknowledgements

This Project was funded by the Spanish Ministry of Science and Innovation, I+D+I funding scheme (Project number HAR2008-04461/HIST) *Recursos olvidados en el estudio de grupos prehistóricos: el caso de la pesca en sociedades meso-neolíticas de la llanura rusa. (Forgotten Resources in Prehistory: Fishing in the Mesolithic-Neolithic of the Russian Plain).*

Notes

1 TA-862: 6.535±60 BP y TA-1746; 6350±60 BP. Radiocarbon ages were calibrated using the curve INTCAL13 (Reimer *et al.* 2013).
2 LE-671: 3870±70 BP and LE-749: 3860±100 BP

References

Berzins, V. 2008. *Sarnate: Living by Coastal Lake during the East Baltic Neolithic.* Oulu University Press.

Burov, G. M. 1969. О поисках древних деревянных вещей и рыболовных сооружений в старичных торфяниках равнинных рек. КСИА -Краткие сообщения института археологии. (In search of ancient fishing structures and other wooden items in the old river bogs of the (Russian) plains. *KSIA, Short Notes from the Institute of Archaeology* 117, 130–4

Burov, G. M. 1972. Археологические памятники Верхней Свияги. Ульяновск. Приволжское книжное издательство. (*The Archaeological Monuments of Upper Svijagi*). Ulianovsk: Приволжское книжное издательство

Burov, G. M. 1988. Запорный лов рыбы в эпоху неолита в Восточной Европе. *Советская археология*, No. 3. Москва.С. 145–160. (Fishing with traps in the East European Neolithic. *Sovietskaya Arjeologia*, No. 3, pp. 145–160. Moscow)

Chaix, L. 2003. A short note on the Mesolithic Fauna from Zamostje 2 (Russia). In Larson, L., Kindgren, H., Knutsson, K., Loeffler, D. and Åkerlund, A. (eds), *Mesolithic on the Move*, 645–48. Oxford: Oxbow Books

Clemente Conte, I., Gyria, E. Y., Lozovskaya, O. V. and Lozovski, V. M. 2002. Análisis de instrumentos en costilla de alce, mandíbulas de castor y caparazón de tortuga de Zamostje 2 (Rusia). In Clemente, I., Gibaja, J. F. and Risch, R. (eds), *Análisis Funcional: su aplicación al estudio de sociedades prehistóricas*, 187–96. Oxford: British Archaeological Report S1073

Clemente Conte, I. and Gyria, E. Y. 2003. Анализ орудий из ребер лося со стоянки Замостье 2 (7 слой, раскопки 1996–97 гг.) Археологические Вести. – СПб (Analysis of tools made on elk ribs from Zamostje (level 7, from the 1996–1997 excavations.) *Arheologicheskie Vesti* 10, 47–59

Clemente Conte, I., Maigrot, Y., Gyria, E., Lozovskaya, O. and Lozovski, V. 2013. Aperos para pesca e instrumentos para el procesado de pescado en Zamostje 2 (Rusia): una experimentación para reconocer los rastros de uso. In Palomo, A., Piqué, R. and Terradas, X. (eds), *Experimentación en arqueología. Estudio y difusión del pasado*, 63–71. Girona: Sèrie Monogràfica del MAC-Girona 25(1)

Dolukhanov, P. M., Gey, N. A., Miklyaev, A. M. and Mazurkevich, A. N. 1989. Rudnya – Serteya, a stratified dwelling – site in the upper Duna basin (a multidisciplinary research project). *Fennoscandia arhaeologica* 6, 23–7

García Diaz, V. and Clemente Conte, I. 2011. Procesando pescado: reproducción de las huellas de uso en cuchillos de sílex experimentales. In Morgado, A., Baena, J. and García, D. (eds), *La investigación experimental aplicada a la arqueología*, 153–9. Granada: Universidad de Granada, Universidad Autónoma de Madrid, Asociación Experimenta. Málaga

Gyria, E. Y., Maigrot, Y., Clemente Conte, I., Lozovski, V. and Lozovskaya, O. 2013. From bone fishhooks to fishing techniques. The example of Zamostje 2 (Mesolithic and Neolithic of the central Russian plain). In Lozovski, V. M., Lozovskaya, O. V. and Clemente Conte, I. (eds), *Zamostje 2. Lake Settlement of the Mesolithic and Neolithic Fisherman in Upper Volga Region*, 111–19. St Petersburg: Russian Academy of Science, Institute for the History of Material Culture

Kondakov, N. D. 1983. Коми: Охотники и рыболовы во второй половине XIX- начале XX в. Наука, Москва. (*Comi, Hunter-Fishers from the Second Half of the 19th and Early 20th Centuries*). Moscow: Nauka

Krajnov, D. A. 1991.Рыболовство у неолитических племен Верхнего Поволжья. Рыболовство и морской промысел в эпоху мезолита-раннего металла в лесной и лесостепной зоне Восточной Европы. – Л.: Наука – С.129–152. (*Fishing in the Neolithic tribes of the Upper Volga Region . Fisheries and marine exploitation from the Mesolithic period to the beginning of the Metal Ages in the forest and steppes of Eastern Europe*), 129–52. Leningrad: Nauka

Levenkov, V. P. 1969. Новые раскопки стоянки Подзорово. Краткие сообщения института археологии (КСИА*). New Excavations at the Site of Podzorovo.*) Вып.117. *Москва. с. 84–90,* 84–90. Moscow: KSIA-Kratkie soobschenia Instituta Arkheologii 117

Loze, I. A. 1979. Поздний неолит и ранняя бронза Лубанской низины. Рига, Зинатне. (*The Late Neolithic, and Early Bronze Ages in the Lubans Plain.*) Riga: Szinatne

Loze, I. A. 1986. Рыболовный закол эпохи неолита на поселении Звидзе. Краткие сообщения Института археологии (КСИА). Вып.185. с. 78–81. Москва. (*A Neolithic fish fence at the site of Zvidze*) , 78–81. Moscow: KSIA-Kratkie soobschenia Instituta Arkheologii 185

Lozovskaya, O. V. 2011. Деревянные изделия позднего мезолита – раннего неолита лесной зоны европейской части России: комплексные исследования (по материалам стоянки Замостье 2). Диссертация на соискание ученой степени канд.ист.наук. СПб. ИИМК РАН. (*Wooden implements from the Late Mesolithic-Early Neolithic of European Russia: Detailed investigations of the items from Zamostje 2.* PhD Dissertation, Institute of the History of Material Culture, Russian Academy of Science, St Petersburg

Lozovskaya, O. V., Lozovski, V. M., Mazurkevich, A. N., Clemente Conte, I. and Gassiot, E. 2012. Деревянные конструкции на стоянке каменного века Замостье 2: новые данные. КСИА, №227, М.: Языки славянской культуры, с. 250–259. (Wooden objects from the prehistoric site of Zamostje 2; new information). *Languages of the Slavic Cultures, KSIA 227*, 250–9

Lozovskaya, O. V., Lozovski, V. M. and Mazurkevich, A. N. 2013. Палеоландшафт рубежа мезолита-неолита на стоянке Замостье 2 (бассейн Верхней Волги)/VIII всероссийское совещание по изучению четвертичного периода: «Фундаментальные проблемы квартера, итоги изучения и основные направления дальнейших исследований». Сб. статей (г. Ростов-на-Дону, 10–15 июня 2013 г.). – Ростов-на-Дону: Издательство ЮНЦ РАН, 2013. С.379–381 (Paleolandscape dynamics during the Mesolithic-Neolithic transition at the site Zamostje 2 (Volga-Oka region). *VIII All-Russian Conference on Quaternary Research: Fundamental problems of the Quaternary, results and main trends of future studies. Collection of papers Rostov-on-Don, 10–15 June 2013*), 379–81. Rostov-on-Don: SSC RAS Publishers

Lozovski, V. M. 1996. *Zamostje 2. Les derniers chasseurs-pêcheurs préhistoriques de la Plaine Russe.* Guides archéologiques du « Malgré-Tout ». Treignes: Editions de CEDARC

Lozovski, V. M. 1997. Рыболовные сооружения на стоянке Замостье-2 в контексте археологических и этнографических данных. Древности Залесского края. Материалы к международной конференции «Каменный век европейских равнин: объекты из органических материалов и структура поселений как отражение человеческой культуры», 1–5 июля 1997, Сергиев Посад, сс.52–65. (Fishing structures at the site Zamostje 2 in the context of archaeological and ethnographic data). In Manushina, T. N., Masson, V. M., Vishnevski, V. I., Lozovski, V. M. and Lozovskaya, O. V. (eds), *Zalessky Region Antiquities. Proceedings of the International Conference 'Stone Age of the European Plains: Objects Made from Organic Materials and Settlement Structure as a Reflection of Human Culture'*, 52–65. Possad: Sergiev

Lozovski, V. M. 1999. Archaeological and ethnographic data for fishing structures. Coles, B., Coles, J. and Jorgensen, M. S. (eds), *Bog Bodies, Sacred Sites and Wetland Archaeology*, 139–45. Exeter: WARP Occasional Papepr 12

Lozovski, V., Lozovskaya, O. and Clemente Conte, I. (eds), 2013. *Zamostje 2. Lake Settlement of the Mesolithic and Neolithic Fisherman in the Upper Volga Region.* St Petersburg: IHMC RAS

Lozovski, V., Lozovskaya, O., Mazurkevich, A., Hook, D. and Kolosova, M. 2014. Late Mesolithic–Early Neolithic human adaptation to environmental changes at an ancient lake shore: The multi-layer Zamostje 2 site, Dubna River floodplain, Central Russia. In Bronnikova, M. and Panin, A. (eds), *Human Dimensions of Palaeoenvironmental Change: Geomorphic Processes and Geoarchaeology*, 146–61 *Quaternary International* 324 special edition

Maigrot, Y., Clemente Conte, I., Gyria, E. Y., Lozovskaya, O. V. and Lozovski, V. M. 2014a. Des hameçons en os aux techniques de pêche: le cas de Zamostje 2 (Mésolithique et Néolithique de la plaine centrale de Russie). In Arbogast, R. M. and Greffier-Richard, A. (eds), *Entre archéologie et écologie, une Préhistoire de tous les milieux. Mélanges offerts à Pierre Pétrequin*, 243–53. Besançon: Presses universitaires de Franche-Comté, Annales Littéraires de l'Université de Franche-Comté 928; série «Environnement, sociétés et archéologie » 18

Maigrot, Y., Clemente Conte, I., Gyria, E. Y., Lozovskaya, O. V. and Lozovski, V. M. 2014b. From bone fishhooks to fishing techniques: the example of Zamostje 2 (Mesolithic and Neolithic of the Central Russian Plain). In Manur, M. E., Lima, M. A. and Maigrot, Y. (eds), *Traceology Today: methodological issues in the Old World and the Americas*, 55–60. Oxford: British Archaeological Report S2643

Mazurkevich, A. N. and Miklyaev, A. M. 1998. О раннем неолите междуречья Ловати и Западной Двины. Археологический сборник Государственного Эрмитажа. Вып. 33. 7–32. (*The Early Neolithic between the Lovat and Western Dvina Rivers*), 7–32. St. Petersburg: Archaeological Collection of the State Hermitage 33.

Morales Muñiz, A. 2010. Inferences about prehistoric fishing gear based on archaeological fish assemblages. In Bekker-Nielsen, T. and Bernal Casasola, D. (eds), *Ancient Nets and Fishing Gear*, 25–53. Cádiz:. Universidad de Cádiz, Monographs of the Sagena Project 2

Pedersen, L. 1995. 7000 years of fishing: stationary fishing structures in the Mesolithic and afterwards. In Fischer, A. (ed.), *Man and Sea in the Mesolithic. Coastal Settlement Above and Below Present Sea Level*, 75–86. Oxford: Oxbow Monograph 53

Radu, V. and Desse-Berset, N. 2012. The fish from Zamostje and its importance for the last hunter-gatherers of the Russian Plain (Mesolithic–Neolithic). In Lefèvre, C. (ed.), *Proceedings of the General Session of the 11th International Council for Archaeozoology Conference (Paris, 23–28 August 2010)* , 147–161. Oxford: British Archaeological Report S2354

Radu, V. and Desse-Berset, N. 2013. Fish and fishing at the site of Zamostje 2. In Lozovski *et al.* (eds) 2013, 194–213.

Reimer, P. J., Bard, E., Bayliss, A., Beck, J. W., Blackwell, P. G., Bronk Ramsey, C., Grootes, P. M., Guilderson, T. P., Haflidason, H., Hajdas, I., HattŽ, C., Heaton, T. J., Hoffmann, D. L., Hogg, A. G., Hughen, K. A., Kaiser, K. F., Kromer, B., Manning, S. W., Niu, M., Reimer, R. W., Richards, D. A., Scott, E. M., Southon, J. R., Staff, R. A., Turney, C. S. M. and van der Plicht, J. 2013. IntCal13 and Marine13 Radiocarbon Age Calibration Curves 0–50,000 Years cal BP, *Radiocarbon* 55(4), 1869–87

Rimantene, R. K. 1979. *Sventoji. Narvos kulturos gyvenvietes.* (*Sventoji: Settlements of the Narva Culture.*) Vilnius: Mokslas

Rimantene, R. K. 1980. *Sventoji. Pamariu kulturos gyvenvietes.* (*Sventoji: Settlements of the Pomeranian Culture.*). Vilnius: Mokslas

Rimantene, R. K. 1991. Озерное рыболовство и морская охота в каменном веке Литвы // Рыболовство и морской помысел в эпоху мезолита - раннего металла. Ленинград. : Наука, 1991. – C.65–86. (Lake fishing and marine hunting in the Stone Age in Lithuania. (ed.), *Fishing and Marine Exploitation in the Mesolithic–Metal Ages*, 65–86. Leningrad: Nauka

Rimantene, R. K. 1992. Neolithic hunter-gatherers at Sventoji in Lithuania. *Antiquity* 66, 367–76

Sementsov, A. A., Romanova, E. N. and Dolukhanov, P. M. 1969. Радиоуглеродные даты лабораторииЛОИА. Советская археология, No. 1, c. 251–261. Москва. (Radiocarbon dates from the LOIA laboratory. *Sovietskaya Arjeologia*, No. 1, pp. 251–261. Moscow)

Tolstov, S. P. (ed.) 1956. Народы Сибири. под ред. С.П.Толстова Серия Народы мира. Изд. АН СССР. Москва. (*The Peoples of Siberia*). Moscow: Academy of Sciences, Peoples of the World Series

Vankina, V. I. 1970. Торфяниковая стоянка Сарнате. Рига. «Зинатне». *The Bog Site of Sarnate.* Riga: Szinatne

Vasil'ev, V. I. 1962. Проблемы происхождения орудий запорного рыболовства обских угров // Труды Института этнографии. Новая серия. Т. 78. Сибирский этнографический сборник. Вып.4. Москва, с. 137–152. (*Problems of the Origin of Stationary Fishery Constructions Among the Ugric Population of the Ob River,* 137–52. Siberian Ethnographical Materials, Contributions of the Institute of Ethnography New Series 78. Moscow: Academy of Sciences

Zelenin, D. K. 1991. Восточнославянская этнография, Москва, "Наука". (*Ethnography of Western Slavs*). Moscow: Nauka

Zhilin, M. G. 2004. Природная среда и хозяйство мезолитического населения центра и северо-запада Восточной Европы. Москва, Наука. (*Environment and Economy of the Mesolithic populations from Central and North West Europe*). Moscow: Nauka

14. Plants and archaeology in Australia

Sally Brockwell, Janelle Stevenson and Anne Clarke

The Aboriginal people of Australia remained hunter-gatherers until the contact period. Therefore all evidence of plant use is pre-agricultural although there are examples of eco-system manipulation. Evidence of plant use in Australia is ancient, going back over 40,000 years. Analysis of plant remains and a variety of methodologies have been used to address a range of key questions in relation to settlement of the continent from the Pleistocene until the Holocene. Results have been intrinsic to a number of well-known themes and debates in Australian archaeology. Given the size of Australia and the long period of time involved, the following synthesis is a brief overview only, and further reading can be found in the reference list. However, some of the key issues and methodologies employed have come together in one study area, Kakadu National Park, and these are presented here in some detail. We are fortunate in Australia that many Aboriginal people still retain detailed knowledge of plants and traditional uses and ethnobotany has been a significant consideration in archaeological interpretation.

Plants can help us interpret the past. They can give us information on diet, tool use, site use, niche exploitation, seasonality, environment, and climate. In Australia, plants and plant remains have been used in archaeological interpretation to address a wide variety of questions, often in an inter-disciplinary way.

Although evidence of plant use in Australian sites does not always survive, it is ancient, dating back to 40,000 years in the Kimberley in Western Australia (McConnell and O'Connor 1997; Frawley and O'Connor 2010). A 1989 edited volume, *Plants in Australian Archaeology* (Beck *et al.* 1989), examined all aspects of plant based archaeological studies in Australia at that time, including taphonomy of plants in archaeological sites (Beck 1989); macroscopic plant remains (Clarke 1989); carbonised plant macrofossils (Denham and Donoghue 1989); pollen (Head 1989); plant residues on stone tools (Hall *et al.* 1989); phytolith analysis (Bowdery 1989); ethnohistory and ecology (Gott 1989); and plant use in a contemporary Aboriginal community and implications for archaeology (Meehan 1989). In the last 20 years, there have been further macro-, micro- and molecular plant fossil studies in the Australasian region, many summarised in a review article by Tim Denham *et al.* (2009a) 'Archaeobotany in Australia and New Guinea: practice, potential and prospects'.

In Australia, plants have been used in archaeological interpretation using the techniques provided by four main disciplines:

ARCHAEOBOTANY is the study of botanical remains in archaeological sites and uses macro-, micro- and more recently molecular botanical remains to interpret the archaeological record.

- *Macroscopic plant remains,* when preserved, are the best primary evidence of plant exploitation. Remains such as charcoal, wood, seeds, fibres, roots and resin, not only provide data on diet, habitat exploitation, material culture, and technology, they can provide more precise indications of seasonality than fauna, as plants flower and fruit at specific times. These studies are under-developed in Australia, with a few well known exceptions (e.g., McConnell and O'Connor 1997; Frawley and O'Connor 2010; Martin 2011; Byrne *et al.* 2013; Dotte *et al.* 2014 and see below).

- *Microbotanical residues,* such as starch, phytoliths, and silica polish on stone tools can aid in the interpretation of plant exploitation, food processing techniques, tool and site use, as well as providing a record of environmental change as background to the archaeology (e.g., Kamminga 1982; Fullagar 1991; Bowdery 1998; Wallis 2001; Clarkson and Wallis 2003; Fullagar *et al.* 2008). A summary and history of phytolith research in Australia is provided by Hart and Wallis (2003; Wallis 2003), and starch research by Torrence and Barton (2006), while an edited volume by Haslam *et al.* (2009) includes recent papers on Australian plant residue research (e.g., Field *et al.* 2009). An important example of archaeobotany contributing to broader archaeological research questions is the study of seed-grinding tools in arid central Australia that elucidated the nature of change in Mid–Late Holocene desert economies (for a summary see Hiscock 2008, 207–9; Smith 2013, 198–202). Smith (1986) argued that an increase of millstone fragments found in desert rockshelter deposits indicated increased grinding of grass seeds to support increased population in the Mid–Late Holocene, although other authors disagree about the timing (e.g., Veth and O'Connor 1996; Fullagar and Field 1997; Gorecki *et al.* 1997; Balme *et al.* 2001). However, the fragmentary nature of up to 40% of samples has made typological identification difficult (Smith 2013, 201). Residue analysis (Fullagar and Field 1997; Gorecki *et al.* 1997; Veth *et al.* 1997; Balme *et al.* 2001; Fullagar *et al.* 2008) has provided promising results to solve this problem although the methodology needs refining (Smith 2013: 201).

- *Molecular archaeobotany* identifies and analyses ancient DNA of plants that can aid in the interpretation of plant domestication, plants within archaeological sites and therefore past environments and Aboriginal subsistence strategies. This technique, which utilises the analysis of bulk sediment is particularly useful for sites where plant preservation is poor, as is often the case in Australian sites (e.g., Dáithí *et al.* 2012; Murdoch University 2011). Denham *et al.* (2009a, 4) summarise other molecular studies currently being undertaken in Australia to identify plant species in archeological contexts.

ETHNOBOTANY is the study of how plants are used by living and historic peoples. With the help of Aboriginal informants, ethnobotanists record local names of plants and their uses and build reference collections that can be used by archaeobotanists (e.g., Levitt 1981; Russell-Smith 1985; Latz 1995; Russell-Smith *et al.* 1997; Head *et al.* 2002). They also utilise ethnohistoric and ethnographic resources and museum collections.

Contact in Australia is comparatively recent (mostly 19th century) and we are fortunate in having a rich ethnographic record and detailed historic accounts of plant use by Aboriginal people, as well as extensive collections of material culture and photographs (e.g., the Spencer and Thomson collections held in Museum Victoria). All this can aid site interpretation even in the absence of plant remains. In 1988, Betty Meehan and Rhys Jones published an edited volume *Archaeology with Ethnography: an Australian perspective* that includes a number of papers, dealing with 'the problem of invisible plants' (Meehan and Jones 1988, iii–iv). A well-known example in Australia is Beth Gott (1989; 1999; 2008) who argued convincingly on the basis of 19th century observation that prehistoric Aboriginal populations in south-eastern Australia relied on the tubers of daisy yam (*Microseris scapigera*) and rhizomes of the wetland reed *Typha* sp. as staples; this type of information is mainly absent from the archaeological record. However recently Sarah Martin (2011) presented evidence from macroscopic charcoal consistent with *Typha* rhizome that supports this theory (cf. also Pardoe and Martin 2011). Ethnobotanical research into the kinds of raw plant materials that are used to make cultural objects can unravel the relationship between form and function of the archaeological tool kit (cf. Kamminga 1988). As Kamminga (1988, 26) points out, woody plants form the basis of most of the Aboriginal tool kit. The nature of the raw material can influence both form and style of artefacts used to modify it, for example the stone adzes 'tulas' used to scrape the hardwoods of the arid region (Kamminga 1988, 28). Ethnobotany can also provide the basis for exploring broader questions, such as the hunter-gatherer versus horticulturist debate and why Aboriginal Australians did not adopt agriculture although they were aware of and to some extent employed horticultural techniques and used fire as a land management tool (cf. Jones 1969; Golson 1971; Harris 1977; Hynes and Chase 1982; Cribb *et al.* 1988; Jones and Meehan 1989; Cribb 1996; Levitus 2005; 2009; Bowman *et al.* 2007; Denham 2007; Petty and Bowman 2007; Denham *et al.* 2009b; Gammage 2011). We are also fortunate that to this day many Aboriginal people maintain their interest in gathering plant foods 'bush tucker' and continue to manufacture traditional arts and crafts. Useful sources of information on this topic include Clarke's (2012) encyclopaedic and lavishly illustrated book on contemporary Aboriginal plant use entitled *Australian Plants as Aboriginal Tools*, Brock's (1988) *Top End Native Plants* that lists plants of northern Australia by community and describes Aboriginal uses, the Australian National Botanic Gardens that maintain a website on Aboriginal plant use (ANBG 2013, http://www.anbg.gov.au/aboriginal-resources/index. html) and Hamby's (2005) beautifully illustrated book on historical and contemporary fibre practice in northern Australia, *Twined Together*.

ETHNOARCHAEOLOGY is the practice of archaeologists recording the subsistence patterns, settlement strategies, and material culture of Indigenous societies in order to elucidate patterns in the archaeological record. In Australia, there are several famous examples that include detailed recording of plant exploitation. Betty Meehan spent a year in central Arnhem Land living with the An-barra people (Meehan 1982; Jones and Meehan 1989); Richard Gould had an on-going project in the Western Desert over some 30 years (Gould 1980; Allen and Holdaway 2011); Jim O'Connell worked with the Alyawarra people of central Australia (O'Connell *et al.* 1983; O'Connell and Hawkes 1984); and

Scott Cane worked in the Western Desert in Western Australia (Cane 1984). Along with ethnobotanical research, these studies have exponentially increased our knowledge of the use of plants in Aboriginal society. As mentioned above, they have mainly emphasised the large role played by plants and their invisibility in the archaeological record. However, they are still immensely valuable as they can aid the interpretation of tool use, site use and seasonality, as well as target species for inclusion in reference collections for interpretation of residues and macrobotanical remains. Gould's (1977) excavations at Puntutjarpa and Smith's (2004) at Puritjarra in central Australia revealed carbonized plant remains that can be compared directly with ethnobotanical data collected by Gould in the same area (Gould 1969; 1977).

- *Experimental archaeology* is a sub-discipline of ethnoarchaeology and focuses on replication, life histories of particular tools, taphonomy of residues and compilation of reference collections. For example, Kamminga (1982, 53–80), Akerman (1998), Fullagar *et al.* (1999) and Berehowyj (2013) conducted extensive experimental work on usewear and residues produced by working plants with a variety of stone raw materials and bone, including fibres, plant food, wood, and bark. As part of her PhD research, Bowdery (1998) compiled pioneering reference collections of phytoliths from the Australian arid zone. Yates *et al.* (2014) conducted experimental work to determine the feasibility of dating ancient plant residues on stone tools that lack a chronological context.

PALAEOECOLOGY informs on past environments and can contextualise archaeological and historical records. It provides data for understanding environmental change particularly between present day ecosystems and pre-European landscapes. Through the analysis of micro-botanical remains (pollen, phytoliths, diatoms, charcoal) found in sedimentary deposits, we can gain insight into the changing floristics of a landscape over long periods of time and explore the relative shifts in biodiversity in association with climate change and human impact. Palaeoecological research also offers a significantly improved understanding of the implications of natural and anthropogenic fires in the context of landscape history through the analysis of high resolution charcoal records. In the Australian context debate still exists around the human transformation of landscapes prior to European arrival. The most noteworthy example is Lynchs Crater in north-east Queensland, an intensively studied sedimentary deposit going back some 130,000 years (Kershaw 1986; 1995; Turney *et al.* 2001; 2004; Rule *et al.* 2012). Significant changes to rainforest elements at around 40,000 years ago in association with fire, have been attributed alternatively to changes in climate and the arrival of people in the area. The most recent evidence and interpretation suggests that people may well have been responsible for these vegetation changes as well as significant impact on megafauna (Rule *et al.* 2012).

Case study: Alligator Rivers region

We will now turn to the case study of the Alligator Rivers region in western Arnhem Land where plant based evidence has been crucial in the interpretation of landscape evolution and the archaeology. The region contains Kakadu National Park and lies 12° south of the Equator in the tropical savannah environment of northern Australia, approximately 200 km east of Darwin (Fig. 14.1). In the north it is bounded by the van Diemen Gulf and in the east by the Arnhem Land plateau. The region comprises a number of landforms including floodplains, lowland plains, sandstone escarpment and the dissected Arnhem Land plateau; the five major rivers are the East, South and West Alligator, Wildman, and upper Katherine Rivers (Fig. 14.1). The climate is characterised by high temperatures and two major seasons, the dry season from May to November and the wet season from December to April when up to 1600 mm of rain falls. This

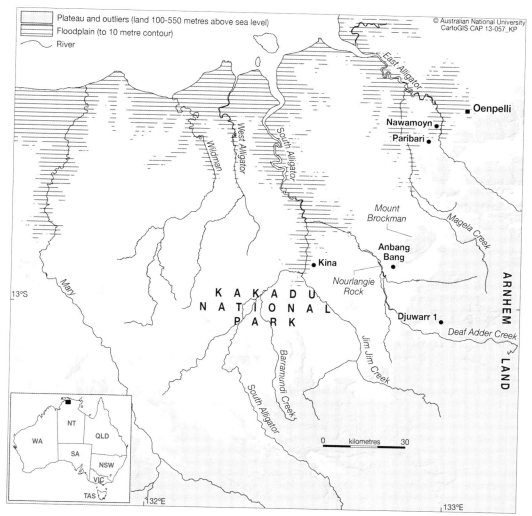

Fig. 14.1. Map of Alligator Rivers region (CartoGIS, Australian National University)

marked seasonality has a major effect on hydrology, and thus on vegetation and animal communities (Department of the Environment 2014).

A number of cross-disciplinary plant focused studies have been undertaken in this region of western Arnhem Land that has one of Australia's most significant archaeological records; from Pleistocene rockshelter sequences containing the oldest edge-ground axes in the world, spectacular rock art more than 10,000 years old, dense concentrations of stone artefacts and earth mounds on the edges of the vast freshwater wetlands of its river systems, and contact sites containing remnants of early European and Chinese occupation. Plant based studies have been key in the interpretation of landscape evolution and the archaeological record. Through the analysis of macro- and microscopic plant remains, rock art motifs, and residue studies, prehistoric subsistence patterns have been elucidated, as well as seasonality and settlement patterns.

Holocene landscape evolution and climate

The coastal plains of northern Australia are part of a young dynamic landscape that has undergone rapid transformation over the last 10,000 years. The evolutionary history of the Alligator Rivers floodplains were unravelled through a combination of palaeoecology and geomorphology (e.g., Chappell and Woodroffe 1985; Hope *et al.* 1985; Woodroffe *et al.* 1985; Clark and Guppy 1988) (Fig. 14.2). This research revealed that initiation of the floodplains began with post-Pleistocene sea level rise flooding down-cut river valleys in the region, with subsequent siltation leading to vast mangrove swamps dominated by *Rhizophora* sp. *c.* 8000–6000 BP. This phase is known colloquially as the 'Big Swamp Phase'. Between 5000 and 2000 BP, further siltation and coastal progradation led to the cutting off of tidal influence, the retreat of the mangroves to the river channels and coast, and a period of transition when a mosaic of estuarine and freshwater environments existed on the floodplains. The pollen evidence shows that during this time *Rhizophora*, which requires tidal inundation, declined and *Avicennia marina*, which grows on landward fringes and needs only intermittent tidal inundation, became more common. With the ponding of freshwater from the annual monsoon against the cheniers (old beach ridges), freshwater wetlands with their exceedingly rich floral and faunal resources became widely established on the floodplains from 2000 BP. This is reflected in the floodplain sediment cores as black organic rich muds and containing the pollen of vegetation dominated by freshwater grasses.

Although the evolution of the Alligator Rivers floodplains was largely dictated by sea level rise, the changing climate throughout this period was also influencing broader landscape change and as a consequence, human interaction; in particular a change from low seasonality in the early Holocene to increased seasonality in the late Holocene, with a general trend toward increasing aridity and climatic variability after 4000 BP (see Kershaw and van der Kaars 2013 for the most recent review of northern Australia). This scenario is supported by data extracted from pollen records, corals, and foraminifera from lake and sea bottom sediments (e.g., Hiscock and Kershaw 1992; Lees 1992; Schulmeister 1992; Nott *et al.* 1999, 233; Prebble *et al.* 2005; Luly *et al.* 2006; Moss and Kershaw 2007; Rowe 2007; Moss *et al.* 2012).

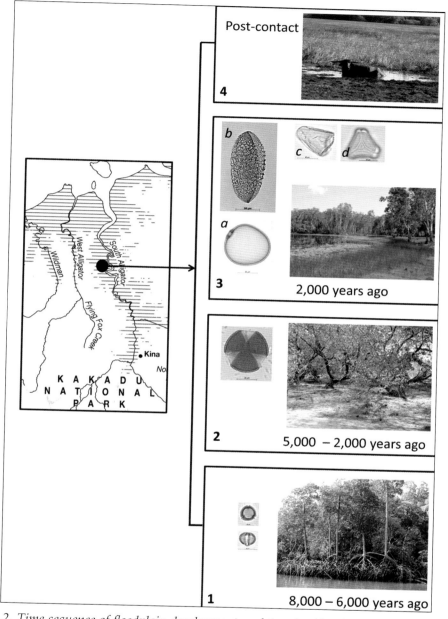

Fig. 14.2. Time sequence of floodplain development as determined by the palynological record. Box 1: Rhizophora *mangrove forest. Pollen record dominated by* Rhizophora *pollen. Box 2:* Avicennia *mangrove forest. Pollen record dominated by* Avicennia *pollen (pictured). Box 3: Freshwater wetlands. Pollen from this period is a mixture of freshwater taxa including; grass (a), Liliaceae (b), sedge (c) and Melaleuca (d). Box 4: Impact of feral animals has led to a loss of integrity of sequence and poor pollen preservation*

Holocene archaeology

Archaeologists have documented the key importance of the Alligator Rivers floodplains to the prehistoric Aboriginal economy (cf. Schrire 1982; Meehan *et al.* 1985; Allen and Barton 1989; Brockwell 1996; Hiscock *et al.* 1992; Hiscock 1996; 1999; Brockwell *et al.* 2001). With the arrival of the Big Swamp Phase about 7000 BP, settlement was concentrated in rock shelters close to the northern floodplains around the East Alligator River and Magela Creek, exploiting the rich estuarine resources of the mangroves (Fig. 14.1). Occupation continued here until the Transition Phase when the rockshelters were abandoned *c.* 3000 BP, then reoccupied with the establishment of freshwater wetlands on the floodplains *c.* 1500 BP. A hypothesis yet to be tested is that populations may have relocated to the recently emerged, resource-rich sub-coastal wetlands from the Arnhem Land plateau 40 km to the east, and from a coastline 70 km to the north that became increasingly less productive after climatic and environmental changes some 500 years ago (Jones 1985; Hiscock 1999; Brockwell *et al.* 2001; 2013; Bourke *et al.* 2007).

In later prehistory, the South Alligator wetlands were of central significance within the pattern of Indigenous adaptation pursued in western Arnhem Land. Their resources provided year-round sustenance to local groups of Aboriginal people, with rich plant resources providing food and raw materials for the manufacture of cultural items and extractive tools, and major protein harvests in the form of waterfowl, waterfowl eggs, turtles, and barramundi, which attracted large numbers of seasonal visitors from surrounding areas at particular times of the year (Brockwell *et al.* 1995; 2001; Brockwell 1996). In the dry season plants were gathered from the wetlands themselves, including bamboo (*Bambusa arnhemica*) and *Phragmites* (*P. karka*) reeds both used to manufacture spearshafts, paperbark (*Melaleuca* spp.) used in a multitude of ways, for example to construct baskets (Fig. 14.3), food wrapping for earth ovens, cladding for shelters and

Fig. 14.3. Baskets from the Alligator Rivers region (Spencer Collection, Museum Victoria)

food items, such as water chestnuts (*Eleocharis* spp.) and water lily seeds (*Nymphaea* spp.) that were ground and used to make bread (Russell-Smith *et al.* 1997). Both these items were present in such quantities in the late dry season that along with the wetlands fauna they could support large numbers of people. It is proposed that regional ceremonies were carried out at this time of year. In the wet season, foraging was focused on the open woodland. Yams, tubers, and wet season fruits were gathered from monsoon rainforests and open woodland on higher ground abutting the floodplains (cf. Brockwell *et al.* 2001) (Table 14.1).

Archaeobotany, ethnobotany, residue studies, and archaeological experiments have been key in the reconstruction of the archaeological record of the Alligator Rivers region regarding diet, subsistence strategies, material culture, technology, seasonality, and regional mobility. Of note are three sites in Kakadu that have remarkable preservation of macro-botanical remains, including uncarbonised fruits, nuts, flowers, and corms from seasonally productive plant species, and preserved remains of wooden and plant fibre artefacts. Anbangbang 1 is located in the Nourlangie-Mt Brockman massif, an outlier of the western Arnhem Land escarpment, in open woodland plains about 30 km east of the wetlands (Fig. 14.1). A large lagoon, also called Anbangbang, lies close to the foot of the outlier (Jones and Johnson 1985a, 39–64). At Anbangbang 1, macroscopic plant remains were recovered from the surface and upper levels of the deposit (dated to 800 BP; Clarke 1985; 1989). On the basis of seasonal availability of these plants, Clarke (1987, 310, 314) has argued that the site was occupied mainly during the early dry season and wet season (Table 14.1). This conclusion did not imply that occupation was continuous year-round, but rather occupation at times when particular resources were available (Clarke 1987, 310). The fibre artefacts consist of 35 remnants of string showing different techniques of knotting and tying, including a bundle of string from a dilly bag (Clarke 1985, 90–2). This collection has been analysed by archaeological textile expert, Dr Judith Cameron (Australian National University) (Cameron *et al.* in prep.). The wooden artefact collection from Anbangbang includes firesticks, hardwood points, and barbed fishing spears made from bamboo (*Bambusa arnhemenica*) and *Phragmites* reeds, both freshwater wetland species (Clarke 1985, 91–5; 1987, 300; Brockwell *et al.* 2001, 371). According to the ethnographic evidence, bamboo and *Phragmites* reeds were also used in the manufacture of composite spears to hunt magpie geese (*Anseranus semipalmata*) which gather in large numbers in billabongs and wetlands in the late dry season (Leichhardt 1847, 505; Spencer 1914, 358 (Fig. 14.4); Russell-Smith 1985a, 248). This evidence, together with faunal remains of Collett's rat (*Rattus colletti*) and big quail (*Coturnix chinensis*), both species found in floodplain sedge habitat (Foley 1985, 100–1) indicates contact with the Alligator Rivers wetlands, as does the rock art described below.

Djuwarr I is a small rockshelter situated about half way up the escarpment in Deaf Adder Gorge (Kamminga and Allen 1973, 94; Jones and Johnson 1985b, 224–7) (Fig. 14.1). Like Anbangbang, it has remarkably good macroscopic plant preservation with a similar range of plants that indicate it was occupied at a similar time of the year. Likewise the presence of bamboo and *Phragmites* indicated contact with the wetlands of the South Alligator River (Clarke 1985; 1987; 1988).

Parbari rockshelter is a site located in an outlier close to the freshwater floodplains of Magela Creek and the East Alligator River (Kamminga and Allen 1973, 23–36, 45–52;

Table 14.1: Plants and their uses Alligator Rivers Region

Season	Species	Type	Habitat	Use	Site location Wetlands (Pari-bari)	Plains (Anbang-bang)	Escarpment (Djuwarr)
EDIBLE SPECIES							
Early dry–early wet	*Nymphaea violacea*	aquatic	F, L	roots, seeds, stems	+++	+++	+++
Dry	*Canarium australianum*	tree	M	nut		+	+
Dry	*Cycas armstrongii*	palm	W	seed	+		+
Dry	*Terminalia carpentariae*	tree	W	fruit		+	
Dry	*Terminalia ferdinandiana*	tree	W	fruit	++		
Late dry	*Pandanus spiralis*	tree	W	kernels		+	+
Late dry–early wet	*Eleocharis dulcis*	aquatic	F	corm	+++		
Late dry–early wet	*Nelumbo nucifera*	aquatic	F	seeds, rhizomes	+++		
Early wet	*Buchanania obovata*	tree	W	fruit		+	+
Early wet–wet	*Persoonia falcata*	shrub	W	fruit		++	++
Early wet–wet	*Vitex glabrata*	tree	M, R	fruit		+	+
Early wet–wet	*Syzgium sp.*	tree	M, W, R	fruit		++	++
Wet–dry	*Triglochin procera*	aquatic	F, R	tuber	+++	+++	+++
All year	*Livistona humilis*	palm	W	heart of palm	+		

Season

MATERIAL CULTURE

Species	Type	Habitat	Use	Wetlands (Pari-bari)	Plains (Anbang-bang)	Escarpment (Djuwarr)
Bambusa arnhemica	graminoid	F	spear shaft	+	+	+
Barksia dentata	tree	F, W	fire light		+	+
Brachychiton paradoxum	tree	W	fibre	+		
Callitris intratropica	tree	W	spear shaft/thrower			
Denhamia obscura	tree	M	fish poison		+	
Eucalyptus sp.	tree	W	fuel		+	
Hibiscus tiliaceus	tree	M	spear shaft	+		
Melaleuca sp.	tree	F	paperbark		+	+
Pandanus spiralis	tree	W	fibre		+	+
Phragmites karka	graminoid	F	spear shaft	+	+	
Phyllanthus sp.	vine		rope		+	
Owenia vernicosa	tree	W	fish poison	+		
Planchonella arnhemica	tree	W	spear shaft/thrower		+	+
Polyalthia holtzeana	tree	M	spear thrower	+		

Key: + present ++ important +++ staple; Habitat: F floodplain; L lagoon; M monsoon rainforest; R riverine; W woodland; Sources: Schrire 1982; Russell–Smith 1985; Brock 1988; Clarke 1988; Russell-Smith *et al.* 1997

Fig. 14.4. Spears from the Alligator Rivers region (Spencer Collection, Museum Victoria)

Schrire 1982, 146–226) (Fig. 14.1). Macroscopic plant remains were found in the upper level. This level was not dated but the presence of contact material (glass, iron, and cloth) suggests that the site was occupied as recently as 150 years ago (Schrire 1982, 51, 60). The plant remains indicate occupation in all seasons, with an emphasis on dry-late dry season exploitation, which accords well with its proximity to the floodplains (Schrire 1982, 58–60, 223; Clarke 1987, 311–13) (see Table 14.1). Paribari also contained a number of wooden artefacts as well as wood shavings, suggesting that the artefacts were made *in situ*. Because of the recent nature of the deposit, Schrire (1982, 63–5) assigned a typology according to the ethnographic literature. The items included spear shafts, firesticks, a message stick, and a spear thrower knob. There were also the remains of a paperbark mat, woven fibres, which may be the remains of baskets that are common to the area, string rolled from plant material, and resin that was commonly used to attach composite weapons (Schrire 1982, 66).

Most recently macrobotanical studies by two University of Queensland PhD students at Madjedbebe (formerly Malakanunja II), a Pleistocene/Holocene age rockshelter in the north of Kakadu, have successfully revealed much about past plant foraging strategies. An anthracological study by Xavier Carah (Dotte-Sarout *et al.* 2014) has reconstructed and traced changes over time of approaches to fuel selection. Anna Florin (2014) has tracked the diversity of the macrobotanical food remains from first settlement of the site, through the Last Glacial Maximum and into the Holocene.

There has been relatively little research of plant remains at the microscopic level within the archaeological sites of the Alligator Rivers region. Wild rice (*Oryza* spp.), *Phragmites*, and bamboo are common throughout the freshwater wetlands and a pilot study that examined sediment samples from the bed of the freshwater swamp at Kina showed that phytoliths are well preserved within the black organic clays (Hope *et al.* 1985). There are numerous examples of *Oryza australiensis* at levels dated to 1370±70 BP and some tentative identification of *Phragmites* in the same levels (Hope *et al.* 1985, 157–61). Analysis of phytoliths from auger samples taken from open archaeological sites (artefact concentrations and mounds) on the nearby South Alligator wetlands revealed the presence of wild rice (*Oryza* spp.), *Phragmites* and perhaps bamboo. These sites were undated but a nearby site at Kina has been dated to 280±140 BP. It was concluded that small but significant quantities of these plant species were brought onto the sites in prehistoric times presumably to be processed as food and material culture (Fujiwara *et al.* 1985, 161–3).

Residue and use wear analysis was undertaken by Fullagar (1988; Fullagar *et al.* 1999) on wooden artefacts from Anbangbang (Clarke 1985, 91–5). He found evidence that some of the wooden artefacts from Anbangbang had been worked with metal tools. Several of the points and spear shafts had traces of resin that suggested hafting (Fullagar 1988, 18–22). Ethnographically, resin is known to have been obtained from the gum of *Terminalia* sp., which was present in the macrobotanical assemblage from Anbangbang (Clarke 1988, 127; Fullagar *et al.* 1999, 19).

Stone flakes with glossy silica polish along both sides of one edge were found exclusively at the wetlands sites of the South Alligator River and in the upper levels of the Paribari and Nawamoyn rockshelters, located adjacent to the Magela Creek wetlands, and from sites in Oenpelli close to a large freshwater lagoon on the East Alligator River (McCarthy and Setzler 1960, 278–9; Kamminga and Allen 1973, 12–13; Schrire 1982, 70, 131–2; Brockwell 1996) (Fig. 14.5). These stone flakes were made on a number of different raw materials which indicates that the nature of the stone type was not essential to function and these tools were made from whatever stone was available (Brockwell 1989, 82). The presence of these tools suggests that they were being used to process the silica rich plants that grow abundantly in the freshwater wetlands. Kamminga (1982, 93–5) has examined residue and usewear on a number of these specimens. The small size of the flakes, extensive polish and traces of resin on six flakes indicates that they were almost certainly hafted (Kamminga 1982, 95; Brockwell 1989, 87). Indeed a hafted specimen was found in sites at Oenpelli by McCarthy and Setzler (1960, 269) (Fig. 14.6). That all the flakes show extensive polish, have acute angled and fragile working edges and lack edge damage indicates that they were used to work a yielding, but not strongly fibrous or elastic vegetable material. Kamminga (1982, 95) has concluded that as these tools were recovered either on the surface or upper levels of open and rockshelter wetlands sites, that the usewear and residue are consistent with processing silica rich plants in or near swamps. Subsequently, on the basis of ethno-archaeological experiments, Akerman (1998) suggested that the polish was caused by stripping the spines of spike rush (*Eleocharis* sp., a wetland species) prior to weaving. There is some ethnographic evidence that, in the absence of other suitable materials in floodplain areas, dilly bags, baskets, skirts, pubic fringes, tassels, and mats were manufactured from *Eleocharis* sp. (Berndt and Berndt 1970, 39; Chaloupka and Guiliani 1984, 63).

Fig. 14.5. Polished flake South Alligator River (Dragi Markovic)

The rock art of the region also provides insights into subsistence and technology of prehistoric plant use through depictions of floral species, e.g., grass prints, yams, and water lilies (Figs 14.7–14.9) and plant-based material culture, canoes, dilly bags, spears, etc. (e.g., Lewis 1988; Chaloupka, 1993, 146–56) (Fig. 14.10). Rock art is found exclusively in the escarpment, plateau valleys, and outliers of the Kakadu region, where the walls of the rockshelters offer suitable surfaces for painting and afford protection from the weather.

Chaloupka (1984, 42–7; 1985; 1993) placed western Arnhem Land rock art in chronological sequence by relating its subject matter to Holocene evolutionary changes in floodplain conditions, and the consequent changes in the nature of the resource base. An 'Estuarine Period' has been defined, characterised by depictions of fauna, especially fish that flourished in the rivers and floodplains of the region with the emergence of estuarine conditions *c.* 7000 years ago. Many of these animals were portrayed with the decorative and descriptive representation of what is known as the 'X-ray style' (Fig. 14.10). Clarity of presentation allows many animals to be identified to species level. The Estuarine Period is succeeded by the 'Freshwater Period' about 2000 years ago, in which freshwater faunal and floral species are depicted, e.g., water lilies (*Nymphaea* sp.) (Fig. 14.9).

Lewis (1988) disputed this analysis. His alternative sequence divided the recent art of the region according to technological changes in material culture. Thus, he defined a 'Boomerang Period', 'Hooked Stick/Boomerang Period', 'Broad Spearthrower Period', and 'Long Spearthrower Period'. He placed these categories in a temporal sequence by relating them to dates for the extinction in northern Australia of the thylacine (*Thylacinus cynocephalus*) and Tasmanian Devil (*Sarcophilus harrsii*) (both depicted in the art), regional archeological changes in stone artefact technology, as well as environmental evolution of the floodplains. On this basis, Lewis argued that the Broad Spearthrower Period can be dated from between 6000 and 3000 years ago, and the Long Spearthrower Period from 3000 BP onwards. Lewis considered that changes in

Fig. 14.6. Hafted polished flake (Australian Museum)

Fig. 14.8. Yam figure, western Arnhem Land (Daryl Wesley)

Fig. 14.9. Water lilies, western Arnhem Land (Daryl Wesley)

Fig. 14.7. Grass stencils, western Arnhem Land (Daryl Wesley)

Fig. 14.10. (right) Hunter with multi-pronged spear and barramundi (Lates calcarifer) depicted in the X-ray style (Paul Taçon)

spear technology were directly related to the changes in resource base. Taçon (1987) attempted to reconcile these two approaches by examining the X-ray art of the region. He divided this style in to 'Early' and 'Late' periods, based on the form and complexity of the subject matter. Taçon suggested a 'Transition' period dated to about 3000 years BP where both freshwater and terrestrial species are represented. About 90% of X-ray art, mostly depicting freshwater species, is dated to the period after 2000 years BP. The depiction of estuarine and freshwater animal and plant motifs in the art of the outliers and the escarpment provides evidence that people inhabiting inland areas had contact with the floodplains. Chippindale and Taçon (1998) refined the chronological phases yet again based on a combination of dating methods.

Yams are a staple food source in western Arnhem Land and feature powerfully in the rock art with enigmatic images of so-called yam figures that started as a depiction of the tuber itself and later developed into symbolic representations of humans and animals (Fig. 14.8). The Rainbow Snake, a highly significant ancestral being of the Dreaming, is depicted in early representations as a yam figure. Yams feature in other mythologies and are central to ceremonies recorded by anthropologists Baldwin Spencer in 1912 and the Berndts in the late 1940s (Chaloupka 1993, 138–45). The yam, *Dioscorea* spp., requires a wet/dry tropical climate with at least 1200 mm of rainfall. The advent of warmer and wetter conditions and the beginning of the modern monsoon regime *c.* 6000 BP suggests these mythologies and rock-paintings may have originated then (Chaloupka 1993, 138–45; Taçon *et al.* 1996).

Historical period – contact

The 'Contact Period' in northern Australia can be separated roughly into two overlapping phases, the south-east Asian period from at least the 1660s (probably older) through to the early 1900s and the European contact period beginning in the early 1800s (Macknight 1976; Mitchell 1996; Clarke 2000a, 2000b; Clarke and Frederick 2008; May *et al.* 2010; Taçon *et al.* 2010). The latter part of the Freshwater Phase was dominated by environmental changes brought about by European contact, with impacts from feral animal and exotic weed species adversely affecting productivity, especially of the floodplains.

Buffalo and pigs were originally introduced by the British through settlements on the Coburg Peninsula to the north east of the study area in the 1840s. The grazing, wallowing, and trampling habits of buffalo and the rooting habits of pigs are particularly destructive to wetland plants. Buffaloes also create swim channels that break down levees and allow salt water to poison fresh water areas. In addition, they tend to return to the same paths and pads, which hastens the drying out of swamps and causes erosion. Likewise exotic weeds (*Mimosa pigra* and *Salvinia molesta* are particularly problematic) choke freshwater floodplains creating similar effects (Petty *et al.* 2007; Bradshaw 2008; Walden and Gardener 2008). This destruction is reflected in the palaeoecological record (Hope *et al.* 1985) (Fig. 14.2). It has been suggested by Meehan (1988; 1991) that before Aboriginal residents acquired shotguns, the presence of the buffaloes posed a threat that deterred them from foraging in floodplain areas.

Historical investigations in the Alligator Rivers region have identified a sharp disruption of prehistoric patterns in the late 19th century. The one overwhelming impact of contact was radical population decline, estimated at 96% of pre-contact numbers across the sub-coastal plains between the Adelaide and East Alligator Rivers, and probably caused by the introduction of epidemic diseases. This seems to have occurred quickly and direct evidence for what happened is fragmentary. Some survivors moved to the towns and mines along the railway line to the west, while others, their numbers supplemented by incoming groups, established new dry-season relations with white buffalo shooters in the bush from the 1880s. Social history and European ethnographic research has shown that, alongside these changes, Aboriginal groups sustained aspects of late prehistoric floodplains human ecology into the early decades of contact with Europeans (Chaloupka 1981; Meehan 1988; 1991; Russell-Smith *et al.* 1997), with later disruption to traditional Aboriginal landscape management, probably most pronounced in the shift of fire management practices (Levitus 2005; 2009; Bowman *et al.* 2007; Petty and Bowman 2007).

Ethnobotanical studies in western Arnhem Land have also contributed detailed data on traditional plant use and fire management practices in the Alligator Rivers region (cf. Specht 1958; Berndt and Berndt 1970; Chaloupka and Guiliani 1984; Russell-Smith 1985; Russell-Smith *et al.* 1997). In cooperation with Traditional Owners over many years, these studies have compiled extensive lists of all dietary and material culture plant from all biogeographic regions of Kakadu including, floodplains, riverine, lowland woodland/open forest, Arnhem Land escarpment and plateau, and lowland and sandstone rainforest. They include plant species name (Linnean, common and Aboriginal), taxa, parts of the plant utilised, habitat, season of harvest, and whether the food was of minor/major importance or a staple. They have revealed that prior to the arrival of European settlers in the region in the 1860s, the freshwater floodplains and riverine habitats provided the major plant food sources, especially in the late dry season when the corms and seeds of water lilies (*Nelumbo nucifera*) and water chestnut (*Eleocharis dulcis*) could support large groups of people (Table 14.1). Edible plant resources were less productive in woodlands, escarpment, plateau and rainforest, although they could be critical at certain times of year, e.g., yams from the monsoon rainforests in the wet season when floodplains and rivers were flooded and resources dispersed. Studies of fire regimes have shown that it was a carefully managed tool used for controlled mosaic burning in the early dry season to avoid devastating fires in the late dry season (cf. Russell-Smith *et al.* 1997, 159).

Studies of plant-based material culture have also provided important insights into prehistoric subsistence and regional patterns. Plant materials were the mainstay of Aboriginal material culture in the Alligator Rivers region (Kamminga 1988, 26; Hodgson 1995, 81). They included wood used for dugout canoes, spear shafts and throwers, digging sticks, hafted handles for tools, didgeridoos amongst other items; fibres from leaves, reeds, and bark used to make dilly bags, mats, fish traps, items of clothing, and mosquito nets. Bark was used to construct shelters, rafts, bark for paintings, torches (Hodgson, 1995, 81–4). Hodgson's (1995) study of material culture from the Alligator Rivers Region held in museum based collections has demonstrated significant inter-regional differences between raw materials used and types of artefacts manufactured, especially of spears. For example, bamboo (*Bambusa arnhemica*) spear shafts were found in greater quantities near to the source of bamboo (1995, 192–3). Regional differences are also reflected in the macro-botanical plant remains found in archaeological contexts at the sites of Paribari, Djuwarr I, and Anbanbang (Schrire 1982; Clarke 1985; 1988; 1989; Hodgson 1995).

Traditional production of plant-based material culture is continued in the region today. Artists create fibre mats and baskets using plant-based dyes and traditional techniques and paint traditional designs on sheets of bark with red, white and yellow ochre sourced locally, taking much of their inspiration taken from local rock art motifs (Hamby 2005; Injaluk Arts and Crafts 2013 http://www.injalak.com).

Summary and future directions

In summary, the plant-based data gathered from the Alligator Rivers region has aided the interpretation of prehistoric plant use, tool use, site use, and seasonality, with one of the most significant stories being one of Late Holocene seasonal mobility and regional contact (cf. Brockwell *et al.* 2001). This model is supported by other forms of archaeological evidence as well as ethnographic evidence. Berndt (1951, 171) for example reports that bundles of goose spears were traded from the South Alligator River to Oenpelli 80 km away in exchange for serrated stone and shovel nose spears from the escarpment (Brockwell *et al.* 2001, 376), borne out by thousands of stone artefacts made from quartzite and chert, and other raw materials only available from the plateau valleys of the escarpment and further south, in South Alligator River wetland sites (Brockwell 1996).

The archaeological, historical, and ethnographic evidence, and importantly arch-aeobotanical data and detailed ethnobotany of plant use and material culture, have enabled a reconstruction of a Late Holocene seasonal round in the Alligator Rivers region. The region was divided into clan territories owned by different groups. Stone country people from the Arnhem Land plateau valleys travelled to the wetlands to join the wetland owners in the large regional ceremonies where trading of stone from the plateau valleys and wetlands for products such as spears made from bamboo and *Phragmites* reeds. Large semi-sedentary camps on the edge of the floodplains allowed exploitation of the rich wetlands resource in the late dry season. With the coming of the wet season these camps were flooded. Wetlands owners moved to higher ground adjacent to the wetlands and the visitors returned to their camps in the outliers and plateau valleys (Leichhardt 1847; Berndt and Berndt 1970; Chaloupka 1981; Jones 1985; Brockwell *et al.* 2001).

The most informative archaeological plant based evidence from the Alligator Rivers region is the macro-botanical remains. This primary level of information reveals not only what plants were being used but importantly seasonality of site use that elucidates past patterns of subsistence, mobility and settlement. Secondary lines of evidence such as phytoliths and residue analyses have been little explored as yet. While the tertiary lines of evidence, such as the rock art, illustrate changing plant based technology in a landscape where the preservation of plant remains in archaeological contexts is extremely poor and highlights the plants that were 'spiritually' significant.

However, many lines of evidence remain be explored. Ethnographic and rock art studies still have much to reveal. Detailed studies of plant-based ethnographic material culture from collections held in museums (e.g., the Spencer and Thomson collections at Museum Victoria), as well as depictions in the rock art of weaponry, basketry, and personal ornaments, have the potential to reveal much information regarding ephemeral prehistoric technology and to elucidate technological changes to plant based material culture over time (cf. Lewis 1988; Chalopuka 1993, 146–56; Hodgson 1995, 181–5).

Apart from the studies of Fullagar, Kamminga, and Fujiwara *et al.* in the 1980s, microfossils, such as phytoliths and residues, have barely been explored and molecular evidence not at all. There is potential to continue use wear and residue studies on wooden artefacts from archaeological contexts at Anbangbang and Djuwarr 1. Fullagar's (1988; Fullagar *et al.* 1999) findings indicate that with appropriate experimental reference collections it will be possible to investigate subtle shifts in subsistence within a predominantly hunting and gathering society through the study of patterns and changes in the use of tools (Fullagar 1988, 32; Fullagar *et al.* 1999).

Investigation of palaeofire histories in the challenging fire-prone tropical savannahs of northern Australia has not been dealt with in any depth. However recently funded work has the potential to contribute to the long term history of Aboriginal burning, including an investigation of how the disconnection of Aboriginal people from their traditional burning practices may have impacted on the composition of vegetation through time. Palaeoecology can contribute to these discussions by analysing the charcoal fragments from within sedimentary deposits as a proxy for fire activity (cf. Whitlock and Larsen 2001). These issues are currently being investigated by Stevenson through two Australian Research Council (ARC) Linkage Projects: *Enhancing Cultural Heritage Management for Mining Operations: a multi-disciplinary approach* (LP110100180); and *From Prehistory to History: landscape and cultural change on the South Alligator River, Kakadu National Park* (LP110201128).

Concluding remarks

The status of archaeobotanical research in the Alligator Rivers region is typical of elsewhere in Australia. The study of plants within an archaeological context is still in its infancy in Australia, despite its proven power as an interpretive tool elsewhere and remains peripheral to mainstream archaeology in Australia today (Denham *et al.* 2009a). In part this is due to the preservational contexts that we have to work with, but also to the small research community generally here in Australia. Denham *et al.* (2009a) have argued for a more central role of archaeobotany within the Australian setting and goes on to suggest that some of the big research questions that often pre-occupy archaeological enquiry in Australia may be answered using plant based studies; 'human adaptation to environmental change', 'colonisation of Sahul', 'interpreting environmental management', 'emergence and transformation of agriculture and aboriculture in New Guinea', 'Aboriginal diets and health'. To this end they have established a working group in a bid to co-ordinate the technical management of data, as well as develop several initiatives to redefine standards and acceptable practice, curate reference collections, disseminate knowledge through on-line databases, and rationalise resources (see http://archaeobotany.net/ for working group webpage).

Another network of interest for archaeobotanists is Palaeoworks (2013), formed by a group of palaeoecologists from the Australian National University. Palaeoworks hosts a website that includes the Indo-Pacific Pollen Site Database and the Australasian Pollen and Spore Atlas. This group of researchers and several others from around the country are currently investigating the long term history of people and fire in the Australian landscape. While, the requisite high resolution records for the Australian environment are still in their infancy, there is a growing movement amongst local palaeoecologists to develop such records as a standard aspect of their research program and is leading to a better understanding of fire frequency for several environmental settings within the Australian landscape (e.g., Black *et al.* 2007; Haberle *et al.* 2010).

This chapter has demonstrated the importance of plants in the interpretation of Australian archaeological sites and the enormous potential for further research, especially into microbotanical remains. Given the critical role of plants as food and material culture in Aboriginal society, more attention needs to be focused on archaeobotanical research and integrating it within mainstream archaeological disciplines.

References

Akerman, K. 1998. A suggested function for Western Arnhem Land use of polished flakes and eloueras. In Fullagar, R. (ed.), *A Closer Look: recent Australian studies of stone tools*, 179–188. Sydney: Sydney University Archaeological Methods Series 6

Allen, H. and Barton, G. 1989. *Ngarradj Warde Jobkeng: White Cockatoo Dreaming and the prehistory of Kakadu*. Sydney: University of Sydney Oceania Monograph 37

Allen, H. and Holdaway, S. 2011. A retrospective review of Richard A. Gould's Living Archaeology. *Ethnoarchaeology* 3(2), 203–20

ANBG 2013. Australian National Botanical Gardens. *Aboriginal Plant Use* http://www.anbg.gov.au/aboriginal-resources/index.html (17 May 2013)

Balme, J., Garbin, G. and Gould, R. 2001. Residue analysis and palaeodiet in arid Australia. *Australian Archaeology* 53, 1–6

Beck, W. 1989. The taphonomy of plants. In Beck *et al.* (eds) 1989, 31–53

Beck, W., Clarke, A. and Head, L. 1989. *Plants in Australian Archaeology*. St Lucia: Tempus 1, Anthropology Museum, University of Queensland

Berehowyj, L. 2013. Can use-wear be used to identify tuber processing on siliceous stone? An experimental study from Australia. *Australian Archaeology* (77), 9–19

Berndt, R. M. 1951. Ceremonial exchange in western Arnhem Land. *Southwestern Journal of Anthropology* 7(2), 156–76

Berndt, R. M. and Berndt, C. H. 1970. *Man, Land and Myth in North Australia: the Gunwinggu People*. Sydney: Ure Smith

Black, M. P., Mooney, S. D. and Haberle, S. G. 2007. The fire, human and climate nexus in the Sydney Basin, eastern Australia. *Holocene* 17(4), 469–580

Bourke, P., Brockwell, S., Faulkner, P. and Meehan, B. 2007. Climate variability in the mid to late Holocene Arnhem Land region, north Australia: archaeological archives of environmental and cultural change. In Lape, P. (ed.), *Climate Change and Archaeology in the Pacific*, 91–101. *Archaeology in Oceania* 42(3)

Bowdery, D. 1989. Phytolith analysis: introduction and applications. In Beck *et al.* (eds) 1989, 161–96

Bowdery, D. 1998. *Phytolith Analysis Applied to Pleistocene-Holocene Archaeological Sites in the Australian Arid Zone.* Oxford British Archaeological Report S695

Bowman, D. M. J. S., Dingle, J. K., Johnston, F. H., Parry, D. and Foley, M. 2007. Seasonal patterns in biomass smoke pollution and the mid-20th-century transition from Aboriginal to European fire management in northern Australia. *Global Ecology and Biogeography* 16, 246–56

Bradshaw, C. J. A. 2008. Invasive species. Feral animal species in northern Australia: savvy surveillance and evidence-based control. In Walden, D. and Nou, S. (eds), *Landscape Change Overview. Kakadu National Park Landscape Symposia Series 2007-2009, Symposium 1*, 58–65. Darwin: Supervising Scientist

Brock, J. 1988. *Top End Native Plants.* Darwin: John Brock

Brockwell, C. J. 1989. *Archaeological Investigations of the Kakadu Wetlands, Northern Australia.* Unpublished MA thesis, Australian National University, Canberra

Brockwell, S. 1996. Open Sites of the South Alligator River Wetland, Kakadu. In Veth, P., Hiscock, P. (eds), *Archaeology of northern Australia*, 90–105. St Lucia: Tempus 4. Anthropology Museum, University of Queensland

Brockwell, S., Clarke, A. and Levitus, R. 2001. Seasonal movement in the prehistoric human ecology of the Alligator Rivers region, north Australia. In Allen, J., Ambrose, W., Anderson, A. and Andrews, A. (eds), *Histories of Old Ages. Essays in Honour of Rhys Jones*, 361–80. Canberra: Pandanus Publications, Centre for Archaeological Research, Australian National University

Brockwell, S., Levitus, R., Russell-Smith, J. and Forrest, P. 1995. Aboriginal heritage. In Press, A. J., Lea, D. A. M., Webb, A. and Graham, G. (eds), *Kakadu: natural and cultural heritage and its management*, 15–63. Darwin: Australia Nature Conservation Agency and the North Australia Research Unit, Australian National University

Brockwell, S., Marwick, B., Bourke, P., Faulkner, P. and Willan, R. 2013. Late Holocene climate change and human behavioural variability in the coastal wet-dry tropics of northern Australia: evidence from a pilot study of oxygen isotopes in marine bivalve shells from archaeological sites. *Australian Archaeology* 76, 21–33

Byrne, C., Dotte-Sarout, E. and Winton, V. 2013. Charcoals as indicators of ancient tree and fuel strategies: An application of anthracology in the Australian Midwest. *Australian Archaeology* 77, 94–106

Cameron, J., Martens, T., Keany, B., Brockwell, S. and Clarke, A. in prep. *Fibre technology from Anbangbang I, western Arnhem Land: Unraveling links to remote ancestors*

Cane, S. 1984. *Desert Camps.* Unpublished PhD thesis. Australian National University, Canberra.

Chaloupka, G. 1981. The traditional movement of a band of Aborigines, Kakadu. In T. Stokes, T. (ed.), *Kakadu National Park: education resources.* Appendix 1. Canberra and Darwin: Australian National Parks Wildlife Service and the Northern Territory Department of Education

Chaloupka, G. 1984. *From Palaeoart to Casual Paintings.* Darwin: Northern Territory Museum of Arts & Sciences Monograph 1

Chaloupka, G. 1985. Chronological sequence of Arnhem Land plateau rock art. In Jones (ed.) 1985, 269–80

Chaloupka, G. 1993. *Journey in Time. The World's Longest Continuing Art Tradition. The 50,000 Year Story of the Australian Aboriginal Rock Art of Arnhem Land.* Chatswood: Reed Books

Chaloupka, G. and Guiliani, P. 1984. *Gunbulk abel gundalg: Mayali flora.* Darwin: Unpublished report on the use of plants in Kakadu. Northern Territory Museum of Arts and Sciences. Copies are available at the Northern Territory and Museum libraries in Darwin, NT

Chappell, J. and Woodroffe, C. 1985. Morphodynamics of Northern Territory tidal rivers and floodplains. In Bardsley, K., Davie, J. D. S. and Woodroffe, C. D. (eds), *Coasts and Tidal Wetlands of the Australian Monsoon Region*, 85–96. Darwin: North Australia Research Unit, Australian National University, Mangrove Monograph 1

Chippindale, C. and Taçon, P. 1998. *The Archaeology of Rock-Art*. Cambridge: Cambridge University Press

Clark, R. L. and Guppy, J. C. 1988. A transition from mangrove forest to freshwater wetland in the monsoon tropics of Australia. *Journal of Biogeography* 15, 665–84

Clarke, A. 1985. A preliminary archaeobotanical analysis of the Anbangbang 1 site. In Jones (ed.) 1985, 77–96

Clarke, A. 1987. *An Analysis of Archaeobotanical Data from two sites in Kakadu National Park*. Unpublished MA thesis, University of Western Australia, Perth

Clarke, A. 1988. Archaeological and ethnobotanical interpretations of plant remains from Kakadu National Park, Northern Territory. In Meehan, B. and Jones, R. (eds), *Archaeology with Ethnography*, 123–36. Canberra: Department of Prehistory, Research School of Pacific Studies, Australian National University

Clarke, A. 1989. Macroscopic plant remains. In Beck *et al.* (eds) 1989, 54–89

Clarke, A. 2000(a). Time, tradition and transformation: the archaeology of intercultural encounters on Groote Eylandt, northern Australia. In Torrence, R. and Clarke, A. (eds), *The Archaeology of Difference: negotiating cross-cultural engagements in Oceania*, 142–81. London: Routledge One World Archaeology 38

Clarke, A. 2000(b). 'The Moorman's trowsers': Aboriginal and Macassan interactions and the changing fabric of Indigenous social life. In O'Connor, S. and Veth, P. (eds), *East of Wallace's Line. Modern Quaternary Research in South-East Asia*, 315–35. Leiden: A.A. Balkema

Clarke, A. and Frederick, U. 2008. The mark of marvellous ideas: Groote Eylandt rock art and the performance of cross-cultural relations. In Veth, P., Neale, M. and Sutton, P. (eds), *Strangers on the Shore: early coastal contacts with Australia*, 148–164. Canberra: National Museum of Australia

Clarke, P. 2012. *Australian Plants as Aboriginal Tools*. Dural: Rosenberg Publishing

Clarkson, C. and Wallis, L. 2003. The search for El Nino/Southern Oscillation in archaeological sites: recent results for Jugali-ya Rockshelter, Wardaman Country. In Hart and Wallis (eds) 2003, 137–52

Cribb, R. 1996. Shell mounds, domiculture and ecosystem manipulation on western Cape York Peninsula. In Veth, P. and Hiscock, P. (eds), *Archaeology of Northern Australia*, 150–173. St Lucia: Tempus 4, Anthropology Museum, University of Queensland

Cribb, R., Walmberg, R., Wolmby, R. and Taisman, C. 1988. Landscape as cultural artefact: Shell mounds and plants in Aurukun, Cape York Peninsula. *Australian Aboriginal Studies* 2, 60–73

Dáithí, C. M, Pearson, S. G., Fullagar, R., Chase, B. M., Houston, J., Atchison, J., White, N. E. Bellgard, M. I., Clarke, E., Macphail, M., Thomas, M., Gilbert, P., Haile, J. and Bunce, M. 2012. High-throughput sequencing of ancient plant and mammal DNA in herbivore middens. *Quaternary Science Reviews* 58, 135–45

Denham, T. 2007. Traditional forms of plant exploitation in Australia and New Guinea: the search for common ground. *Vegetation History and Archaeobotany* 17, 245–8

Denham, T. and Donoghue, D. 1989. Carbonised plant macrofossils. In Beck *et al.*(eds) 1989, 90–110

Denham, T., Atchison, J. Austin, J., Bestel, S., Bowdery, D., Crowther, A., Dolby, N., Fairbairn, A., Field, J., Kennedy, A., Lentfer, C., Matheson, C., Nugent, S., Parr, J., Prebble, M., Robertson, G., Specht, J., Torrence, R., Barton, H., Fullagar, R., Haberle, S., Horrocks, M., Lewis, T. and Matthews, P. 2009a. Archaeobotany in Australia: practice, potential and prospects. *Australian Archaeology* 68, 1–10

Denham, T., Donohue, M. and Booth, S. 2009b. Horticultural experimentation in northern Australia reconsidered. *Antiquity* 83, 634–48

Department of the Environment, 2014. *Welcome to Kakadu National Park*. Australian Government website: http://www.environment.gov.au/topics/national-parks/kakadu-national-park (accessed 16 April 2014)

Dotte-Sarout, E., Carah, X. and Byrne, C. 2014. Not just carbon: Assessment and prospects for the application of anthracology in Oceania. *Archaeology in Oceania*. DOI: 10.1002/arco.5041

Field, J., Cosgrove, R., Fullagar, R. and Lance, B. 2009. Survival of starch residues on grinding stones in private collections: A study of morahs from the tropical rainforests of NE Queensland. In Haslam, M. and Robertson, G. (eds), *Archaeological Science under a Microscope: Papers in Honour of Tom Loy*, 218–228. Canberra: Terra Australis 30

Florin, S. A. 2014. Thesis abstract 'Archaeological investigations into plant food use at Madjedbebe (Malakununja II)'. *Australian Archaeology* 78, 115

Foley, D. 1985. Faunal analysis of Anbangbang and Djuwarr I. In Jones (ed.) 1985, 97–102

Frawley, S. and O'Connor, S. 2010. A 40,000 year wood charcoal record from Carpenter's Gap 1: new insights into palaeovegetation change and Indigenous foraging strategies in the Kimberley, Western Australia. In Haberle, S., Stevenson, J. and Prebble, M. (eds), *Altered Ecologies*. Canberra: ANU EPress http://epress.anu.edu.au/titles/terra-australis/ta32

Fujiwara, H., Jones, R. and Brockwell, S. 1985. Plant opals (phytoliths) in Kakadu archaeological sites. In Jones (ed.) 1985, 155–64

Fullagar, R. 1988. *Microscopic Study of Aboriginal Artefacts from Arnhem Land*. Unpublished report prepared for Betty Meehan. Available from Division of Anthropology, Australian Museum, Sydney

Fullagar, R. 1991. The role of silica in polish formation. *Journal of Archaeological Science* 18(1), 1–24

Fullagar, R. and Field, J. 1997. Pleistocene seed grinding implements from the Australian arid zone. *Antiquity* 71(272), 300–7

Fullagar, R., Field, J. and Kealhofer, L. 2008. Grinding stones and seeds of change: starch and phytoliths as evidence of plant food processing. In Rowan Y. M. and Ebling, J. R. (eds), *New Approaches to Old Stones: recent studies of ground stone artifacts*, 159–172. London: Equinox

Fullagar, R., Meehan, B. and Jones, R., 1999. Residue analysis of ethnographic plant-working and other tools from northern Australia. In Anderson-Gerfaud, P. (ed.), *Prehistory of Agriculture; new experimental and ethnographic approaches*, 15–23. Berkeley CA: Institute of Archaeology, UCLA Monograph 40

Gammage, B. 2011. *The Biggest Estate on Earth: how Aborigines made Australia*. Sydney: Allen and Unwin

Golson, J. 1971. Australian Aboriginal food plants: some ecological and culture-historical implications. In Mulvaney, D. J. and Golson, J. (eds), *Aboriginal Man and Environment in Australia*, 196–238. Canberra: Australian National University Press

Gorecki, P., Grant, M., O'Connor, S. and Veth, P. 1997. The morphology, function and antiquity of Australian grinding implements. *Archaeology in Oceania* 32(2), 141–50

Gott, B. 1989. The uses of ethnohistory and ecology. In Beck *et al.*(eds) 1989, 197–213

Gott, B. 1999. Cumbungi, *Typha* species: a staple Aboriginal food in southern Australia. *Australian Aboriginal Studies* (1), 33–50

Gott, B. 2008. Indigenous use of plants in south-eastern Australia. *Telopea* 12(2), 215–26

Gould, R. 1969. Subsistence behaviour among Western Desert Aborigines of Australia. *Oceania* 34(4), 253–74

Gould, R. 1977. Puntutjarpa Rockshelter and the Australian desert culture. *Anthropological Papers of the American Museum of Natural History* 54(1), 1–187

Gould, R. 1980 (reissued 2009). *Living Archaeology*. Cambridge: Cambridge University Press

Haberle, S., Stevenson, J. and Prebble, M. (eds), 2010. *Altered Ecologies: fire, climate and human influence on Terrestrial Landscapes*. Terra Australis 32. ANU Epress, Canberra. http://epress.anu.edu.au/titles/terra-australis/ta32

Hall, J., Higgins, S. and Fullagar, R. 1989. Plant residues on stone tools. In Beck *et al.* (eds) 1989, 136–60

Hamby, L. (ed.), 2005. *Twined Together: Kunmadj Njalehnjaleken.* Gunbalanya, NT: Injalak Arts and Crafts

Harris, D. 1977. Subsistence strategies across Torres Strait. In Allen, J., Golson, J. and Jones, R. (eds), *Sunda and Sahul: prehistoric studies in southeast Asia, Melanesia and Australia*, 421–63. London: Academic

Hart, D. and Wallis, L. (eds), 2003. *Phytolith and Starch Research in the Australian-Pacific-Asian Regions: the state of the art.* Canberra: Terra Australis 19

Haslam, M., Robertson, G., Crowther, A., Nugent, S. and Kirkwood, L. (eds), 2009. *Archaeological Science under a Microscope.* Canberra: Terra Australis 30 http://epress.anu.edu.au/titles/terra-australis/ta30_citation

Head, L. 1989. Pollen. In In Beck *et al.* (eds) 1989, 111–35

Head, L., Atchison, J. and Fullagar, R., 2002. Country and garden: ethnobotany, archaeobotany and Aboriginal landscapes near the Keep River, northwestern Australia. *Journal of Social Archaeology* 2, 173–96

Hiscock, P. 1996. Mobility and Technology in the Kakadu Coastal Wetlands. *Bulletin of the Indo-Pacific Prehistory Association* 15, 151–7

Hiscock, P. 1999. Holocene coastal occupation of western Arnhem Land. In Hall, J. and McNiven, I. J. (eds), *Australian Coastal Archaeology*, 91–103. Canberra: Research Papers in Archaeology and Natural History 31

Hiscock, P. 2008. *Archaeology of Ancient Australia.* London and New York: Routledge

Hiscock, P. and Kershaw, A. P. K. 1992. Palaeoenvironments and prehistory of Australia's tropical top end. In Dodson, J. (ed.), *The Naïve Lands: human/environment interactions in Australia and Oceania*, 43–75. Melbourne: Longman Cheshire

Hiscock, P., Guse, D. and Mowat, F. 1992. Settlement patterns in the Kakadu wetlands: Initial data on size and shape. *Australian Aboriginal Studies* 2, 84–9

Hodgson, R. 1995. *Variation in Aboriginal Material Culture of the Alligator Rivers Region.* Unpublished MA thesis, Northern Territory University, Darwin

Hope, G., Hughes, P. J. and Russell-Smith, J. 1985. Geomorphological fieldwork and the evolution of the landscape of Kakadu National Park. In Jones (ed.) 1985, 229–40

Hynes, R. and Chase, A. 1982. Plants, sites and domiculture: Aboriginal influence upon plant communities in Cape York Peninsula. *Archaeology in Oceania* 17, 38–49

Injaluk Arts and Crafts 2013. Website http://www.injalak.com/ (accessed 16 April 2014)

Jones, R. 1969. Fire-stick farming. *Australian Natural History* 16, 224–8

Jones, R. (ed.) 1985. *Archaeological Research in Kakadu National Park.* Canberra: Australian National Parks and Wildlife Service, Special Publication 13

Jones, R. and Johnson, I., 1985a. Rockshelter excavations: Nourlangie and Mt Brockman massifs. In Jones (ed.) 1985, 39–76

Jones, R. and Johnson, I. 1985b. Deaf Adder Gorge: Lindner site, Nauwalabila I. In Jones (ed.) 1985, 165–227

Jones, R. and Meehan, B. 1989. Plant foods of the Gidjingali: ethnographic and archaeological perspectives from northern Australia on tuber and seed exploitation. In Harris, D. and Hillman, G. (eds), *Foraging and Farming: the evolution of plant exploitation*, 120–35. London: One World Archaeology 13

Kamminga, J., 1982. *Over the Edge: functional analysis of Australian stone Tools.* St Lucia: Anthropology Museum, University of Queensland Occasional Papers in Anthropology 12

Kamminga, J. 1988. Wood artefacts: a checklist of plant species utilised by Australian Aborigines. *Australian Aboriginal Studies* 2, 26–59

Kamminga, J. and Allen, H. 1973. *Report of the Archaeological Survey. Alligator Rivers Environmental Fact-Finding Study*. Darwin: Government Printer

Kershaw, P. 1986. Climatic change and Aboriginal burning in north-east Australia during the last two glacial/interglacial cycles. *Nature* 322, 47–9

Kershaw, A. P. 1995. Environmental change in greater Australia. *Antiquity* 69, 656–75

Kershaw, A. P. and Kaars, S. van der, 2013. Tropical Quaternary climates in Australia and the southwest Pacific. In Metcalfe, S. E. and Nash, D. J. (eds), *Quaternary Environmental Change in the Tropics*. Oxford: Blackwell Scientific

Latz, P. 1995. *Bushfires and Bush Tucker: Aboriginal plant use in central Australia*. Alice Springs: IAD Press

Lees, B. G. 1992. Geomorphological evidence for late Holocene climatic change in northern Australia. *Australian Geographer* 23, 1–10

Leichhardt, L. 1847. *Journal of an Overland Expedition in Australia from Moreton Bay to Port Essington*. London: T. & W. Boone

Levitt, D. 1981. *Aboriginal Plants and People: Aboriginal uses of plants on Groote Eylandt*. Canberra: Australian Institute of Aboriginal Studies

Levitus, R. 2005. Management and the model: burning Kakadu. In Minnegal, M. (ed.), *Sustainable Environments, Sustainable Communities: Potential Dialogues between Anthropologists, Scientists and Managers*, 29–35. Melbourne: University of Melbourne SAGES Research Paper 21

Levitus, R. 2009. Change and catastrophe: adaptation, re-adaptation and fire in the Alligator Rivers region. In Russell-Smith, J., Whitehead, P. and Cooke, P. (eds), *Culture, Ecology and Economy of Fire Management in North Australian Savannas. Rekindling the Wurrk Tradition*, 41–67. Canberra: CSIRO Publishing

Lewis, D. 1988. *Rock Paintings of Arnhem Land, Australia: Social, Ecological and Material Culture Change in the Post-Glacial Period*. Oxford: British Archaeological Report S415

Luly, J. G., Grindrod, J. F. and Penny, D. 2006. Holocene palaeoenvironments at Three-Quarter Mile Lake. *Holocene* 16, 1085–94

McCarthy, F. D. and Setzler, F. M. 1960. The archaeology of Arnhem Land. In Mountford, C. P. (ed.), *Records of the American-Australian Scientific Expedition to Arnhem Land* 2, 215–95. Melbourne: Melbourne University Press

McConnell, K. and O'Connor, S., 1997. 40,000 year record of food plants in the southern Kimberley ranges, Western Australia. *Australian Archaeology* 45, 20–31

Macknight, C. C. 1976. *The Voyage to Marege: Macassan Trepangers in Northern Australia*. Melbourne: Melbourne University Press

Martin, S., 2011. Palaeoecological evidence associated with earth mounds of the Murray Riverine plain, southeastern Australia. *Environmental Archaeology* 16(2), 162–172. http://www.maneyonline.com/doi/abs/10.1179/174963111X13110803261056 (accessed 16 April 2014)

May, S. K., Taçon, P. S. C., Wesley, D. and Travers, M. 2010. Painting history: Indigenous observations and depictions of the other in northwestern Arnhem Land, Australia. *Australian Archaeology* 71, 57–65

Meehan, B. 1982. *Shell Bed to Shell Midden*. Canberra: Australian Institute of Aboriginal Studies

Meehan, B. 1988. Changes in Aboriginal exploitation of wetlands in northern Australia. In Wade-Marshall, D. and Loveday, P. (eds), *Floodplains Research. Northern Australia: Progress and Prospects* 2, Appendix 2, 1–23. Darwin: North Australia Research Unit, Australian National University

Meehan, B. 1989. Plant use in a contemporary Aboriginal community and prehistoric implications. In Beck *et al.* (eds) 1989, 14–30.

Meehan, B. 1991. Wetland hunters: some reflections. In Haynes, C. D., Ridpath, M. G. and Williams, M. A. J. (eds), *Monsoonal Australia: landscape, ecology and man in the northern lowlands*, 197–206. Rotterdam: A.A. Balkema

Meehan, B. and Jones, R., 1988. *Archaeology with Ethnography: an Australian perspective*. Canberra: Department of Prehistory, Australian National University

Meehan, B., Brockwell, S., Allen, J. and Jones, R. 1985. The wetlands sites. In Jones (ed.) 1985, 103–53

Mitchell, S. 1996. Dugongs and dugouts, sharptacks and shellbacks: Macassan contact and Aboriginal marine hunting on the Cobourg Peninsula, north western Arnhem Land. In Glover, I. and Bellwood, P. (eds), *The Chiang Mai Papers* 2, 181–91. Canberra: Bulletin of the Indo-Pacific Prehistory Association 15

Moss, P. T. and Kershaw, A. P. K. 2007. A late Quaternary marine palynological record (oxygen isotope stages 1 to 7) for the humid tropics of northeastern Australia based on ODP site 820. *Palaeogeography, Palaeoclimatology, Palaeoecology* 251, 4–22

Moss, P. T., Cosgrove, R., Ferrier, A. and Haberle, S., 2012. Holocene environments of the sclerophyll woodlands of the wet tropics of northeastern Australia. In Haberle, S. and David, B. (eds), *Peopled Landscapes. Archaeological and Biogeographic Approaches to Landscapes*, 329–343. Canberra: Terra Australis, ANU Press

Murdoch University, 2011. *Murdoch researchers to probe ancient DNA at Devil's Lair* http://media. murdoch.edu.au/murdoch-researchers-to-probe-ancient-dna-at-devil%E2%80%99s-lair (accessed 3 February 2014)

Nott, J., Bryant, E. and Price, D. 1999. Early Holocene aridity in tropical northern Australia. *Holocene* 9(2), 231–6

O'Connell, J. F. and Hawkes, K. 1984. Food choice and foraging sites among the Alyawara. *Journal of Anthropological Research* 40 (4), 504–35

O'Connell, J. F., Latz, P. K. and Barnett, P. 1983. Traditional and modern plant use among the Alyawara of central Australia. *Economic Botany* 37 (1), 83–112

Palaeoworks, 2013. http://palaeoworks.anu.edu.au/ (26 March 2013)

Pardoe, C. and Martin, S. 2011. *Murrumbidgee Province Aboriginal Cultural Heritage Study*. AACAI (Australian Association of Consulting Archaeologists) Monograph Series http://www.aacai. com.au/publications/Monograph/

Petty, A. and Bowman, D. M. J. S. 2007. A satellite analysis of contrasting fire patterns in Aboriginal and Euro-Australian managed lands in tropical north Australia. *Fire Ecology* 3, 32–47

Petty, A. M., Werner, P. A., Lehmann, C. E. R., Riley, J. E., Banfai, D. S. and Elliott, L. P. 2007. Savanna responses to feral buffalo in Kakadu National Park, Australia. *Ecological Monographs* 77(3), 441–63

Prebble, M., Sim, R., Finn, J. and Fink, D. 2005. A Holocene pollen and diatom record from Vanderlin Island, Gulf of Carpentaria, lowland tropical Australia. *Quaternary Research* 64, 357–71

Rowe, C. 2007. A palynological investigation of Holocene vegetation change in Torres Strait, seasonal tropics of northern Australia. *Palaeogeography, Palaeoclimatology, Palaeoecology* 25, 83–103

Rule, S., Brook, B. W., Haberle, S. G. Turney, C. S. M., Kershaw, A. P. and Johnson, C. N. 2012. The aftermath of megafaunal extinction: ecosystem transformation in Pleistocene Australia. *Science* 335, 1483–6

Russell-Smith, J. 1985. Studies in the jungle: people, fire and monsoon forest. In Jones (ed.) 1985, 241–67

Russell-Smith, J., Lucas, D., Gapindi, M., Gunbunuka, B., Kapirigi, N., Namingum, G., Lucas, K., Giuliani, P. and Chaloupka, G. 1997. Aboriginal resource utilization and fire management practice in western Arnhem Land, monsoonal northern Australia: notes for prehistory, lessons for the future. *Human Ecology* 25, 159–95

Schrire, C. 1982. *Alligator Rivers Prehistory: prehistory and ecology in Western Arnhem Land*. Canberra: Terra Australis 7

Schulmeister, J. 1992. A Holocene pollen record from lowland tropical Australia. *Holocene* 2, 107–16

Smith, M. 1986. The antiquity of seed-grinding in central Australia. *Archaeology in Oceania* 21(1), 29–39

Smith, M. 2004. The grindstone assemblage from Puritjarra rockshelter: Investigating the history of seed-based economies in arid central Australia. In Murray, T. (ed.), *Archaeology from Australia*, 168–86. Melbourne: Australian Scholarly Publishing

Smith, M. 2013. *The Archaeology of Australia's Deserts*. Cambridge: Cambridge World Archaeology

Specht, R. L. 1958. An introduction to the ethno-botany of Arnhem Land. In Specht, R. L. and Mountford, C. P. (eds), *Botany and Plant Ecology* 3, 479–503. Melbourne: Melbourne University Press

Spencer, W. B. 1914. *The Native Tribes of the Northern Territory of Australia*. London: Macmillan

Taçon, P. S. C. 1987. Internal-external: a re-evaluation of the 'x-ray' concept in western Arnhem Land rock art. *Rock Art Research* 4 (1), 36–50

Taçon, P. S. C., May, S. K., Fallon, S. J., Travers, M., Wesley, D. and Lamilami, R. 2010. A minimum age for early depictions of Southeast Asian praus in the rock art of Arnhem Land, Northern Territory. *Australian Archaeology* 71, 1–10

Taçon, P. S. C., Wilson, M. and Chippindale, C. 1996. Birth of the Rainbow Serpent in Arnhem Land rock art and oral history. *Archaeology in Oceania* 31, 103–24

Torrence, R. and Barton, H. (eds), 2006. *Ancient Starch Research*. Walnut Creek CA: Left Coast Press

Turney, C. S. M., Kershaw, A. P., Clemens, S. C., Branch, N., Moss, P. T. and Fifield, L. K. 2004. Millennial and orbital variations of El Niño/Southern Oscillation and high-latitude climate in the last glacial period. *Nature* 428, 306

Turney, C. S. M., Kershaw. A. P., Moss, P., Bird, M. I., Fifield, L. K., Cresswell, R. G., Santos, G. M., di Tada, M. L., Hausladen, P. A. and Zhou, Y. 2001. Redating the onset of burning at Lynch's Crater (north Queensland): implications for human settlement in Australia. *Journal of Quaternary Science* 16, 767–71

Veth, P. and O'Connor, S. 1996. A preliminary analysis of basal grindstones from the Carnavon Range, Little Sandy Desert. *Australian Archaeology* 43, 20–2

Veth, P., Fullagar, R. and Gould, R. 1997. Residue and use-wear analysis of grinding implements from Puntutjarpa Rockshelter in the Western Desert: current and proposed research. *Australian Archaeology* 44, 23–5

Walden, D. and Gardener, M. 2008. Invasive species: weed management in Kakadu National Park. In Walden, D. and Nou, S. (eds), *Landscape Change Overview*, 66–83. Darwin: Kakadu National Park Landscape Symposia Series 2007–2009. Internal Report 532. Available on-line: http://www.environment.gov.au/system/files/resources/e71dc68a-edb7-4986-ae2f-498d4a873691/files/ir532.pdf

Wallis, L. 2001. Environmental history of northwest Australia based on phytolith analysis at Carpenter's Gap 1. *Quaternary International* 83–5, 103–17

Wallis, L. 2003. The history of phytolith researchers in Australia. In Hart and Wallis (eds) 2003, 1–17

Whitlock, C. and Larsen, C. 2001. Charcoal as a fire proxy. In Smol, J. P., Birks, H. J. B. and Last W. M. (eds), *Tracking Environmental Change Using Lake Sediments 3: Terrestrial, Algal and siliceous indicators*, 75–97. Dordrecht: Kluwer

Woodroffe, C. D., Thom B. G. and Chappell, J. 1985. Development of widespread mangrove swamps in mid-Holocene times in northern Australia. *Nature* 317, 711–13

Yates, A., Smith, A. M., Parr, J., Scheffers, A. and Joannes-Boyau, R. 2014. AMS dating of ancient plant residues from experimental stone tools: a pilot study. *Journal of Archaeological Science* 49, 595–602

15. Plentiful scarcity: plant use among Fuegian hunter-gatherers

Marian Berihuete Azorín, Raquel Piqué Huerta and Maria Estela Mansur

When European settlers arrived in the 19th century, the Isla Grande of Tierra del Fuego was inhabited by the Yamana and Selknam peoples. Ethnographic descriptions of those populations are focused largely on the importance of hunting as the main source of subsistence and plant food was considered to have had a marginal role in their economy. This general idea was also supported by the lack of archaeobotanical studies. Nevertheless, recent research at historic Selknam and Yamana sites, and the analysis of linguistic and ethnographic data, has shed new light on the role of plants among Fuegian populations.

Even so, the Fuegian hunter-gatherers may be somewhat different to other groups living in comparable latitudes. The reason is the low plant diversity, which comprises only 417 native species. Moreover, geographical and seasonal constraints related to the species growing in the region, also conditioned the exploitation of plant resources. Plants however, had their particular role within the framework of the Fuegian economy: they were a constant source of raw materials, as well as a valuable source of food which complemented the other dietary resources including meat and seafood.

This chapter focuses on the relationship of those groups with the plant resources available in their region. Ethnographic and archaeological data will be combined in order to provide a better understanding of plant consumption and its role within the broader context of the other available food resources.

Rapoport *et al.* (1998) estimate that we currently use fewer than 1% of all plant species available worldwide. According to the FAO, the current worldwide economy is based on slightly over 100 cultivated species (Rapoport *et al.* 1998), while traditional knowledge about edible or useful wild plants is fast disappearing. The perspective on which resources are available, and which of these may be useful, is quite different among present day hunter-gatherer groups, as it probably was for past hunter-gatherers who used a greater number of species in a wider variety of ways, than it is for modern agriculturally dependent societies.

Based on ethnographic descriptions, we know that plant resources were important to people everywhere. For example, the Ju/'hoansi Bushmen of the Kalahari Desert relied on plants for around 85% of their dietary requirements (Kelly 1995, 65). According to Lee and De Vore (1971), they could name around 200 plant species, 85 of which were classified as edible, including one species (*Ricinodendron rautanenii*) considered 'essential'. The Nukak populations in the Amazon used around 57 plant species, 41 of these were considered edible (Politis 1999, 106).

In the circumpolar regions, taxonomic diversity, density, and availability is relatively low in comparison to lower latitudes with more temperate climatic conditions. The lower solar radiation in high latitudes diminishes productivity, resulting in a limited quantity of available biomass. However, in spite of the reduced variability in these regions, the systematic use of plants, such as the berries of some recurrent species like *Ribes* spp., *Empetrum* spp., *Fragaria* spp. or *Rubus* spp., by, for instance, the people of the Aleutian islands from Alaska to Kamchatka (Veltre *et al.* 2006), the Selknam in Tierra del Fuego (Martínez-Crovetto 1968) and the Sami in Fennoscandia (Qvarnström 2006), is well documented. More recently, Nestle (2000) identified over 1000 potentially useful plant species, of which at least 550 are known to have been used by Arctic populations. Porsild, in reference to the Chukchi, states that:

> when available, vegetable food constitutes a regular part of at least their principal meals, eaten eagerly, and certain kinds even with avidity; furthermore, they consider these foods important enough each year to gather supplies that last them through the long, grim winter. (Porsild 1953, 16)

The traditional societies of Tierra del Fuego have been widely cited in ethographical and archaeological literature (De Agostini 1956; Lothrop 2002 [1928]; Bridges 2000 [1948]; Gusinde 1982 [1931]; Chapman 1982; 1986; Borrero 1991). However, due to a lack of archaeological data and the scarcity of information, particularly in the early ethnographic literature, little attention has been paid to the use of plants. In this chapter, we combine the available ethnographic documentation with the archaeological data, to examine the relationship of Fuegian hunter-gatherer groups to their plant resources and we examine the way this information can be used to obtain a better understanding of subsistence strategies.

Fuegian territory and societies

Tierra del Fuego is an archipelago situated in the southernmost extreme of South America (latitudes 52°–56°). It is formed by the Isla Grande (large island) of Tierra del Fuego and a series of smaller islands surrounded by the Pacific and Atlantic Oceans and the Magellan Strait. The landscape of the Isla Grande is characterised by two main areas, separated by the foothills of the Andes. The north part of the island is dominated by a plateau, while the abrupt relief of the 'Cordillera Fueguina' mountain range is found in the south and west.

The present vegetation is heterogeneous mosaic. Towards the south of the Isla Grande and the adjacent areas surrounding the Beagle Channel, evergreen and deciduous Magellanic forests dominate (Pisano 1977). These forests occupy about a third of the surface of the Isla Grande; *Nothofagus* is the main genus though smaller quantities of *Drimys winteri* (winter's bark) and *Maytenus magellanica* (Magellan's mayten) are also present. The shrubby stratus is poor and only becomes abundant at the forest margins. The principal shrub species are *Berberis buxifolia* (box-leaved barberry), *Berberis illicifolia* (Magellan barberry), *Ribes magellanica* (wild currant), *Embothrium coccineum* (Chilean firetree), and *Chiliotrichum diffusum* (fachine). The *Nothofagus pumilio* (lenga beech) grows best in well-drained soils at low altitude, but is also present on mountain slopes up to the tree line. The forest of *Nothofagus antarctica* (low beech) becomes dominant in areas with higher aridity and shallow soils, or a high water table; it is found most abundantly in the hills of the north and centre of the Isla Grande. In the south and west where the annual rainfall exceeds 800–850 mm, the evergreen forest spreads out and here the dominant tree is *Nothofagus betuloides* (Magellan's beech). In some regions the transition between deciduous and evergreen forest is known as the 'mixed Magellan forest'. Towards the north and east of the Isla Grande, increasing numbers of small meadows and bogs can be found in the forests. The vegetation of the northern plains is limited to a low diversity of shrubs such as *Chiliotrichum diffusum* (fachine).

When Europeans arrived in Tierra del Fuego, it was inhabited by several hunter-gatherer groups including the Yamana, Alakalufs, Haush and Selknam (Fig. 15.1). Each group was specialised in exploiting different resources. None of these groups survived the pressure of European colonisation, due to a combination of the introduction of diseases, the destruction of their way of life and even hunted and killed, and within a few years of the arrival of Europeans the native population had become almost extinct. Nevertheless, during the brief but intense period in which they were in contact with people of European origin, Fuegian populations were studied and described by travellers, scientists and ethnographers (De Agostini 1956; Gusinde 1982 [1931]; Chapman 1982; 1986); Bridges 2000 [1948]; Lothrop 2002 [1928]). The first accounts date to the 16th century and comprise a variety of written and photographic information which can be used together with the archaeological data.

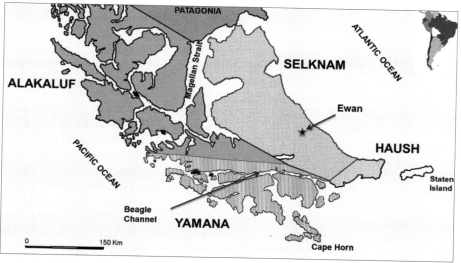

Fig. 15.1. Map of Tierra del Fuego showing the approximate territories occupied by each native group and the archaeological sites mentioned in the text

The traditional interpretation of the Fuegian hunter-gatherer economy, based on ethnographic descriptions, is that they were sea lion and guanaco hunters, on the coast and in the hinterland, respectively. As these ethnographic accounts have been the main source of information about the native Fuegians until recently, they have had a significant impact on how archaeologists approached the study and interpretation of these societies. Most ethnographic accounts characterized the Fuegian communities according to their predominant subsistence base, and how their food was obtained. The maritime groups comprised the Yamana and Alakaluf, who lived on the coast and depended almost entirely on canoes for transport. These people travelled among the Magellan-Fuegian channels; principally exploiting marine mammals and other coastal resources as fish and shellfish. Yamana society has been the object of extensive archaeological, ethnographic, and ethnoarchaeological research (Gusinde 1986 [1937]; Estévez and Vila 1995; Orquera and Piana 1999a; 1999b). The Yamana occupied the shores of the Beagle Channel and the adjacent islands. They moved in bark canoes and exploited a wide range of coastal resources, in particular marine mammals, fish, birds, shellfish and other seafood including crustaceans. They moved frequently and returned to the same places repeatedly. These locations are highly visible archaeologically, due to the large quantity of material remains (Orquera and Piana 1999b).

Yamana society was based on a strictly gendered division of labour. Women were responsible for managing the canoe and gathering molluscs, fungi and plants, as well as firewood collection and the management of fire. Men hunted marine mammals, although this activity required the displacement by canoe and therefore the collaboration of women (Gusinde 1986 [1937]).

The Selknam, who inhabited the north part of the Isla Grande, were based in the interior. They focused largely on hunting terrestrial herbivores, predominantly guanaco, and only exploited coastal resources seasonally. The Haush inhabited in the south-eastern part of the island and shared many customs with the neighbouring Selknam.

A detailed examination of the ethnographic literature, together with the available archaeological data, has permitted a better understanding of the role plants played in Selknam society and how these were organized within the wider food procurement strategies. Selknam society was organised into territories called *haruwen*. According to Chapman's (1982, 20) study, there were 82 *haruwen*, though there appears to have been some instability both in the number of *haruwen* and their boundaries. Some *haruwen* bordered the coast, while others were exclusively inland. Several related families of between 40 and 120 people lived in each *haruwen*. Within the *haruwen*, mobility was high; however, access to different *haruwen*, was restricted by kinship. Although ethnographic accounts emphasize the role of guanaco hunting in subsistence, other food resources are also mentioned including rodents, birds, freshwater fish, marine shellfish, marine mammals, beached cetaceans, fungi, and plants. As among the Yamana, collection of food was governed by a strictly gendered division of labour: hunting and fishing were the responsibility of men, the gathering of molluscs, fungi and plants were female activities (Gusinde 1982 [1931]; Chapman 1982).

Mobility was determined by guanaco movements. In the summer, people moved into the forests and mountain foothills as this is where the guanaco grazed, while in winter (May to November), people moved to the coast if their territories incorporated coastal regions, or to valleys and plains if their territories were inland. The availability of other resources, such as eggs and fruits, was seasonal, while molluscs, rodents and other larger mammals were available all year round. The higher availability of resources in late spring, summer and autumn favoured aggregation into larger groups. In winter, however, the population split into smaller groups (Chapman 1982).

Archaeological investigations confirm the seasonal occupation of some sites and the gathering together of large groups (Mansur and Piqué 2009; 2012). Ewan is located in the centre of the Isla Grande, 12 km from the Atlantic coast. It is a ceremonial site that was occupied during the spring–summer of 1905–1906 to perform a *hain* ritual, a rite of passage for the initiation of young men into adulthood. The site consists of two areas, one area contains a large ritual hut, and the other area, which is located about 200 m away, contains four hearths; these are thought to relate to the many dwelling structures found in this area. The investigations at Ewan have shown not only actual evidence for the seasonal gathering of groups, but also the wide geographical area of resource collection. This included coastal resources (molluscs, fish), terrestrial animals, notably guanaco and sheep, as well as resources from the surrounding forest, in particular a range of fruits and other plant parts were also identified.

Use of plants among the Fuegian societies: ethnography and archaeobotany

Among the ethnographers, the general impression was that the role of plants in the diet was marginal, and that plant gathering was a secondary economic activity. Martin Gusinde, possibly the most significant ethnographer of the region, describes the Selknam use of plants for food as follows: 'The yield from the plant kingdom is so small and meagre that it does not count in the housekeeping' (Gusinde 1982 [1931], 407). About Yamana plant consumption he writes: 'Mother Nature voluntarily offers him fungi and berries of various kinds, and these, as we realize, are mere condiments' (Gusinde 1986 [1937], 291). These comments reflect the general thinking that dominated Fuegian ethnographies.

Plant diversity in Tierra del Fuego is low even in comparison with other circumpolar regions. Moore (1983) recorded 545 species, of which around 417 are considered native to Tierra del Fuego. In Alaska for example, which has a similar latitudinal range, 1550 species have been recorded (Hultén 1968). This low plant biodiversity helped to reinforce the idea that the diet was almost exclusively meat-based. Equally, while the lack of focus on plants in the diet may in part be due to the low plant biodiversity and the highly seasonal availability of these resources, gathering activities were usually carry out by women. It is likely that the European ethnographers of the late 19th and early 20th centuries – all of whom were male – may have been more interested in accompanying men on their hunts, or in other activities such as tool manufacture and house building (Berihuete 2010, 35). However, plants were consumed on a regular basis, and in spite of their apparently low quantitative dietary contribution, a significant effort was placed on plant collection.

We began by analysing ethnographic publications including Gallardo (1998 [1910]), Beauvoir (1998 [1915]), Gusinde (1982 [1931]), Bridges (2000 [1948]), and Martínez-Crovetto (1968), for the Selknam group, and Bridges (1933), Gusinde (1986 [1937]), and Orquera and Piana (1999b), for the Yamana. The selection of works was based on their contemporaneity with the occupation of the studied archaeological sites, except in the case of Martínez-Crovetto (1968), as this is the only publication comprising ethnobotanical research among the last Selknam people who lived in the late 1960s. Based on these publications, we were able to document which species were consumed, which parts of the plants were used, and the processes of production and consumption.

Table 15.1 lists 39 plant, fungi and lichen taxa which were identified in the ethnographic literature as used by Selknam groups, (Berihuete 2010; 2013). Different parts of these plants were gathered and consumed, including the fruits of prickly heath (*Pernettya mucronata*), mustard seeds (*Descurainia canenscens/antarctica*), dandelion leaves (*Taraxacum magellanicum*), the stems of some rushes (*Marsippospermum grandiflorum*), 'clavelito' roots (*Hypochoeris incana*), lenga beech sap (*Nothofagus pumilio*), and the bark of the winter bark tree (*Drimys winteri*) (Berihuete 2013; 2014). Different methods of preparation include direct raw consumption (fruits of red crowberry *Empetrum rubrum*), ash baking (for example the roots and tubers of *Bolax caespitosa* and *B. gummifera*) and the 'complex' preparation of *Descurainia antarctica* seeds which were roasted, ground into a paste with a cylindrical shaped stone, and mixed with seal fat (Gusinde 1982 [1931]; Chapman 1982, 25). Fruits and other plant parts were normally gathered in sacks and baskets and carried to the settlement (Figs 15.2–15.4).

Fig. 15.2. Yamana woman gathering berries (photo: De Agostini, canoeros Rosa Yagan o Lukataia, ?1917)

Fig. 15.3. Selknam woman picking plants (photo) Martín Gusinde, published in Antropología física T IV–Vol. II, 1989, 617)

Fig. 15.4. Selknam women using baskets to collect their gathered fruit on the Atlantic coast (photo: De Agostini, 1923. Museo Nazionale Della Montagna 'Duca degli Abruzzi' Club Alpino Italiano – Sezione di Torino. Torino, Italia)

Table 15.1: Main species used by Yamana and Selknam native people (taken from ethnographic texts)

Species	English name	Used part	Preparation way	Use	Used by
Acaena ovalifolia Ruiz & Pavón	2-spined Acaena	Roots	Boiled, applied with a bandage to wounds	Medicinal	Selknam
Adesmia lotoides Hooker f.	Leguminosae family	Rhizomes	Direct consumption	Food	Selknam
Agaricus pampeanus Speg.	–	Fruiting body (mushroom)	Raw	Food	Selknam
Agropyron patagonicum (Speg.) Parodi	Couch grass family	Flower wearing stalks	Little baskets	Technology	Selknam
Apium australe Thouars	Wild celery	Leaves & roots	Direct consumption or boiled	Food	Selknam Yamana
Arjona patagonica Dcne	Santalaceae family	Roots & tubers	Direct consumption	Food	Selknam
Armeria maritima (Miller) Willd.	Sea thrift	Roots	Baked in ashes	Food	Yamana
Azorella filamentosa Lam.	Azorella	Roots & tubers	Direct consumption or baked in ashes	Food	Selknam
Azorella lycopodioides Gaudich, *A. monantha* Clos, *A. selago* Hooker f., *A. trifurcata* (Gaertner) Hooker f.	Azorella	Roots & tubers	Direct consumption or baked in ashes	Food	Selknam Yamana
Berberis buxifolia Lam.	Box-leaved barberry	Berries	Direct consumption	Food	Selknam
Berberis empetrifolia Lam.	Crowberry-leaved barberry	Berries	Direct consumption	Food	Selknam
Berberis ilicifolia L.f.	Michay, barberry sp.	Berries	No data	Food	Yamana
Bolax caespitose Hombron & Jacquinot	Apiaceae family	Roots & tubers	Direct consumption or baked in ashes	Food	Selknam
Bolax gummifera (Lam.) Sprengel	Balsam bog	Roots & tubers	Direct consumption or baked in ashes	Food	Selknam
Boopis australis Dcne	Calyceraceae family	Roots & tubers	Baked in ashes	Food	Selknam
Calvatia bovista var. *magellanica* (L.) Pers.	Burst puffball	Fruiting body (mushroom)	Dried as tinder for starting fires	Technology	Selknam

		Used part	Preparation way	Use	Used by
Calvatia lilacina (Mont. & Berk.) Henn.	Puffball	Fruiting body (mushroom)	Burnt: smoke inhaled to cure colds	Medicinal	Selknam
Cardamine glacialis (Forster f.) DC	Bittercress sp.	Leaves (?)	No data	Food	Yamana
Chiliotrichum diffusum (Forster f.) O. Kuntze	Fachine	Branches	Tattoos. Flowers were rubbed on eyes to clear sight	Personal ornament. Medicinal	Selknam
Cladonia laevigata Vain.	Lichen species	Whole plant	For body washing, before getting dry with ánhuel (*Usnea* sp.) (a type of lichen)	Hygiene	Selknam
Cyttaria darwinii Berkeley; *C. Harioti* Fischer; *C. Hookeri* Berkeley	–	Fruiting body (mushroom)	Raw or baked	Food	Selknam
Descurainia canescens auct., non (Nutt) Prantl; *D. antarctica* (E. Fourn.) O. E. Schultz	Tansy mustard genus	Seed	Ground & toasted, mixed with guanaco fat	Food	Selknam
Drimys winteri Foster & Foster f.	Winter's Bark	Bark	Decoction against dandruff	Hygiene	Selknam
Embothrium coccineum Foster & Foster f.	Chilean firetree	Nectar	No data	Food	Yamana
Empetrum rubrum Vahl ex Willd.	Red crowberry (Diddle-dee)	Berries	Direct consumption. Raw.	Food	Selknam
Festuca gracillima Hooker f.	Tussac	Grass	Stuffing for leather shoes	Clothing	Selknam Yamana
Fistulina hepatica (Schaeff.) With.	Beefsteak	Fruiting body (mushroom)	Raw	Food	Selknam
Fragaria chiloensis (L.) Mill.	Chilean strawberry	Fruits	Direct consumption	Food	Selknam Yamana
Hypochoeris incana (Hooker & Arn.) Macloskie; *H. incana* var. *integrifolia* (Sch. Bip. ex Walp.) Cabrera	Asteraceae family	Roots & tubers	Grilled or baked in the ashes	Food	Selknam
Hypochoeris radicata L.	Hairy cat's ear	Leaves	Direct consumption	Food	Selknam
Marsippospermum grandiflorum (L. f.) Hooker f.	Juncaceae family	Stalk	Roasted & flattened by hand to weave baskets	Technology	Selknam
Mysodendron punctulatum Banks ex D.C.	Mistletoe genus	Whole plant	Body rubbing against muscular pains	Medicinal	Selknam
Nothofagus antarctica (Forster f.) Oersted	Antarctic beech/low beech	Wood	Tools & hut building	Technology	Selknam

Species	English name	Used part	Preparation way	Use	Used by
Nothofagus betuloides (Mirbel) Oersted	Magellan's beech	Bark	Bird hunting torches	Technology	Selknam
Nothofagus pumilio (Poeppig & Endl.) Krasser	Lenga	Sap	Direct consumption	Food	Selknam Yamana
Oreomyrrhis andicola auct., non (Kunth) Hooker f.	Apiaceae family	Roots & tubers	Direct consumption	Food	Selknam
Osmorhiza chilensis Hooker & Arn.	Sweet cicely	Root	Baked in ashes	Food	Yamana
Pernettya mucronata (L. f.) Gaudich. ex G. Don.	Ericaceae family	Berries	Direct consumption	Food	Selknam Yamana
Pernettya pumila (L. f.) Hooker	Ericaceae family	Berries	Direct consumption	Food	Selknam Yamana
Philesia sp.	–	Flowers	No data	Food	Yamana
Poa flabellata (Lam.) Raspail	Tussac	Stems	No data	Food	Yamana
Polyporus eucalyptorum Fr.*	–	Fruiting body (mushroom)	Direct consumption	Food	Selknam
Polyporus aff. *Gayanus* Lév.	–	Fruiting body (mushroom)	Direct consumption	Food	Selknam
Ribes magellanicum Poiret	Wild currant	Berries, tea of leaves and infusion of bark	Direct consumption raw or boiling of some parts	Food	Selknam Yamana
Rubus geoides Sm.	Rainberry	Berries	Direct consumption, raw	Food	Selknam Yamana
Taraxacum magellanicum, Comm. ex Sch. Bip.; *T. gilliesii* Hooker & Arn. and *T. officinale* Weber	Dandelion	Flowers, leaves and roots	Direct consumption, raw	Food	Selknam Yamana
Usnea magellanica (Mont.) Motyka	Old's man beard	Whole plant	As towel	Hygiene	Selknam

Among the Yamana, 17 plant and fungi species were identified as regularly used (see Table 15.1) (Berihuete 2010). These include berries, roots, tubers, and leaves, though little information has survived regarding cooking and preparation methods. The collection of some plants could be labour intensive, for example underground storage organs, had to be dug for. Some roots were most likely roasted whole; these include the Chilean sweet-cicely (*Osmorhiza chilensis*), sea thrift (*Armeria maritima*), and wild celery (*Apium australe*). However, the tubers of *macachi* (*Arjona patagonica*) were so highly appreciated by the Selknam, that during 'harvest' time, people moved to coastal areas where they were particularly abundant (Martínez-Crovetto 1968). This species seems to have been important for other indigenous people living in Patagonia who also used it as a food (Ochoa and Ladio 2011, 275). Other information includes the drying of fungi and some plant resources that required a series of specific steps in their preparation; for example to obtain plant sap, the bark had to be detached and then the inner bark scratched with a spoon or a shell (Martínez-Crovetto 1968). Another example of complex processing is the mixing of prickly heath berries with whale or sea lion fat. This type of preparation is also known elsewhere, the Selknam used *Descurainia antarctica* in the same way, and the Inuit of Alaska also mixed crowberries (*Empetrum nigrum*) with whale or sea lion fat (Moerman 1998, 210).

Among both the Selknam and the Yamana, different words related to plant management and plant biological cycles, such as 'to reap', 'to boil', 'to fry', 'to roast', 'well cooked', 'half cooked', 'to gather', 'to gather for oneself' or 'to gather for storing and later use' and others like 'to germinate', 'to sprout', 'to bloom', 'first fruits', 'early fruit', 'green fruit' or 'to ripe', which suggest an applied botanical understanding, were used to describe the reproductive cycles of plants and management of their use of this resource (Berihuete 2010; 2014).

The retrieval and analysis of plant remains are slowly being incorporated into archaeological excavations in the region. Yamana sites such as Túnel VII and Lanashuaia, and Selknam sites such as Ewan, Kami 1 and Kami 7, have had an archaeobotanical component incorporated into their excavations (Berihuete 2010). Among these sites, Ewan stands out due to the quantity of recovered remains (over 12,000 seeds). Consequently, the archaeobotanical analysis from this site has provided a more precise picture of plant use among Selknam communities (Berihuete 2010; 2013; Mansur and Piqué 2012).

All recovered plants are interpreted as food remains, specifically because the predominant plants identified are the red crowberry (*Empetrum rubrum*) and cleavers (*Galium* sp.). Both are edible, and red crowberry (*Empetrum rubrum*) is also referred to as food in the ethnographic literature for both the Selknam and Yamana. Although only very few remains have been recovered, we have also documented the presence of these species at all other sites.

Discussion and conclusions

While other authors have focused their attention on the nutritional values of different traditional plant foods consumed in circumpolar regions (i.e., Kuhnlein 1989; Kuhnlein and Turner 1991; Kuhlein and Soueida 1992), for example, kelp, berries, and sorrel were

Table 15.2 Nutritional values of some plant foods growing in Tierra del Fuego

Scientific name	Common name	Part Used	Food energy (kcal)	Water g	Protein g	Fat g	Carbohydrate g	Fiber g	Ash g	Thiamine mg	Riboflavin mg	Niacin mg	Vit. C mg	Vit. A RE	Calcium mg	Phosphorus mg	Sodium mg	Potassium mg	Magnesium mg	Copper mg	Zinc mg	Iron mg	Manganese mg	Molybdenum mg	Chloride mg
Berberis aquifolium Berberidaceae	Tall Oregon-grape	Berry	–	–	–	–	–	–	–	–	–	–	–	–	–	–	–	–	–	–	–	–	–	–	–
Berberis nervosa Berberidaceae	Low Oregon-grape	Berry	72	81	3.4	1.5	13.4	–	0.7	–	–	–	27.6	–	36	–	–	–	16.2	–	0.9	0.7	–	–	–
Berberis thunbergii Berberidaceae	Japanese barberry	Berry	–	–	3.2	2.0	–	2.7	1.0	–	–	–	–	–	83	83	–	–	–	–	–	–	–	–	–
Chenopodium species Chenopodiaceae	Goosefoot/pigweed	Greens	43	–	2.4	1.1	–	2.4	0.7	–	–	–	80.0	1160	309	–	–	–	–	–	–	1.2	–	–	–
Descurainia pinnata Brassicaceae	western tansy mustard	Seeds	–	–	24.4	38.4	63.3	–	–	–	–	–	–	–	–	–	–	–	–	–	–	–	–	–	–
Empetrum nigrum Ericaceae	Black crowberry/curlewberry	Fruit	35	89	0.2	0.7	9.5	5.9	0.7	<0.01	<0.01	0.1	51.0	–	90	11	2.5	87	7.9	1.0	0.1	0.4	0.4	–	–
Gallium aparine Rubiaceae	Common bedstraw/cleavers	Greens	–	–	1.6	–	–	2.1	1.8	–	–	–	–	–	145	65	39	517	13.0	0.1	–	3.2	0.7	–	97.0
Phalaris canariensis Poaceae	Canary grass	Grains	–	9	18	5.5	–	1.5	2.9	0.65*	–	1.2*	–	–	27	580	18	363	181	0.1	4.5	6.4	3.6	0.4	–
Phalaris canariensis Poaceae	Canary grass	Roots	–	–	4.6	–	–	–	–	–	–	–	–	–	272	136	181	523	90.7	0.5	4.0	181	8.7	4.2	–
Polygonum species Poligonaceae	Knotweed	Greens	–	–	3.3	0.4	1.4	1.1	0.7	–	–	–	–	–	99	50	2.8	240	80.0	–	–	4.1	0–7	–	–
Polygonum aviculare Poligonaceae	Common knotweed	Greens	–	–	–	–	–	–	–	–	–	–	77.9	–	–	–	–	–	–	–	–	–	–	–	–

Scientific name	Common name	Part Used	Food energy (kcal)	Water g	Protein g	Fat g	Carbohydrate g	Fiber g	Ash g	Thiamine mg	Riboflavin mg	Niacin mg	Vit. C mg	Vit. A RE	Calcium mg	Phosphorus mg	Sodium mg	Potassium mg	Magnesium mg	Copper mg	Zinc mg	Iron mg	Manganese mg	Molybdenum mg	Chloride mg
Ribes species Saxifragaceae	Currant	Fruit	50	86	1.4	0.2	12.1	34	0.6	0.04	0.05	0.1	41.0	72	32	23	20	257	13.0	0.1	0.2	1.0	0.2	–	–
Ribes species Saxifragaceae	Gooseberry	Fruit	44	88	0.9	0.6	10.2	1.9	0.5	0.04	0.03	0.3	27.1	29	25	27	1.0	198	10.0	0.1	0.1	0.3	0.1	–	–
Rubus species Rosaceae	Raspberry	Berry	49	86	0.9	0.6	11.6	3.0	0.4	0.03	0.09	0.9	25	13	22	12	0.0	152	18.0	0.1	0.5	0.6	1.0	–	–
Rubus species Rosaceae	Blackberry	Berry	52	86	0.7	0.4	12.7	4.1	0.5	0.03	0.04	0.4	21.0	16	32	21	1.0	196	20.0	0.1	0.3	0.9	1.3	–	–
Rubus species Rosaceae	Wild dewberry	Berry	–	84	0.9	0.8	14.0	–	0.6	–	–	–	–	–	54	31	–	–	–	–	–	–	–	–	–
Rumex species Polygonaceae	Dock	Greens	28	91	2.1	0.3	5.6	0.8	1.1	0.09	0.22	0.5	0.48	400	44	63	4.0	390	103	–	–	2.4	–	–	–
Stellaria media Caryophyllaceae	Chickweed/common starwort	Leaves	–	90	1.6	0.2	5.3	1.8	1.3	0.02	0.14	0.5	34.8	613	91	56	122	585	41.7	0.1	–	3.5	1.7	–	70.0
Stellaria media Caryophyllaceae	Chickweed/common starwort	Seeds	–	–	17.5	0.5	51.7	8.8	16.5	–	–	–	–	–	–	–	–	–	–	–	–	–	–	–	–
Taraxacum officinale Asteraceae	Dandelion	Flowers	–	–	–	–	–	–	–	–	–	–	–	9.2	–	–	–	–	–	–	–	–	–	–	–
Taraxacum officinale Asteraceae	Dandelion	Greens	45	85	2.7	0.7	9.2	1.6	1.8	0.19	0.28	–	35.0	1400	209	64	73	422	51.5	0.3	–	4.1	0.7	–	329
Taraxacum officinale Asteraceae	Dandelion	Buds	–	86	3.1	–	–	–	–	–	–	–	30.0	80	–	–	–	–	–	–	–	–	–	–	–

Values from Kuhnlein and Turner 1991, and (*)http://www.botanical-online.com/

major carbohydrate sources in the traditional Inuit diet (Kuhnlein and Soueida 1992, 115), there is still a lack of information concerning the species indigenous to Tierra del Fuego. Table 15.2 lists some of the species/genera for which data are available, and which are comparable to the species identified at Ewan. As we can see in the table, some of the species collected by the Fuegians have significant amounts of nutrients. However, in addition to their role in providing essential nutrients including carbohydrates, vitamin C and folic acid, plants could have been gathered by Fuegian people for other purposes. Some plants may have had a social or ritual significance, or they could have been selected for their taste. Fuegian people invested a considerable amount of work preparing their fat cakes with *Descurainia antarctica* seeds. The small seeds had to be gathered, ground, and mixed with seal fat. This suggests that the cakes were a valuable food source in terms of their cultural/ culinary as well as their nutritional qualities.

In addition to their role as providers of essential carbohydrates, plant resources in Tierra del Fuego were also important in terms of their micronutrients, including minerals and vitamins. High levels of vitamin C are found in many berries that grow in this region and that were used by the native populations. For native *Rubus* species, the quantity of vitamin C ranges between 43.4–89.4 mg/100 g; for *Berberis*, 36.6–198.8 mg/100 g; for *Ribes magellanicum* 111.6 mg/100g and for *Empetrum rubrum* the mean value is 87.6 mg/100 g (Pino *et al.* 2011). If we consider that the mean value of vitamin C in oranges is 53 mg/100 g, it places into perspective the value of these fruits.

While a diet that consists of lean meat alone can lead to 'rabbit starvation' (Porsild 1953, 15; World Health Organisation 2002, 230), a diet that is very high in animal-based products is possible, by including fats and other resources obtained from animals such as chyme (Milton 2000, 665). However, although the variety of edible plant resources is scarce and Fuegian societies developed elaborate strategies to exploit terrestrial and marine animal and fish resources, it is clear that plants were also significant. The scarcity of fruits and fresh vegetables throughout much of the year may have turned these into a desired and awaited food when they became available in terms of their nutritional qualities and the variation that they would have offered in the diet.

Many of the plant resources were only available seasonally. For instance leaves are more suited to be gathered in spring, before flowering occurs, berries usually ripen between summer and early autumn, while roots are typically gathered in late autumn – winter, when plants build up their carbohydrate energy sources. Ethnobotanical descriptions from other regions illustrate how some groups leave crowberries (*Empetrum nigrum*) on the plant through the winter for collection fresh or frozen into the early spring, when they can be gathered even from beneath snow (Porsild 1953: 20). This type of resource management offered an opportunity for access to a range of plant-based nutrients over extended periods; however there is no ethnographic data on plant resource management strategies for Tierra del Fuego. Being a highly mobile society, transporting plant foods is likely to have been reduced to a minimum. However, preservation techniques have been documented for fungi as these were dried over the fire and carried around while travelling (Gusinde 1982 [1931]).

While it appears that in Tierra del Fuego plant resource use may have been lower than in other regions of similar latitude, it is clear that Fuegian communities had developed highly efficient subsistence strategies that maximised the available resources, including plants, and allowed them to thrive within their environmental context (Nestle 1999, 211)

Despite their relatively low abundance and diversity, plants played an important role in subsistence, while possibly also fulfilling roles in cultural and social aspects of life. Using a combination of ethnographic literature and ethnobotany, Ochoa and Ladio (2011, 272) claim that the use of roots and tubers decreased over recent time as contact with farmers and the availability of domesticated products increased, until they became a marginal resource. Though no firm data exist on relative proportions of other traditional foods, it is very likely that their use also altered after contact.

Ethnographic and archaeobotanical data demonstrate that plants were gathered and consumed in a variety of ways by the inhabitants of Tierra del Fuego though as we have outlined, Fuegian ethnography is marked by the lack of accurate descriptions on the use of plants. However, while ethnographic descriptions are relatively scarce particularly when compared to the rich ethnobotanical studies available for other regions (e.g., Moerman, 1998 for North America), they provide enough information to demonstrate that plants were an important complement to the diet.

Acknowledgements

Part of this research was carried out under the framework of several wider projects: 'Ritual en grupos cazadores-recolectores. Espacios rituales y espacios domésticos entre los Selknam de Tierra del Fuego (Argentina)' (IPHE/AMN/CMM, Arqueología Exterior 2005); 'Estudi de la variabilitat en el consum de recusos litorals i aquàtics des d'una perspectiva arqueològica' (2006EXCAVA00021) and 'Arqueologia de la gestión de los recursos sociales y el territorio (AGREST)' (2009 SGR 734). 'Explotación de recursos y circulación humana en la zona central de Tierra del Fuego' (CONICET PIP 0452), 'Ambiente, recursos y dinámica poblacional en sociedades cazadoras-recolectoras de Tierra del Fuego' (PICT ANPCYT 2648). Marian Berihuete is a post-doctoral fellow of the Alexander von Humboldt Foundation.

References

Beauvoir, J. M. 1998 [1915]. *Diccionario Shelknam. Indigenas de Tierra del Fuego*. Ushuaia: Sus tradiciones, costumbres y lengua. Zagier and Urruty Publications

Berihuete Azorín, M. 2010. *El Papel de los Recursos Vegetales No Leñosos en las Economías Cazadoras-Recolectoras: Propuesta Para el Estudio de su Gestión: El Caso de Tierra de Fuego (Argentina)*. PhD dissertation. Departament de Prehistòria, Universitat Autònoma de Barcelona. http://hdl.handle.net/10803/32064

Berihuete-Azorín, M. 2013. First archaeobotanical approach to plant use among Selknam hunter-gatherers (Tierra del Fuego, Argentina). *Archaeological and Anthropological Sciences* 5(3), 255–266. DOI 10.1007/s12520-013-0137-4

Berihuete Azorín, M. 2014. Las plantas en las economías fueguinas: una perspectiva etnoarqueológica. In Oria, J. and Tivoli, A. M. (eds), *Cazadores de Mar y Tierra. Estudios Recientes en Arqueología Fueguina*, 389–408. Ushuaia: Editora Cultural Tierra del Fuego

Bridges, T. 1933. *Yamana–English: A Dictionary of the Speech of Tierra del Fuego.* Mödling: Editado por Ferdinand Hestermann y Martin Gusinde. Missionsdruckerei St. Gabriel

Bridges, E. L. 2000 [1948]. *El último confín de la Tierra. (Uttermost Part of the Earth).* Buenos Aires: Editorial Sudamericana

Chapman, A. 1982. *Drama and Power in a Hunting Society: the Selk'nam of Tierra del Fuego.* Cambridge: Cambridge University Press

Chapman, A. 1986. *Los Selk'nam. La vida de los Onas.* Buenos Aires: Emecé

De Agostini, A. M. 1956. *Treinta años en la Tierra del Fuego.* Buenos Aires: Peuser

Estévez, J. and Vila, A. (Coords), 1995. *Encuentros en los conchales fueguinos.* Barcelona: Treballs d'Etnoarqueologia 1. CSIC/UAB

Gallardo, C. R. 1998 [1910]. *Los Onas de Tierra del Fuego.* Ushuaia: Zagier and Urruty Publications

Gusinde, M. 1982 [1931]. *Los indios de Tierra del Fuego. Tomo 1: Los selk'nam.* Buenos Aires: Centro argentino de etnología americana, Consejo Nacional de Investigaciones Científicas y Técnicas

Gusinde, M. 1931. *Die Feuerland Indianer. Band I: Die Selk'nam.* Mödling bei Wien: Anthropos

Gusinde, M. 1986 [1937]. *Los indios de Tierra del Fuego. Tomo II: Los Yamana.* Buenos Aires: Centro argentino de etnología americana, Consejo Nacional de Investigaciones Científicas y Técnicas,. (originally published as *Die Feuerland Indianer. Band II: Die Yamana.* Mödling bei Wien: Anthropos)

Hultén, E. 1968. *Flora of Alaska and Neighbouring Territories: a manual of the vascular plants.* Stanford CA: Stanford University Press

Kelly, R. L. 1995. *The Foraging Spectrum. Diversity in Hunter-Gatherer Lifeways.* Washington DC & London: Smithsonian Institution Press

Kuhnlein, H. V. 1989. Nutrient values in indigenous wild berries used by the Nuxalk People of Bella Coola, British Columbia. *Journal of Food Composition and Analysis* 2, 28–36

Kuhnlein, H. V. and Turner, N. J. 1991. *Traditional Plant Foods of Canadian Indigenous People. Nutrition, Botany and Use.* Philadelphia PA: Food and Nutrition in History and Anthropology 8

Kuhnlein, H. V. and Soueida, R. 1992. Use and nutrient composition of traditional Baffin Inuit foods. *Journal of Food Composition & Analysis* 5, 112–26

Lee, R. B. and De Vore, I. 1971. The Bushmen of the Kalahari Desert, record and filmstrip. In *Studying Societies: Patterns in Human History.* New York: Macmillan, developed for the Anthropology Curriculum Study Project, American Anthropological Association

Lothrop, S. K. 2002 [1928]. *The Indians of Tierra del Fuego.* New York: Museum of the American Indian, Heye Foundation

Mansur, M. E. and Piqué, R. 2009. Between the forest and the sea: hunter-gatherer occupations in the subantarctic forests in Tierra del Fuego (Argentina). *Arctic Anthropology* 46(1–2), 144–57

Mansur, M. E. and Piqué, R. 2012. *Arqueología del Hain. Investigaciones Etnoarqueológicas en un Sitio Ceremonial Selknam de Tierra del Fuego. Implicancias Teóricas y Metodológicas para los Estudios Arqueológicos.* Madrid: CSIC

Martínez-Crovetto, R. 1968. Nombres de plantas y sus utilidades según los indios Onas de Tierra del Fuego. *Revista de la Facultad de Agronomía y Veterinaria de la Universidad del Nordeste, Estudios Etnobotánicos* 4(3) 1–20 http://www.tierradelfuego.org.ar/museo/virtual/botanica.htm#%281%29

Milton, K. 2000. Hunter-gatherer diets-a different perspective. *American Journal of Clinic Nutrition* 71, 665–7

Moerman, D. E. 1998. *Native American Ethnobotany.* Portland OR: Timber Press

Moore, D. M. 1983. *Flora of Tierra del Fuego.* Oswestry: Anthony Nelson & Missouri Botanical Garden

Nestle, M. 1999. Animal v. plant foods in human diets and health: is the historical record unequivocal? *Proceedings of the Nutrition Society* 58, pp 211–218. doi:10.1017/S0029665199000300

Nestle, M. 2000. Paleolithic diets, a skeptical view. *Nutrition Bulletin* 25, 43–7

Ochoa, J. J. and Ladio, A. H. 2011. Pasado y presente del uso de plantas silvestres con órganos subterráneos de almacenamiento comestibles en la Patagonia. *Bonplandia* 20(2), 265–84

Orquera, L. A. and Piana, E. L. 1999a. *Arqueología de la región del canal Beagle (Tierra del Fuego, República Argentina)*. Buenos Aires: Publicaciones de la Sociedad Argentina de Antropología

Orquera, L. A. and Piana, E. L. 1999b. *La vida social y material de los Yámana*. Buenos Aires: EUDEBA

Pino, M. T., Obando, L. and Torres, J. 2011. Propiedades Antioxidantes de algunos Frutales Nativos Magallánicos. Especial INIA y Los Alimentos. *Tierra Adentro* 95, 55–60. http://www.inia.cl/wp-content/uploads/revista_tierra_adentro/TA95.pdf

Pisano, E. 1977. Fitogeografía de Fuego-Patagonia chilena. Comunidades vegetales entre latitudes 52° y 56°. *Anales del Instituto de Punta Arenas* 8, 121–247

Politis, G. 1999. Plant exploitation among the Nukak hunter gatherers of Amazonia: between ecology and ideology. In Gosden, G. and Hather, J. (eds), *The Prehistory of Food. Appetites for Change*, 99–125. London: Routledge

Porsild, A. E. 1953. Edible plants of the Arctic. *Arctic* 6, 15–34

Qvarnström, E. 2006. *De tycka emellertid av gammal vana att det smakar gott, och tro dessutom att det är bra för hälsan. Samiskt växtutnyttjande från 1600-talet fram till ca 1950 (Sami Plant Use from the 1600s to 1950)*. Master-thesis, Swedish University of Agricultural Sciences, Umeå

Rapoport, E. H., Ladio, A., Raffaele, E., Ghermandi, L. and Sanz, E. H. 1998. Malezas Comestibles. Hay yuyos y yuyos. *Revista de Divulgación Científica y Tecnológica de la Asociación Ciencia Hoy* 9, 49 http://www.cienciahoy.org.ar/hoy49/malez01.htm

Veltre, D. W., Pendleton, C. L., Schively, S. A., Hay, J. A. and Tatarenkova, N. 2006. *Aleut/ Unangax Ethnobotany: An annotated Bibliography*. Akureyri: CAFF International Secretariat

World Health Organization, 2002. *Protein and amino acid requirements in human nutrition*. Report of a joint FAO/WHO/UNU expert consultation. Geneva: WHO Technical Report Series, 935

16. Ethnobotany in evolutionary perspective: wild plants in diet composition and daily use among Hadza hunter-gatherers

Alyssa N. Crittenden

Human–plant interactions are often fundamental to evolutionary models of nutrition, medicine, behavioural ecology, and aspects of cultural transmission. Here, I explore ethnobotany, defined as the study of the relationship between humans and plants and their evolutionary significance, among the Hadza hunter-gatherers of Tanzania. The diets of foraging populations are often conscripted as referential models of the Palaeolithic diet, yet few quantitative nutritional studies of hunter-gatherer diet composition are available. I provide data on annual diet composition among the Hadza, including species targeted and per cent contribution to diet. In addition to diet, I outline Hadza use of wild plants as extraction and/or processing tools (e.g., bows and arrows and digging sticks), and non-nutritive daily uses of plants (e.g., medicinal purposes, body adornment, and household activities). The evolutionary significance of each category of plant use is examined. Data on Hadza ethnobotany not only provides invaluable and timely cross-cultural information on plant use, but also enhances our understanding of early hominin paleoecology.

The utilisation of wild plants is at the core of evolutionary perspectives on nutrition, medicine, and aspects of cultural transmission. Increasingly, plant-human interactions are associated with human adaptive physiology, morphology, and behaviour (Etkin 1996; Johns 1996; Bridges and Lau 2006).

This relationship between people and wild plants, termed 'ethnobotany', has many historic manifestations and draws widely from varied disciplines such as ecology, anthropology, botany, and ethnopharmacology (Clement 1988; de Albuquerque and Hanazaki 2009). Despite this interdisciplinary academic ancestry, many ethnobotanical studies ignore the evolutionary significance of human plant exploitation. Recently there has been a push to expand the purview of ethnobotany to include the study of the 'direct interrelations between humans and plants and their evolutionary consequences' (Johns 1996, 10). Here, I will operationalise this definition of ethnobotany by placing the use of wild plants by the Hadza hunter-gatherers of Tanzania in evolutionary perspective. I provide data on annual diet composition (including identification and description of species targeted and per cent contribution to diet), the use of wild plants as extraction and/or processing tools (e.g., bows and arrows, digging sticks), and non-nutritive daily uses of plants (e.g., medicinal purposes, body adornment, household activities). A greater understanding of ethnobotany among the Hadza informs our interpretation of evolutionary models of nutrition, foraging behaviour, and paleoecology.

Hunter-gatherers and the evolution of the human diet

Shifts in diet composition have been linked to many key milestones in human evolution such as family formation, pair bonding, tool making, brain expansion, cooperation, and increased longevity (Lee and DeVore 1968; Washburn and Lancaster 1968; Pennisi 1999; Aiello and Wheeler 1995; Leonard *et al.* 2007; Wrangham 2012). In the past decade, our understanding of the diets of early hominins has been greatly augmented with evidence from the analysis of morphology, stone tools, butchery patterns, and fossil dentition – including microwear, stable isotope analysis, and the extraction of organic compounds (Hardy *et al.* 2012) and phytoliths (Henry *et al.* 2012; Lucas *et al.* 2013) from dental calculus. These lines of direct evidence suggest that approximately 2.5 million years ago, the earliest members of the genus *Homo* had a generalised diet composed of plant and animal resources – foods with both C_3 (trees and bushes) and C_4 (grasses and sedges) photosynthetic pathways (Lee-Thorp *et al.* 2001; Sponheimer *et al.* 2006; Ungar and Sponheimer 2011). Building upon these findings using indirect evidence, such as referential modelling of hunter-gatherer diet composition, greatly supplements our understanding of the paleoecology of early hominins.

The diets of contemporary hunting and gathering populations are often recruited not only as a reference standard for the evolution of human nutrition but also as a potential representation of Palaeolithic diet (Eaton and Konner 1985; Eaton *et al.* 1996; Cordain *et al.* 2000). Hunter-gatherers are by no means models of Palaeolithic populations. They do, however, practise a nomadic lifestyle and subsistence regime that characterises the bulk of human history, making them ideal populations in which to study evolutionary aspects of behavioural and nutritional ecology. Despite attempts to characterise a universal hunter-gatherer diet, no such composite diet exists, as there is wide variation in terms of ecology, climate, and overall diet composition between populations (Eaton 2006; Strohle and Hahn 2006; Strohle 2010); reliance on plant foods ranges from 6–15% of the annual diet in tundra regions and 46–55% in grasslands (Cordain *et al.* 2000). Despite the significance placed on the diets of foraging peoples, few quantitative nutrition studies are available (see Kaplan *et al.* 2000 and Cordain *et al.* 2002).

Below, quantitative annual diet composition data for the Hadza of Tanzania are presented and their implications for the evolution of the human diet are explored. The Hadza are a particularly relevant foraging population in which to study the evolution of the human diet because they occupy an area of East Africa that is often referred to as 'the crucible of human evolution' (Johanson and Wong 2009; Roberts *et al.* 2012) and harbours resources that are likely similar to those exploited by our hominin ancestors.

The Hadza

The Hadza hunter-gatherers live in a 4000 km² area around the shores of Lake Eyasi in northern Tanzania, East Africa. Of the total population size, which numbers roughly 1000, only approximately 200 individuals practise a predominantly hunting and gathering way of life, meaning that the bulk of their diet is derived from wild plant foods and game animals. The 200 Hadza who continue to practise a foraging lifestyle live in small nomadic camps with fluid residential composition (Marlowe 2010). Camp size, which fluctuates depending on seasonality and/or resource availability, can swell up to 100 members when resources are freely available and contract to a mere 5–10 individuals when resources are scarce. There is frequent movement between camps, which may be linked to Hadza notions of land rights, as they do not traditionally recognise control over natural resources (Woodburn 1968; Kaare 1994). Increasingly, however, they are staying in larger camps for longer periods of time in direct response to ecological pressure from the influx of other tribes moving into the area (Fosbrooke 1956; Crittenden 2014).

The Hadza, like most foraging populations, are central-place provisioners (Marlowe 2006) who collect food on a daily basis and return to camp to distribute the food to weanlings, dependent children, elderly, injured, or disabled camp members. Food is widely shared both within and outside of the household (Marlowe 2004). Camps move approximately every 2–3 months in response to the seasonal availability of resources and distinct wet and dry seasons associate with differential subsistence behaviours and social arrangements. During the dry season, which lasts roughly from June to October, camp size is relatively large. The larger aggregation of people during the dry season may be due to the limited availability of water and/or the greater concentration of hunted animals near watering holes (Bunn *et al.* 1988). During the wet season, which lasts roughly from November to May, camp size is smaller and may be linked to more freely available drinking water (Woodburn, 1968).

The Hadza diet, which is very well balanced, includes a wide variety of plant foods, small to large sized game, a vast array of bird species, and the larvae and honey of both stingless and stinging bees. Significant sex differences exist in both diet composition and foraging behaviour. Men consume a greater amount of meat, whereas women consume a great proportion of plant foods (Berbesque *et al.* 2011; Schnorr *et al.* 2014). This maps onto foods targeted while foraging, as women typically forage in groups and target plant foods, while men tend to hunt solo or in pairs and focus on game hunting and honey collection. When unsuccessful on a hunt, either for meat or honey, men will focus collection efforts on baobab fruit. Female foraging may be constrained by the fact that they are carrying nursing infants (Marlowe 2003), who accompany their mothers on daily foraging trips until weaned. Once a child is being weaned (at approximately 2 or 3 years of age) he or she is typically left in camp with older children or an elderly caregiver (Crittenden and Marlowe 2008), being too large to be carried easily and too young to walk long distances to the locations where the women forage. Children are active foragers in their own right and are able to collect approximately half of their daily caloric intake by the time that they are 5 years old (Jones *et al.* 1989; 1997). Children tend to focus on resources that are relatively easy to collect (for example berries, fruit, nuts, and/or small game animals and birds) and located close to camp (Crittenden *et al.* 2013; Crittenden and Zes 2015).

Diet composition

Despite long-standing interest in hunter-gatherer diet composition, relatively few studies of the nutrient composition of wild plant foods have been undertaken. Four published studies to date (Vincent 1985; Murray *et al.* 2001; Schoeninger *et al.* 2001; Crittenden 2009) provide data on the macronutrient composition of several species of plant foods consumed by the Hadza. The diet composition data presented here is extracted from Crittenden (2009) and represents all food taken back to camp during a study period in five camps over 1 calendar year, 2005–2006. The most accurate means of estimating energy contribution by food type is to use kilocalorie dry weight values for each food. The energy values for all plant foods including baobab, berries, honey, legumes, drupes, nuts, and tubers were determined using standard analytical methods for wild foods (Conklin Brittain *et al.* 2006). The energy values for birds and game meat were determined based on published values (Prange *et al.* 1979; Clum *et al.* 1996; USDA 2008). Resource availability and rainfall patterns during the study year map onto the greater resource availability and general seasonal trends in this region of Lake Eyasi over the past two decades (Marlowe 2010), allowing us to interpret these data as an accurate snapshot of diet composition among the Hadza.

The contribution of each food type (measured in %kilocalories) to the overall annual diet is shown in Figure 16.1. Birds, small, medium, and large sized game meat comprise 32% of the annual diet. During the study period, targeted animals included several species of birds and the following game animals: hyrax, aardvark, serval, baboon, bushbaby, dik dik, impala, kudu, zebra, giraffe, and cape buffalo. As the primary focus here is plant foods and their contribution to the diet, a detailed discussion of the types of avian and game meat consumed will not be covered. See Marlowe (2010) and Peterson *et al.* (2012) for an exhaustive list of species targeted. Outlined below is a detailed discussion of each type of plant food consumed. All of the plant foods listed below, except baobab, are exclusively collected by women. The other notable exception is honeycomb, collected only by men, which is included here as a 'plant' food because the liquid honey is the product of plant resources.

Baobab

Baobab fruit (*Adansonia digitata*), or n//obabe, is consumed throughout the year and comprises 14% of the annual diet. The fruit has an inedible hard, green outer shell that accounts for approximately 50% of the total weight of the fruit (Nour *et al.* 1980). The inside of the fruit is composed of approximately 15–20 seeds which are covered with dry, white, chalky pulp that may be consumed in four ways: (1) directly out of the shell, discarding the hard seed inside the pulp, (2) pounded into a flour, removing the seed husks by winnowing on the surface of small piece of animal hide, (3) combining the flour with water and/or berry juice to create a sweet paste, or (4) removing the intact seeds from the dung of baboons, sun drying, and then pounding into flour. The fruit pulp alone is low in fat, protein, and fibre whereas the pulp flour (seed and pulp combined) is relatively high in fat, protein, and fibre. The fruit pulp and pulp flour are both high in simple carbohydrates (Crittenden 2009).

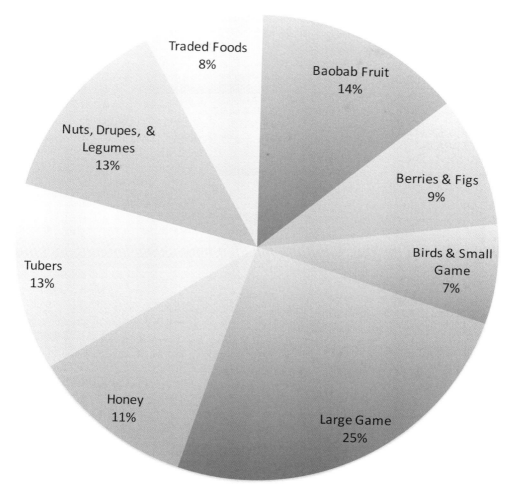

Fig. 16.1. Hadza annual diet composition

Berries and figs

The Hadza consume several species of berries (Fig. 16.2) and one species of fig throughout the year, comprising 9% of the annual diet. The berry species represented here include undushabe (*Cordia senensis*), tafabe (*Salvadora persica*), nguilabe (*Grewia ectosicarpa*), and kongolobi (*Grewia bicolor*). Additional berry species are consumed, however they were not consumed during the study year and, therefore, not analysed for nutrient content. For an exhaustive list of all fruit species targeted by the Hadza, see Peterson *et al.* (2012). The majority of the berries consumed during the study period have proportionately large seeds with very little pulp, and when consumed, the seeds are either expectorated or passed through the digestive system with little or no physical

Fig. 16.2. Hadza women with berries

alteration (i.e., no mastication). All of the berry species listed are low in fat, relatively low in protein, and high in simple carbohydrates (Crittenden 2009). All berries are consumed raw, either fresh off of the plant or slightly dried.

Figs (*Ficus sycomorus*), or ogoyo, are a popular and highly desirable food – they are low in protein and high in fat, simple carbohydrates, and fibre (Crittenden 2009). While mainly collected and consumed by children and adolescents, figs are a very popular food item and are routinely shared with adults back in camp. They are collected in the early morning or late afternoon, in order to avoid competition with neighbouring baboon troops who also consume this fruit. The figs are collected by picking them up off of the ground beneath the tree once they have fallen, or after shaking the tree vigorously to encourage the figs to drop to the ground. They are typically consumed whole, including the small seeds inside of the pulpy fruit.

Nuts, drupes, and legumes

Nuts, drupes, and legumes comprise 13% of the annual diet. The marula nut (*Sclerocarya birrea*), called pawe, is pounded with a rock and consumed raw. Marula nut is high in protein and fat (Crittenden 2009) and is a highly valued food item for the short time that it is available during the year.

The one species of drupe consumed during the study year, mashalobe (unknown Genus species), contains a very hard seed surrounded by dense pulp. They are low in fat and protein and high in fibre and simple carbohydrates (Crittenden 2009). Directly after collection, this drupe is boiled to render the pulp soft and pliable. The pulp, once boiled, falls readily off of the hard seed and is eaten, while the seed is inedible and discarded. This is a food that is typically targeted by young female foragers, although it is widely shared and consumed by all members of camp.

Legumes, mangwala, are largely collected and consumed by juvenile foragers. The primary species targeted (*Acacia nilotica*) looks similar to domestic soybean pods. They are high in protein and fibre and low in fat and simple carbohydrates (Crittenden 2009). The pods are boiled after collection and the pulse (inner seed) is extracted from the outer covering and consumed.

Honeycomb (liquid honey and bee larvae)

Hadza honey collection is exclusively a men's foraging task, yet honeycomb (liquid honey and larvae) is listed by all foragers (men, women, and children) as the top ranked food (Marlowe and Berbesque 2009). A Hadza honey hunter leaves camp on his daily 'walkabout' targeting game animals and/or honey – he will target the honey of stingless bees (*Meliponinae* sp.), called Na'ateko and Kanoa, and stinging bees (*Apis mellifera*), called Ba'alako. If the honey hunter encounters the honey guide bird (*Indicator indicator*), he will soon encounter the very lucrative hive of the African killer bee (*A. mellifera adansonii*). The honey guide bird and the honey hunter communicate back and forth via a series of whistles and chatters while the bird leads the man to the hive (Crittenden 2011; Wood *et al.* 2014). Once at the tree, typically a baobab, the honey hunter chops into the tree with his axe creating a large hole. He then smokes the hive, which functions to pacify the bees by reducing the electroantennograph response of the guard bees, who typically release the alarm pheromone, iso-pentyl acetate, when threatened (Boch and Shearer 1962; Visscher *et al.* 1995). Once the bees are pacified, the honey hunter then reaches into the hive and retrieves the honeycomb while the honey guide bird is left to consume the wax from the comb and bees exiting the hive.

Combined, the honeycomb of all three species of bees comprises 11% of the Hadza diet. Liquid honey is a concentrated source of fructose and glucose and contains approximately 80–95% sugar (Murray *et al.* 2001; Bogdanov *et al.* 2008) and trace amounts of several essential vitamins and minerals (Iskander 1995; Terrab *et al.* 2004). Bee larvae, also consumed and highly valued, is a good source of protein, fat, and several essential minerals and B vitamins (Murray *et al.* 2001; Finke 2005).

Tubers

Tubers, or underground storage organs, are routinely listed as a key Hadza food (Vincent 1985; Marlowe 2010) and in this study, comprise 13% of the annual diet. Higher estimates of the contribution of tubers to Hadza diet have been previously published (Marlowe and Berbesque 2009); however these higher values are likely due to differences in measurement – kilograms (Marlowe and Berbesque 2009) versus kilocalories (Crittenden 2009). A compounding issue with estimating the contribution of tubers to the diet is in regard to the analytical measurement of fibre, where differences in chemical composition analysis can cause the energetic value of tubers to be relatively low (Schoeninger *et al.* 2001) or relatively high (Galvin *et al.* n.d.). The energy values used here are extracted from Crittenden (2009) and represent an intermediate value.

The tuber species represented in this study are matukwayako (*Coccinea surantiaca*), //ekwa (*Vigna frutescens*), shakeako (*Vigna macrorhyncha*), and shumuwako (*Vatoraea pseudolablab*). The majority of tubers that are collected by the Hadza are located up to 3 m below the surface of the ground and are accessed using the sharpened tip of a digging stick. Tubers may be peeled and consumed raw or roasted unpeeled on a fire (Fig. 16.3). When consuming //ekwa (*Vigna frutescens*), the tuber is chewed for up to 3 minutes and then a quid is expectorated (Schoeninger *et al.* 2001). Before expelling the quid, the fibrous mass is sucked free of all moisture while still in the mouth. All tuber species are low in fat and protein and relatively high in fibre and simple carbohydrates (Crittenden 2009).

Fig. 16.3. Roasting //ekwa tubers

The Hadza routinely consume a wide variety of animal and plant resources, as outlined above. Detailed studies on the types of foods targeted, their nutritional properties, and their contribution to the overall diet are key in elucidating cross-cultural ecological differences in diet composition. Collecting quantified nutritional data among hunter-gatherers while they are still actively foraging is critical for informing our understanding of the dietary complexity and diversity of contemporary foraging populations as well as the evolution of the human diet. The Hadza data presented here can be used to interpret assumptions about the Palaeolithic diet.

Hadza nutrition and evolution of the human diet

Studies of diet composition among the Hadza can inform our understanding of the paleoecology of early *Homo* living in a savannah mosaic ecosystem. The Hadza, who live in an environment similar to that exploited by our hominin ancestors (Marlowe 2010), target a wide range of resources year round, where plants make up the majority of the diet (over 60%). This maps onto predictions about Palaeolithic diet that suggest early members of the genus *Homo* had a generalised diet composed of variable amounts of plant and animal resources (Sponheimer *et al.* 2006; Ungar and Sponheimer 2011).

Tubers, a significant year-round Hadza food, have received a great deal of attention as a key food in hominin evolution (Hawkes *et al.* 1989; O'Connell *et al.* 1999; Pennisi 1999). The ability to exploit underground resources, such as tubers, is argued to have been a significant part of the early *Homo* foraging repertoire that would have allowed for survival in a savannah environment. Specifically, the roasting of tubers has been linked with the evolution of the controlled use of fire in early members of genus *Homo* (Wrangham *et al.* 1999; Wrangham 2012). Cooking breaks down indigestible compounds for more efficient fermentation in the colon and also denatures toxins (Wrangham *et al.* 1999). Hadza tuber roasting is often employed as potential theoretical support for the so-called 'cooking hypothesis'. The Hadza flash-roast their tubers before consumption (although they do 'snack' on raw tubers), on surface fires for approximately 3–5 minutes (Fig. 16.3). These fires are ephemeral and recent ethno-geoarchaeological evidence on the sedimentary aspects of the combustion structures of Hadza fires suggests that they would be very difficult to trace archaeologically – if not impossible (Mallol *et al.* 2007). Definitive evidence in regard to whether roasting tubers changes their nutritional bioavailabiliy remains elusive. Recent work on the mechanical properties of tubers, however, suggests that roasting tubers softens their peels, thus expediting manual peeling and lessening the work of mastication – effectively reducing the cost of digestion (Dominy *et al.* 2008).

Further evidence of the significance of tubers to early hominin diet comes from preliminary microwear analysis of replica stone tools used by Hadza women to peel, process, and section tubers. The archaeological microtraces from the experimental use of the replicas matches the microwear of the Oldowan Kanjera stone tools, indicating that the Oldowan tools may have been used to process fibrous plant material in addition to butchering and processing animal protein (Lemorini *et al.* 2014).

The Oldowan tool kit may also have been used to target beehives for honey and bee larvae, which likely supplemented meat and plant foods, providing considerable amounts of energy (Crittenden 2011; Wrangham 2012). The ability of early *Homo* to exploit beehives would have not only provided energy to the expanding hominin brain, but may have been associated with their ability to nutritionally out-compete other species.

In addition to human nutrition, Hadza ethnobotany can also inform our interpretations of the daily non-nutritive uses of plants and the implications of using wild plant resources in food extraction and/or processing techniques.

Plants in medicinal use and material culture

Many plant species are employed for medical use. There are no shamans or medicine men or women among the Hadza, however most adults know what plant species to target for particular ailments. The bark of mondoko (*Entadrophragma caudatum*) is peeled, boiled, and used as a compress. For chest congestion, they smoke the dry root shavings and leaves of the plant pun//upun//u (*Croton menyhartii*) or boil the bark of morongodako (*Zanthoxylum chalybeum*) and drink the tea (Peterson *et al.* 2012). Additional plants (of unknown species) are used to treat sore throats, diarrhoea, general aches and pains, fever, and snake bites (Woodburn 1959; Marlowe 2010).

Jewellery often plays the double role of functioning as body adornment and traditional medicine. It may also serve as a means of status (Marlowe 2010). Traditional jewellery – including earrings, necklaces, bracelets, anklets, waistbands, and headbands – was historically made entirely out of organic components, including plant products (grasses, seeds, and reeds) and items such as small animal bones, bird feathers, pangolin scales, shells, and porcupine quills. While all of these items are still routinely used (see Fig. 16.4), much of the jewellery now incorporates an increasing amount of glass

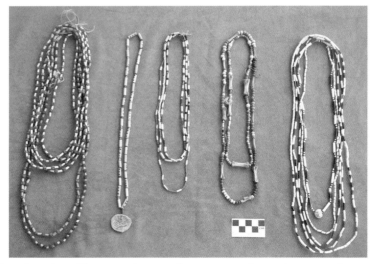

Fig. 16.4. Plant based necklaces made out of grasses, reeds, seeds, porcupine quills, glass, or plastic beads, and small animal bones

beads that are obtained mainly through trade with neighbouring tribes or as gifts from researchers and NGO workers.

The Hadza liberally use a wide variety of plants in the construction of their material culture (Table 16.1). The first descriptions of many of the items listed below can be found in the work of Dorothea Bleek (1931), Ludwig Kohl-Larsen (1958), and James Woodburn (1970), yet all of the items listed are still in wide use today, both in the nomadic bush camps and in the more settled camps located on the periphery of the villages.

Digging sticks, which are used by women and children to dig for tubers, and bows and arrows, which are used by men and boys for hunting, are either made out of the wood of mutateko (*Dombeya kirkii*), kongoloko (*Grewia bicolor*), or ts'apaleko (*Cordia* sp.) (Peterson *et al.* 2012). The shafts of the arrows are straightened during construction by heating the wood over the ashes of a fire and then using the mouth as a clamp to bend the shaft into place (Woodburn 1970; Berbesque *et al.* 2012). Arrows are then fletched by attaching bird feathers, typically guinea fowl or vulture, to the shaft of the arrow with plant resin and binding the fletching with animal sinew. For large game hunting, the Hadza tip their arrows with one of two types of plant poison, shanjo (*Strophanthus eminii*) or panjube (*Adenium obesum*) (Bartram 1997; Marlowe 2010).

Several other men's items also rely on contributions from plants. Axes are critically important tools for Hadza men that allow them to access honey, a main component of their diet (Crittenden 2011) and a highly preferred food item (Marlowe and Berbesque 2009). Axe handles are made from gobandako (*Terminalia brownie*) or mnupeko (*Lonchocarpus eriocalyx*) (Peterson *et al.* 2012). Fire drills, long narrow sticks used to create fire by inserting the tip into a smaller piece of wood and 'drilling' to create friction, are either made out of *Commiphora schimperi* or *Markhamia obtusifolia* (Peterson *et al.* 2012). Men's gambling discs (used for games of chance) were historically made out of the bark of the baobab tree (*Adansonia digitata*) (Woodburn 1970); they now use Tanzanian coins for almost all games; although a handful of old discs remain in circulation. Additional men's items made out of unknown plant species include the handles of knives and climbing pegs hammered into baobab trees to facilitate honey collection.

The construction of houses relies entirely on wild plant resources (Crittenden 2009; Marlowe 2010). During the dry season, Hadza sleep mostly outside, yet during the transition season (from wet to dry) and in the rainy season, they build grass huts. The building of huts is entirely women's work and can be quite rigorous (Fig. 16.5). Women begin the process by selecting tree branches, either from *Grewia bicolor* or another sturdy species, and bending them to fit in holes in the ground that measure approximately two to three inches in diameter. The branches are bent across an area measuring roughly eight to ten feet and then interlocked with one another, creating a dome shaped structure. This hut is then filled in with grass of an unknown species until all of the spaces between the branches are filled (Fig. 16.6).

A wide variety of plants are also used to make children's toys and other miscellaneous household items. Young boys are given small bows and arrows from the time that they can walk independently – the size of the bow and complexity of the arrow mature along with the boy (Crittenden *et al.* 2013) (Fig. 16.7). The very first bow is made of a small twig and the accompanying arrows are made of grass and tipped with beeswax or plant resin, in order to weight them down so that they can be shot from the bow.

Table 16.1: Hadza plants and their uses

Hadza Name	Scientific Name	Part used	Uses
n//obabe	*Adansonia digitata*	Baobab fruit	Pulp flour high in fat, protein, & fibre; both fruit pulp & pulp flour high in simple carbohydrates; hollowed pods also used to make rattles as toys for small children
		Baobab bark	Historically used to make gambling disks
undushabe	*Cordia senensis*	Berries	Consumed raw, either fresh or dried; high in simple carbohydrates; tafabe twigs also used to make useful household items
tafabe	*Salvadora persica*		
nguilabe	*Grewia ectosicarpa*		
kongolobi	*Grewia bicolor*		
ogoyo	*Ficus sycomorus*	Figs	Consumed whole; collected by children; high in fat, simple carbohydrates, & fibre
pawe	*Sclerocarya birrea*	Marula Nut	Pounded & consumed raw; high in protein & fat
mashalobe	unknown	Drupe	Consumed after being boiled; collected by young females; high in fibre & simple carbohydrates
mangwala	*Acacia nilotica*	Legumes	Inner seed consumed after being boiled; collected by juveniles; high in protein & fibre
matukwayako	*Coccinea surantiaca*	Tubers	Consumed raw or roasted (additionally, //ekwa must be masticated for up to 3 minutes – a quid is then expelled); high in fibre & simple carbohydrates
//ekwa	*Vigna frutescens*		
shakeako	*Vigna macrorhyncha*		
shumuwako	*Vatoraea pseudolablab*		
mondoko	*Entadrophragma caudatum*	Medicines	Peeled, boiled & used as a compress
pun//upun//u	*Croton menyhartii*		Dry root shavings & leaves smoked for chest congestion
morongodako	*Zanthoxylum chalybeum*		Bark boiled & consumed as tea; also used to make useful household items
mutateko	*Dombeya kirkii*	Bows & Arrows, Digging Sticks	Digging sticks used by women to collect tubers; bows & arrows used by men & teenage boys for hunting
kongoloko	*Grewia bicolor*		
ts'apaleko	*Cordia* sp.		
shanjo	*Strophanthus eminii*	Hunting poisons	Used on tips of arrows for hunting large game
panjube	*Adenium obesum*		
gobandako	*Terminalia brownie*	Axe handles:	Important in the collection of honey by men
mnupeko	*Lonchocarpus eriocalyx*		
	Commiphora schimperi	Fire drilling sticks:	Used to create fire by use of friction with a smaller piece of wood
	Markhamia obtusifolia		
	Grewia bicolor	Housing construction:	Tree branches bent & interlocked into domes by women; used for shelter during transition & rainy seasons
tangako	unknown	Children's toys:	Used to make spinning tops for children

Fig. 16.5. Women carrying grass back to camp to construct huts

Fig. 16.6. Hadza hut during the transition between the wet and dry season

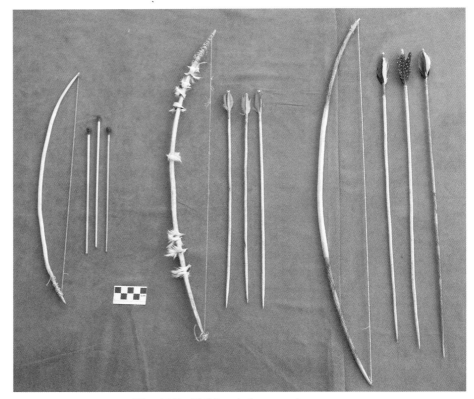

Fig. 16.7. Children's bows and arrows

Young children are often given spinning tops (made from tangako, an unidentified plant species) or rattles (made from hollowed out baobab fruit pods) as first toys (Woodburn, 1970). Useful household items include toothbrushes, made out of frayed twigs of *Salvadora persica* or *Zanthoxylum chalybeum* (Peterson *et al.* 2012), and gourds and hollowed out baobab pods are used for transporting water, honey, or embers (Marlowe 2010). Hadza women occasionally weave baskets out of plant fibres; historically, they created baskets far more regularly before the introduction of aluminum cooking pots and plastic buckets.

Hadza ethnobotanical material culture and human evolution

The exploitation of wild plants by the Hadza not only informs our understanding of the evolution of the human diet, as explored above, but may also inform our interpretation of the links between foraging techniques, material culture, and behaviour in human evolution.

Much of Hadza material culture is essential in key foraging strategies, such as the digging stick, bows and wooden tipped arrows, climbing pegs for honey collection, and fire drills. All of these items would not easily survive the archaeological record, which has intriguing implications for interpretations of early hominin behaviour.

Archaeological findings are habitually used to determine when 'modern' *Homo* behaviour arose (Lombard and Haidle 2012), yet a lithic bias persists (Waguespack *et al.* 2009). In Oswalt's classic compendia of forager technology around the world, he emphasises the importance of 'simple' foraging technology, such as the stick, and how it laid the foundation for more elaborate implements (Oswalt 1973; 1976). This acknowledgement of the significance of organic tools has been extended to the digging stick (Bartholomew and Birdsell 1953; Zihlman and Tanner 1978; Laden and Wrangham 2005), yet given how much attention has been paid to tuber extraction and the related implications for human evolution, it is surprising that the digging stick has not received more consideration. The fact that Hadza digging sticks, critical foraging implements, do not persist for years (let alone decades) in the savannah mosaic environment, should influence the way that we perceive early hominin tuber extraction and allow us to push back this foraging behaviour much earlier than a strict interpretation of the fossil record would allow. The same logic applies to other plant based tools.

Finding bow and arrow technology, primarily consisting of lithic arrowheads, in the fossil record is considered to be an indicator of complex behaviour and cognition in human evolution (Wadley and Mohapi 2008; Coolidge and Wynn 2009; Ambrose 2010; Lombard and Parsons 2011). Although there remains debate regarding where and when composite tools first appear (ranging from the Acheulean to the Middle Palaeolithic), there appears to be consensus on the interpretation that bow and arrow technology is linked to key strategic innovations in human foraging behaviour (Shea and Sisk 2010; Lombard and Haidle 2012). The production of mechanically projected weaponry is argued to require hierarchical thought (Barham 2010) and the integration of working memory with prospective and constructive memory (Ambrose 2010). This argument can be reasonably extended to include wooden tipped arrows. Although there is a dearth of wooden tipped arrows in the archaeological record, there is an overwhelming abundance of wood only hunting technology in the ethnographic record, including the Hadza (Ellis 1997; Waguespack *et al.* 2009). It has been suggested that this 'prehistory paucity' of wooden tipped arrows might be an artefact of poor preservation and not necessarily due to technological, economic, or cognitive factors (Waguespack *et al.* 2009). The Hadza consume large quantities of birds and small to medium sized game – all hunted with wooden tipped arrows. This is consistent with the ethnographic record, which suggests that nearly all foraging populations who use bows and arrows, although preferring stone (or now metal) tipped arrows for large game, use wooden tips to target small and medium sized game animals and birds (Ellis 1997). Using this knowledge, it can be argued that the use of mechanically assisted weaponry and the advent of hierarchical thinking may have emerged much earlier in human evolution than previously predicted, perhaps originating with the genus *Homo*.

The construction of climbing pegs for honey collection and the use of the fire drill pose similar issues. The climbing pegs hammered into baobab trees that allow the Hadza honey hunter to reach the beehive and the drills used to create fire to smoke the hive would not survive into the archaeological record. These tools, which facilitate honey collection, may have likely been part of the early hominin tool kit (Crittenden 2011; Wrangham 2012).

Conclusion

Investigation of the ways in which the Hadza utilise wild plants contributes to our cross-cultural knowledge of human–plant interactions and provides intriguing insight into early hominin paleoecology. Hadza ethnobotany, as explored here, includes diet composition, medical uses, and non-nutritive uses of plants. The analysis of Hadza diet reveals that they target a vast array of both plant and animal resources, supporting the notion that the diet of early hominins was generalised and varied. Hadza consumption of tubers and honeycomb (liquid honey and bee larvae) has intriguing implications for interpretations of the Palaeolithic diet. The majority of early *Homo* dietary reconstructions have historically emphasised the vital role played by meat (Eaton *et al.* 1988; Cordain *et al.* 2001; Stanford and Bunn 2001; Bunn 2007), yet dietary studies of hunter- gatherers continue to provide rich data that highlight the importance of alternative foods.

Knowledge of Hadza use of wild plants for non-nutritive daily tasks (medicine, body adornment, and household maintenance) increases our understanding of the extent to which contemporary foraging populations exploit plant resources. Furthermore, the use of plants during foraging for extraction and/or processing (bows and arrows, digging sticks, climbing pegs, and fire drills), provides insight into early hominin behavioural ecology. The majority of these items would not survive the archaeological record, yet remain vital aspects of Hadza material culture. The ability of the Hadza to forage full time, participating in a predominantly hunting and gathering subsistence regime, is declining. Geopolitical, ecological, and social changes are occurring rapidly in Tanzania – as throughout the world – and these changes have significant effects on the amount of land that the Hadza have access to and the types of resources housed on this land. It is likely that within the next decade there will be very few Hadza practicing a nomadic hunter-gatherer lifestyle, making studies of ethnobotany opportune and critical.

Acknowledgements

I would like to thank the University of Nevada, Las Vegas and the University of California, San Diego for research funding. I would also like to extend my gratitude to the following people: Dr Nancy Lou Conklin-Brittain for lab training and tremendous collegiality; Dr Frank Marlowe and Dr Audax Mabulla for research support; Lene and Johannes Kleppe for their gracious hospitality; Daudi Peterson for his support in the field and his generosity; Happy Msofe, Ephraim Mutukwaya, Golden Ngumbuke, and Pastory Bushozi for their hard work and dedication; and to the Hadza, for welcoming me into their homes and for making this type of research so thoroughly enjoyable.

References

Aiello, L. C. and Wheeler, P. 1995. The expensive-tissue hypothesis: The brain and the digestive system in human and primate evolution. *Current Anthropology* 36, 199–221

Ambrose, S. H. 2010. Coevolution of composite-tool technology, constructive memory, and language. *Current Anthropology* 51, S135–47

Barham, L. 2010. A technological fix for 'Dunbar's dilemma'? *Proceedings of the British Academy* 158, 367–89

Bartholomew, G. A. and Birdsell, J. B. 1953. Ecology and the protohominids. *American Anthropologist* 55, 481–98

Bartram, L. E. 1997. *A Comparison of Kua (Botswana) and Hadza (Tanzania) Bow and Arrow Hunting. Projectile Technology*, 32–43. New York: Plenum

Berbesque, J. C., Marlowe, F. W. and Crittenden, A. C. 2011. Sex differences in Hadza eating frequency by food type. *American Journal of Human Biology* 23, 339–45

Berbesque, J. C., Marlowe, F. W., Pawn, I., Thompson, P., Johnson, G. and Mabulla, A. 2012. Sex differences in Hadza dental wear patterns. *Human Nature* 23, 270–282

Bleek, D. F. 1931. The Hadzapi or Watindega of Tanganyika territory. *Africa* 4, 273–86

Boch, R. and Shearer, D. A. 1962. Identification of geraniol as the active component in the Nassanoff pheromone of the honey bee. *Nature* 194, 704–6

Bogdanov, S., Jurendic, T., Sieber, R. and Gallman, P. 2008. Honey for nutrition and health: A review. *American College of Nutrition* 27, 677–89

Bridges, K. W. and Lau, Y. H. 2006. The skill acquisition process relative to ethnobotanical methods. *Ethnobotany Research & Applications* 4, 115–18

Bunn, H. T. 2007. Butchering backstraps and bearing backbones: Insights from Hadza foragers and implications for Paleolithic archaeology. In Pickering, T. R., Toth, N. and Schick, K. (eds), *Breathing Life into Fossils: taphonomic studies in honor of CK (Bob) Brain*, 269–79. Bloomington IN: Stone Age Institute Press

Bunn, H. T., Bartram, L. E. and Kroll, E. M. 1988. Variability in bone assemblage formation from Hadza hunting, scavenging, and carcass processing. *Journal of Anthropological Archaeology* 7, 412–57

Clement, C. R. 1988. The potential use of the pejibaye palm in agroforestry systems. *Agroforestry Systems* 7, 201–12

Clum, N. J., Fitzpatrick, M. P. and Dierenfeld, E. S. 1996. Effects of diet on nutritional content of whole vertebrate prey. *Zoo Biology* 15, 525–37

Conklin-Brittain, N. L., Knott, C. D. and Wrangham, R. W. 2006. *Energy Intake by Wild Chimpanzees and Orangutans: Methodological Considerations and a Preliminary Comparison.* Cambridge: Cambridge Studies in Biological and Evolutionary Anthropology 48

Coolidge, F. L. and Wynn, T. 2009. *The Rise of Homo Sapiens: the evolution of modern thinking.* Chichester: Wiley-Blackwell

Cordain L, Watkins, B. A. and Mann, N. J. 2001. Fatty acid composition and energy density of foods available to African hominids: evolutionary implications for human brain development. *World Review of Nutrition and Dietetics* 90, 144–61

Cordain, L., Miller, J. B., Eaton, S. B. and Mann, N. 2000. Macronutrient estimations in hunter-gatherer diets. *American Journal of Clinical Nutrition* 72, 1589–90

Cordain, L., Eaton, S. B., Brand Miller, J., Mann, N. and Hill, K. 2002. Original communications-the paradoxical nature of hunter-gatherer diets: Meat-based, yet non-atherogenic. *European Journal of Clinical Nutrition* 56, S42

Crittenden, A. N. 2009. *Allomaternal Care and Juvenile Foraging among the Hadza: Implications for the Evolution of Cooperative Breeding in Humans.* PhD Dissertation, University of California San Diego

Crittenden, A. N. 2011. The importance of honey consumption in human evolution. *Food and Foodways* 19, 257–73

Crittenden, A. N. 2014. *Etnografía de los Hadza: su importancia para la evolución humana.* In Baquedano, E. (ed.), *La Cuna de la Humanidad,* 209–20. Madrid: Instituto de Evolución en **África** (IDEA) and Museo Nacional de Antropología

Crittenden, A. N. and Marlowe, F. W. 2008. Allomaternal care among the Hadza of Tanzania. *Human Nature* 19, 249–62

Crittenden, A. N. and Zes, D. A. 2015. Food sharing among Hadza hunter-gatherer children. *PloS One* 10, e0131996 (doi: 10.1371/journal.pone.0131996)

Crittenden, A. N., Conklin Brittain, N. L., Zes, D. A., Schoeninger, M. J. and Marlowe, F. W. 2013. Juvenile foraging among the Hadza: implications for human life history. *Evolution & Human Behavior* 34, 299–304

de Albuquerque, U. P. and Hanazaki, N. 2009. Five problems in current ethnobotanical research – and some suggestions for strengthening them. *Human Ecology* 37, 653–61

Dominy, N. J., Vogel, E. R., Yeakel, J. D., Constantino, P. and Lucas, P. W. 2008. Mechanical properties of plant underground storage organs and implications for dietary models of early hominins. *Evolutionary Biology* 35, 159–75

Eaton, S. B. 2006. The ancestral human diet: What was it and should it be a paradigm for contemporary nutrition? *Proceedings of the Nutrition Society* 65, 1–6

Eaton, S. B. and Konner, M. 1985. Paleolithic nutrition: A consideration of its nature and current implications. *New England Journal of Medicine* 312, 283–9

Eaton, S. B., Eaton, S. B. III, Konner, M. J. and Shostak, M. 1996. An evolutionary perspective enhances understanding of human nutritional requirements. *Journal of Nutrition* 126, 1732–40

Eaton, S. B., Shostak, M. and Konner, M. 1988. *The Paleolithic Prescription.* New York: Harper and Row

Ellis, C. J. 1997. Factors influencing the use of stone projectile tips. In Knecht, H. (ed.), *Projectile Technology,* 37–74. New York & London: Plenum

Etkin, N. L. 1996. Medicinal cuisines: diet and ethopharmacology. *Pharmaceutical Biology* 34(5), 313–26

Finke, M. D. 2005. Nutrient composition of bee brood and its potential as human food. *Ecology of Food & Nutrition* 44(4), 257–70

Fosbrooke, H. A. 1956. A Stone Age tribe in Tanganyika. *South African Archaeological Bulletin* 11, 3–8

Galvin, K. A., Hawkes, K., Maga, J. A., O'Connell, J. F. and Jones, N. G. B. n.d. *The Composition of Some Wild Plant Foods used by East African Hunter Gatherers*

Hardy, K., Buckley, S., Collins, M. J., Estalrrich, A., Brothwell, D., Copeland, L., García-Tabernero, A., García-Vargas, S., de la Rasilla, M. and Lalueza-Fox, C. 2012. Neanderthal medics? Evidence for food, cooking, and medicinal plants entrapped in dental calculus. *Naturwissenschaften* 99, 617–26

Hawkes, K., O'Connell, J. F. and Blurton Jones, N. G. 1989. *Hardworking Hadza Grandmothers. Comparative Socioecology,* 341–66. Oxford: Blackwell Scientific

Henry, A. G., Ungar, P. S., Passey, B. H., Sponheimer, M., Rossouw, L., Bamford, M. and Berger, L. 2012. The diet of *Australopithecus sediba. Nature.* doi:10.1038/nature11185

Iskander, F. Y. 1995. Trace and minor elements in four commercial honey brands. *Journal of Radioanalytical & Nuclear Chemistry* 201, 401–8

Johanson, D. C. and Wong, K., 2009. *Lucy's legacy: the quest for human origins.* New York: Harmony Books

Johns, T. 1996. *The Origins of Human Diet and Medicine.* Tucson AZ: University of Arizona Press

Jones, N. G. B., Hawkes, K. and O'Connell, J. F. 1989. Modeling and measuring costs of children in two foraging societies. In Standen, V. and Foley, R. A. (eds), *Comparative Socioecology,* 367–90. Oxford: Blackwell Scientific

Jones, N. G. B., Hawkes, K. and O'Connell, J. F. 1997. Why do Hadza children forage? In Segal, N., Weisfeld, G. E. and Weisfeld, C. C. (eds), *Uniting Psychology and Biology: Integrative Perspectives on Human Development*, 279–313. Washington DC: American Psychological Association

Kaare, B. T. M. 1994. The impact of modernization policies on the hunter-gatherer Hadzabe: the case of education and language policies of post-independence Tanzania. In Burch, E. S. and Ellanna, L. J. (eds), *Key Issues in Hunter-Gatherer Research*, 315–31. Oxford: Berg

Kaplan, H., Hill, K., Lancaster, J. and Hurtado, A. M. 2000. A theory of human life history evolution: diet, intelligence, and longevity. *Evolutionary Anthropology Issues News & Reviews* 9, 156–85

Kohl-Larsen, L. 1958. *Wildbeuter in Ostafrika: Die Tindiga ein Jäger- und Sammlervolk*. Berlin: Dietrich Reimer Verlag

Laden, G. and Wrangham, R. 2005. The rise of the hominids as an adaptive shift in fallback foods: plant underground storage organs (USOS) and Australopith origins. *Journal of Human Evolution* 49, 482–98

Lee, R. B. and DeVore, I. 1968. *Man the Hunter*. Chicago: Aldine

Lee-Thorp, J. A., Holmgren, K., Lauritzen, S. E., Linge, H., Moberg, A., Partridge, T. C., Stevenson, C. and Tyson, P. D. 2001. Rapid climate shifts in the southern African interior throughout the mid to late Holocene. *Geophysical Research Letters* 28, 4507–10

Lemorini, C., Plummer, T. W., Braun, D. R., Crittenden, A. N., Ditchfield, P. W., Bishop, L. C., Hertel, F., Oliver, J. S., Marlowe, F. W., Schoeninger, M. J. and Potts, R. 2014. Old stones' song: Use-wear experiments and analysis of the Oldowan quartz and quartzite assemblage from Kanjera South (Kenya). *Journal of Human Evolution* 72, 10–25

Leonard, W. R., Snodgrass, J. J. and Robertson, M. L. 2007. Effects of brain evolution on human nutrition and metabolism. *Annual Review of Nutrition* 27, 311–27

Lombard, M. and Haidle, M. N. 2012. Thinking a bow-and-arrow set: cognitive implications of Middle Stone Age bow and stone-tipped arrow technology. *Cambridge Archaeological Journal* 22, 237–64

Lombard, M. and Parsons, I. 2011. What happened to the human mind after the Howiesons Poort? *Antiquity* 85, 1433–43

Lucas, P. W., Omar, R., Al-Fadhalah, K., Almusallam, A. S., Henry, A. G., Michael, S., Thai, L. A., Watzke, J., Strait, D. S. and Atkins, A. G. 2013. Mechanisms and causes of wear in tooth enamel: implications for hominin diets. *Journal of the Royal Society Interface* 10(80), 20120923

Mallol, C., Marlowe, F. W., Wood, B. M. and Porter, C. C. 2007. Earth, wind, and fire: Ethnoarchaeological signals of Hadza fires. *Journal of Archaeological Science* 34, 2035–52

Marlowe, F. W. 2003. A critical period for provisioning by Hadza men: implications for pair bonding. *Evolution and Human Behavior* 24, 217–29

Marlowe, F. W. 2004. What explains Hadza food sharing? *Research in Economic Anthropology* 23, 69–88

Marlowe, F. W. 2006. Central place provisioning, the Hadza as an example. In Hohmann, G., Robbins, M. and Boesch, C. (eds), *Feeding Ecology in Apes and Other Primates*, 359–77. Cambridge: Cambridge University Press

Marlowe, F. W. 2010. *The Hadza: Hunter-gatherers of Tanzania*. Berkeley CA: University of California Press

Marlowe, F. W. and Berbesque, J. C. 2009. Tubers as fallback foods and their impact on Hadza hunter-gatherers. *American Journal of Physical Anthropology* 140, 751–8

Murray, S. S., Schoeninger, M. J., Bunn, H. T., Pickering, T. R. and Marlett, J. A. 2001. Nutritional composition of some wild plant foods and honey used by Hadza foragers of Tanzania. *Journal of Food Composition and Analysis* 14, 3–13

Nour, A. A., Magboul, B. I. and Kheiri, N. H. 1980. Chemical composition of baobab fruit (*Adansonia digitata*). *Tropical Science* 22, 383–8

O'Connell, J. F., Hawkes, K. and Blurton Jones, N. G. 1999. Grandmothering and the evolution of *Homo erectus*. *Journal of Human Evolution* 36, 461–85

Oswalt, W. H. 1973. *Habitat and Technology: the evolution of hunting*. New York: Holt, Rinehart & Winston

Oswalt, W. H. 1976. *An Anthropological Analysis of Food-Getting Technology*. New York: Wiley

Pennisi, E. 1999. Did cooked tubers spur the evolution of big brains. *Science* 283(5410), 2004–5

Peterson, D., Baalow, R. and Cox, J. 2012. *By the Light of a Million Fires*. Dar es Salaam: Mkuki na Nyota

Prange, H. D., Anderson, J. F. and Rahn, H. 1979. Scaling of skeletal mass to body mass in birds and mammals. *American Naturalist* 113, 103–22

Roberts, E. M., Stevens, N. J., O'Connor, P. M., Dirks, P. H. G. M., Gottfried, M. D., Clyde, W. C., Armstrong, R. A., Kemp, A. I. S. and Hemming, S. 2012. Initiation of the western branch of the East African Rift coeval with the eastern branch. *Nature Geoscience* 5, 289–94

Schnorr, S. L., Candela, M., Rampelli, S., Centanni, M., Consolandi, C., Basaglia, G., Turroni, S., Biagi, E., Peano, C., Severgnini, M., Fiori, J., Gotti, R., De Bellis, G., Luiselli, D., Brigidi, P., Mabulla, A., Marlowe, F. W., Henry, A. G. and Crittenden, A. N. 2014. Gut microbiome of the Hadza hunter-gatherers. *Nature Communications* 5, doi:10.1038/ncomms4654

Schoeninger, M. J., Bunn, H. T., Murray, S. S. and Marlett, J. A. 2001. Composition of tubers used by Hadza foragers of Tanzania. *Journal of Food Composition and Analysis* 14, 15–25

Shea, J. J. and Sisk, M. L. 2010. Complex projectile technology and *Homo sapiens* dispersal into western Eurasia. *PaleoAnthropology* 2010, 100–22

Sponheimer, M., Passey, B. H., De Ruiter, D. J., Guatelli-Steinberg, D., Cerling, T. E. and Lee-Thorp, J. A. 2006. Isotopic evidence for dietary variability in the early hominin *Paranthropus robustus*. *Science* 314, 980–2

Stanford, C. B. and Bunn, H. T. (eds) 2001. *Meat-Eating and Human Evolution*. New York: Oxford University Press

Ströhle, A. and Hahn, A. 2006. Evolutionary nutrition science and dietary recommendations of the Stone Age – ideal answer to present day questions or reason for criticism? Part 2: Ethnographic results and scientific implications. *Ernahr-Umsch* 52, 53–8

Ströhle, A., Hahn, A. and Sebastian, A. 2010. Estimation of the diet-dependent net acid load in 229 worldwide historically studied hunter-gatherer societies. *American Journal of Clinical Nutrition* 91, 406–12

Terrab, A., Recamales, A. F., Hernanz, D. and Heredia, F. J. 2004. Characterisation of Spanish thyme honeys by their physicochemical characteristics and mineral contents. *Food Chemistry* 88, 537–42

Ungar, P. S. and Sponheimer, M. 2011. The diets of early hominins. *Science* 334, 190–3

US Department of Agriculture. 2008. *Agricultural Research Service USDA National Nutrient Database for Standard Reference, 2008*. Release 21. Nutrient Data Laboratory Home Page. http://www.ars. usda.gov/ba/bhnrc/ndl

Vincent, A. S. 1985. Plant foods in savanna environments: A preliminary report of tubers eaten by the Hadza of northern Tanzania. *World Archaeology* 17, 13–48

Visscher, P. K., Vetter, R. S. and Robinson, G. E. 1995. Alarm heromone perception in honeybees is decreased by smoke (Hymenoptera: Apidae). *Journal of Insect Behavior* 8, 11–18

Wadley, L. and Mohapi, M. 2008. A segment is not a monolith: evidence from the Howiesons Poort of Sibudu, South Africa. *Journal of Archaeological Science* 35, 2594–605

Waguespack, N. M., Surovell, T. A., Denoyer, A., Dallow, A., Savage, A., Hyneman, J. and Tapster, D. 2009. Making a point: wood-versus stone-tipped projectiles. *Antiquity* 83, 786–800

Washburn, S. and Lancaster, J. B. 1968. The evolution of hunting. In Lee and DeVore (eds) 1968 293–303

Wood, B. M., Pontzer, H., Raichlen, D. A. and Marlowe, F. W. 2014. Mutualism and manipulation in Hadza–honeyguide interactions. *Evolution and Human Behavior* 35, 540–546

Woodburn, J. 1959. *Hadza Conceptions of Health and Disease*, 89–94. Kampala: East African Institute of Social Research

Woodburn, J. 1968. Stability and flexibility in Hadza residential groupings. In Lee and DeVore (eds) 1968, 103–10

Woodburn, J. 1970. *Hunters and Gatherers: the material culture of the nomadic Hadza*. London: British Museum

Wrangham, R. W. 2012. Honey and fire in human evolution. In Sept, J. and Pilbeam, D. (eds), *Casting the Net Wide: papers in honor of Glynn Isaac and his approach to human origins research*, 149–67. Oxford: Oxbow Books

Wrangham, R. W., Jones, J. H., Laden, G., Pilbeam, D. and Conklin-Brittain, N. 1999. The raw and the stolen. *Current Anthropology* 40, 567–94

Zihlman, A. L. and Tanner, N. 1978. Gathering and the hominid adaptation. In Tiger, L. and Fowler, M. (eds), *Female Hierarchies*, 53–62. Chicago: Aldine

17. Wild edible plant use among the people of Tomboronkoto, Kédougou region, Senegal

Mathieu Guèye and Papa Ibra Samb

This chapter outlines the results of a research project to identify and record the indigenous food plants traditionally consumed in the rural community of Tomboronkoto, Senegal, and their methods of consumption. Here, the Malinké people harvest plant parts from 92 species as part of their diet. Fruits then leaves are the most common plant parts collected. Some plants are considered famine food and represent fallback foods. The high diversity of food species used reflects the important contribution of forest products to the local diet in the rural community of Tomboronkoto.

The importance of wild plants in the diet of non-industrialised peoples is widely recognised. While the botanical knowledge that underpins this has largely been lost in 'western' countries, it survives elsewhere. Today, a reduction in biological diversity and ecological degradation is being caused by a combination of several factors including intense human pressure and ecosystem degradation due to profound climate-induced changes in soil and drought (Clement 1997). However, the use of wild plants, for raw materials, as part of the diet, and as medicine, is still widespread across much of Africa and offers an insight into the level of applied ecological knowledge in many traditional communities.

Significant efforts have been made to record the use of dietary and medicinal plants in Africa. Examples include Baumer (1995) who identified around 350 wild, edible species still used habitually in West Africa. Malaisse (1997) identified edible wild products in the Bemba territory, Democratic Republic of Congo, Vivien and Faure (1996) recorded the wild fruit still collected and eaten in Cameroon, and Ambé (2001) recorded 75 species of wild fruit that are still collected and eaten by the Malinké from the region of Séguéla, Côte d'Ivoire, and the Guinean savannah. Guèye and Diouf (2007) identified 40 species of plant that are still collected for consumption of their leaves in Senegal, while 38 species have been identified in Cameroon (Stevels 1990). This level of indigenous knowledge is retained today largely by communities who live in the most vulnerable ecosystems.

The south and south-east regions of Senegal had the largest phyto-biodiversity of the west African region (Ba and Noba 2001). The rural population of Tomboronkoto on the edge of the Niokolo Koba park is an area of high phyto-biodiversity and inhabitants still draws heavily on the forest for resources that are essential to their survival. Many species, particularly those that are lesser known, and the knowledge surrounding their uses, are currently at risk, as gold mining and industrial development compete for the same space. Within this framework, we set up a research project to record and identify the indigenous food plants traditionally consumed in the rural community of Tomboronkoto, and their methods of consumption. While many plants are also used for their medicinal properties, these are not included in this chapter.

Materials and methods

Study area

The rural community of Tomboronkoto, on the border of the Niokolo Koba National Park, lies around 660 km from Dakar, in the region of Kédougou, Bandafassi district. It covers 2267.9 km^2 (Fig. 17.1). The park is bordered to the north by the rural communites of Khossanto and Dialakoto, to the south by the community of Bandafassi, to the east by the community of Kédougou and to the west by the Niokolo Koba National Park. The population is predominantly Malinké, and comprises 7877 inhabitants (PLD 2003) spread among 28 villages. It is in the Soudano-Guinéen climatic zone, and its proximity to the National Park offers the population a wide variety of native plants. The people practice animal husbandry, raising cattle, sheep and goats, and cultivate millet, maize, and peanuts.

Data collection

Based on the information we obtained from previous visits, we selected 20 villages for the investigation. We used the technique of open semi-structured interviews, which are participatory tools, to collect several types of information from the local people. The semi-structured interviews were conducted using open, indirect, and direct questions in order to learn about the different food plants. The principal aims of our research were to identify and characterise species and/or varieties, local names and peoples perceptions about the different species and/or varieties. Secondly we have developed a resource map that showed the distribution of inventoried species and their varieties in the study area, as well as the traditional knowledge linked to them.

When we arrived at a village, we began by having a group session. These group interviews were extremely useful to explore and circulate information from across the socio-economic strata of the village, and to provide a platform for generating fruitful discussion among participants. This direct contact with many of the villagers led to the establishment of a focus group in each village. Since we did not speak the local language, we sought the help of an interpreter-guide who not only knew the language well but was also familiar with the plant species. This guide was chosen after discussion with some of the villagers to ensure their botanical and local knowledge. The guide

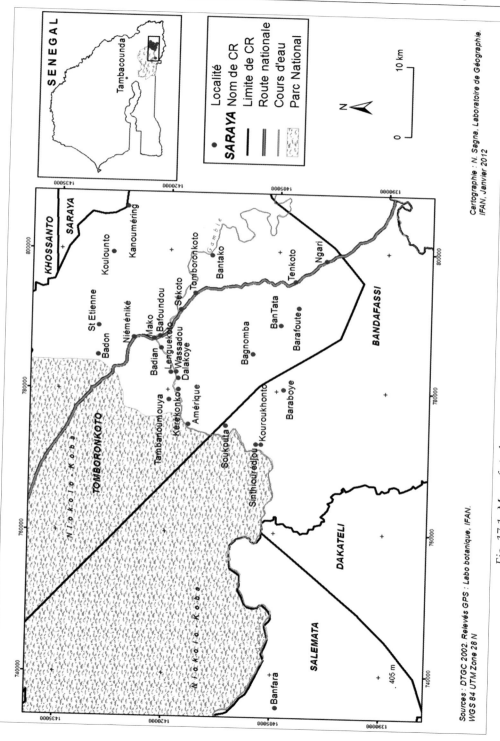

Fig. 17.1. Map of study area, Tomboronkoto, Kédougou region, Senegal

accompanied us on all our visits to the forest, showed us the species that had been identified in the group session, and allowed us to collect samples.

Individual interviews were also conducted with the elderly and with young people. The selection of respondents was made with the help of the villagers who identified people with good knowledge of local plants and their uses. Grenand *et al.* (2004) argue that the only cultural criterion that acts upon the value of an informant is their knowledge relative to the other members of the community or the reputation they enjoy. Once we had selected informants, we met them wherever we could, (in the village, in the fields, panning for gold, etc). Individual interviews are preferably made during walks in the woods, as suggested by Cunningham (2002). In this way, species are indicated directly by the informant and immediately collected. In cases where the informant was too old or busy after the interview, we relied on the knowledge of the guide – interpreter to collect specimens, and then met afterwards with the informant who validated our collections. We also made use of direct observations and conducted casual conversations, these allowed us to estimate knowledge and to elicit responses (Martin 1995).

Species identification

Some species were directly identified in the field, others in the laboratory using Flora (Hutchinson and Dalziel 1954; Berhaut 1967; 1971; 1974; 1975a; 1975b; 1976; 1979; Lebrun 1973), and other published works (Arbonnier 2000; Hawthorne and Jongkind 2006) or directly using reference material from the IFAN (Institut Fondamentale d'Afrique Noir) herbarium collection, based in Dakar. For the selection of valid scientific names, we consulted the Geneva-based database of conservatory and botanical gardens (http://www.villege.ch/musinfo/bd/cjb/africa/ recherche.php) regularly updated updating and the IPNI (International Plant Name Index).

Data processing

The data were processed using a range of descriptive statistical techniques, and consistency was confirmed by the comparison method developed by El Rhaffari *et al.* (2002), in which information is considered consistent when reported at least twice in two different locations and by different informants, in other cases it is called divergent. Only consistent information was included in the data processing.

The Fidelity level (IF) is the percentage of informants who cited the use of a species in a defined use category; this is calculated using the technique described by Begossi (1996) and Trotter and Logan (1986).

IF (%) = (Ip/Iu) ×100

Results

Diversity of food plants

Food plants comprise 92 species, from 65 genera, and 43 families. The most common family represented is the *Tiliaceae* with 12% of all species. This is followed by the *Anacardiaceae*, the *Apocynaceae*, and the *Caesalpiniaceae*, each with 6%. Following this, the *Amaranthaceae*, and the *Rubiaceae* each represent 4.6% of the recorded food taxa (Table 17.1). The most diverse genera represented are the *Grewia* and the *Corchorus* each with five species; *Amaranthus* and *Dioscorea* have four species represented; *Commelina*, *Ficus*, *Lannea*, each have three; *Diospyros*, *Hibiscus*, *Raphionacme*, *Ocimum*, *Boerhaavia* have two. The other genera are represented by one species each. The families with the largest number of species do not always have the greatest diversity; despite this they can be some of the best represented genera. In the vast majority of cases (91%), only one plant part is used. Nevertheless, for 9% of plants, at least two different parts are made use of by the local population (Table 17.1). Fruits are by far the most commonly used food items (60%), followed by leaves (32%). Tubers (4%), gum (3%) and roots (1%) were little used (Fig. 17.2).

Fruits

Among the plant species used as food, 49% are exploited for their fruit. Of these, 93% of species are eaten directly. Only the fruits of three species (*Allophyllus africanus, Kigelia africana,* and *Raphia palma-pinus*) are cooked before consumption while the fruit of only 13% of species are traded (Fig. 17.3).

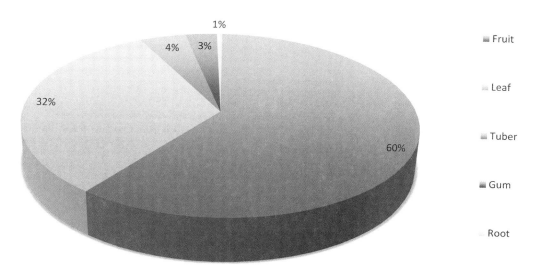

Fig. 17.2 Relative importance of plant parts

Table 17.1. *List of food plants recorded in the rural community of Tomboronkoto, Kédougou, Senegal*

Family	Species	Part eaten
Amaranthaceae	*Amaranthus graecizans* L.	Herb
Amaranthaceae	*Amaranthus hybridus* L.	Herb
Amaranthaceae	*Amaranthus spinosus* L.	Herb
Amaranthaceae	*Amaranthus viridis* L.	Herb
Anacardiaceae	*Lannea microcarpa* (Oliv.) Engl.	Fruit
Anacardiaceae	*Lannea velutina* A. Rich.	Fruit
Anacardiaceae	*Lannea acida* A. Rich.	Fruit
Anacardiaceae	*Spondias monbin* L.	Fruit
Anacardiaceae	*Sclerocarya birrea* (A. Rich.) Hochst.	Fruit
Annonaceae	*Annona senegalensis* Pers.	Fruit
Annonaceae	*Hexalobus monopetalus* (A. Rich.) Engl. et Diels	Fruit
Apocynaceae	*Saba senegalensis (A. DC.)* Pichon	Fruit
Apocynaceae	*Carissa edulis* (Forssk.) Vahl.	Fruit
Apocynaceae	*Landolphia heudelotii* A. DC.	Fruit
Apocynaceae	*Raphionacme brownii* Scott-Elliot	Tuber
Apocynaceae	*Raphionacme splendens* subsp. *bingeri* (A.Chev.) Vent.	Tuber
Araceae	*Amorphophallus Aphyllus*	Tuber
Araceae	*Stylochyton hypogaeus* Lepr.	Herb
Araceae	*Stylochiton lancifolius* Kotschy & Peyr.	Herb
Arecaceae	*Borassus aethiopum* Mart.	Fruit, heart, tuber
Arecaceae	*Raphia palma-pinus* (Gaertn.) Hutch.	Fruit
Asclepiadaceae	*Leptadenia hastata* Decne.	Herb
Bignoniaceae	*Kigelia africana* (Lam.) Benth.	Fruit
Bombacaceae	*Adansonia digitata* L.	Fruit, herb
Bombacaceae	*Bombax costatum* Pellegr. et Vuillet.	Herb
Bombacaceae	*Ceiba pentandra* (L.) Gaertn.	Herb
Boraginaceae	*Cordia myxa* L.	Herb, fruit
Caesalpiniaceae	*Senna obtusifolia* Link.	Leaves, herb
Caesalpiniaceae	*Cordyla pinnata* (Lepr. ex A. Rich.) Milne-Redhead	Fruit
Caesalpiniaceae	*Detarium microcarpum* G. et Perr.	Fruit
Caesalpiniaceae	*Tamarindus indica* L.	Fruit, condiment, herb
Capparaceae	*Crateava adansonii* DC.	Herb
Celastraceae	*Gymnosporia senegalensis* (Lam.) Loes.	
Chrysobalanaceae	*Parinari excelsa*	Fruit
Chrysobalanaceae	*Neocarya macrophylla* (Sabine) G.T. Prance ex White	Fruit, grain
Commelinaceae	*Commelina erecta* L.	Herb
Commelinaceae	*Commelina erecta* L. subsp. *erecta*	Herb

Family	Species	Part eaten
Commelinaceae	*Commelina benghalensis* L.	Herb
Convolvulaceae	*Ipomoea setifera* Poir.	Herb
Convolvulaceae	*Jacquemontia tamnifolia* Griseb	Herb
Cucurbitaceae	*Blastania fimbristipula* Kotschy & Peyr.	Herb
Dioscoreaceae	*Dioscorea cayenensis* Lam.	Air tuber
Dioscoreaceae	*Dioscorea bulbifera* L.	Tuber
Dioscoreaceae	*Dioscorea dumetorum* (Kunth) Pax	Tuber
Dioscoreaceae	*Dioscorea sagittifolia* Pax.	Tuber
Ebenaceae	*Diospyros mespiliformis* Hochst. ex A. DC.	Fruit
Ebenaceae	*Diospyros heudelotii* Hiern	Fruit
Fabaceae	*Pterocarpus santaloides* DC.	Fruit
Ficoidaceae	*Trianthema portulacastrum* L.	Herb
Icacinaceae	*Icacina senegalensis* Juss.	Fruit
Labiatae	*Ocimum gratissimum* L.	Condiment, herb
Labiatae	*Ocimum* sp. L.	Condiment, herb
Labiatea	*Hyptis suaveolens* Poit.	Condiment
Loganiacaea	*Strychnos spinosa* Lam.	Herb, fruit
Malvaceae	*Hibiscus asper* Hoek. F.	Leaves, herb, condiment
Malvaceae	*Pavonia triloba* Guill. & Perr.	Leaves, herb, condiment
Malvaceae	*Hibiscus* spp. L.	Leaves, herb, condiment
Meliaceae	*Trichilia emetica* Vahl.	Fruit
Mimosaceae	*Parkia biglobosa* (Jacq.) R. Br. ex G. Don	Fruit, grain, condiment
Moraceae	*Ficus sycomorus* subsp. *gnaphalocarpa* (Miq.) C.C. Berg	Fruit
Moraceae	*Ficus dicranostyla* Mildbr.	Herb
Moraceae	*Ficus sur* Forssk.	Herb, fruit
Myrtaceae	*Syzygium guineense* (Willd.) DC.	Fruit
Nyctaginaceae	*Boerhaavia diffusa* L.	Herb
Nyctaginaceae	*Boerhaavia erecta* L.	Herb
Olacaceae	*Ximenia americana* L.	Fruit
Pedaliaceae	*Ceratotheca sesamoides* Endl.	Herb
Poaceae	*Chrysopogon nigritanus* (Benth.) Veldkamp	Root, condiment
Rhamnaceae	*Ziziphus mauritiana* Lam.	Fruit
Rubiaceae	*Sarcocephalus latifolius* (Smith) Bruce	Fruit
Rubiaceae	*Spermacoce octodon* (Hepper) J.-P.Lebrun & Stork	Herb
Rubiaceae	*Pavetta crassipes* K.Schum.	Herb
Rubiaceae	*Gardenia erubescens* Stapf et Hutch.	Herb, fruit
Sapotaceae	*Vitellaria paradoxa* Gaertn. f.	Fruit
Spindaceae	*Allophyllus africanus* P. Beauv.	Fruit
Simaroubaceae	*Quassia undulata* (Guill. et Perr.) F. Dietr.	Fruit

Family	Species	Part eaten
Sterculiaceae	Cola cordifolia (Cav.) R. Br.	Fruit
Sterculiaceae	Sterculia setigera Del.	Gum
Tiliaceae	Grewia bicolor Juss.	Fruit
Tiliaceae	Grewia lassiodiscus K. Schum.	Fruit
Tiliaceae	Grewia tenax (Forsk.) Fiori	Fruit
Tiliaceae	Grewia mollis Juss.	Fruit
Tiliaceae	Corchorus fascicularis Lam.	Herb
Tiliaceae	Corchorus olitorius L.	Herb
Tiliaceae	Grewia barteri Burret	Herb
Tiliaceae	Corchorus aestuans L.	Herb
Tiliaceae	Corchorus spp. (Tourn.) L.	Herb
Tiliaceae	Corchorus tridens L.	Herb
Tiliaceae	Triumfetta pentandra A.Rich.	Herb
Ulmaceae	Celtis toka (Forssk.) Hepper & J.R.I.Wood .	Fruit
Verbenaceae	Vitex madiensis Oliv.	Fruit, herb

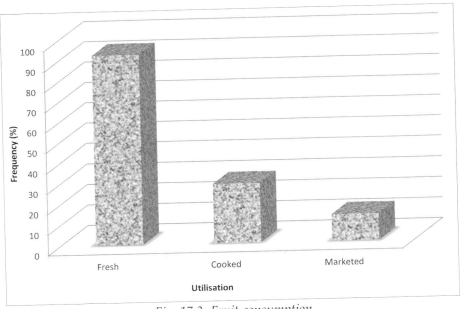

Fig. 17.3. Fruit consumption

Species are classified by combining their abundance and frequency of consumption. This gives three categories of fruits :1) the common or well-known fruit, 2) medium known fruits and 3) little known fruits. Each category is then divided into well used fruit, medium and little used fruit (Fig. 17.4).

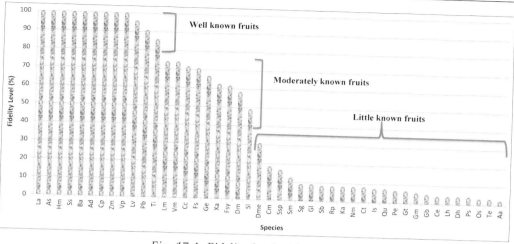

Fig. 17.4. Fidelity level of forest fruits

The first category, the common or well-known fruit, represent 35.5% of those eaten. These are primarily fleshy fruit and the pulp is often eaten directly though in certain cases, they are cooked before consumption. These are heavily exploited fruits and 75% of them are regularly collected (Figs 17.3 and 17.5). Moderately known fruits are the least exploited with only 13.3% collected regularly. In contrast, 51% of the least well known fruits are exploited. Overall, the less common fruits (last two categories) are exploited between 12% and 44% (Fig. 17.3).

Leafy vegetables

Among the species identified as fruits, 49% are exploited for their leaves and consumed in various ways, though in most cases (98%) the leaves are used as flavouring herbs. The use of leaves is less common (15.5%) and they are rarely used as spices (9%) (Fig. 17.6).

The classification of leafy vegetable species is based on three categories: 1) well known, 2) less well known and 3) little known. Each category is further divided into three subsets (well exploited, medium well exploited and little exploited) (Fig. 17.7). Species whose leaves are commonly eaten (Fig. 17.8) and those whose leaves are consumed moderately represent only a small proportion of leafy vegetables, 15.5% and 22% respectively. In contrast, little known leafy vegetables comprise 62% of all the leafy vegetables consumed. The common species are dominated by perennials, followed by annual species.

Tubers

We identified eight species from four genera, belonging to four families whose tubers were collected for consumption by the people of Tomboronkoto (Table 17.2). All tubers are cooked prior to being eaten, except *Raphionacme brownii* and *Raphionacme splendens subsp. bingeri* (Fig. 17.9) which are only ever consumed fresh. When the main root of *B. aethiopum* is collected when it is young, it can be eaten fresh though sometimes it is also cooked (Table 17.2).

Fig. 17.5. Examples of well known fruits: a) Annona senegalensis; *b)* Hexalobus monopetalus *c)* Borassus aethiopum

Borassus aethiopum is the species whose tubers are the best known and most consumed (75%). This is followed by *Dioscorea cayenensis* (58%) and *Dioscorea sagittifolia* (54%) which are eaten moderately (Fig. 17.10). Plants from the genus Raphionacme and other species (Fig. 17.11) are little known and infrequently consumed (12–25%).

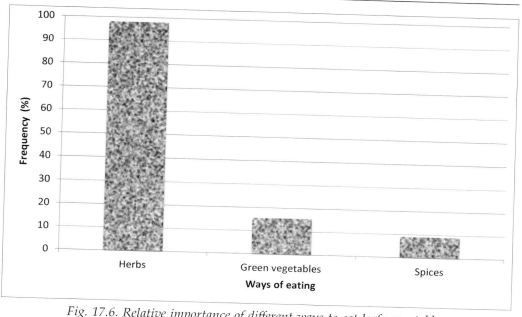

Fig. 17.6. Relative importance of different ways to eat leafy vegetables

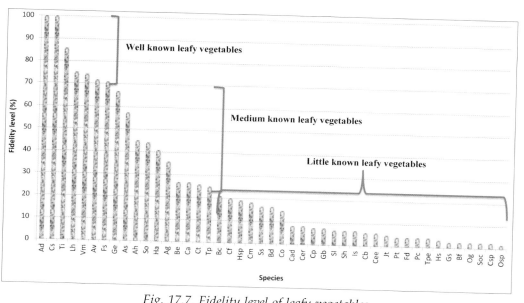

Fig. 17.7. Fidelity level of leafy vegetables

Fig. 17.8. Example of well known leafy vegetable: Ceratotheca sesamoides

Lean-time and famine foods

The edible parts of 20 species are consumed during lean times. These species come from 16 genera and 13 families. These include four species of *Dioscoreaceae* and two species each of *Apocynaceae, Arecaceae, Caesalpiniaceae* and *Rubiaceae* (Table 17.3). All the others are single species. The most harvested plant parts during lean times are fruits (39%), followed by tubers (35%) and leaves (26%) (Table 17.3). They are eaten fresh or cooked. Leaves are most often cooked (71%), followed by tubers (67%) then fruit (62%) (Figs 17.12 and 17.13). Direct food consumption of lean period food is less common. Some plant parts, such as the fruits of *Icacina senegalensis* are exploited in two different ways. The fruit is eaten fresh and the seed is crushed seed to extract the kernel which is then ground into flour and used to make a particular local dish. The fruit of *Cordia myxa* is also used to to sweeten the ubiquitous millet porridge.

Fig. 17.9. Examples of tuber always eaten uncooked: Raphionacme splendens *subsp.* binger

Table 17.2: List of tubers collected and eaten by the rural community of Tomboronkoto

Family	Species	Method of consumption
Apocynaceae	*Raphionacme splendens* subsp. *bingeri* (A.Chev.) Vent.	Fresh
Apocynaceae	*Raphionacme brownii* Scott-Elliot	Fresh
Araceae	*Amorphophallus Aphyllus* (Hook.) Hutch.	Cooked
Arecaceae	*Borassus aethiopum* Mart.	Fresh or cooked
Dioscoreaceae	*Dioscorea cayenensis* Lam.	Cooked
Dioscoreaceae	*Dioscorea sagittifolia* Pax.	Cooked
Dioscoreaceae	*Dioscorea bulbifera* L.	Cooked
Dioscoreaceae	*Dioscorea dumetorum* (Kunth) Pax	Cooked

Table 17.3: List of species eaten during lean periods or famine

Family	Species	Part eaten
Apocynaceae	*Raphionacme splendens* subsp. *bingeri* (A.Chev.) Vent.	Tuber
Apocynaceae	*Raphionacme brownii* Scott-Elliot	Tuber
Araceae	*Amorphophallus Aphyllus* (Hook.) Hutch.	Tuber
Arecaceae	*Raphia palma-pinus* (Gaertn.) Hutch.	Fruit
Arecaceae	*Borassus aethiopum* Mart.	Fruit and tuber
Bignoniaceae	*Kigelia africana* (Lam.) Benth.	Fruit
Boraginaceae	*Cordia myxa* L.	Herb and fruit
Caesalpiniaceae	*Cordyla pinnata* (Lepr. ex A. Rich.) Milne-Redhead	Fruit
Caesalpiniaceae	*Senna obtusifolia* Link.	Herb
Celastraceae	*Gymnosporia senegalensis* (Lam.) Loes.	Herb
Dioscoreaceae	*Dioscorea sagittifolia* Pax.	Tuber
Dioscoreaceae	*Dioscorea bulbifera* L.	Tuber
Dioscoreaceae	*Dioscorea cayenensis* Lam.	Tuber
Dioscoreaceae	*Dioscorea dumetorum* (Kunth) Pax	Tuber
Fabaceae	*Pterocarpus santaloides* DC.	Fruit
Icacinaceae	*Icacina senegalensis* Juss.	Fruit
Moraceae	*Ficus sur* Forssk.	Herb and fruit
Rubiaceae	*Gardenia erubescens* Stapf et Hutch.	Fruit
Rubiaceae	*Pavetta crassipes* K.Schum.	Herb
Verbenaceae	*Vitex madiensis* Oliv.	Herb

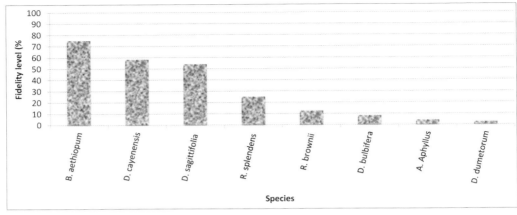

Fig. 17.10. Fidelity level of tubers eaten

It is stirred in boiling water, then filtered and added into the porridge as this is being prepared.

With regards to lean season and famine periods, foods were split into those that are well-known, medium well known and little known (Fig. 17.11). The most commonly used lean season foods comprise four of the least common species. This is followed by seven medium little known and nine medium well-known species. The exploitation of the species in different categories varies little but appears more diverse in terms of little known lean season foods (Fig. 17.11).

Discussion

The level of knowledge and use of the 92 plants known to have been collected for food is highly variable. Those with the highest fidelity level (IF) are the best known and most widely consumed. Food species with low fidelity levels (IF), are

Fig. 17.11. Example of little known tuber: Dioscorea bulbifera

also the least known and least used. Most plants fall into these categories of little used plants. This may be due to various factors including their rarity in the study area. Examples include plants such as *Parinari excelsa, Landolphia heudelotii, Neocarya*

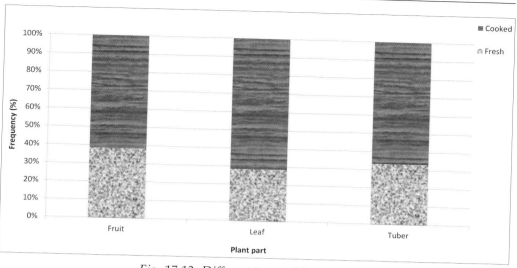

Fig. 17.12. Different types of famine foods

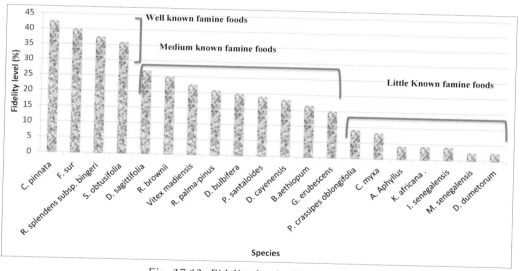

Fig. 17.13. Fidelity level of famine foods

macrophylla, Syzygium guineense. A change in eating habits has also been noted in recent years (Guèye 2012). Many plants that used to be used as food are now being abandoned in favour of imported foods, even though in many cases, these do not necessarily offer better quality. This is the case of the *Cordia myxa* foods, used to sweeten the millet porridge, and the fruits of *Pterocarpus santaloides* which used to comprise the main meal for many families during lean periods (Guèye 2012; Guèye *et al.* 2014). *C. myxa* fruit has been replaced by industrial sugar despite all the health risks associated with this.

Likewise, the use of traditional leafy vegetables is becoming less and less common as the cultivation of exotic vegetables has increased even though it has been clearly demonstrated that traditional leafy vegetables provide greater production per unit area in a shorter time period than imported cereals (Watson and Eyzaguire 2002) and are much richer in vitamins and minerals than most 'European' vegetables (Westphal *et al.* 1985). For much of the year leafy vegetables can provide most of the vitamins and minerals required (Sundriyal *et al.* 2004. Misra *et al.* 2008).

Among the food plants recorded by Thiombiano *et al.* (2012) in Burkina Faso, *Leptadenia hastata, Vitellaria paradoxa, Parkia biglobosa,* and *Lannea microcarpa* are the principal lean season species used. Some of recorded food species are also eaten in other regions, including nine species in Ethiopia (Teklehaymanot and Giday 2010), two species (*Tamarindus indica* and *Amaranthus viridis*) in Malaysia (Ong *et al.* 2011), 30 species in Uganda (Katende *et al.* 1999) and 24 species in Kenya (Maundu *et al.* 1999). In south-west Cameroon, Schippers and Budd (1997) showed that indigenous flavouring plants make up 50% of the dietary plant foods.

Adam (1963) in an outline of useful plants from the Niokolo Koba Park, highlights the way the native plants from here can provide all the plant-based dietary essentials as well as their extensive medicinal properties. Recently Guèye *et al.* (2014) have shown that the vast majority of the most commonly eaten fruits on the edge of Niokolo Koba Park are available from the end of the hot dry season to the middle of the rainy season. This period coincides with the lean or famine season, when yields from the previous rainy season are depleted and the new crops are not yet ripe (Norman and Dixon 1995). Thus, in view of our results, it seems that these dietary plant contribute significantly to the food security of the local population.

Though Gautier-Bréguin (1992) suggested that the quantity and quality of available food products in West Africa could not provide the recommended amounts of nutrition, many forest products are consumed on-site immediately after collection and therefore not included in the overall balance of nutrition. In fact, the people of the forest show few signs of malnutrition

Rural populations have a good knowledge of forest foods and recognise their different qualities (Diouf *et al.* 2007; Guèye and Diouf 2007; Teklehaymanot and Giday 2010; Guèye 2012), some of which have recently been scientifically confirmed (Danthu *et al.* 2001; Ayessou *et al.* 2009; 2011).

Ayessou *et al.* (2009) have demonstrated that *Maerua pseudopetalosa*, a lean season food used in the district of Kéniaba (Senegal), has a protein content similar to *Vigna unguiculata* and may therefore be a useful contribution to dietary diversity during lean times. The fruit *Sarcocephalus latifolius* is also a good source of protein approaching 20% (Ayessou *et al.* 2011). These proteins are paricularly important for people doing heavy physical work and for children during periods of growth and as weaning foods (Ayessou *et al.* 2009).

Analyses performed on other wild fruits such as *Icacina senegalensis, Cordyla pinnata, Ficus sycomorus gnaphalocarapa* and *Sarcocephalus latifolius* show that they are all good sources of minerals (Ayessou *et al.* 2011). However, acidity, vitamin C and total mineral composition is different for each fruit. The fruit of *S. latifolius* stands out for its high acidity and high content of vitamin C while *I. senegalensis* and *C. pinnata* are rich in fructose. Such species can play a vital role in the fight against hunger and malnourishment (Ayessou *et al.* 2011).

The broad range of wild plants still collected for food in Tomborokoto provides a small glimpse into the breadth of applied botanical knowledge and the way these plants can contribute to providing a balanced diet in a traditional dietary setting. Additionally, many species also have known medicinal properties and these continue to be made use of.

Conclusion

The Malinké people harvest plant parts from 92 species as part of their diet. Fruits are the most common plant parts collected, followed by leaves. Certain species are used only in lean times, and represent fallback foods. The high diversity of food species reflects the important contribution of forest products to the local diet in the rural community of Tomboronkoto. However the level of knowledge and consumption of forest foods is highly variable. Collection focuses mostly on the best-known plants while most food species are not well known and are today, underutilized. The consumption of certain species such as *Cordia myxa* or *Pterocarpus santaloides* is very limited in comparison to the more exotic plants. The medicinal properties of some forest plants are also known about. These plants are selected as food but are also used for treating or preventing certain illnesses in which case a double benefit is obtained.

The participatory method has ensured significant contribution from many members of the community. This has allowed us to compile a wealth of information on the various indigenous food plants of the Tomboronkoto population and the range of ways these contribute to the diet, both as mainstream and famine foods. The wide range of plants and plant parts used reflects their sustained importance in rural life here and offers a glimpse into the extent of traditional knowledge and the variety of ways that different plants and plant parts contribute to the diet in non-industrialised economies in West Africa.

References

Adam, J. G. 1963. Les plantes utiles du Parc national du Niokolo-Koba (Sénégal). Comment vivre uniquement avec leurs ressources. *Notes africaines* 9, 5–21

Ambé, G-A. 2001. Les fruitiers sauvages comestibles des savanes guinéennes de la Côte d'Ivoire: état de la connaissance par une population locale, les Malinké. *Biotechnology Agronomy Society & Environment* 5(1), 43–58

Arbonnier, M. 2000. *Arbres, Arbustes et Lianes des Zones Sèches d'Afrique de l'Ouest*. Paris: CIRAD-MNHN-UICN

Ayessou, N. C., Guèye, M., Dioh, E., Konteye, M., Cissé, M. and Dornier, M. 2009. Composition nutritive et apport énergétique du fruit de *Maerua pseudopetalosa* (Gil et Gil-Ben) DeWolf (*Capparidaceae*), aliment de soudure au Sénégal. *Fruits* 64(3), 147–56

Ayessou, C. N., Ndiaye, C., Cissé, M., Guèye, M. and Sakho, M. 2011. Nutritional contribution of some Senegalese forest fruits running across Soudano-Sahelian zone. *Food and Nutrition Sciences* 2(6), 606–12

Ba, A. T. and Noba, K. 2001. Flore et biodiversité végétale au Sénégal. *Sécheresse* 12(3), 149–55

Baumer, M. 1995. *Arbres, Arbustes et Arbrisseaux Nourriciers en Afrique Occidentale*. Dakar: Enda Tiers-Monde

Begossi, A. 1996. Use of ecological methods in ethnobotany: diversity indices. *Ecological Methods in Ethnobotany* 5, 280–9

Berhaut, J. 1967. *Flore du Sénégal Plus Complète Avec les Forêts Humides de la Casamance*. Dakar, Senegal: Clair Afrique

Berhaut, J. 1971. *Flore Illustrée du Sénégal. Dicotyledones. Tome 1 Acanthacées à Avicenniacées*. Dakar: Gouvernement du Sénégal-Ministère du développement Rural et de l'Hydraulique, Direction des Eaux et Forêts

Berhaut, J. 1974. *Flore Illustrée du Sénégal. Dicotyledones. Tome 2 Balanophoracées à Composées*, Dakar: Gouvernement du Sénégal-Ministère du développement Rural et de l'Hydraulique, Direction des Eaux et Forêts

Berhaut, J. 1975a. *Flore Illustrée du Sénégal. Dicotyledones. Tome 3 Connaracées à Euphorbiacées*. Dakar: Gouvernement du Sénégal-Ministère du développement Rural et de l'Hydraulique, Direction des Eaux et Forêts

Berhaut, J. 1975b. *Flore Illustrée du Sénégal. Dicotyledones. Tome 4 Ficoidées à Legumineuses*. Dakar: Gouvernement du Sénégal-Ministère du développement Rural et de l'Hydraulique, Direction des Eaux et Forêts

Berhaut, J. 1976. *Flore Illustrée du Sénégal. Dicotyledones. Tome 5 Legumineuses Papilionacées*. Dakar: Gouvernement du Sénégal-Ministère du développement Rural et de l'Hydraulique, Direction des Eaux et Forêts

Berhaut, J. 1979. *Flore Illustrée du Sénégal. Dicotyledones. Tome 6 Linacées à Nymphéacées*. Dakar: Gouvernement du Sénégal-Ministère du développement Rural et de l'Hydraulique, Direction des Eaux et Forêts

Clement, J. C. 1997. Les variétés traditionnelles de mil en Afrique Sahélienne facteurs de stabilité et de variation. *Actes du Colloque 'Gestion des Ressources Génétiques des Plantes en Afrique des Savanes'*, 133–42

Cunningham, A. B. 2002. *Applied Ethnobotany: people, wild plant use and conservation*. London: Earthscan, People and Plants Conservation Manua.

Danthu, P., Soloviev, P., Totté, A., Tine, E., Ayessou, N., Gaye, A. and Niang, T., 2001. Caractères physico-chimiques et organoleptiques comparés de jujubes sauvages et des fruits de la variété Gola introduite au Sénégal. *Fruits* 57(3), 173–82

Diouf, M., Guèye, M., Faye, B., Dieme, O. and Lo, C. 2007a. The commodity systems of four indigenous leafy vegetables in Senegal. *Water SA*, 33(3), 343–8

El Rhaffari, L., Zaid, A., Hammani, K. and Benlyas, M. 2002. Traitement de la leishmaniose cutanée par la phytothérapie au Tafilalet. *Revue Biologie & Santé* 1(4), 45–54

Gautier-Bréguin, D. 1992. Plantes de cueillette alimentaire dans le Sud du V-Baoulé en Côte d'Ivoire. Description, écologie, consommation et production. *Boissiera* 46, 1–341

Grenand, P., Moretti, C., Jacquemin, H. and Prévost, M. F. 2004. *Pharmacopées Traditionnelles en Guyane*. Paris: IRD

Guèye, M. 2012. *Etude ethnobotanique chez les Malinké de la Communauté rurale de Tomboronkoto (Région de Kédougou) et valorisation des collections historiques de l'Herbier de l'Institut fondamental d'Afrique noire (IFAN) Ch. A. DIOP/UCAD*. Thèse de Doctorat d'État ès Sciences Naturelles, UCAD

Guèye, M., Ayessou, N. C., Koma, S., Diop, S., Akpo, L. E. and Samb, P. I. 2014. Wild fruits traditionally gathered by the Malinke ethnic group in the edge of Niokolo Koba Park (Senegal). *American Journal of Plant Sciences*, 5 (9), 1306–1317

Guèye, M. and Diouf, M. 2007. Traditional leafy vegetables in Senegal: diversity and medicinal uses. *African Journal of Traditional, Complementary and Alternative Medicine (AJTCAM)* 4(4), 469–75

Hawthorne, W. and Jongkind, C. 2006. *Woody Plants of Western African Forests: a guide to the forest trees, shrubs and lianes from Senegal to Ghana*. Kew: Royal Botanic Gardens

Hutchinson, J. and Dalziel, J. M. 1954. *Flora of West Tropical Africa* 1(1), 2nd edn (rev. R. W.J. Keay) London: Crown Agents for Overseas Governments and Administrations

Katende, A. B., Ssegawa, P. and Bernie, A. 1999. *Wild Food Plants and Mushrooms of Uganda*. Nairobi, Kenya: RELMA Technical Handbook 19

Lebrun, J. P. 1973. *Enumération des Plantes Vasculaires du Sénégal*. Maisons-Alfort: IEMVT Étude Botanique 2

Malaisse, F. 1997. *Se Nourrir en Forêt Claire Africaine. Approche Écologique et Nutritionnelle*. Gembloux, Wageningen: CTA

Martin, G. J. 1995. *Ethnobotany. A Method Manual*. London: Chapman & Hal

Maundu, P. M., Ngugi, G. W. and Kabuye, C. H. S. 1999. *Traditional Food Plants of Kenya*. Nairobi, Kenya: Kenya Ressource Centre for Indigenous Knowledge, National Museum of Kenya

Misra, S., Maikhuri, R. K., Kala, C. P., Rao, K. S. and Saxena, K. G. 2008. Wild leafy vegetables: A study of their subsistence dietetic support to the inhabitants of Nanda Devi Biosphere Reserve, India. *Journal of Ethnobiology & Ethnomedicine* 4, 15

Murphy, J. and Sprey, L. H. 1984. *Introduction aux Enquêtes Agricoles en Afrique*. Wageningen, Netherlands: ILRI

Norman, D. W. and Dixon, J. 1995. *Sustainable Dryland Cropping in Relation to Soil Productivity*. FAO Soils Bulletin 72. Food and Agriculture Organisation, United Nations. http://fao.org/docrep/V9926E/V9926E00.htm

Ong, H.-C., Chua, S. and Milow, P. 2011. Traditional knowledge of edible plants among the Temuan Villagers in Kampung Jeram Kedah, Negeri Sembilan, Malaysia. *Scientific Research & Essays* 6(4), 694–7

Rachie, K. O. 1979. *Tropical Legumes: resources for the future*. Washington DC: National Academy of Sciences

Schipper, R. and Budd, L. 1997. *Workshop on African Indigenous Vegetables held in Lime, Cameroon, 13-18 Jan. 1997*. Workshop papers. Nairobi/Chatham: IPGRI

Stevels, J. M. C. 1990. Légumes traditionnels du Cameroun, une *étude* agro-botanique. *Wageningen Agricultural University Papers* No. 90–1, 262p.

Sundriyal, M., Sundriyal, R. C. and Sharma, E. 2004. Dietary use of wild plant resources in the Sikkim Himalaya, India. *Economic Botany* 4, 626–38

Teklehaymanot, T. and Giday, M. 2010. Ethnobotanical study of wild edible plants of Kara and Kwego semi-pastoralist people in Lower Omo River Valley, Debub Omo Zone, SNNPR, Ethiopia. *Journal of Ethnobiology & Ethnomedicine* 6, 23

Thiombiano, D. N. E., Lamien, N., Dibong, D. S., Boussim, I. J. and Belem, B. 2012. Le rôle des espèces ligneuses dans la gestion de la soudure alimentaire au Burkina Faso. *Sécheresse*, 23(6), 86–93

Trotter, R. T. and Logan, M. H. 1986. *Informant Consensus: a new approach for identifying potentially effective medicinal plants*, 91–112. New York: Bedfore Hills

Vivien, J. and Faure, J. J. 1996. *Fruitiers sauvages d'Afrique-Espèces du Cameroun*. Wageningen: CTA

Watson, J. W. and Eyzaguire, P. B. 2002. Home gardens and in situ conservation of plant genetic resources in farming systems. *Proceedings of the Second International Home Gardens Workshop 17–19 July 2001.* Eds Watson, J. W. and Eyzaguire, P. B., 184p. Witzenhausen: Federal Republic of Germany, International Plant Genetic Resources Institute (IPGRI)

Westphal, E., Embrechts, J., FerWerda, J. D., Van Gils-Meeus, H. A. E., Mustsaers, H. J. W. and Westphal-Stevels, J. M. C. 1985. *Cultures Vivrières Tropicales avec Référence Spéciale au Cameroun,* 321–463. Wageningen: Pudoc

Index

Abies alba (see also silver fir) 75
Abora 81, *257*, 259, 262
Aboriginal 3, 159, 251, 273–276, 280, 288, 289, 291, 292
Abri Romanic 75
Abu Hureyra 5, 17, 91, 97, 102, 103, 194
Acacia nilotica 325, *330*
Acaena ovalifolia 308
Acer sp. (see also maple) 265
Acheulean 74, 75, 79, 333
Acorn (see also *Quercus*) 5, 6, 22, 91, 93, 97, 113, 114, 116, 128, 143, 191, 194, 197
Actinoptychus senarius 227, 230
Adansonia digitata (see also baobab) *77,* 322, 329, *330, 346*
Adenium obesum 329, *330*
Adesmia lotoides 308
Adze 266, 275
Africa xi, 3, 32, 57, 58, *60,* 62, *63,* 66, 73–76, 79, 155, 159–164, 171–176, 245, 320, 321, 341, 344, 356, 357
Agaricus pampeanus 308
Agriculturalist 3, 21, 159
Agriculture xi, 1, 3, 9, 17, 21, 32, 43, 91, 103, 191, 193, 275, 291
Agropyron patagonicum 308
Ain Mallaha 96
Alakaluf 303, 304
Al Khiday 5, 247
Alaska 118, 302, 305, 311
Alder (see also *Alnus*) 261, 265
Aleutian islands 302
Alligator River 276–278, 280–282, 284, 285, 289–291
Allium (see also wild onion) 116, 125, 126
Allophyllus africanus 345, *347*
Almond (see also *Amygdalus*) 4, 22, 46, 48, 91, 93, 95–97, 100, *194*
Alnus (see also alder) 265
Alpine bistort 118

Amaranthaceae 345, *346*
Amaranthus 345, *346,* 356
Amud 5, 75
Amygdalus (see also almond) 4, 46, 93, 95–98, *194*
Amylopectin 26, 115
Amylose 26, 115
Anacardiaceae 345, *346*
Anatolia 95, 102, 208
Anbangbang 281–283, 285, 290, 291
An-barra 275
Andes 302
Animal 1, 6–8, 22–24, 26, 32, 33, 43, 48, 55, 56, 58, 61, 64, 65, 71, 75, 79, 81, 95, 101, 102, 113, 114, 129, 144, 155, 164, 172, 174, 182, 191, 193, 195, 203, 241–245, 248, 266, 268, 278, 279, 286, 288, 305, 314, 320–322, 325, 327–329, 333, 334, 342
Annona senegalensis 346, 350
Annonaceae 220, *346*
Anseranus semipalmata (see also magpie geese) 281
Antler 2, 71, 78, 135, 144, 150, 268
Antrea 80
Apium australe (see also wild celery) *308,* 311
Apocynaceae 64, 345, *346,* 352, *353*
Araceae 223, 226, 228, 229, 232, *346, 353*
Araucaria angustifolia 228
Archaeobotanical 4, 91, 92, 94, 96, 97 101– 103, 113, 114, 117, 118, 120, 124, 128, 129, 135, 193, 215, 216, 233, 251, 290–292, 301, 311, 315
Archaeobotany xi, 114, 150, 273, 274, 281, 291, 292, 305
Arctic 6, 7, 81, 302
Ardipithecus 161, 245
Arecaceae 220, 228, *229, 232, 346,* 352, *353*
Arjona patagonica 308, 311
Armeria maritima (see also sea thrift) *308,* 311
Arnhem Land 275–278, 280, 281, 286–290
Arrow 79, 80, 319, 329, *330,* 332–334
Arrowhead (see also *Sagittaria*) 118, 119

Asclepiadaceae 346
Ash (see also *Fraxinus*) 114, 127, 177, 265, 306
Ashes 119, 124, 128, 156, 228, 231, 232, 308–310, 329
Asia 91, 95, 103
Asparagus 97
Atlantic 4, 114, 124, 302, 305, *307*
Atriplex (see also orache) 91, 95, 97, 103, 116
Au. Afarensis 159, 161–165
Au. Anamensis 159, 161–163, 165, 245
Aurignacian 195
Auroch 95, 100, 114
Australia 3, 6, 251, 273–278, 286, 288, 291
Australopithecus (see also *Au. afarensis; Au. anamensis*) 5, 161–163, 245
Avena (see also oat) 91, 98
Avicennia 278, 279
Axe 144, 266, 278, 325, 329, *330*
Azorella 308

Baboon 33, 57, 163, 182, 322, 324
Bacteria 41, 42, 56, 115
bag(s) 79, 80, 82, 281, 285, 286, 289
Bamboo (see also *Bambusa arnhemica*) 280, 281, 284, 285, 290
Bambusa arnhemica (see also bamboo) 280, 281, *283*, 290
Bananas 24, 26
Banksia dentata 283
Bantu 159
Baobab (see also *Adansonia digitata*) 77, 321, 322, 325, 329, *330*, 332, 333
Bark 22, 61, *63*, 71, 73, 76–78, 80, 81, 116, 144–147, 163, 184, 255, 260–262, 265, 268, 276, 289, 290, 302, 304, 306, 309–311, 328, *330*
Barley (see also *Hordeum*) 3, 21, 25, 27, 91, 93, 95–101, 191, *194*, 196, 199, 204–206, 209
Barramundi 280, *287*
Basket *145*, 147, 224, 260, 280, 284, 285, 290, 306–309, 332
Beads 79, *328*, 329
Beaver 268
Bedding 75, 173, 184, 207
Bedrock 3, 96, 112, 191, 192, 195–197, 199–204, 206, 207
Bees 23, 321, 325
Beeswax 77, 329
Beidha 102
Berberis 302, 303, *308*, 314
Berries 7, 91, 93, 94, 113, 114, 116, 141, 302, 305, *306*, *308*–311, 314, 321–324, *330*
Beta vulgaris ssp.*maritima* (see also sea beet) 124, *125*

Betula (see also birch) 265
Big quail (see also *Coturnix chinensis*) 281
Bignoniaceae 346, 353
Birch (see also *Betula*) 81, 114, 116, 265
Birch bark 73, 76–78, 80, 147, 261
Bird cherry (see also *Prunus padus*) 197, 265
Bird 32, 64, 95, 245, 251, 268, 304, *310*, 321, 322, 325, 328, 329, 333
Bitter vetch (see also *Vicia*) 98
Bitumen 17, 71, 76, 82
Blade 96, 100, *139*, 141–146, 148, 149, 264
Blombos Cave 79
Boat (see also canoe) 79, 253, 255
Boerhaavia 345, 347
Bolax 306, *308*
Bolboschoenus maritimus (see also sea club-rush) 97, 100, 197, 207
Bolkilde 81
Bølling-Allerød 92, 96
Bolomor Cave 74
Bombacaceae 346
Bone 1, 2, 31, 33, 34, 40, 44, 71, 73–76, 78, 80, 102, 113, 135, 144, 147–150, 171, 173, 186, 203, 216, 217, 242, 244, 248, 254, 255, 259, 268, 276, 328
Bonobo 55, 72
Boopis australis 308
Borage (see also *Boraginaceae*) 94
Boraginaceae (see also borage) 94, *346, 353*
Borassus aethiopum 346, 350, 353
Border Cave 75, 79
Boreal 80, 114, 124
Borneo 5, 6
Bow 75, 79, 80, 319, 329, *330*, 332–334
Brachychiton paradoxum 283
Brain 19, 20, 22, 23, 31–37, 44, 45, 164, 182, 320, 328
Brassica 100
Brazil 215–219, 228, 231
Brush 74, 93
Buchanania obovata 282
Budongo 58
Bulb 95, 113, 116, 117, 125, *126*, 182
Bulrush (see also *Scirpus*) 80, 116, 142, *145*, 264
Burial 79, 93, 112, 191, 192, 195, *196*, 200, 203, 204, 215, 217, *219*, 221, 222, 225, 227–230, 232
Bushmen 301

Caesalpiniaceae 345, 346, 352, 353
Cafer Höyük 102
Calathea 226, 228, 229
Callitris intratropica 283
Całowanie 115, 118, 119

Caltha palustris 142

Calvatia 308, 309

Cameroon *63*, 163, 341, 356

Canada 8, 118

Canarium australianum 282

Canoe (see also boat) *77, 78*, 286, 289, 304

Capparaceae 346

Capparis 97

Carassius carassius (see also Crucian carp) 254

Carbohydrate 2–7, 17, 19–24, 26, 31–37, 41, 42, 44, 47, 48, 97, 114–116, 125, 128, 172, 224, 231, 242, 247, 248, 314, 322, 324–326, 330

Cardamine glacialis 309

Carpinus betulus (see also hornbeam) 265

Castor seeds 75

Çatalhöyük 102, 103

Catalunya 75

Cave 4–6, 74–76, 79, 80, 95, 96, 98, 112, 172, 174, 176, 177, 180, 184, 191–196, 199, *201*, 204, 206–209, 247

Celastraceae 346, 353

Ceramic 116, 217, 261

Ceratotheca sesamoides 347, 352

Cercopithecoidea 155, 163

Cereal 17, 21, 22, 24, 25, 27, 28, 91, 93, 96–103, 112, 128, 138, 141, 191, 193, 356

Cetaceans (see also whale) 304

Charcoal xi, 1, 3, 44, 74, 93, 97, 117, 127, 171, 173, 174, 177, 216, 218, 219, 231, 258, 274–276, 291

Chenopod 91, 95, 97, 103

Chenopodiaceae 95

Chenopodium (see also fat hen) 101, 116

Chert *139*, 290

Chesowanja 74

Chia Sabz 98, 99

Chickpea (see also *Cicer*) 21, 22, 99, 101

Chiliotrichum diffusum 303, 309

Chimpanzee (see also *Pan troglodytes*) 2, 4, 17, 33, 44, 55–59, 61, 62, 65, 71, 72, 163, 182

China 5

Chisel 144, 266

Chloridoideae 226

Chogha Golan 98, *99*

Cholesterol 27, 37

Christ's thorn (see also *Paliurus spina-christi*) 93

Chrysobalanaceae 346

Chyme 7, *314*

Cicer (see also chickpea) 46, 101

Clacton-on-Sea 75

Cladonia laevigata 309

Clay 44, 79, 80, 93, 127, 177, 198, 199, *201*, 260, 284

Cleaver 311

Climate 7, 22, 24, 96, 97, 113, 163, 244, 273, 276–278, 288, 320, 341

Climatic 7, 92, 96, 155, 159–161, 163, 172, 174, 191, 193, 194, 251, 278, 280, 302, 342

Climbing pegs 329, 333, 334

Clothing 79–81, 289, *309*

Coccinea surantiaca 326, 330

Colic 58

Collett's rat (see also *Rattus colletti*) 281

Colonsay 122

Comí 256, 258

Commelina 345–347

Commelinaceae 346, 347

Commiphora schimperi 329, 330

Compositae (see also sunflower) 61, 94

Congo 59, *63*, 341

Conopodium majus (see also pignut) 125–127

Constipation *63*

Convolvulaceae 229, 232, 347

Cooking 26, 32, 35, 43–47, 73, 75, 93, 125, 128, 129, 159, 172, 173, 231, 232, 311, 327, 332

Corchorus 345, 348

Cord (see also string) 71, 79– 81, 135, 258–261, 264

Cordia myxa 323, 329, 330, 346, 352, 353, 355, 357

Cordyla pinnata 346, 353, 356

Corms 20, 116, 118, 164, 182, 281, 282, 289

Cornus sanguinea (see also dogwood) 77

Coscinodiscus 227, 230

Côte d'Ivoire 341

Coturnix chinensis (see also big quail) 281

Crop(s) 24, 91, 93, 99, 102, 103, 148, 149, 356

Croton menyhartii 328, 330

Crowberry (see also *Empetrum rubrum*) 306, *308*, *309*, 311

Crucian carp (see also *Carassius carassius)* 255

Cruciferae 102

Crustaceans 304

Cucurbitaceae 220, 347

Cultivation 2, 91, 92, 97–100, 102, 103, 191, 193, 218, 231–233 356

Cycas armstrongii 282

Cyperaceae 32, 47, 95, 118, 121, 199, *208*

Cyperus rotundus (see also wild nut-grass) 5, 207, 247

Cyttaria darwinii 309

Daisy yam (see also *Microseris scapigera*) 275

Daub 100

De Bruin 135, 148–150

Death xi, 31, 101, 204, 215, 222, *227*

Dederiyeh Cave 98, 194

Deer 95, 114

Demirkoy 100
Denhamia obscura 283
Denisova Cave 79
Denmark 80, 120, 124–*127*, 129, 135, 150
Derjø 80
Descurainia 306, *309*, 311, 314
Desert 8, 33, 92, 95, 274–276, 301
Detarium 77, 346
Dhra *99*, 100
Diarrhoea 36, 40, 58, 59, 61, 65, 328
Diatom 175, 177, *178*, 180, 215, 220, *221*, 223,
    227, 229, 230, 232, 276
Dicotyledon 4, 124, *172*, 176, 178, 180, 182, 199,
    208, 209
Didgeridoo 289
Diet 1, 3, 6–9, 17, 19–21, 23, 24, 26–28, 31–34,
    36–38, 40–45, 48, 58, 59, 66, 101, 102,
    111–118, 120, 129, 155, 158, 159, 161–164,
    171–174, 176, 182, 184, 186, 191, 194, 206,
    208, 215–217, 220, 222, 226, 228, 231, 232,
    241–249, 251, 252, 273, 274, 281, 291, 305,
    314, 315, 319–327, 329, 332, 333, 341, 357
Digging sticks 75, 289, 319, 326, 329, *330*, 333, 334
Dilly bag 281, 285, 286, 289
Dioscorea (see also yam) 218, 219, 224, 228, *229*,
    288, 345, *347*, 350, *353*, *354*
Dioscoreaceae 229, 232, *347*, 352, *353*
Diospyros 345, *347*
Diploneis ovalis 227, 230
Disease 9, 23, 26, 34, 36, 37, 40, 43, 44, 56–59, 66,
    216, 289, 303
Dja´de *99*, 101
Djuwarr 281–283, 290, 291
DNA 22, 38, 274
Dogwood (see also *Cornus sanguinea*) 77, 78
Dolní Věstonice 4, 79, 117
Dombeya kirkii 329, *330*
Domesticated 101–103, 217, 232, 241, 315
Domesticates 102, 103, 243
Domestication 28, 43, 91, 92, 98–101, 191, 193,
    218, 243, 274
Drimys winteri 302, 306, *309*
Dronten-N23 77
Drupe 322, 324, 325, 330
Dubna River 253, 264–266
Dwelling (see also house, hut) 95–97, 259, 261, 305
Dye 290
Dysentery 58, 62, 65
Dzudzuana Cave 80

Ebenaceae 347
Ecology 6, 155, 159, 166, 273, 289, 319, 320, 334

Egg 61, 62, 64, 119, 280, 305
Egypt 44, 207
Ein Gev I 92
Einkorn 91, 97–99, 101, 102
el Hemmeh 98, *99*
El Sidrón 5, 8, 76, 247
Eleocharis spp. (see also water chestnuts) 47, 281,
    282, 285, 289
Elk 255
Elm (see also *Ulmus*) 114, 256, 259, 261, 265
Embothrium coccineum 303, *309*
Emmer 91, 93, *99*, 102
Empetrum (see also crowberry) 302, 306, *309*, 311,
    314
Enamel 2, 32, 111, 155–158, 161, 164, 244, 245, 248
Entadrophragma caudatum 328, *330*
Enteritis *63*
Environment 6–9, 20, 21, 33, 43–45, 76, 92, 95,
    76, 116, 118, 124, 125, 135, 140, 155, 159–161,
    163, 165, 173, 182, 207, 217, 224, 229, 230,
    232, 241, 243, 246, 247, 251, 273, 274, 276–
    278, 292, 327, 333
Enzyme 23–25, 37, 38, 40, 42, 243
Epipalaeolithic 91, 95, 102, 103, 195
Equisetum (see also horsetail) 116, 120
Ertebølle 80, 113, 116, 124–*127*, 135, 148
Estonia 80
Ethiopia 44, 57, *63*, 356
Ethnobotany 3, 8, 251, 273–275, 281, 290, 315,
    319, 320, 328, 334
Ethnography xi, 3, 251, 275, 305, 315
Ethnohistory 3, 251, 273
Eucalyptus sp. *283*
Euphrates River 92
Eurasia 103, 118, 119, 193
Europe 5, 6, 8, 73, 74, 77, 80, 82, 111, 113–118,
    120–122, 124, 128, 129, 136, 148, 244
European xi, 6, 111, 117, 125, 128, 129, 141–143,
    150, 251, 253, 255, 260, 265, 268, 276, 278,
    288, 289, 301, 303, 305, 356
Euryale ferox (see also prickly water lily) 4
Evenks 258
Ewan 305, 311, 314
Exos lucius (see also pike) 255
Eynan 193, *194*, *205*

Faba bean (see also *Vicia faba*) *99*, 101
Fabaceae 229, *347*, *353*
Fallback food 4, 162–164, 252, 341, 357
Famine 341, 352–357
Fat hen (see also *Chenopodium album*) 116
Festuca gracillima 309

Fever 56, 59, 62, 63, 328

Fibre 16, 17, 19, 20, 24, 27, 31, 32, 34, 41, 42, 48, 71, 72, 76, 78–82, *123*, 144, 146, *147*, 216, 218, 224, 255, 259–262, 264, 274–276, 281, *283*, 284, 289, 290, 322, 324–326, *330*, 332

Ficoidaceae 347

Ficus (see also fig) 94, 98, 324, *330*, 345, *347*, *353*, 356

Fig (see also *Ficus*) 94, 98–100, 323, 324, *330*

Figurine 79

Finland 80

Fippenborg 80

Fire 2, 6, 35, 44, 66, 73–75, 78, 82, 117–119, 142, 172, 173, 186, 255, 275, 276, *283*, 289, 291, 292, 304, *308*, 314, 326, 327, 329, *330*, 333, 334

Fire drill 329, 333, 334

Firestick 281, 284

Firewood 65, 304

Fish 3, 32, 37, 76, 95, 113, 114, 129, 135, 144, 216, 217, 243, 253–*256*, 258–262, 264, 266, 268, *283*, 286, 304, 305, 314

Fish-fence 255, 258

Fishing net 80, 224

Fish-screen 255

Fish trap 3, 76, 135, 251, 253–268, 289

Fistulina hepatica 309

Flake 76, 96, 141, 143, 285, *286*

Flint 76, 92, 135, 136, *139*, 141, 143–146, 148, 149, 266

Floating sweet grass (see also *Glyceria fluitans*) 116

Float 81, 255

Food 1–8, 17, 19–24, 26–28, 31–37, 41–48, 55–59, 61, 63, 66, 73–75, 91, 93–97, 100–103, 111–129, 135, 141–144, 155, 156, 158–165, 172, 173, 180, 182, 184, 191, 193–195, 197, 206, 207, 215–218, 220, 222, 224, 226–228, 230–232, 241–247, 251, 252, 274–276, 280, 281, 284, 285, 288, 289, 292, 301, 302, 304, 305, 308–311, 314, 315, 320–322, 324, 325, 327–329, 334, 341, 342, 345, 346, 352, 354–357

Forage 24, 321, 334

Forager 193, 321, 325, 333

Foraging 35, 58, 64, 180, 209, 281, 284, 288, 319–321, 325, 327, 332–334

Forest 3, 6, 57–59, 65, 92, 96, 97, 114, 125, 159, 161, 191, 194, 217, 220, 233, 246, 276, *279*, 281, *283*, 289, 302–305, 341, 342, 344, *349*, 356, 357

Forstermoor 80

Fragaria chiloensis 302, *309*

Fraxinus (see also ash) 265

Friesack 80

Frugivorous 161–163

Fruit 1, 4, 5, 19, 22, 23, 26–28, 32, 57, 63, 77, 78, 91–93, 95–97, 100, 102, 111, 113, 114, 116, 156, 161, 163, 173, *180*, 182–184, 216, 218–220, 268, 274, 281, 282, 305–309, 311, 314, 321–324, 330, 332, 341, 345–350, 352, 353, 355–357

FTIR 3, 171, 174, 177

Fuel 1, 4, 33, 38, 39, 42, 74, 75, 77, 95, 172, 173, 184, 215, 218, 228, 283, 284

Fungi 23, 56, 304–306, 311, 314

Gazelle 95

GC-MS 241, 242, 246–248

Geophyte 172, 176, 184, *185*

Georgia 80

Germany 75, 76, 80

Gesher Benot Ya'aqov 4, 44, 74, 75, 172

Gibraltar 4

Gilgal I 98–100

Gluconeogenesis 6, 7, 23, 37

Glucose 6, 19, 20, 22–27, 31, 34, 36–38, 41, 47, 115, 325

Glue 71

Glyceria fluitans (see also floating sweet grass) 116

Glycogen 23, 26

Goat 95, 342

Gonnersdorf 79

Gorham's Cave 4

Gorilla 55, 59, 64, 163

Gouge 266

Gourd 332

Grains 17, 21, 22, 24, 25, 27, 28, 36, 91, 93, 95–97, 101–103, 115, 116, 136, 199, 204, 206–208, 225, 226, 347

Grape (see also *Vitis*) 94, 95, *194*

Grass (grasses) 3, 5, 8, 17, 24, 32, 73–75, 80, 91, 93, 95–98, 100–103, 112, 116, 143, *144*, 159, 164, 171, 174, *178*, 180, 182, 191–194, 199, *202*–209, *226*, 228, 231, 243, 245, 246, 274–279, 286, *287*, *308*, *309*, 320, 328, 329, *331*

Grassland 8, 33, 126, 159, 160, 165, 182, 209, 320

Grave 93, 204, 207

Gravettian 5

Greece 102

Grewia 323, 329, *330*, 345, *348*

Grinding (implements, stones, tools) 5, 91, 95–97, 111, 143, 193, 274

Grit 156, 165

Groundstone 194, 195

Guanaco 304, 305, *309*
Gum 28, 76, 157, 285, 345, *348*
Gura Cheii-Râşnov Cave 76
Gusinde, Martin 305, *307*
Gut 19, 20, 23, 24, 26, 27, 31, 34, 41, 43, 47, 62,
    164, 216

H. (Homo) Erectus 173
H. (Homo) Ergaster 165, 184
H. (Homo) Habilis (see also *Homo habilis*) 73, 171,
    174, 182, 184, 186
Habitat 6, 57, 97, *99*, 101, 114, 160, 161, 164, 209,
    274, 281, *282, 283*, 289
Hadza 3, 75, 172, 251, 319–334
Hallan Çemi 100, 208
Halsskov 124–129
Hardinxveld 122, 135, 143, 148–150
Harpoon 79, 255
Harvest 21, 22, 43, 46, 121, 280, 289, 311, 341, 357
Harvesting 2, 5, 31, 75, 91, 96, 101, 138, 141, 193
Hattemerbroek 77
Haush 303, 304
Hayonim Cave 96, 194
Hazel 6, 114, 146, 261
Hazelnut 22, 113, 114, 122, 128, 143
Health 8, 19, 20, 27, 31, 34, 36, 40–42, 55, 56, 61,
    291, 314, 356
Hearth 4, 74, 76, 93, 95, 127, 129, 172–174, 176,
    177, 184, 217, 229, 305
Heinrich event 92
Helianthemum salicifolium 101
Hexalobus monopetalus 346, *350*
Hibiscus 283, 345, *347*
Hide 75, 80, 81, 96, 136, 138, 144, 148, 322
Hindu 159
Hoge Vaart 129
Holocene 2, 91, 92, 98, 113, 121, 159, 184, 194,
    228, 230, 244, 273, 274, 278, 280, 284, 286,
    290
Holy thistle (see also *Silybum marinanum*) 94
Hominin 1, 4, 7, 8, 17, 31–33, 35, 36, 41, 44, 45,
    48, 56, 66, 71–73, 111, 112, 155, 159–161, 164,
    171–174, 182, 241–243, 245–248, 319, 320,
    327, 328, 333, 334
Hominini 32, 155
Hominoidea 32, 155
Homo (see also *H. erectus, H. ergaster, H. habilis,*
    Homo sapiens) 31, 33, 43, 44, 48, 66, 73, 111,
    160, 164, 165, 171, 186, 195, *200, 201*, 203,
    204, 208, 245, 320, 327, 328, 333, 334
Homo Sapiens 32, 171, 186
Honey 23, 251, 321, 322, 325, 327, 329, *330*, 332–334

Hook 255, 259, *353*
Hordeum (see also barley) 91, 93, 96, 101, 103,
    194, 199, 204
Hornbeam (see also *Carpinus betulus*) 265
Horsetail (see also *Equisteum*) 116, 120
Horticulture 112, 217, 232, 233
Horticulturalist 8, 275
House (see also dwellings, huts) 79, 259, 261,
    262, 305, 329
Household 81, 100, 101, 319–321, 329, 330, 332, 334
Hunter-gatherers 1–3, 5, 8, 9, 38, 59, 74, 91–93,
    95, 103, 111, 112, 114, 116–118, 123, 129, 135,
    140, 141, 150, 158, 159, 172, 174, 182, 184,
    193– 195, 207, 231, 251, 273, 275, 301, 302,
    304, 319–322, 327, 334
Hunting 6, 7, 31, 33, 78, 79, 95, 96, 113, 172, 193,
    232, 268, 291, 301, 304, *310*, 320, 321, 329,
    330, 333, 334
Hut (see also dwellings, houses) 75, 93, 95, 96,
    305, 309, 329, *331*
Hypochoeris 306, *309*
Hypoplasia 184

Icacina senegalensis 347, 352, *353*, 356
Icacinaceae 347, *353*
Ide (see also *Leuciscus idus)* 255
Illness 56, 58, 61, 65, 66, 357
India 65, 73
Infant 44, *63*, 200, 321
Infection 55–59, 61–*63*, 65, 66
Insect 21, 23, 27, 32, 43, 56, 246
Inuit 6, 7, 159, 311, 314
Ipomoea 223, 228, *229, 347*
Iran 91, *99*
Iraq 91, 98, *99*
Iron Age 138, 260
Isotopes 3–5, 7, 32, 33, 98, 101, 111, 112, 173, 174,
    193, 217, 241–248, 320
Israel 4, 5, 44, 74, 75, 80, 91, 96, 172, 191, 193, 247
Italy 76, 102, 248

Jabuticabeira 3, 215, 216, 218–221, *223*–232
Japan 5, 6
Jerf el Ahmar 98–100
Jericho 98–100
Jewellery 79, 328
Jomon 6
Jordan 91, 92, 96, 100
Ju/'hoansi 301

Kakadu 3, 273, 277, 281, 284, 286, 289, 291
Kalahari 81, 301

Kalunde, Mohamedi Seifu 65, 66
Kebara Cave 4
Kebaran 195, *201, 204*
Kenya 57, *63*, 74, 182, 356
Kenyanthropus platyops 245
Kets 258
Khant 258
Kharaneh IV 92
Khoe-San 159
Kibale 57, 58
Kigelia africana 345, 346, 353
Kimberley 273
Knives 138, 141, 255, 329
Knotgrass (see also *Polygonum*) 91, 97, 100, 103, 118
Komes 258
Kongsted Lyng 81
Konigsaue 76
Koobi Fora 74
Koriak 258
Korjaki 256, 258
Körtik Tepe 101
Koryals 258
Kostenki 79
Ksar'Akil 79
!Kung 8, 172

Labiatae 347
Lactose 19, 22–24
Łajty 120
Lake Eyasi 174, *175*, 321, 322
Lake Hula 194
Lake Kinneret 93
Lake Lubans 258
Lake Zvidze 258
Landolphia heudelotii 346, 354
Language 66, 342
Lannea 345, 346, 356
Lascaux 80
Last Glacial Maximum (see also LGM) 2, 91, 92, 284
Latvia 80, 81, 253, 257–259
Lauraceae 228, 229
Leakey, Jonathan 174
Leakey, Mary 174
Leaves 24, 57, 59, 61–*63*, 65, 71, 72, 74, 75, 77, 92, 95, 96, 118, 124, 125, 156, 163, *172*, 180, 184, 199, 228, 242, 289, 306, *308*–311, 314, 325, 328, *330*, 341, 345–347, 349, 352, 357
Lebanon 79, 91
Legume 4, 5, *21*, 28, 46, 91, 96, 97, 100, 208, 322, 324, 325, *330*
Lehringen 75

Lentil *21*, 22, 24, 97–*99*, 101, *194*
Leptadenia hastata 346, 356
Lesser celandine (see also *Ranunculus ficaria*) 116, 117, 121, *122*
Leuciscus idus (see also ide), 255
Levant 2, 17, 91, 92, 96, 98–103, 112, 191–195, 208, 209
LGM (see also Last Glacial Maximum) 2, 91–93
Lichen 306, *309*
Light microscopy 137
Liliaceae 279
Lime 80, 81, 114, *145*, 146
Lipids 19, 22, 26, 34, 38, 242, 244, 248
Lithic (see also stone tools) 1, 78, 92, 96, 100, 103, 111, 160, 164, 175, 195, 217, 218, 230, 232, 261, 333
Lithuania 80, 81, 253, 259, 260
Livistona humilis 282
Loganiacaea 347
Lonchocarpus eriocalyx 329, 330
Lugovskoje 261
Lupin (see also *Lupinus pilosus*) 96
Lupinus pilosus (see also lupin) 96, *194*

Macaques 73
Macrofossils 175, 273
Maerua pseudopetalosa 356
Magellan 302, 303, 304, *310*
Magpie geese (see also *Anseranus semipalmata*) 281
Mahale 58, 59, 61, 62, 65
Maize (see also *Zea mays*) 215, 228, 231, 232, 241, 342
Malaria 58, 61–63
Malinké 3, 341, 342, 357
Mallow (see also *Malva*) 94
Malva (see also mallow) 94, 96
Malvaceae 347
Mangabey 57, 182
Maple (see also *Acer* sp.) 265
Marantaceae 226, 229
Marine mammals 304
Markhamia obtusifolia 329, 330
Marmugino 260, 262
Marsippospermum grandiflorum 306, 309
Maya 8
Meat 6, 7, 17, 31–33, 36–38, 40–43, 48, *63*, 129, 159, 164, 165, 171–173, 245, 247, 301, 305, 314, 321, 322, 327, 334
Medicinal 2, 5, 7–9, 17, 55, 56, 58, 59, 61, 62, 65, 66, 73, 94, 95, 216, 247, *308, 309*, 319, 320, 328, 341, 342, 356, 357

Medicine 1, 4, 8, 9, 20, 32, 58, 64–66, 120, 224, 228, 251, 319, 328, *330*, 334, 341
Mediterranean 92, 93, 96, 176, 191–193
Melaleuca spp. (see also paperbark) *279*, 280
Meliaceae 347
Mesolithic xi, 3, 5, 6, 73, 75–81, 111, 113–129, 135, 136, 141–143, 148, 149, 251, 253, 254, 268
Mezhirich 117
Microlith 92, 96
Microseris scapigera (see also daisy yam) 275
Microwear 1, 2, 4, 9, 111, 135–138, 140, 141, 143, 144, 148–150, 155–159, 161–166, 173, 184, 193, 320, 327
Middle Stone Age 5, 44, 171
Millet 21, 241, 243, 342, 352, 355
Mimosaceae 347
Miocene 3
M'lefaat 100
Møllegabet 80, 124
Mongolia 8
Monkey 55, 57
Monocotyledon 4, 33, 75, 176, 178, 180, 199, 218
Monosaccharides 19, 20, 22–24, 28, 116
Moose 268
Moraceae 347, 353
Morocco 5
Morphology 3, 26, 155, 161, 162, 173, 199, 245, 319, 320
Mortality (see also death) 61
Mortar 3, 91, 93, 96, 112, 191, 192, 195–197, 199–201, 203, 206, 207, 219
Mosquito net 289
Mossel Bay 172, 174, 175, 180, 184–186
Mount Carmel 191, 192, 195, 208, 209
Mozambique 5, *63*
Mud brick 101
Muller 96
Mureybet 97–100
Museum 218, 259, 274, 275, *280, 284, 286,* 289, 291
Myrtaceae 220, 223, 228, *229, 347*
Mysodendron punctulatum 309

Nahal Soreq Cave 193
Nahal Oren *194*, 196
Nålebinding 80, 81
Natufian 3, 91, 96, 97, 112, 191–196, *201,* 203–209
Ndoki forest 59
Neanderthal 4–8, 44, 73, 75, 76, 79, 115, 244, 245, 247
Needle 79, 147–150, 255
Nelumbo nucifera (see also water lilies) *282,* 289
Nematode 56–59, 61, 65

Nentser 258
Neocarya macrophylla 346
Neolithic 5, 17, 75, 77, 80, 81, 91–93, 97–99, 101–103, 116, 124–126, 129, 141, 142, 148, 149, 195, 253, 254, *257–262,* 264, 265, 268
Net 78–80, 135, *145,* 150, 224, 253, 255, 260, 261, 289
Netherlands 77, 78, 120–124, 129, 138, 148
Netiv Hagdug 98–100
Net-sinker 261
Nettle (see also *Urtica*) 81, 145, 147
Nevali Çori 101
New Guinea 8, 82, 273, 291
Niah Cave 5, 6
Nothofagus 302, 303, 306, *309, 310*
Nukak 302
Nuphar (see also water lily) 116, 128, 129, 142, 197
Nutrition 17, 19, 20, 26, 36, 216, 220, 232, 319, 320, 327, 328, 356
Nutritional 2, 7–9, 17, 19, 21, 28, 32, 34, 43, 45, 46, 55, 57, 58, 115, 128, 171, 173, 194, 311, *312,* 314, 319, 320, 327
Nuts 4, 5, 19, 22, 26, 32, 46, 95, 96, 100, 115, 128, 141, 143, 161, 163, 194, 218, 281, 321, 322, 324
Nyctaginaceae 347
Nymphaea (see also water lily) 116, 128, 142, 197, 281, *282,* 286

Oak (see also acorn, *Quercus*) 4, 93, 114, 116, 128
Oar (see also paddle) 266, 268
Oat (see also *Avena*) 21, 27, 91, 98, *99*
Ob Ugrians 258
Occlusal 111, 155–158, 161–164, 166
Ochre 76, 290
Ocimum 345, *347*
Ohalo 4, 17, 75, 80, 91–93, 95, 102, 209
Oka River 253
Öküzini Cave 95
Olacaceae 347
Oldowan 327
Olduvai 3, 111, 171, 173–176, 178, 182, 184, 186
Oligosaccharides 19, 20, 23, 24, 42
Olives 91, 93
Orache (see also *Atriplex*) 116
Orang utan 72
Oreomyrrhis andicola 310
Oryza spp. (see also rice) *284,* 285
Osmorhiza chilensis (see also sweet-cicely) *310,* 311
Oven 95, 123, 231, 232, 246, 280
Owenia vernicosa 283

Pacific 73, 292, 302
Paddle 253, 255, 264, *267*
Palaeolithic xi, 4, 5, 8, 34, 44, 74–76, 79–81, 100, 117, 173, 195, 209, 247, 319, 320, 327, 333, 334
Palestinian authority 91
Paliurus spina-christi (see also Christ's buckthorn) 94
Palm 21, 22, 112, 171, *178*, 180, 182–184, 186, 199, 215, 218, 226, 228, *282*
Pan troglodytes (see also chimpanzee) 33, 57, 163
Pandanus spiralis 282, 283
Panicoideae 226
Paperbark (see also *Melaleuca* spp.) 280, *283*, 284
Paralia sulcata 227, 230
Paranthropus 160, 164, 171, 174, 245
Parasite 55–59, 61, 62, 64, 65
Parbari rockshelter 281
Parenchyma 2, 5, 111, 113–115, 117–121, 123–125, 127–129
Parinari excelsa 346, 354
Parkia biglobosa 347, 356
Pastoralist 159
Patagonia 311
Pavlov 79
Pea (see also *Pisum*) 21, 22, 24, 46, 96, 97, 99, 143, *194*
Peanut 21, 22, 342
Pectin 27, 28
Pedaliaceae 347
Perca fluviatilis (see also perch) 255
Perch (see also *Perca fluviatilis)* 255
Pernettya 306, 310
Persoonia falcata 282
Pestle 96, 195, 197, 219
Philesia sp. *310*
Phoenix reclinata (see also wild date palm) 182, *183*
Phragmites (see also reed) 116, 120, 141, 199, 264, 280, 281, *283–285*, 290
Phyllanthus sp. *283*
Phytolith 2, 3, 5, 75, 96, 111, 112, 155, 156, 159, 171–178, 180, 182–184, 186, 191, 192, 194, 198–200, 202–204, 206, 215, 220, 221, 223–226, 228, *229*, 231, 232, 273, 274, 276, 284, 289–291, 320
Picea sitchensis (see also Sitka spruce) 73
Pig 64, 65, 95, 114, 288
Pigment 71
Pignut (see also *Conopodium majus)* 125–127
Pike (see also *Exos lucius)* 255, 258, 259
Pine (see also *Pinus*) 5, 22, 73, 75, 77, 81, 114, 116, 228, 255, 256, 258, 260–262, 264–266, 268
Pinnacle Point 171, 173–176
Pinus (see also pine) 75, 265

Pistacia (see also pistachio) 4, 93, 95–98, 193
Pistachio (see also *Pistacia atlantica*) 4, 22, 91, 93, 95, 100, 194
Pisum (see also pea) 96, *194*
Pitch 17, 71, 73, 76, 78
Pith 22, 59, 61, 62, 65, 121
Planchonella arnhemica 283
Pleistocene 2, 4, 32, 34, 48, 76, 113, 155, 159, 171, 173, 174, 194, 207, 244–246, 273, 278, 284
Pliocene 112, 159, 161, 162, 164, 242, 245, 246
Poa flabellata 310
Poaceae 215, 219, *223*, 226, 228, 229, 232, *347*
Podzorovo 262
Poland 76, *115*, 116, 118–120
Pollen 6, 98, 115, 118, 125, 126, 173, 225, 273, 276, 278, 279, 292
Polyalthia holtzeana 283
Polygonum (see also knotgrass) 91, 97, 100, 103, 118
Polyporus 310
Polysaccharides 19, 20, 22, 24, 26–28, 30, 34, 41, 42, 115
Pooideae 226
Poplar (see also *Populus*) 80, 81, 265
Populus (see also poplar) 80, 265
Porosphaera globularis 79
Post-mortem 156–159
PPN 91, 93, 191, 193
PPNA 91, 98, 100–103
PPNB 91, 98, 101–103
Prebiotics 19, 20, 42
Pre-Boreal 98, 114, 118
Prickly water lily (see also *Euryale ferox*) 4
Primate 4, 31–33, 55–59, 66, 72, 73, 155, 158, *162*, 182
Protozoa 56, 57
Prunus padus 197
Pterocarpus santaloides 347, 353, 355, 357
Pulses 22, 24, 34, 91, 93, 96–103, 191, 325
Pygmy 159
Pyrus 95

Qarassa 96, 102
Qermez Dere 100
Qesem Cave 5, 74, 247
Quartzite 156, 290
Quercus (see also oak) 4, 93, 96, 97, 128, 193, 197
Quern 100, 141

Rainfall 57, 288, 303, 322
Rambling vetch (see also *Vicia*) 91, 98, 99
Ranunculus ficaria (see also lesser celandine) 116, 117, 121–123
Raphia palma-pinus 345, 346, 353

Raphionacme 345, *346*, 349, 350, *352*, *353*
Raqefet 3, 191, 192, *194–197*, *205*, 206, 208, 209
Raspberry (see also *Rubus*) 94
Rattus colletti (see also Collett's rat) 281
Reed (see also *Phragmites*) 6, 116, 120, 136, 138,
    141, *145*, 148, 199, 257, 264, 275, 280, 281,
    289, 290, 328
Reindeer 65, 114
Residue 5, 9, 28, 96, 116, 137, 184, 198, 199, 207,
    208, 220, 241, 242, 244, 248, 273, 274, 276,
    278, 281, 285, 290, 291
Resin 71, 73, 75, 76, 157, 274, 284, 285, 329
Rhamnaceae 347
Rhizome 2, 20, 113, 115–119, 123, 124, 128, 178,
    228, *262*, 275, *282*, *308*
Rhizophora sp. 278
Ribes 302, 303, *310*, 314
Rice (see also *Oryza* spp.) 21, 284, 285
Roach (see also *Rutilus rutilus*) 255
Rock art 278, 281, 286, 288, 290, 291
Rodent 304, 305
Romania 76
Root 2, 4, 17, 19, 20, 24, 25, 28, 47, 63–66, 81,
    93, 95, 111–117, 121, 122, 124–129, 142, 143,
    163, 175, 182, 206, 231, 246, 274, *282*, 306,
    308–311, 314, 315, 328, *330*, 345, *347*, 349
Rope 79–81, 144, *145*, 147, 224, 262, *283*
Rubiaceae 345, *347*, 352, *353*
Rubus (see also raspberry) 94, 302, *310*, 314
Russia 3, 75, 76, 79, 80, 135, 251, 253, 255, 258,
    260, 261
Rutilus rutilus (see also roach) 255
Rye (see also *Secale*) 5, *21*, 25, 27, 28, 91, 97–99,
    103, 204

Sagittaria (see also arrowhead) 118, *119*
Sahul 291
Sakhtysh 261, 262
Salix (see also willow) 80, 93, *146*, *147*, 262, 265
Saloum Delta 74, *77*, *78*, 82
Saltbaek Vig 120
Salvadora persica 323, *330*, 332
Sambaqui 3, 215–220, 228–231
Sami 302
Sap *63*, 71, 306, *310*, 311
Sapotaceae 347
Sarcocephalus latifolius *347*, 356
Sarcophilus harrsii (see also Tasmanian devil) 286
Sārnate 81, *257*, 259, 262
Saudi Arabia 57
Savannah 32, 33, 112, 159–161, 164, 182, 241,
    246, 277, 291, 327, 333, 341

Scandinavia 73, 81, 113, 124, 148
Scanning electron microscopy (see also SEM) 1,
    2, 4, 9, 111, 113, 114, 157
Scheemda-Scheemderzwaag 77
Schipluiden 77, 124, 125, 148
Schistosomiasis 62, *63*
Schistosoma mansoni (blood fluke) *57*
Schöningen 75
Scirpus (see also bulrush) 5, 80, 91, 97, 103, 141,
    145, 264
Sclerocarya birrea 324, *330*, *346*
Scotland 122
Sea beet (see also *Beta vulgaris* ssp. *maritima*)
    124, 125
Sea club-rush (see also *Bolboschoenus maritimus*)
    103
Sea lion 304, 311
Seal 204, 306, 314
Seasonality 92, 95, 159, 251, 273, 274, 276, 278,
    281, 290, 321
Sea thrift (see also *Armeria maritima*) *308*, 311
Seaweed 7
Secale (see also rye) 5, 91, 97, 103
Sedge 32, 75, 95, 100, 112, 118, 121, 164, 171, 178,
    180, 182, 199, 207, 243, 245–247, *279*, 281, 320
Seed 1, 4, 5, 19, 20–25, 27, 28, 32, 36, 46, 47, *63*,
    71, 75, 77, 78, 91–103, 111, 115–117, 128,
    129, *144*, 161, 163, 173, 182, 184, 191–194,
    203–209, 216, 218–220, 228, 274, 281, *282*,
    289, 306, 309, 311, 314, 322–325, 328, 330,
    352
Selknam 3, 301–315
SEM (see also scanning electron microscopy) 4,
    25, 114–*119*, 121, 123, *125*, *127*, 129, 156, 157,
    165
Semenov, Sergei 136–138
Senegal 3, 74, 77, *78*, 82, 252, 341, *343*, *346*, 356,
    357
Serteya 81, 261
Serteya 81, 261
Shanidar 5
Sheep 64, 305, 342
Shell 3, 4, 17, 27, 71, 72, 75, 79, 112, 122, 128, *194*,
    216–219, *227*, 243, 264, 311, 322, 328
Shellfish *74*, 180, 217, 243, 244, 304
Shell mound 3, 216, 218, 219
Shrotses 258
Shvjantoji 260
Siberia 253, 255–258
Sickle 96, 100–102, 138, 141, 193
Sidubu 75, 76
Sigersdal Mose 80
Siiversti 80

Silica 2, 62, 120, 155, 156, 178, 199, 274, 285
Silver fir (see also *Abies alba*) 75
Silybum marianum (see also holy thistle) 94
Simaroubaceae 347
Sitka spruce (see also *Picea sitchensis*) 73
Skoldnaes 80
South Africa 3, 74–76, 79, 160–162, 164, 171–176
South America 302
Spain 5, 9, 75, 247
Spear 75, 281, 283–290
Spindaceae 347
Sponge spicules 175, 177
Spore 180, 292
Spy 5
Stable isotopes 3–5, 7, 101, 111, 173, 217, 241–248, 320
Stanovoje 4, 262
Staosnaig 122
Star Carr 6, 120, 140
Starch 5, 7, 17, 19, 20, 22–28, 31–35, 41, 42, 45–47, 95, 96, 111–120, 124, 128, 129, 164, 173, 215, 220–232, 274
Stem 4, 20, 22, 24, *63*, 80, 94, 118, 120, 126, 141, 145–147, 184, 199, *202*, 204, 207, 208, 282, 306, 310
Steppe 92, 95, 97
Sterculiaceae 348
Stipa (see also grasses) 91, 95, 97, 103
Stipagrostis (see also grasses) 91, 97, 103
Stomach 7, 61–*63*, 65
Stone tool (see also lithic) 5, 33, 44, 71, 76, 78, 80, 91, 93, 100, 135, 136, 138, 141, 143, 149, 150, 173, 194, 273, 274, 276, 320, 327
Striation 135, 136, 138, 142, 143, 147, 155–159, 163–165
String (see also cord) 71, 78–82, 258, 261, 281, 284
Strophanthus eminii 329, *330*
Sucrose 19, 23, 24
Sunflower (see also *Compositae*) *21*, 22, 94
Sungir 79
Šventoji 81, 260, 262
Svijaga River 261
Swan 147, *148*
Swartkrans 74
Sweet-cicely (see also *Osmorhiza chilensis*) 311
Swifterbant 148
Syria 91, 92, 96, 98, 102
Syzgium sp. *282*

Taeniatherum 101, 102
Taforalt 5

Tamarindus indica 346, 356
Tamarisk 93
Tanzania 3, 58, *63*, 65, 172, 319–321, 329, 334
Tar 73, 76–78, 82
Taraxacum magellanicum, 306, *310*
Tartars 258
Tasmanian devil (see also *Sarcophilus harrsii*) 286
Taurus mountains 92
Taxus baccata (see yew) 75
Teeth (see also tooth) 1, 4, 73, 77, 79, 95, 155, 157, 158, *165*, 173, 220, 224, 226, 242, 268
TEK 9
Tell Aswad 101
Tell 'Abr 98, *99*
Tell Halula 102
Terminalia 282, 285, 329, *330*
Termiteria 175
Textile 80, 81, 135, 281
Thomandersia laurifolia 59
Thylacine (see also *Thylacinus cynocephalus*) 286
Thylacinus cynocephalus (see also Thylacine) 286
Tierra del Fuego 3, 251, 301–303, 305, *312*, 314, 315
Tigris River 100
Tiliaceae 345, 348
Tłokowo 76
Tolpis virgata 102
Tomboronkoto 341–343, *346*, 349, *353*, 357
Tongwe 61, 65
Tooth (see also teeth) 5, 32, 72, 73, 156–158, 161, 164, 165, 173, 220, 221, 244, 245, 248
Toothache 72, 73, 76
Toothbrush 220, 332
Toothpick 73
Torralba 75
Toxic 21, 31, 43, 48, 57, 64
Toys 329, *330*, 332
Trapa natans (see also water chestnut) 4
Trematode *63*
Trifolieae 96
Triglochin procera 282
Triticum (see also wheat) 5, 91, 93, 97, 103, *194*, 199, 224
Tubercles (see also tubers) 163
Tubers (see also tubercles) 2, 4, 5, 7, 17, 19, 23–25, 46, 47, 75, 92, 93, 95, 111–122, 125–127, 129, 138, 141, 142, 182, 196, 207, 218, 219, 231, 251, 275, 281, 282, 288, 306, *308–311*, 315, 322, 325–327, 329, *330*, 333, 334, 345–347, 349, 350, 352–354
Tulstrup Mose 80
Tunnel-Drontermeer 77

Turkey 8, 79, 91, *99*, 101
Turtles 280
Tuva 258
Tybrind Vig 80, 124, *125*, 135
Typha 116, 141, 142, 145, 180, 275

Ucagızlı Cave 79
Ug River 260
Uganda 57, 58, *63*, 356
Ukagir 258
Ukraine 117
Ulkestrup Mose 80
Ulmaceae 348
Ulmus (see also elm) 265
Umm el Tlel 76
Underground storage organs (see also USO) 2,
    4, 20, 32, 33, 97, 113–115, 161, 172, 173, 182,
    228, *229*, 231, 311, 325
Uragiri 258
Urtica (see also nettle) 141, *145–147*
Usnea 309, *310*
USO 2, 4, 20, 21, 31–35, 46, 161, 164, 172–174,
    182, 228

Vatoraea pseudolablab 326, *330*
Veenkoloniën 120, 124
Vegetation 1, 6, 19, 57, 92, 111, 113, 120, 171–174,
    176, 178, 180, 182, 209, 224, 276, 278, 291,
    302, 303
Verbascum 95
Verbenaceae 348, 353
Vernonia amygdalina 61, *63*
Vicia faba (see also faba bean) 101
Vicia (see also bitter vetch, rambling vetch) 91,
    98, *99, 194*
Vigna 63, 326, *330*, 356
Vines 71
Vinkelmore 80
Vis 75, 80, 260, 262
Vistula River 118
Vitellaria paradoxa 347, 356
Vitex 282, 348, 353
Vitis (see also grape) 94, 95, 114, *194*
Voronezh River 262

Wadi Jilat 6 95
Wadi Kubbaniya 5, 207
Water chestnut (see also *Eleocharis* spp., *Trapa
    natans*) 4, 47, 113, 114, 281, 289
Water lily (see also *Nymphaea, Nuphar, Nelumbo
    nucifera*) 4, 129, 142, 197, 281
Wedge 266
Whale (see also cetacean) 311
Wheat (see also *Triticum*) xi, 3, 5, 21, 22, 25–28,
    91, 93, 95, 97, 98, 101–103, 191, 194, 199,
    204– 206, 209
Wicker (see also willow, *Salix*) 251, 262
Wild celery (see also *Apium australe*) *308*, 311
Wild date palm (see also *Phoenix reclinata*) 182
Wild nut-grass (see also *Cyperus rotundus*) 5, 207
Wild onion (see also *Allium*) 116
Willow (see also *Salix*) 80, 81, 93, 101, *145*, 146,
    255, 256, 258, 259, 265
Wonderwerk Cave 74, 172
Wood xi, 1, 4, 65, 71–77, 81, 97, 117, 122, 143,
    144, 161, 173, 177, 184, 199, 216, 218, 220,
    228, 255, 259, 265, 266, 274, 276, 284, 289,
    304, *309*, 329, *330*, 333, 344
Woodland 33, 93, 97, 126, 159, 163, 164, 193, 209,
    281, *283*, 289
Worm 58, 59, 61–63

Yam (see also *Dioscorea*) 20, 112, 215, 218, 219,
    228, 275, 281, 286–289
Yamana 301, 303–306, *308–* 311
Yangtze Harbour 6, 78, 120–124, 129
Yew (see Taxus baccata) 75
Younger Dryas 97, 98, 193, 194, 209

ZAD 2 98, *99*
Zagros 92, 102
Zamostje xi, 3, 80, 251, 253–*256*, 262, *263*,
    265–268
Zanthoxylum chalybeum 328, *330,* 332
Zaraisk 79
Zea mays (see also maize) *223*, 228, *229*
Ziziphora 98
Zvidze 258, 262